Principles of Cyber-Physical Systems

This unique introduction to the foundational concepts of cyber-physical systems (CPSs) describes key design principles and emerging research trends in detail. Several interdisciplinary applications are covered, with a focus on the wide-area management of infrastructures, including electric power systems, air transportation networks, and health care systems.

Design, control, and optimization of cyber-physical infrastructures are discussed, addressing security and privacy issues of networked CPSs, presenting graph-theoretic and numerical approaches to CPS evaluation and monitoring, and providing readers with the knowledge needed to operate CPSs in a reliable, efficient, and secure manner. Exercises are included at the end of each part of the book.

This is an ideal resource for researchers and graduate students in electrical engineering and computer science, as well as for practitioners using CPSs in aerospace and automotive engineering, medical technology, and large-scale infrastructure operations.

Sandip Roy is Professor at the School of Electrical Engineering and Computer Science at Washington State University, where he teaches systems and control engineering and directs the Network Controls Group.

Sajal K. Das is Professor of Computer Science and Daniel St. Clair Endowed Chair at the Missouri University of Science and Technology, where he is also the director of CReWMaN lab. He is a recipient of the IEEE Computer Society Technical Achievement Award for pioneering contributions to sensor networks and mobile computing, and is a Fellow of the IEEE.

Principles of Cyber-Physical Systems

An Interdisciplinary Approach

Edited by

SANDIP ROY
Washington State University

SAJAL K. DAS
Missouri University of Science and Technology

CAMBRIDGE
UNIVERSITY PRESS

CAMBRIDGE
UNIVERSITY PRESS

University Printing House, Cambridge CB2 8BS, United Kingdom

One Liberty Plaza, 20th Floor, New York, NY 10006, USA

477 Williamstown Road, Port Melbourne, VIC 3207, Australia

314–321, 3rd Floor, Plot 3, Splendor Forum, Jasola District Centre, New Delhi – 110025, India

79 Anson Road, #06–04/06, Singapore 079906

Cambridge University Press is part of the University of Cambridge.

It furthers the University's mission by disseminating knowledge in the pursuit of education, learning, and research at the highest international levels of excellence.

www.cambridge.org
Information on this title: www.cambridge.org/9781107066618
DOI: 10.1017/9781107588981

First published 2020

A catalogue record for this publication is available from the British Library.

ISBN 978-1-107-06661-8 Hardback

Contents

Contributors

Ravi Akella
DENSO International America Inc., San Jose, CA, USA

Erik Blasch
Information Directorate, Air Force Research Laboratory, Rome, NY, USA

Paul Bogdan
Electrical and Computer Engineering Department, University of Southern California, Los Angeles, CA, USA

Alex Borden
Electrical and Computer Engineering Department, University of Wisconsin – Madison, Madison, WI, USA

Airlie Chapman
Mechanical Engineering Department, University of Melbourne, Australia

Brijesh Kashyap Chejerla
Electrical and Computer Engineering Department, Missouri University of Science and Technology, Rolla, MO, USA

Genshe Chen
Intelligent Fusion Technology, Inc., Germantown, MD, USA

Vasu Chetty
Information and Decision Algorithms Laboratories, Department of Computer Science, Brigham Young University, Provo, UT, USA

Patrick Corrigan
School of Electrical Engineering and Computer Science, Washington State University, Pullman, WA, USA

Sajal K. Das
Computer Science Department, Missouri University of Science and Technology, Rolla, MO, USA

Rahul Dhal
School of Electrical Engineering and Computer Science, Washington State University, Pullman, WA, USA

Rahav Dor
Department of Computer Science & Engineering, School of Engineering & Applied Science, Washington University in St. Louis, Saint Louis, MO, USA

Chinwendu Enyioha
Department of Electrical Engineering, Harvard University, USA

James J. Filliben
NIST, Gaithersburg, MD, USA

Thoshitha Gamage
Department of Computer Science, Southern Illinois University Edwardsville, Edwardsville, IL, USA

Shuo Han
Department of Electrical and Systems Engineering, University of Pennsylvania, Philadelphia, PA, USA

Gerry Howser
Department of Computer Science, Missouri University of Science and Technology, Rolla, MO, USA

Ali Jadbabaie
Institute for Data, Systems, and Society, Massachusetts Institute of Technology, Cambridge, MA, USA

S. Jagannathan
Department of Electrical and Computer Engineering, Missouri University of Science and Technology, Rolla, MO, USA

James Krueger
Integrative Physiology & Neuro, Spokane, WA, USA

Yu Lei
Department of Computer Science and Engineering, University of Texas at Arlington, Arlington, TX USA

Bernard Lesieutre
Electrical and Computer Engineering Department, University of Wisconsin – Madison, Madison, WI, USA

Chenyang Lu
Department of Computer Science & Engineering, School of Engineering & Applied Science, Washington University in St. Louis, Saint Louis, MO, USA

Sanjay Madria
Department of Computer Science, Missouri University of Science and Technology, Rolla, MO, USA

Radu Marculescu
Electrical and Computer Engineering, Carnegie Mellon University, Pittsburgh, PA, USA

Nuno Martins
Electrical and Computer Engineering, University of Maryland, College Park, MD, USA

Bruce McMillin
Department of Computer Science, Missouri University of Science and Technology, Rolla, MO, USA

Mehran Mesbahi
Department of Aeronautics and Astronautics, University of Washington, WA, USA

Kevin Mills
NIST, Gaithersburg, MD, USA

Paul Moulema
Department of Computer Science and Information Technology, Western New England University, Springfield, MA, USA

Cameron Nowzari
Department of Electrical and Systems Engineering, University of Pennsylvania, Philadelphia, PA, USA

Masaki Ogura
Department of Electrical and Systems Engineering, University of Pennsylvania, Philadelphia, PA, USA

George Pappas
Department of Electrical and Systems Engineering, University of Pennsylvania, Philadelphia, PA, USA

Khanh Pham
Space Vehicles Directorate, Air Force Research Laboratory, Kirtland, NM, USA

Victor M. Preciado
Department of Electrical and Systems Engineering, University of Pennsylvania, Philadelphia, PA, USA

Parameswaran Ramanathan
Electrical and Computer Engineering Department, University of Wisconsin – Madison, Madison, WI, USA

Tom Roth
NIST, Gaithersburg, MD, USA

Sandip Roy
School of Electrical Engineering and Computer Science, Washington State University, Pullman, WA, USA

Avimanyu Sahoo
Department of Electrical and Computer Engineering, Missouri University of Science and Technology, Rolla, MO, USA

Eric Schoof
Electrical and Electronic Engineering Department, University of Melbourne, Australia

Dan Shen
Intelligent Fusion Technology, Inc., Germantown, MD, USA

Marcos M. Vasconcelos
Electrical and Computer Engineering Department, University of Southern California, Los Angeles, CA, USA

Yan Wan
Department of Electrical Engineering, University of Texas at Arlington, Arlington, TX, USA

Sean Warnick
Information and Decision Algorithms Laboratories, Department of Computer Science, Brigham Young University, Provo, UT, USA

Junfei Xie
Department of Electrical and Computer Engineering, San Diego State University, San Diego, CA, USA

Guobin Xu
Department of Computer Science and Information Technologies, Frostburg State University, Frostburg, MD, USA

Hao Xu
Department of Electrical and Computer Engineering, Missouri University of Science and Technology, Rolla, MO, USA

Mengran Xue
School of Electrical Engineering and Computer Science, Washington State University, Pullman, WA, USA

Wei Yu
Department of Computer and Information Sciences, Towson University, Towson, MD, USA

Michael Zargham
Department of Electrical and Systems Engineering, University of Pennsylvania, Philadelphia, PA, USA

Yi Zhou
Computer Science and Engineering, University of North Texas, Denton, TX, USA

Preface

Cyber technologies are becoming pervasive in engineered physical systems ranging in scale from micro-robots that travel in the bloodstream to multi-continent transportation networks. The tight interfacing of cyber and physical components in engineered systems permits profound (and sometimes magical) advancements in their functionalities, but at the same time brings forth unique challenges in system operation, analysis, and design. The potentially transformative benefits that come from melding cyber and physical capabilities, and the attendant challenges, are rapidly driving research and development efforts in wide-ranging fields (e.g., medicine, transportation, power transmission, retail). As these cyber-physical systems (CPSs) become increasingly common in our daily lives, tools and techniques for characterizing and designing their operations are increasingly needed. This is because such tools will (1) permit keen analysis and improvement of the intricate interactions between cyber and physical aspects of a particular complex system; and yet (2) transcend system-specific details to provide general insight into CPSs. In short, a foundational theory for CPSs is spanning important disciplines but yielding concrete advances in deployment and operation of particular systems.

The interfacing of cyber technologies with physical systems is not new: Computers have been used in conjunction with electromechanical systems almost from the inception of the digital age, and such fields as telecommunications and manufacturing have a rich history of exploiting digital technologies to enhance physical-system function. What is new, however, is (1) the increasingly wide penetration of cyber-capabilities into physical devices; (2) the growing diversity of uses of cyber-capabilities in these systems; (3) the increasing sophistication of the computer–machine interface; and (4) the advent of integrated technologies whose functionalities supersede those of physical and cyber systems in isolation. It is this growing pervasiveness and sophistication of CPSs that motivates the development of a new foundational theory, one which provides generic insight into interfaced cyber and physical capabilities across disciplines. At its essence, such a foundation requires understanding of how information-transmission and processing capabilities (the hallmarks of cyber systems) can be interfaced with physics-based functionalities. A wide array of interesting and difficult questions arise in engineering these dual-purpose systems, including analysis of their dynamics, stability analysis, performance design, robustness and security characterization, equity, implementation of the cyber-physical interface, and deployment, among others.

The analysis and design of CPSs has drawn wide attention in academia, government, and industry in recent years. In particular, industrial and governmental organizations in several domains – automobile manufacturing, surgical/medical device design, and transportation-system management, to name just a few – are engaged in research and development efforts to build and optimize CPSs. At the same time, university engineering and computer-science programs are beginning to foster scholarship on CPSs, through introduction of new courses, involvement in and support of research projects, faculty-hiring initiatives, etc. As research and development on CPSs continues to expand, there is a growing recognition that knowledge in several fields needs to be integrated and enhanced to develop foundational methods/tools for CPSs and to take these methods to industrial practice. Such foundational integrated work on CPSs crucially requires forums that: (1) provide comprehensive tutorial content on basic CPS principles and concepts; and (2) introduce state-of-the-art research on focused topics that are necessary for advancing the field.

The purpose of this publication is to meet these dual needs for tutorial content and focused research advancement, in a way that is useful for both academic researchers and industrial practitioners. With this goal in mind, we have sought to develop a text that can serve both as a compendium of interesting and important research results, and as a textbook for a graduate or advanced undergraduate course on CPSs. To this end, we have invited leading researchers on CPSs from several disciplines to contribute chapters to the book that contain up-to-date research progress. While these chapters capture new research, we have also asked the authors to include tutorial content to provide a comprehensive introduction to their direction of research; this content includes significant literature review, examples and exercises, illustrations, and homework problems, so that the book is instructional for classroom use. Additionally, we as editors have included sectional overview materials, a scoping chapter for future work on cyber-physical infrastructures, and some discussions and annotations, so as to achieve a cohesive story with smooth transitions and completeness. The two of us hail from different disciplines (controls engineering and computer science) and are collaborating on multidisciplinary research on CPSs; we have sought to draw on our ongoing collaboration to compose an interdisciplinary yet cohesive perspective. In short, it is our intention that the textbook covers the state-of-the-art with regard to influential CPS research, while also meeting the need for a cohesive framework for CPS concepts and principles.

The book is divided into four parts which each address a major conceptual theme in the design and operation of CPSs. Specifically, the chapters in Part I discuss challenges in uncertainty modeling, analysis, and mitigation for CPSs. Part II explores the notion of CPS structure, broadly defined, and explores how structure can be exploited in control and management of CPSs. The chapters in Part III address the growing need for threat assessment and resolution across the cyber and physical components of CPSs, and hence introduce principles related to security, privacy, and vulnerability. Finally, Part IV describes the development of frameworks and testbeds for addressing CPS challenges. Core conceptual and analytical results are presented throughout the book, but are also supported by myriad application case studies that reflect the authors' individual specializations. Besides the four main parts, the editors have included a concluding chapter, which overviews major challenges in management and control of large-scale infrastructure networks. This concluding chapter shows how the major themes of the book can tie together in addressing an engineering grand challenge, namely management of terrestrial-scale infrastructures, and serves as a call-to-action to CPS researchers to focus on wide-area management of infrastructures.

The field of CPSs has grown extremely rapidly over the last 10 years or so: An extraordinary number of conference and journal articles have been published on the topic, new conferences and journals have been initiated, several companies have introduced focused research and development efforts in the area, and – importantly – several other books have been written on the subject. The wealth of recent work on the subject is a strong motivation for this book, but at the same time means that we should distinguish the contributions of this book from other texts. Because of the huge scope and newness of CPS research, intrinsically each text and survey manuscript has a different perspective on the field. We believe that this book encompasses a wide range of core concepts as well as applications of CPSs, and should be of interest widely to academics and practitioners in the field. At the same time, however, we would like to highlight two distinguishing features of our book.

First, this book diverges from the bulk of the literature on CPSs in that there is a central focus on large-scale infrastructures like power-transmission networks and global transportation systems. CPSs as a field grew out of the embedded computing literature, and hence naturally the field's original focus was on physical devices with embedded cyber systems, and in turn on networks of such devices (which constitute the Internet of Things). This book overviews research in these directions, but also recognizes the complementary challenge of operating legacy infrastructures with integrated cyber systems as a keystone CPS problem. We believe that the need for focused research and tutorial content is particularly pronounced in the management of cyber-physical critical infrastructures, like transportation, electric power, and water-distribution systems, among many others. The efficient operation of such infrastructures in a complex and adversarial world constitutes a grand challenge problem for engineers today. Diverse cyber technologies are being integrated into these infrastructures, and are enabling revolutionary changes in their operations. Yet fully harnessing these new technologies requires detailed co-modeling and analysis of the infrastructures' physical-world processes and its cyber systems. For most infrastructures, modeling, analysis, and management are undertaken by domain experts who are well-versed in the physical processes and legacy systems involved in the infrastructure, but typically have limited technical knowledge of computing systems. On the other hand, most computer scientists lack specialized knowledge about the physical processes in infrastructures. Also, the historical development of infrastructure systems often dictates that there is limited interface between computer scientists and domain experts. Given this lack of comprehensive expertise, conceptual and mathematical frameworks are needed that facilitate co-analysis of the cyber and physical systems and allow efficient operation/management. This book aims to provide the conceptual and mathematical frameworks needed for operation and management of CPSs.

Second, a special focus of this book is on coalescing research in computer science and systems/controls engineering to address CPS challenges. Several of the chapters introduce techniques that leverage both systems/controls engineering concepts and computer science constructs, or develop new algorithmic capabilities that merge the fields. These efforts show that coalesced methods can support advanced control and design of CPSs, including for performance-shaping and threat assessment and mitigation. The coalesced methods are crucial for the analysis and design of cyber-physical *networks* (whether infrastructures like power and transportation networks, or communicating-agent systems like multi-vehicle teams or networked devices in a smart building/home). This book demonstrates the application of coalesced methods in these diverse network-focused applications.

It is our hope that the material in the book will be useful for researchers and practitioners across many areas of specialization, who encounter analysis and engineering design problems that involve tightly interfaced cyber and physical components. Target audiences include:

- practitioners in industry engaged in building cyber-physical capabilities, who require a comprehensive overview of foundational methods in the field;
- academic and industrial researchers interested in developing interdisciplinary techniques for the analysis and management of CPSs, such as epidemiologists seeking to integrate cyber decision-making capabilities into hospital-management procedures or public-health initiatives;
- graduate and advanced undergraduate students in engineering and computer science, especially those focused on modeling, algorithms, and systems/control. Specifically, we envision that these students would read the text as part of a graduate or advanced undergraduate course on CPSs.

The book has been designed to allow readers to easily port the introduced concepts and principles to their application area of interest.

Part I

Overcoming Uncertainty

Cyber-physical systems (CPSs) are subject to uncertainties, which may arise from both external disruptions and intrinsic internal variability. These uncertainties may be of many types, affecting the physical or cyber elements of the system (or both), acting persistently or impulsively, impacting knowledge of system parameters or altering state dynamics, admitting statistical models or being completely unmodeled, etc. These many uncertainties profoundly impact the operation of CPSs, because they place fundamental limits on model identification and state forecasting, which in turn complicates control, management, and design. Beyond these essential operational challenges, uncertainties in many cases also bring forth difficult computational challenges in system analysis and engineering, requiring extremely high-dimensional modeling, repetitive simulation of time-consuming programs, and representation of complex and hybrid uncertainties. Addressing the operational and computational challenges that arise due to uncertainty is a keystone challenge in CPS research. Part I of this text introduces problems, principles, and analytical as well as numerical methods related to addressing uncertainty in CPSs.

Modeling and evaluation of uncertainties is already a central part of the engineering and computing sciences. Of particular note, an extensive toolbox for evaluating and mitigating uncertainties in physical-world systems is available, which builds on the classical literature on probability theory and stochastic processes, as well as more recent (but still classical) work on estimation and stochastic control. Meanwhile, the modeling and analysis of uncertainties has been a central focus of the computer and information sciences since their inception: information theory, Markov modeling, and graphical models are some of the typical constructs used to capture uncertainty in these domains. While many of these approaches are being carried over to CPS applications, uncertainty analyses in CPSs also entail new challenges which require new principles and methods:

1. Uncertainties in CPSs may have propagative impacts across cyber and physical elements. Thus, methods are needed for characterizing hybrid, networked, multi-time-scale processes that are subject to uncertainties. Such methods require stochastic modeling frameworks that can encompass physical and cyber elements and metrics for evaluating impacts across the cyber and physical worlds. They also require analysis techniques for stochastic constructs that marry traditional physical-world

models/principles (e.g., Kalman filtering, linear stochastic control) with cyber-world models (e.g., hidden Markov models, random fields, etc.).
2. The complexity of CPSs often dictates that analysis and forecasting are only possible using time-consuming simulation models. Techniques are needed for evaluating parametric and process uncertainties for models with high-dimensional simulations in the loop.
3. Particularly for cyber-physical infrastructures, uncertainties are high-dimensional and multifaceted. For example, the air transportation management infrastructure may be subject to spatially and temporally evolving severe weather, demand uncertainty, rare unmodeled cyber- or physical- world attacks on traffic-control facilities, and failures by human operators resulting from, e.g., fatigue.

The chapters in Part I identify and address challenges in these directions, and apply the developed principles and methods in illustrative examples and real-world applications. Chapter 1, by Bogdan and Marculescu, models cyber-physical processes in systems-on-chip that are intrinsically uncertain, and develops the stochastic calculus required for analyzing such processes. Chapter 2, by Xie et al., pursues evaluation of parametric uncertainties in computationally intensive simulation models for CPSs, and demonstrates application to air-transportation-focused applications. Chapter 3, by Dhal and Roy, describes a stochastic graph-partitioning algorithm that combines uncertainty analysis methods from systems and control engineering and computer science, as is required for CPS applications. It also envisions how such stochastic algorithms can be used dually as models and computational tools for CPS applications. Throughout the development, the authors refer to open challenges and complementary directions of interest related to uncertainty analysis/mitigation in CPSs.

1 From Physical Processes to Theoretical Foundations of Cyber-Physical Systems Design and Optimization

Paul Bogdan and Radu Marculescu

1.1 Introduction

Cyber-physical systems (CPS) refers to a new generation of networked embedded systems that bring together sensing, computation, communication, control, and actuation in order to sustain a continuous interaction with the physical world [1–3]. Simply speaking, CPS consist of networks of embedded computation and communication devices [4], as well as sensors [5, 6] that monitor and measure various physical processes taking place on electrical power grids, transportation and traffic roads, communication and financial networks, medical devices, smart buildings, and more. Hence, CPS need to provide real-time efficiency, dependability, and robustness (i.e., reliable, safe, secure) to natural disasters or adversarial attacks [2, 7–9].

Cyber-physical systems are meant to not only build on our understanding of various physical processes, but also enable more accurate decision-making (e.g., resource allocation, flow regulation) in the presence of uncertain behavior. For instance, aiming at understanding physiological complexity, a rich sensing technology has been developed for both measuring physiological processes [10–12], as well as spatiotemporal biochemistry cellular phenomena [13]. Along the same lines, in order to provide a more comprehensive understanding of traffic flow dynamics across large road network structures, several research efforts investigate how cell phone technology can be harnessed for social tracking and knowledge fusion [5, 13, 14]. In addition, in order to allow more robust decision-making strategies, CPS must not only provide congestion information about the network state, but also enable the development of mitigation strategies that account for uncertain behavior of physical and cyber components. Along these lines, Xue et al. [15] use probabilistic influence modeling to construct a weather-impact simulation environment that generates stochastic trajectories of weather state (presence or absence of convection) which helps gauge the impact of weather dynamics on airspace-relevant parameters.

While sensing technology can offer rich data about physical processes, the theoretical foundations of CPS design and optimization are mostly lacking. Figure 1.1 summarizes these challenges and shows how tools from nonlinear dynamics and statistical physics can help with modeling physical processes and CPS design. Numerous physical processes are sensed at a very fine level of granularity (see the bottom of the pyramid), yet few of them are accurately modeled. Consequently, building the

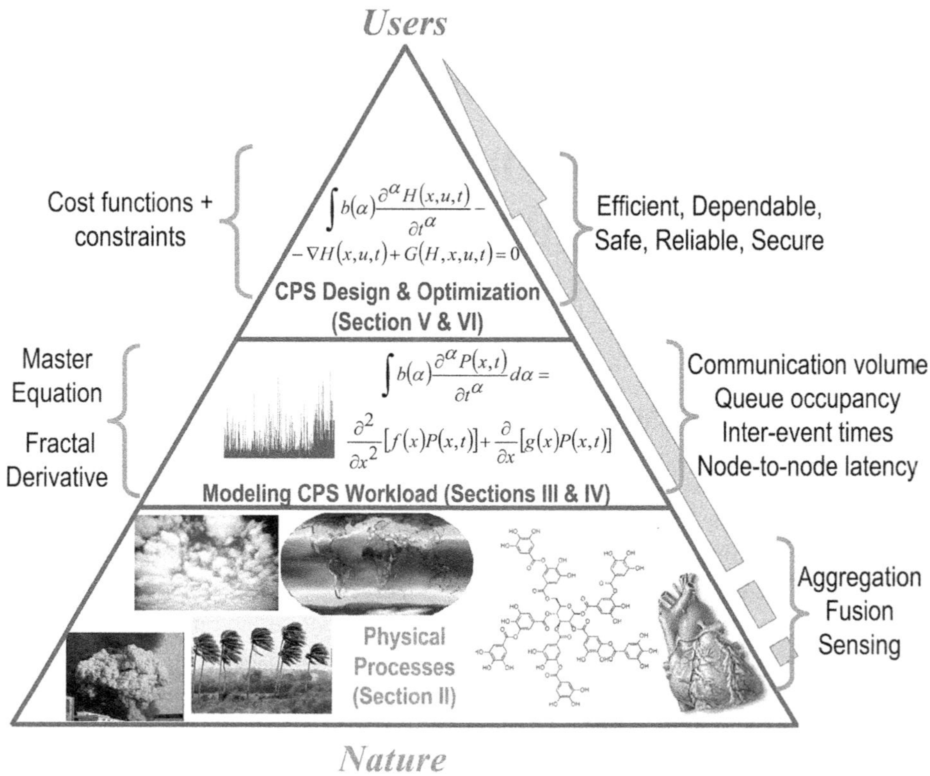

Figure 1.1 CPS pyramid. The pyramid foundation summarizes the large set of physical processes of interest to the CPS community. Insights about physical processes gained through sensing, information fusion, and aggregation lead to a complex heterogeneous CPS workload. Accurate modeling of workloads enables the design of optimal CPS architectures that may improve our lives. Similarly to statistical physics which has been successful in explaining nature, the CPS workloads are modeled via master equations which later are used in defining various optimization problems of interest to the end-users. © 2011 IEEE. Reprinted, with permission, from [17].

appropriate mathematical models of physical processes requires comprehensive signal processing algorithms, advanced mathematical techniques, and efficient computing platforms that can allow for meaningful data aggregation, fusion, and mining. One such example is represented by the need for long-distance patient monitoring via heart-rate sensing. Simply recording each heartbeat or magnitude of R–R intervals and transmitting this information to medical centers would involve significant performance penalty, energy consumption, and security issues.

To address these problems, various CPS components need to be designed to solve specific tasks such that some estimate the risk factors locally, actuate parameters in real time, and eventually transmit remotely only the critical information. Deciding which piece of information to send and which to compress (or discard) requires accurate mathematical modeling and data mining. This way, once these measurements are locally analyzed, some are digitized and further communicated to various decision centers as CPS workloads that can vary in magnitude over time depending on the state of the physical process (the middle part of the pyramid in Figure 1.1). Designing the CPS infrastructure and the routing protocols while accounting for CPS workload characteristics can lead to more efficient solutions [16]. From this perspective, the workload variability represents the Achilles' heel of CPS design since relying on fixed and restrictive protocols not only can lead to poor performance, but also to catastrophic events when CPS workloads had to cope with various abnormalities and abrupt changes in physical processes. Addressing high performance while ensuring a high degree of robustness/reliability/dependability cannot be done without accurate characterization of workloads (e.g., self-similarity, nonstationarity). In addition, there is a stringent need to couple the nature of CPS workloads with novel mathematical methodologies that allow for the design of self-organizing architectures that consist of heterogeneous wireless sensor networks [16] and embedded systems. Such a new science cannot rely solely on reductionism and linear control paradigms that represent the norm nowadays; this represents a big challenge for CPS design and optimization since it needs to account for end-users' satisfaction (top of the pyramid).

In the light of these challenges, we argue that several recurrent CPS design problems can be solved in an elegant manner by combining fractional calculus (e.g., fractional integrals/derivatives), statistical physics (e.g., master equation), and optimal control (e.g., dynamic programming) concepts. Indeed, as shown in Figure 1.1, the statistical physics approaches can help by allowing us to describe the CPS workload and its intermediate states (e.g., queue occupancy, node-to-node latency) via fractal-type master equations. Consequently, taking into consideration the space and time features of the CPS workloads via such master equations can help us formulate various optimization problems such as resource allocation, task mapping, and scheduling (the top of the pyramid in Figure 1.1). By carefully allocating the buffer sizes and the bandwidth between sensors and data centers, we can also accelerate the decision-making process and prevent data loss or delays, which can cause life-threatening situations or increase economic costs. These issues are discussed next.

1.2 Characteristics of Physical Processes: Self-Similar, Fractal, and Nonstationary Dynamics

Many of the physical processes sensed by the CPS (e.g., pollution concentration, temperature, and wind speed for weather prediction; body temperature and heart rate for medical diagnosis; flow densities and velocities for road transportation network management) can be described by a series of concurrent actions that represent the system dynamical state (i.e., behavior) in response to changing environmental stimuli. One classical example is weather prediction, which cannot reach a steady-state regime due to the complex activity of the Sun; indeed, the Sun heats the atmosphere, oceans, biosphere, and land and so it triggers numerous physicochemical processes that lead to various gradients and complex currents of air circulation. Another example of a physical process is human crowd dynamics, which directly influence vehicular traffic. Such physical processes are rarely evolving toward an equilibrium and, even when they reach a steady-state behavior, this typically happens only for short intervals [20]. In spite of their complex behavior, physical processes are characterized by a few fundamental characteristics like self-similarity, fractality, and nonstationarity.

Self-similarity represents the property of geometric objects or time series (stochastic process) to look similar at various scales in space or time [21]. Self-similarity is not necessarily an artifact of some random actions, but rather seems to be the effect of a careful optimization realized by evolution. Indeed, natural objects and biological systems have "learned" over the course of evolution to exploit the self-similar properties to improve their energetic efficiency and robustness. Probably the most intriguing example where self-similarity plays a crucial role is the human body: Our body would not be able to sustain our weight without harnessing the self-similar properties of the bone structure, our muscles would not be able to deliver the strength we need without complicated branching blood flow, and, finally, our intellectual capacity would not be of much value without self-similar neuronal interconnections [22, 23]. From a mathematical perspective, the self-similarity of a stochastic process $x(t)$ is defined by the following relationship:

$$(x(bt_1), x(bt_2), \ldots, x(bt_m)) = (b^{\alpha} x(t_1), b^{\alpha} x(t_2), \ldots, b^{\alpha} x(t_m)), 0 < \alpha < 1 \qquad (1.1)$$

where α is the Hurst exponent, $b > 0$ is a positive scaling factor, and $t_1, t_2, \ldots, t_m$ are successive time stamps. Simply speaking, the self-similarity implies that the joint distribution of the random variables $(x(bt_1), x(bt_2), \ldots, x(bt_m))$ is identical with the joint distribution of the rescaled random variables $(b^{\alpha} x(t_1), b^{\alpha} x(t_2), \ldots, b^{\alpha} x(t_m))$. Moreover, the rescaled versions of the stochastic trajectories (time series) of $x(bt)$ tend to look similar to the original time series $x(t)$. This self-similar behavior is also referred to as scale invariance. For a dynamical system described by multiple variables, the self-similarity is a form of symmetry in the sense that a change of scale in some variables can be compensated by a change in scale of the remaining variables.

Figure 1.2 shows two examples of self-similar behavior. More precisely, Figure 1.2a illustrates the rescaling of the daily wind speeds [24]. Figure 1.2b describes the dynamics of R–R intervals (i.e., time duration between two R waves in the ECG signal[1]) as a function of the number of heart beats for a healthy individual.[2] Under the rescaling operation done across several time scales with respect to the initial time series, one can notice the spiky dynamics of various intervals and subintervals. This spiky behavior has important implications from both practical and theoretical perspectives. For instance, from a physician's perspective, the absence or tendency of

1 An electrocardiogram (ECG) signal represents the electrical activity of the heart over time and consists of P wave (atrial depolarization), QRS complex (ventricles depolarization), T wave (ventricles repolarization), and U wave. Because ventricles contain more muscles than atria, the spike corresponding to an R wave (heart beat) in the QRS complex is more visible than other waves. The difference between two consecutive R waves is called the R–R interval and provides information about heart-rate variability.

2 The R–R interval data sets were obtained from the National Institute of Biomedical Imaging and Bioengineering website (i.e., www.physionet.org).

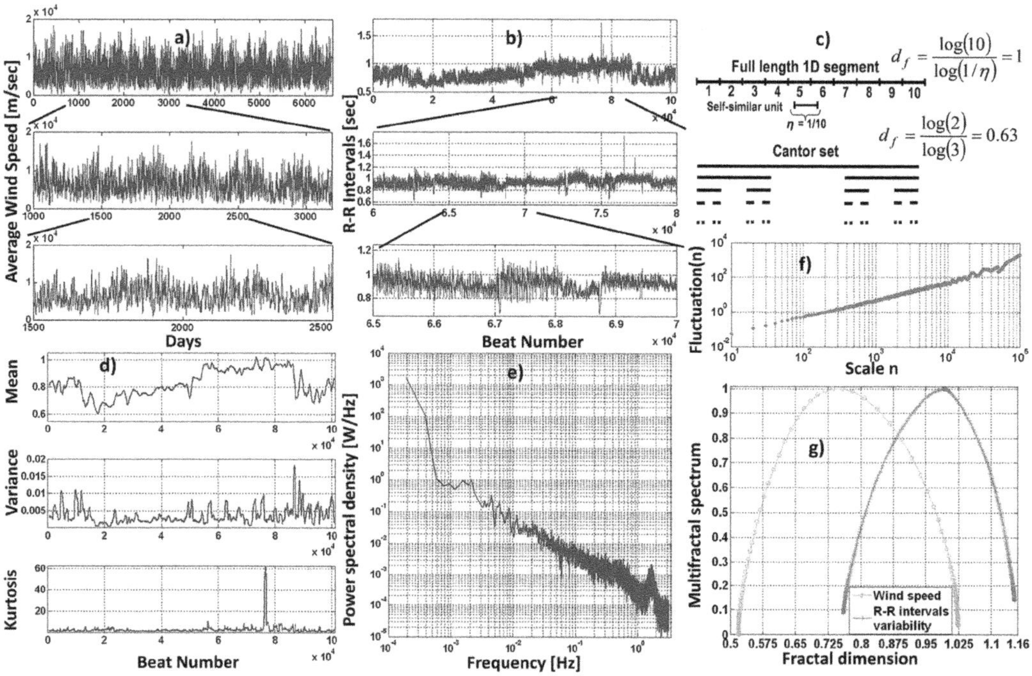

Figure 1.2 (a) Daily average wind speed recorded under the Hubbard Brook Experimental Forest Project [18] from October 27, 1965 through December 31, 2009. (b) R–R intervals collected via electrocardiogram recording of data from a healthy subject exhibit self-similarity. (c) Two geometric objects exemplifying the computation of fractal dimension: a Euclidean one-dimensional object having a fractal dimension equal to 1 (top) and a Cantor set exhibiting a fractal dimension of approximately 0.63 (bottom). (d) Mean, variance, and kurtosis plots of the R–R intervals (as a function of the beat number) for a healthy individual exhibit nonstationary behavior and deviate from Gaussian statistics. (e) Power spectral density of the R–R intervals exhibits a $1/f^{\beta}$ ($\beta = 1.246$) behavior confirming the presence of self-similarity. (f) Average fluctuations of the R–R intervals as a function of the scale size n obtained via detrended fluctuation analysis [19]. (g) Multifractal spectrum of the daily average wind speed and R–R intervals.

losing the fractality is synonymous with the presence of some abnormalities, which can trigger congestive heart failure [25]. On the other hand, from a mathematical standpoint, the fractal pattern exhibited in Figure 1.2a is not differentiable when considered as a time-varying function of a physical process. As we shall see later, this means that the integer order derivatives need to be replaced by derivatives of fractional orders when analyzing the system behavior [21]. We note that there exist many other physical processes that also exhibit similar behavior over long periods of time. For instance, another example of a self-similar process is the cumulative concentration of cloud condensation nuclei (CCN) collected via a CCN spectrometer [17]. From a practical standpoint, the CCN measurements are used to study the ice nucleation process in clouds, assess the impact of industrial pollution on climate change, as well as quantifying cloud dynamics for the purpose of weather forecasting. In order to study the complexity of the self-similar behavior, one can further investigate the existence of fractal behavior.

Fractality, encountered either in space (i.e., fractal geometry) or in time (i.e., fractal dynamics), describes a scale invariance phenomenon characterized by a non-integer fractal dimension [21]. The fractal dimension represents a statistical measure of complexity of an object or process, which describes how a fractal pattern varies with changes in space or time. From a mathematical perspective, the fractal dimension is defined by the following relationship:

$$d_f = \lim_{\varepsilon \to 0} \frac{\log(B(\varepsilon))}{\log\left(\frac{1}{\varepsilon}\right)} \tag{1.2}$$

where $B(\varepsilon)$ represents the minimum number of nonempty boxes (or spheres in 3D representations) of size ε required to cover in space or in time the object or stochastic process under investigation, and ε is the scale size at which the measurement is performed. The intuition from defining this statistical complexity measure comes from the fact that the object or stochastic process is sparsely distributed in space (i.e., it does not fill the Euclidean space entirely) and so it can be encapsulated into a space of volume $L^{d'_f}$, where L is the unit size of the D-dimensional Euclidean space and $d_f < D$, with D being the minimum integer of the Euclidean space covering the set completely.

Simply speaking, a fractal object or stochastic process is one that displays some degree of pattern repetition (self-similarity) over a wide range of scales and the degree to which it covers the space/time domain is sparse, being quantified by a fractional number (see the Cantor set in Figure 1.2c). For instance, while Euclidean objects may present self-similarity under magnification operation (see the one-dimensional segment in Figure 1.2c), they cannot be considered as fractal examples because they can be precisely measured via integer fractal dimensions. It is important to note that perfect self-similarity is proper to only deterministic fractals. In reality, a perfect self-similar behavior is rarely preserved at all scales, so we may encounter stochastic self-similarity characterized by a distribution of fractal dimensions. These stochastic objects or processes are often called multifractals. Figure 1.2g shows the multifractal spectrum of the daily wind speeds and the R–R interval time series. One can clearly notice the rich set of exhibited fractal dimensions and so the importance of accounting for this complex multifractal dynamics in modeling physical processes.

Nonstationarity of a stochastic process implies that its probability density function (PDF) and implicitly its statistical moments (e.g., mean, variance, skewness, kurtosis) vary with time. In mathematical terms, for any short time increment δt, the nonstationarity can be expressed as follows:

$$P(x,t) \neq P(x,t+\delta t) \tag{1.3}$$

To exemplify the existence of nonstationary behavior, Figure 1.2d shows the time-dependency of various high-order moments (i.e., mean, variance, and kurtosis[3]) of

[3] Kurtosis (i.e., the ratio between the fourth-order moment and the standard deviation of a probability distribution) captures the frequency of rare events.

the R–R intervals in Figure 1.2b. The existence of nonstationary behavior can be observed from the moving average plots of the mean and variance of R–R intervals in Figure 1.2d. This implies that, at every point in time, the physical process can be characterized by some (local) fractal exponent which encapsulates the $1/f^{\beta}$ type of behavior (where f is the frequency measured in hertz and β is the power law exponent of the power spectral density[4]) exhibited in Figure 1.2e. In addition to the intrinsic nonstationarity of data, a nonzero kurtosis provides evidence that a Gaussian type of approach does not fit the data well and that a higher-order moment analysis is needed in order to properly characterize this type of behavior.

Although many physical processes display nonstationary and fractal behavior over a wide range of scales, their impact on modeling and optimization is often neglected. For instance, mathematical modeling of wind speeds, precipitation formation, and cloud movement, to name just a few, has not only impact on the CPS workloads (e.g., type and amount of data communicated between CPS components), but also an immediate and enormous economic impact on air, road, and rail traffic. Similarly, relying on continuous monitoring of physiological processes in order to detect abnormalities can modify the structure and type of CPS workloads that need to be communicated between wearable CPS medical devices, on one hand, and physicians and medical data centers, on the other. Starting from these premises, in the next section we discuss the main features of CPS workloads that often share common mathematical features with their corresponding physical processes.

1.3 Workloads in Cyber-Physical Systems

Many of the physical processes relevant to CPS exhibit a systematic relationship at different scales in space and time. This relationship between various scales is important since the statistical properties (e.g., correlations) from one scale, do not vanish, but instead contribute in an additive or multiplicative manner to the statistics of the process at another scale and affect the type and amount of computational and communication workloads. This intrinsic property can also be thought of as one of the main causes for observing self-similarity in CPS workloads. A powerful approach for investigating the existence of self-similarity is to move the investigation from the time- to frequency-domain and analyze the power spectrum[5] of CPS workloads. If the power spectrum of a CPS workload flattens as a horizontal line on a log-log scale, then this implies that its corresponding stochastic process does not display any correlations because all frequencies play an equally important role.[6] By contrast, if the power spectrum diverges for high frequencies, then the CPS workload is said to display long-term memory

[4] As a word of caution, relying on power spectrum analysis is not always robust to investigate the fractal behavior of time series. A more appropriate approach implies measuring the average fluctuations exhibited as a function of the scale size (time interval within the time series) via the detrended fluctuation analysis method [26].

[5] The power spectrum characterizes the contribution of each frequency to the overall signal.

[6] This would be similar to having a white-noise type of behavior, which appears as a flat line when represented on a log-log scale.

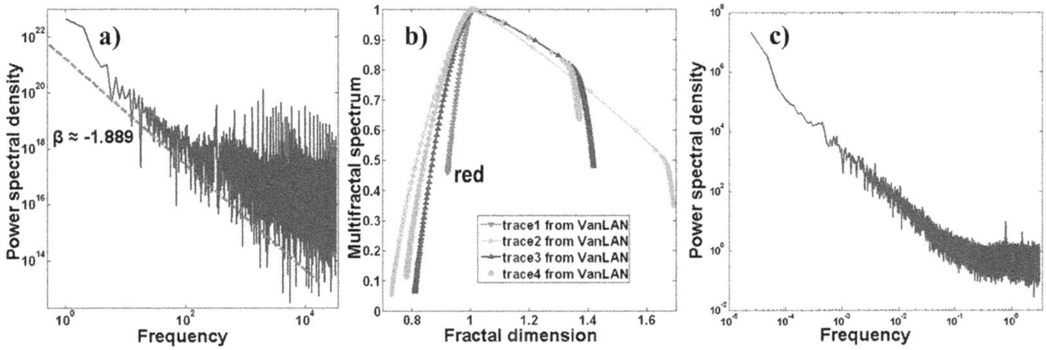

Figure 1.3 (a) Power spectral density of short-range communication in local area network (LAN) established via wireless links between moving vehicles. (b) Multifractal spectrum of information transaction events (i.e., sent/received packets) in a LAN where communication connectivity is established via wireless links and access points in an urban environment. (c) Power spectral density of the packet received signal strength obtained in a wireless network. © 2011 IEEE. Reprinted, with permission, from [17].

effects [19, 21, 25]. This also implies that the CPS workload possesses $1/f^{\beta}$ *type of scaling*, where f and β are the frequency and power law coefficient, respectively. For instance, Figure 1.3a shows the power spectral density of the communication throughput in a heterogeneous network with a frequency exponent of approximately 1.8; this behavior confirms the existence of self-similarity in this workload. A similar $1/f^{\beta}$ scaling can also be observed in Figure 1.3c for the packet received signal strength at the receiver directional antenna in a wireless network [10].

The existence of $1/f^{\beta}$ type of fluctuations indicates that the CPS workloads actually consist of a *mixture* of long packets containing data and control information and short packets consisting of control flags. This is similar to other critical phenomena in nature (e.g., crystal growth, granular flow) and implies that CPS workloads *cannot* be simply described accurately using average values at *fixed* space–time scales.

Due to the intertwined relationship with physical processes, CPS workloads also exhibit nonstationary behavior. Indeed, CPS workloads exhibit more often a heterogeneous set of scaling exponents (see traces 2, 3, and 4 in Figure 1.3b), rather than a homogeneous (or monofractal) one shown as trace 1 in Figure 1.3b. Such a nonstationary workload can also be understood by recalling the time dependency of the fluctuations in the R–R intervals (see Figure 1.2b). The existence of various heterogeneous scaling exponents indicates that some CPS workloads exhibit multifractal properties. Speaking simply, the multifractal spectrum offers a compact mathematical representation of the intricate interactions among CPS components. For instance, Figure 1.3b shows the multifractal spectrum of four communication traces collected from a LAN among several vehicles communicating via wireless links.[7] One can notice that while trace 1 exhibits a monofractal behavior (i.e., narrow vertical spike), the remaining three traces exhibit a multifractal behavior, shown as wide bell-like shapes.

[7] For detailed information about the measurement instrumentation, the reader can consult [27].

1.4 New Formalism for Modeling Cyber-Physical Workloads

Traditionally, the system design methodologies have assumed that workloads are either Markovian (i.e., memoryless) or that they can be modeled via linear time-invariant equations. These assumptions are in disagreement with the observed mathematical characteristics (self-similarity, fractality, nonstationarity) of real processes. One explanation for the observed fractal behavior is that many soft real-time systems adopt an event-based triggered execution which is dictated by events or changes in the physical process in order to save energy or cope with bandwidth constraints over large spatial areas. For hard real-time networked systems that must monitor and communicate vast amounts of data about physical processes, the network is pushed toward criticality, and packet flow dependencies induce correlations contributing to the observed fractality. Consequently, in what follows, we advocate for a paradigm shift concerning the *mathematical modeling* of CPS workloads, one that does not make very restrictive assumptions, but rather derives state dynamical equations from observations (measurements) and leads to mathematical laws concerning system dynamics. Borrowing inspiration from nonequilibrium statistical physics, we propose a master equation description of CPS workloads that enables both probabilistic and deterministic analysis. Rather than postulating deterministic or stochastic dynamical equations, the statistical physics-inspired formalism allows us to assign a probability of finding the CPS component and/or workload in a particular configuration state. Introducing a probabilistic description of CPS states has a double role: (1) it enables the quantification of the likelihood of an event to happen or a state to be reached; and (2) it measures our information about the system states and dynamics. By identifying the possible transitions in the CPS dynamics (i.e., the transition probabilities can represent the intrinsic interactions between the CPS components or the effect of the environment on the CPS) and accounting for the uncertainty/perturbations/noise sources, we can construct, as will be outlined in the following, master equations that quantify the evolution of the CPS state probabilities. The master equation can be of a deterministic nature when no uncertainty/noise effects are considered, or stochastic when specific assumptions about uncertainty are made. In addition, this master equation-based modeling of CPS workloads allows not only for encapsulating network state (e.g., flow intensity, resource availability) dependencies, but also captures the effects of complex protocols among CPS components. Another benefit of the master equation approach is that it enables the derivation and analysis of the higher-order statistical moments associated with the dynamics of various CPS states via fractional or integer ordinary differential equations governing their dynamics. Similarly, the stochastic version of the master equation can also help at deriving the stochastic fractional or integer differential equations describing the dynamics of the statistical moments or related metrics (risk of rare events). These deterministic/stochastic fractional/integer differential equations can be encapsulated into optimization strategies. For instance, the master equation allows us to derive the dynamical equation governing the dynamics of the expectation (mean/average) values of the CPS properties or states.

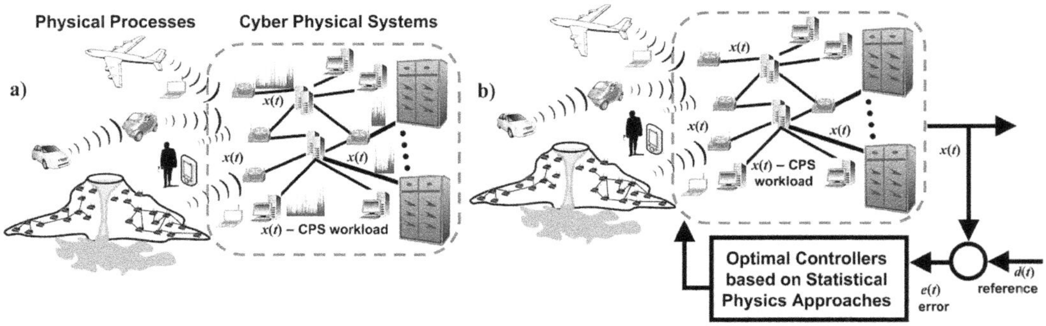

Figure 1.4 (a) A generic representation of the CPS operation from physical processes all the way up to application workloads. Such an itinerary may consist of volcanic activity monitoring, cloud movement, precipitation formation, air pollution, traffic conditions, and physiological processes, followed by compressing the data measurements and communicating them to data centers for further analysis. (b) Feedback control diagram that ensures the QoS reference via statistical physics approaches. Distributed controllers can dynamically estimate the workload and decide, based on specific QoS metrics (e.g., latency), on prioritizing data transmission or allocating more efficiently the communication bandwidth. © 2011 IEEE. Reprinted, with permission, from [28].

Before getting into the mathematical modeling of CPS workloads, let us see first how physical processes contribute to computational and communication workloads. Figure 1.4a shows various types of sensors monitoring diverse physical processes (e.g., volcanic activity, heart rate, car density, CCN concentration). This measured information is then aggregated into packets and communicated as a workload to specialized data centers for further analysis. In the medical domain, a bio-implantable system-on-a-chip can monitor the heart rate by recording the heart electrical activity as a time series. Based on this physiological information, the bio-implantable device can locally analyze the behavior exhibited by the fluctuations in the R–R intervals (as in Figure 1.2) and decide to provide more or less electrical stimuli to the heart. In parallel, abnormalities are communicated to data centers for further data mining and forecasting the likelihood of life-threatening events. Whether the medical device sensing the heart rate is designed to communicate after each heart beat or a fixed number of heart beats, the interval of time between communication packets may exhibit a fractal behavior. By the generalized central limit theorem, the aggregation of multiple independent identically distributed heavy-tailed random variables converges to a new heavy-tailed random variable. Consequently, both the microscopic (from a local medical device) and macroscopic (from multiple medical devices) may contribute to the observed fractality of CPS workloads.

Of note, for medical devices, the CSP workloads can encompass a multitude of multimodal sensing of physiological processes leading to a very heterogeneous CPS workload with varying requirements. For better management of transportation networks, the CPS workloads can arise from both passive mobile sensing of road traffic and active sensing of clouds movement via in-flight airplane measurement. Consequently, the CPS workload communicated through the network consists of traffic metrics and weather measures (e.g., CO_2, CCN concentration). Consequently, from a single type of measurements involving single physical process to those encompassing multimodal sensing, CPS workloads get sent over the shared communication infrastructure (e.g., wireless networks, Internet) and typically consist of a mix of many types of data exhibiting complex spatiotemporal features. Consequently, in what follows, we denote by variable $x(t)$ a stochastic process corresponding to a CPS workload (e.g., communication volume, packet delays).

Relying on total probability theorem and aiming to capture the mathematical characteristics (e.g., self-similarity, nonstationarity) of CPS workloads, we describe the dynamics of the PDF $P(x,t|\alpha)$ representing the stochastic process $x(t)$ as follows:

$$P(x,t|\alpha) = P(x_0,t_0) + \int_0^t dt \int_0^\infty w(x-y,t-t')P(y,t'|\alpha)dy \tag{1.4}$$

where $P(x,t|\alpha)$ denotes the conditional probability of finding the system at time t in a particular state x (e.g., x can represent the level of buffer occupancy, the communication volume between two nodes, the time it takes until a computation request is completed) for a given fractal exponent α, $P(x_0,t_0)$ is the initial condition, and $w\left(x-y,t-t'\right)$ is the kernel probability capturing the dependency of the evolution of probability $P(x,t|\alpha)$ on the memory of the stochastic process $x(t)$. For example, the atmospheric measurements made during flights can be aggregated with road traffic information from cars into heterogeneous workloads and then sent via satellite or intermediate nodes to data centers for further processing. In this case, the stochastic process $x(t)$ denotes the amount of information communicated at a specific time. The aggregation process, which implies changes in both magnitude and timing patterns of $x(t)$, is captured via the joint probability $w\left(x-y,t-t'\right)$. Of note, this distribution depends on the *inter-event times* $t-t$ (i.e., the difference between two consecutive time stamps when the process $x(t)$ takes different values). For power law distributed inter-event times, the kernel takes the following mathematical form $w\left(x-y,t-t'\right) = \frac{v(x-y)\left(t-t'\right)(\alpha-1)}{\Gamma(\alpha)}$, where $v\,(x-y)$ denotes the state transition probabilities and $\Gamma(\alpha)$ is the gamma function [29]. Of note, the worst-case analysis can be retrieved from this formulation by computing the probabilities of rare events (e.g., obtaining the probability distribution for the maximum of a random variable) via high-order moments analysis. Of note, the mathematical relationship in (1.4) represents the rigorous way of constructing a master equation[8] [15, 30] governing the dynamics of a physical phenomenon in nonequilibrium statistical physics.

Due to the inherent multifractal nature of many physical processes, the CPS stochastic processes $x(t)$ can also exhibit a complex self-similar behavior that can be captured via a distribution function $b(\alpha)$[9] of scaling exponents. By multiplying with

[8] The master equation is used in statistical physics to characterize through a dynamical equation the evolution of the probability distribution $P(x,t)$ over a set of states.

[9] The multifractal spectrum $b(\alpha)$ can be estimated either as a histogram of fractal dimensions where each fractal dimension is obtained via box-counting algorithm [31, 32] or by the method of moment estimation via the Legendre formula [31].

$b(\alpha)$ both sides of (1.4) and integrating over the set of fractal exponents, we can get the following relation for the probability $P(x,t)$ of reaching state x at time t:

$$P(x,t) = P(x_0,t_0) + \int_{\alpha_{\min}}^{\alpha_{\max}} b(\alpha)d\alpha \int_0^t \frac{(t-t')^{(\alpha-1)}}{\Gamma(\alpha)}dt' \int_0^\infty v(x-y)P(y,t'|\alpha)dy \tag{1.5}$$

By recalling the definition of the fractional Riemann–Liouville integral of order α [26, 33, 34]:

$${}_0I_t^\alpha P(x,t) = \int_0^t P(x,t')\frac{(t-t')^{(\alpha-1)}}{\Gamma(\alpha)}dt' = {}_0D_t^{-\alpha} \tag{1.6}$$

and using the duality between the fractional Riemann–Liouville integral and fractional Riemann–Liouville derivative

$${}_0D_t^\alpha P(x,t) = {}_0D_t^n\,{}_0I_t^{n-\alpha} = \frac{d^n}{dt^n}\int_0^t P(x,t')\frac{(t-t')^{(n-\alpha-1)}}{\Gamma(n-\alpha)}dt' \tag{1.7}$$

we obtain the following relationship:

$$\int_{\alpha_{\min}}^{\alpha_{\max}} b(\alpha)\frac{\partial^\alpha P(x,t|\alpha)}{\partial t^\alpha}d\alpha = \int_0^\infty v(x-y)P(y,t)dy \tag{1.8}$$

Next, we assume that the transition probabilities $v(x-y)$ has only local dependency on the state of stochastic process expressed via nonlinear mathematical functions $f_1(x-1)$ and $f_2(x+1)$ and, thus, the right-hand side of (1.8) takes the form:

$$\begin{aligned}\int_0^\infty v(x-y)P(y,t)dy &= a_1 f_1(x-1)P(x-1,t)\\ &\quad + a_2 f_2(x+1)P(x+1,t) - [a_1 f_1(x) + a_2 f_2(x)]P(x,t)\\ &\quad + \int_0^\infty \frac{\gamma(x-y)}{H(y,t)}P(y,t)dy \end{aligned}\tag{1.9}$$

where $a_1 f_1(x-1)$ and $a_2 f_2(x+1)$ are the transition probabilities from states $(x-1)$ and $(x+1)$ to state x, $a_1 f_1(x) + a_2 f_2(x)$ represents the probability of remaining in the same state x, and $\gamma(x-y)/H(y,t)$ denotes the memory kernel reflecting the intrinsic multiplicative noise that may affect the evaluation of transition probabilities [35]. In a hypothetical situation when the stochastic evolution of the stochastic process $x(t)$ is perfectly characterized, the multiplicative noise (i.e., $\frac{\gamma(x-y)}{H(y,t)}$) becomes zero.

Based on the relations in (1.8) and (1.9), and using the finite difference method [29], the generalized master equation (1.5) characterizing the evolution of the CPS workload $x(t)$ can be expressed as follows:

$$\begin{aligned}\int_{\alpha_{\min}}^{\alpha_{\max}} b(\alpha)\frac{\partial^\alpha P(x,t|\alpha)}{\partial t^\alpha}d\alpha &= \frac{\partial}{\partial x}[a_1 f_1(x)\\ &\quad + a_2 f_2(x)]P(x,t) - \frac{\partial}{\partial x}[a_1 f_1(x) - a_2 f_2(x)]P(x,t)\\ &\quad + \int_0^\infty \frac{\gamma(x-y)}{H(y,t)}P(y,t)dy \end{aligned}\tag{1.10}$$

The first term in (1.10) is introduced to capture the multifractal temporal features of the stochastic process $x(t)$, and represents the dynamics of CPS workload as a weighted sum of fractional-order derivatives. One can investigate the existence of multifractal behavior by studying the higher-order moments $M_k(t)$ associated with the stochastic process $x(t)$ and checking whether they satisfy the following relation:

$$M_k(t) \approx t^{const-\tau(k)} \tag{1.11}$$

If the mass exponent $\tau(k)$ depends linearly on the kth-order moment of the stochastic process $x(t)$, then this process is called monofractal. Instead, if the mass exponent $\tau(k)$ exhibits a nonlinear relationship with respect to the kth-order moment, then the process is called *multifractal*, i.e., a bell-like distribution of fractal dimensions can be observed (as shown in Figure 1.3b). Generally speaking, this implies that the distribution of $x(t)$ consists of a superposition of multiple power laws. Of note, the fractional order derivative implies that the underlying inter-event times $(t - t')$ associated with the stochastic process $x(t)$ are power law, rather than exponentially, distributed. For a memoryless type of dynamics characterized by exponential inter-event times, the first term reduces to the classical (integer) order time derivative of $P(x,t)$ with respect to time. The second and third term are meant to encompass the magnitude of fluctuations in the stochastic process $x(t)$. Finally, the last term in (1.10) is introduced to model the multiplicative noise that comes from a system in contact with the environment. One can note that if $b(\alpha) = \delta(\alpha - 1)$ and $\gamma(x) = 0$, (1.10) reduces to the normal diffusion equation for the evolution of $P(x,t)$ [20] and allows for deriving classical time-dependent solutions of the queuing models (e.g., M/M/1, M/G/1).

The exhibited multifractal behavior requires new control strategies based on *nonlinear* state equations. To provide some insight on how the master equation can help at building these *nonlinear* state equations and develop new control strategies, one can derive from (1.10) a Langevin-like dynamical equation describing the state of stochastic process with multiplicative noise as follows:

$$\int_{\alpha_{\min}}^{\alpha_{\max}} b(\alpha)\frac{\partial^{\alpha} P(x,t|\alpha)}{\partial t^{\alpha}} d\alpha = f(x,t) + H(x,t)\eta(t) \tag{1.12}$$

where $f(x,t)$ is given by the following relation:

$$f(x,t) = a_2 f_2(x) - a_1 f_1(x) - \int_0^{\infty} \frac{\partial}{\partial x}[a_1 f_1(x) + a_2 f_2(x)]P(x,t)dx \tag{1.13}$$

One important feature of the above-mentioned statistical physics-inspired modeling approach is that it allows us to characterize not only the dynamics of the PDF of a particular stochastic process, but also to derive fractional- or integer-based ordinary differential equations that can be used to describe physical and cyber processes. In addition, this newly introduced mathematical modeling approach can be seen as a general framework that encapsulates the fractal and non-fractal models as particular cases. For instance, one can clearly observe that by considering $b(\alpha) = \delta(\alpha - 1)$ (i.e., no fractal behavior in time), a linear state space dependence $f(x,t) = ax$ and

$H(x,t) = 0$ (zero noise source), (1.11) reduces to a classical linear control state equation. However, in the next section, we will show how (1.12) can be used formulate a new problem of optimal control for CPS.

1.5 CPS Design under Uncertainty Conditions

So far we have seen how master equations can be used to provide both a probabilistic and deterministic description of the CPS workload dynamics and how their mathematical features can be accurately captured in simple state equations. In this section, we discuss how these mathematical models can actually enable the formulation of control approaches for fractal CPS processes (see Figure 1.4b). As in any optimization, the end goal of a CPS design task is to minimize (or maximize) a particular relevant cost function (e.g., power consumption, fuel consumption, thermal profile, failure rate) subject to various resource limitations, performance, and reliability constraints.

Figure 1.5 shows that due to the intrinsic continuous interactions of CPS with the physical environment, the main tasks of computation, communication, and control *cannot* be modeled, analyzed, and optimized in isolation, but instead the uncertainty and noise sources need to be considered as well. Consequently, we formulate the CPS design problem (also described in Figure 1.5) as a general (stochastic) optimization problem:

$$\min_{u(t)} \int_{x \in X} \int_{t_{initial}}^{t_{final}} c(x(t), u(t), d(t), t) p(x(t_{initial})|u(t_{initial}))dt$$

$$s.t.\ u_{\min} \leq u(t) \leq u_{\max}, k = 1 \div N, x_i(0) = x_i^*, i = 1 \div M$$

$$\int_{\alpha_i^{\min}}^{\alpha_i^{\max}} b(\alpha_i) \frac{\partial^{\alpha_i} x(t)}{\partial t^{\alpha_i}} d\alpha_i = f(x_i, x, d, t) + g(u,t) + h(x_i, t)$$

where $c(x(t), u(t), d(t), t)$ represents the cost objective as a function of the states of the system $x_i(t) \in \Xi, i = 1 \div M$, the input control signals $u_k(t) \in \Psi, k = 1 \div N$, and

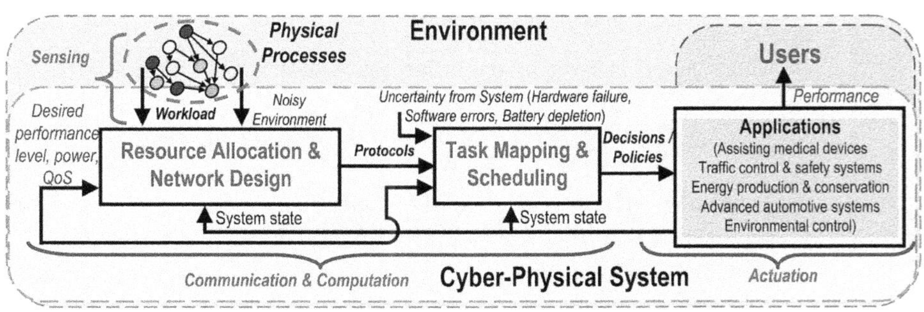

Figure 1.5 The main challenges of CPS design: resource allocation and network design to accurately sense a set of physical processes within a noisy environment, task mapping, and computation scheduling to determine the optimal control strategies that allow satisfying the performance constraints. © 2011 IEEE. Reprinted, with permission, from [28].

the reference/demands signals $d_j\,(t) \in \Lambda,\ j = 1 \div L$; $p(x(t)|u(t))$ is the conditional probability of finding the CPS in state $x(t)$, given the control scenario $u(t)$; u_{min} and u_{max} are the lower and upper bounds on the control signals $u_k\,(t) \in \Psi,\ k = 1 \div N$; α_i and $b(\alpha_i)$ denote the fractional-order derivatives and the distribution of fractal dimensions characterizing the dynamics of $x_i\,(t) \in \Xi,\ i = 1 \div M$; $f(x_i(t),d(t),t)$ is a generic function describing the evolution of system states $x_i\,(t) \in \Xi,\ i = 1 \div M$; $g(u(t),t)$ represents the dependency between the control signals $u_k\,(t) \in \Psi,\ k = 1 \div N$ and the stochastic process $x_i\,(t) \in \Xi,\ i = 1 \div M$ of interest; $h(x_i(t),t)$ is the multiplicative noise amplitude that depends on the actual state of the system; and $\eta(t)$ is the noise coming from a fluctuating environment (e.g., white noise). Of note, in the above formulation, we only consider a fractional differential equation governing the dynamics of the mean of the processes of interest. One can generalize this formulation by considering higher-order moments. Also, one can retrieve classical non-fractional formulations by considering that $b\,(\alpha) = \delta\,(\alpha - 1)$, which leads to a first-order time derivative.

To provide some insight about the sources of this noise, we can, for instance, imagine that under the impact of some disruptive news, a patient whose heart is controlled via a pacemaker experiences a ventricular arrhythmia which results in less blood pumped through the arteries. These changes in the blood flow can only be modeled as some multiplicative noise [36]. Moreover, this arrhythmia indirectly affects not only the heart rate, but also the workload communicated between the pacemaker and the decision center. Similar cases in which the multiplicative noise can affect the CPS workload directly or indirectly can be found in mobile communication channels due to air turbulence, reflection, refraction, or multipath effects (multiplicative interference) [37, 38], in optical channels due to amplitude vector rotation [32].

In order to determine the minimum cost and its associated control signals $u_k\,(t) \in \Psi$, $k = 1 \div N$, we define the following functional:

$$H(x,t) = \min_{u(t)} \int_{x \in X} \int_{t_{initial}}^{t_{final}} c(x(t),u(t),d(t),t)p(x(t_{initial})|u(t_{initial}))dtdx \qquad (1.14)$$

The goal of the stochastic optimization problem defined in (1.14) is to determine the admissible control signals $u_k^*\,(t) \in \Psi,\ k = 1 \div N$, which cause the system or stochastic process of interest (whose dynamics is captured either by (1.10) or (1.12)) to follow an admissible trajectory $x_i^*\,(t) \in \Xi,\ i = 1 \div M$ that minimizes the cost functional in (1.10).[10] Simply put, an admissible trajectory for utilization $x(t)$ of a queue means to keep it within a predefined range such that chances of high blocking probability (queue overflows) are minimized.

The functional $H(x,t)$ in (1.14) can be expressed as a sum of two terms by considering first the minimization over the interval $\left[t_{initial}, t_{initial} + \delta t\right]$, and then over the interval $\left[t_{initial} + \delta t, t_{final}\right]$ as follows:

[10] Note that, because we assume that the state of the system undergoes a stochastic dynamics, we choose to minimize the expectation of the cost function. For highly nonlinear processes, we may also consider the kurtosis of the cost function such that the deviations (or chances) of rare (cost) events are minimized.

$$H(x,t) = \min_{u(t_{initial})} \left\{ \int_{x\in X} \int_{t_{initial}}^{t_{initial}+\delta t} c(x,u,d,t)p(x(t_{initial})|u(t_{initial}))dtdx + \int_{x\in X} \int_{t_{final}+\delta t}^{t_{final}} c(x,u,d,t)p(x(t_{initial})|u(t_{initial}))dtdx \right\} \quad (1.15)$$

In (1.15), the first term on the right-hand side considers the space of *all* possible control signals $u_k(t) \in \Psi$, $k = 1 \div N$ for a time window $\left[t_{initial}, t_{initial} + \delta t\right]$ and encodes all the optimal control signals from time $t_{initial} + \delta t$ onwards. The second term in (1.15) denotes the cost to be minimized over $\left[t_{initial} + \delta t, t_{final}\right]$ interval. Equation (1.15) also shows that the best approach to optimality, given the complex fractal dynamics of the workload, is to search the entire time horizon instead of starting with some random initial configuration and then searching for the optimum.

Using the optimality principle [39], we can distribute the minimization operation in (1.15) over two paths, one that depends on the control signals $u_k(t_{initial}) \in \Psi$, $k = 1 \div N$ at time $t_{initial}$, and the second one which depends on the controls $u_k(t_{initial} + \delta t) \in \Psi$, $k = 1 \div N$ at time $t_{initial} + \delta t$:

$$H(x,t) = \min_{u(t_{initial})} \left\{ \int_{x\in X} \int_{t_{initial}}^{t_{initial}+\delta t} c(x,u,d,t)p(x(t_{initial})|u(t_{initial}))dtdx \right\} + \min_{u(t_{initial}+\delta t)} \left\{ \int_{x\in X} \int_{t_{initial}}^{t_{final}} c(x(t),u(t),d(t),t)p(x(t_{initial})|u(t_{initial}))dtdx \right\} \quad (1.16)$$

One approach to simplify (1.16) is to use the Taylor expansion formula and expand the $H(x(t_{initial} + \delta t), t_{initial} + \delta t)$ term as a function of the previous $H(x(t_{initial}), t_{initial})$, leading to the following relation:

$$H(x, t_{initial} + \delta t) = H(x, t_{initial}) + \sum_{i=1}^{M} \int_{\alpha_i^{\min}}^{\alpha_i^{\max}} b(\alpha_i) \frac{\partial^{\alpha_i} x(t)}{\partial t^{\alpha_i}} d\alpha_i + \sum_{i=1}^{M} \frac{\partial^{\alpha_i} H(x, t_{initial})}{\partial x^{\alpha_i}} d^{\alpha_i} x + O(d^{\alpha_i} x, dt^{\alpha_i}) \quad (1.17)$$

where we make use of the fractal behavior that may be exhibited by the stochastic process $x(t)$ (see (1.10)) and neglect the higher-order terms, i.e., $O\left(d_i^{\alpha} x, dt_i^{\alpha}\right)$. Substituting (1.17) into (1.16) and rearranging the terms leads to a multifractal partial differential equation for the stochastic problem in (1.14):

$$-\sum_{i=1}^{M} \int_{\alpha_i^{\min}}^{\alpha_i^{\max}} b(\alpha_i) \frac{\partial^{\alpha_i} x(t)}{\partial t^{\alpha_i}} dt^{\alpha_i} d\alpha_i = \min_{u(t_{initial})} \left\{ \sum_{i=1}^{M} \frac{\partial^{\alpha_i} H(x, t_{initial})}{\partial x^{\alpha_i}} d^{\alpha_i} x + \int_{x\in X} \int_{t_{initial}}^{t_{initial}+\delta t} c(x,u,d,t)p(x|u)dtdx \right\} \quad (1.18)$$

Equation (1.18) allows the study and stochastic control of monofractal processes by assuming a point-like distribution of fractal exponents $b(\alpha_i) = \delta(\alpha_i - \alpha)$ for all $i = 1 \div M$, then dividing by dt^α and taking the limit $\lim_{dt^\alpha \to 0}$. This algebraic strategy leads to the following fractional partial differential equation:

$$-\frac{\partial^{\alpha_i} H}{\partial t^{\alpha_i}} = \min_{u(t)} \left\{ \sum_{i=1}^{M} \frac{\partial^\alpha H(x, t_{initial})}{\partial x_i^\alpha} \int_{x \in X} \frac{d^\alpha}{dt^\alpha} p(x|u)dx + \int_{x \in X} \lim_{\delta t^\alpha \to 0} \frac{1}{\delta t^\alpha} \int_t^{t+\delta t} c(x,u,d,t)p(x|u)dtdx \right\} \quad (1.19)$$

where, for simplicity, we replaced the $t_{initial}$ with t.

Using the fractal dynamical equation characterizing the evolution of $x_i(t) \in \Xi$, $i = 1 \div M$, (1.19) can be further reduced to the following relation:

$$-\frac{\partial^{\alpha_i} H}{\partial t^{\alpha_i}} = \min_{u(t)} \left\{ \int_{x \in X} \lim_{\delta t^\alpha \to 0} \frac{1}{\delta t^\alpha} \int_t^{t+\delta t} c(x,u,d,t)p(x|u)dtdx + \sum_{i=1}^{M} \frac{\partial^\alpha H(x, t_{initial})}{\partial x_i^\alpha} \int_{x \in X} [f(x_i,t) + g(u,t) + h(x_i,t)\eta_i]p(x|u)dx \right\} \quad (1.20)$$

Equation (1.20) can model the behavior of CPS components by taking into account the characteristics of the workloads through the fractional derivatives of order αu. For instance, the last term in (1.20) captures the interactions between CPS components (e.g., $x_i(t) \in \Xi$, $i = 1 \div M$) and physical processes (e.g., $u_k(t) \in \Psi$, $k = 1 \div N$). Incorporating fractional derivatives as shown in (1.10) and (1.18) to account for the exhibited multifractal behavior that is typical of the CPS workload is meant to allow for more accurate estimates of chances of rare events, unsafe operation modes, or failures with devastating consequences. Of note, the cost function in the optimization problem does *not* rely on memoryless properties (like the ones presented in [39]), but instead it is directly influenced by the multifractal behavior of the CPS workload. Ignoring these multifractal properties can result in optimistic bounds or predictions and therefore compromise the design of the entire CPS system.

1.6 Mathematical Implications of the Fractal Formalism on Bio-implantable CPS Medical Devices

In what follows, we describe one example that links the mathematical characteristics of heart rate variability (e.g., fractal exponent of the mean R–R intervals) with the design of the control algorithm of a bio-implantable pacemaker [40]. To make the discussion more concrete, we consider the time series of R–R intervals shown in Figure 1.6a, which correspond to an individual experiencing bradycardia. Bradycardia is a medical condition in which the heart rate slows down below 60 beats per minute

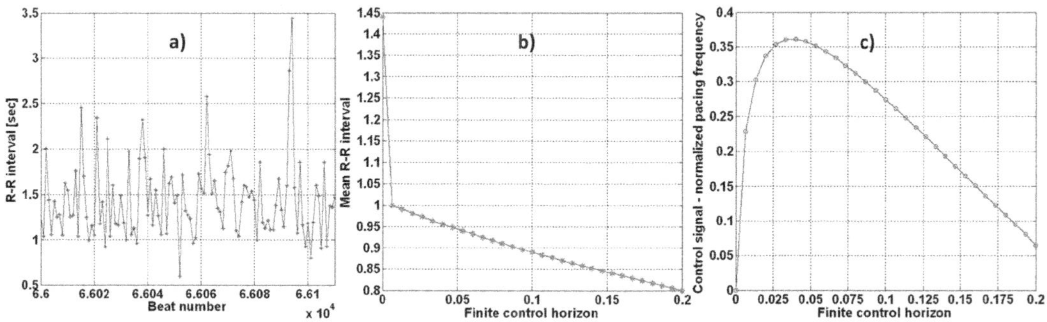

Figure 1.6 (a) A time series of R–R intervals collected via ECG recording from an individual suffering from bradycardia. The maximum and mean of the R–R intervals observed in this time series are 3.4 and 1.4, respectively. (c) Mean of the R–R interval as a function of the control horizon by applying a fractal controller with a fractal exponent of 0.64 (i.e., the order of the fractional-order derivative is equal to 0.64). (c) The fractal control signal as a function of the finite time horizon required for the controller to be able to bring the mean R–R interval from a dangerous region of extreme bradycardia to a normal heart rate of 75 beats per minute or an R–R interval of 0.8 seconds.

and so the mean of R–R intervals starts exceeding 1 second. Of note, the R–R intervals of a healthy individual should range between 0.6 and 1 second, which corresponds to a heart rate of 60 to 100 beats per minute.

As one can observe from Figure 1.6a, the R–R intervals not only exhibit multiple peaks that exceed 1 second, but also display some very large values (see the 3.44 seconds R–R interval on the right side of the graph in Figure 1.6a). Although the mean of the R–R intervals is approximately 1.44 seconds, the large values of a few R–R intervals can cause significant harm to an individual. Large R–R intervals cannot only induce dizziness, fatigue, and weakening, they can also contribute to fainting and life-threatening situations. By carefully analyzing the R–R intervals, we observe that the time series is characterized by a fractal exponent of $\beta = 0.64$ (with an estimation error of $\pm$ 0.26). One can clearly see that this fractal exponent requires a new type of modeling of the R–R interval dynamics, which can be achieved by using a fractional-order differential equation of the following form:

$$\frac{d^{\beta} m_{RR}(t)}{dt^{\beta}} = a m_{RR}(t) + b u(t) \tag{1.21}$$

where $m_{RR}(t)$ and $u(t)$ are the mean of the R–R intervals and the pacing frequency, respectively, $a = 0.09$ and $b = 0.23$ are proportionality coefficients estimated via regression methods. Besides the estimation of parameters, the goodness-of-fit analysis for the model in (1.21) reveals a p-value of 0.95, which confirms that the fractional order differential equation (1.21) cannot be rejected as a good model for the dynamics of R–R intervals in Figure 1.6a.

Following the mathematical methodology in (1.14), the fractal controller seeking to pace the heart and bring the heart rate within a healthy range can be obtained by solving the following fractal optimal control problem:

$$\min_{u(t)} \int_0^T \left\{ \frac{q}{2}(m_{RR}(t) - m_{RR}^{ref})^2 + \frac{r}{2}(u(t) - u^{ref})^2 \right\} dt \tag{1.22}$$

$$\frac{d^\beta m_{RR}(t)}{dt^\beta} = am_{RR}(t) + bu(t),\ m_{RR}(t=0) = 1.44,\ m_{RR}(t=T) = 0.80$$

$$m_{RR}^{\min} \leq m_{RR}(t) \leq m_{RR}^{\max},\ u_{\min} \leq u(t) \leq u_{\max} \tag{1.23}$$

where T represents the magnitude of the finite horizon, m_{RR}^{ref} and u^{ref} denote the reference values imposed for both the mean of the R–R intervals and the pacing frequency, m_{RR}^{min} and m_{RR}^{max} are the minimum and maximum bounds allowed for the mean of the R–R intervals $m_{RR}(t)$, u_{min} and u_{max} are the minimum and maximum bounds allowed on the pacing frequency $u(t)$, q and r are proportionality coefficients weighting the importance of the deviations from the reference value in both state (mean of R–R intervals) and control (pacing frequency) variables, respectively. In addition, $m_{RR}(t = 0) = 1.44$ and $m_{RR}(t = T) = 0.8$ represent the initial and final conditions imposed on the mean of the R–R intervals.

Employing optimal control arguments such as the ones described in Section 1.5, the constrained fractal optimal control problem in (1.22) and (1.23) can be reduced to solving a linear program [40]. Starting from the bradycardia premises summarized in Figure 1.6a, we solve the constrained fractal optimal control problem for the following set of parameters: $T = 0.2$, $q = 0.1$, $r = 0.01$, $m_{RR}^{min} = 0.6$, $m_{RR}^{max} = 1$, $u_{min} = 0$, and $u_{max} = 1$. We employ $N = 30$ as the number of discretization steps between $t = 0$ and $t = T$. Figures 1.6b and 1.6c show the evolution of: (1) the mean of the R–R intervals as a function of the finite control horizon; and (2) the pacing frequency (control signal) as a function of the finite control horizon T. Figure 1.6c shows that a nonlinear control sequence is required to bring the mean of R–R intervals within defined bounds, a goal that cannot be achieved with high accuracy and stability with linear control methods. A similar strategy has been employed in [4] to compute the control signal for a process in which the dynamics are characterized by a time-dependent fractal dimension. In addition, we showed that fractal optimal control algorithms can be efficiently implemented in hardware and can be employed in CPS.

1.7 Conclusion and Future Work

In this chapter, we have presented a radically new view of CPS design, which captures the complex characteristics of CPS workload via a fractal-based master equation. This formalism can serve as a basis of a generalized stochastic (optimal) control framework, which can be further particularized to various CPS design problems like resource allocation, task scheduling and mapping, and more. From ecosystems dynamics all the way down to blood circulation in the heart, many natural processes display complex regulatory and self-organization schemes that reduce to nonlinear stochastic optimization problems. Accurate characterization of CPS workloads through master equations with memory kernels can not only encompass their complex statistical features into nonlinear stochastic optimal control problems, but also open new pathways for online design and optimization algorithms. Future work should target online algorithms for solving (1.9) by combining various computational methods (e.g., dynamic programming, barrier methods).

1.8 Acknowledgments

P. B. acknowledges the support by the US National Science Foundation (NSF) under Grants 1331610 and 1453860. R. M. acknowledges the generous support by the NSF grant CNS-1128624 and SRC grants 2011-HJ-2168 and 2012-HJ-2232.

References

[1] P. Mohan, V. N. Padmanabhan, and R. Ramjee, "Nericell: rich monitoring of road and traffic conditions using mobile smartphones," presented at the *6th ACM Conf. on Embedded Network Sensor Systems*, Raleigh, NC, November 5–7, 2008.

[2] A. D. Polyanin and A. V. Manshirov, *Handbook of Mathematics for Engineers and Scientists*. London: Taylor and Francis, 2007.

[3] R. Sugihara and R. K. Gupta, "Programming models for sensor networks: A survey," *ACM Trans. Sen. Netw.* vol. 4, no. 2, April 2008.

[4] M. Ghorbani and P. Bogdan, "A cyber-physical system approach to artificial pancreas design," presented at the *Proc. of Ninth IEEE/ACM/IFIP Int. Conf. on Hardware/Software Codesign and System Synthesis (CODES+ISSS)*, Montreal, October 2013.

[5] D. E. Kirk, *Optimal Control Theory: An Introduction*. New York: Dover Publications, 1998.

[6] J. A. Stankovic, I. Lee, A. Mok, and R. Rajkumar, "Opportunities and obligations for physical computing systems," *Computer*, vol. 38, no. 11, pp. 23–31, Nov. 2005.

[7] E. A. Lee and D. G. Messerschmitt, *Digital Communication*. Dordrecht: Kluwer Academic Publishers, 1995.

[8] S. G. Samko, A. A. Kilbas, and O. I. Marichev, *Fractional Integrals and Derivatives: Theory and Applications*. London: Gordon & Breach, 1993.

[9] Y. Xu, F. Ren, T. He, et al., "Real-time routing in wireless sensor networks: A potential field approach," *ACM Trans. Sen. Netw.*, vol. 9, no. 3, article 35, Jun. 2013.

[10] J. Allansona and S. H. Fairclough, "A research agenda for physiological computing," *Interact. Comput.*, vol. 16, no. 5, pp. 857–878, 2004.

[11] P. Ch. Ivanov, M. G. Rosenblum, C.-K. Peng, et al. "Scaling and universality in heart rate variability distributions," *Physica A*, vol. 249, pp. 587–593, 1998.

[12] E. Jovanov and A. Milenkovic, "Body area networks for ubiquitous healthcare applications: opportunities and challenges," *J. Med. Syst.*, vol. 35, no. 5, pp. 1245–1254, Oct. 2011.

[13] E. Miluzzo, N. D. Lane, K. Fodor, et al., "Sensing meets mobile social networks: The design, implementation and evaluation of the CenceMe application," presented at the *6th ACM Conf. on Embedded Network Sensor Systems*, Raleigh, NC, November 5–7, 2008.

[14] B. B. Mandelbrot, *The Fractal Geometry of Nature*. New York: W.H. Freeman and Co., 1982.

[15] M. Xue, S. Roy, Y. Wan, and S. K. Das, "Security and vulnerability of cyber-physical infrastructure networks: a control-theoretic approach," in *Handbook on Securing Critical Cyber-Physical Infrastructures: Foundations and Challenges* (S. K. Das, K. Kant, and N. Zhang, eds.). Burlington, MA: Morgan-Kauffman, 2012.

[16] B. J. West, *Fractal Physiology and Chaos in Medicine*. Singapore: World Scientific, 2013.

[17] P. Bogdan and R. Marculescu, "Cyberphysical systems: Workload modeling and design optimization," *IEEE Des. Test. Comput.*, vol. 28, no. 4, pp.78–87, Jul.–Aug. 2011.

[18] K. Park and W. Willinger, *Self-Similar Network Traffic and Performance Evaluation*. Chicester: Wiley, 2000.

[19] K. B. Oldham and J. Spanier, *Fractional Calculus: Theory of Differentiation and Integration to Arbitrary Order*. New York: Academic Press, 2006.

[20] R. Balescu, *Aspects of Anomalous Transport in Plasmas*. Bristol: Institute of Physics Publishing, 2005.

[21] R. Mahajan, J. Zahorjan, and B. Zill, "Understanding WiFi-based connectivity from moving vehicles," in *Proc. of the 7th ACM SIGCOMM Conf. on Internet Measurement*, San Diego, CA, October 24–26, 2007.

[22] J. B. Bassingthwaighte, L. S. Liebovitch, and B. J. West, *Fractal Physiology*. Oxford: Oxford University Press, 1994 .

[23] USDA Forest Service, Wind speed data-set for Hubbard Brook Experimental Forest, at www.hubbardbrook.org/data/dataset.php?id=56.

[24] T. Vicsek, *Fractal Growth Phenomena*. Singapore: World Scientific, 1992.

[25] R. Hilfer and L. Anton, "Fractional master equations and fractal time random walks," *Phys. Rev. E*, vol. 51, no. 2, Feb. 1995.

[26] National Science Foundation, "Cyber-physical systems," at www.nsf.gov/publications/pub_summ.jsp?ods_key=nsf08611.

[27] E. Lee, "CPS Foundations," in *Proc. of the 47th IEEE/ACM Design Automation Conf.*, Anaheim, June 2010.

[28] P. Bogdan and R. Marculescu, "Towards a science of cyber-physical systems design," presented at the *IEEE/ACM Int. Conf. on Cyber-Physical Systems (ICCPS)*, Chicago, IL, April 12–14, 2011, pp. 99–108.

[29] C.-K. Peng, S. V. Buldyrev, S. Havlin, et al., "Mosaic organization of DNA nucleotides," *Phys. Rev. E*, vol. 49, pp. 1685–1689, 1994.

[30] R. Gupta, P. Le Guernic, S. Shukla, and J. P. Talpin, *Formal Methods and Models for System Design: A System Level Perspective*. Dordrecht: Kluwer Academic Publishers, 2004.

[31] C. Lu, J. Stankovic, G. Tao, and S. Son, "Feedback control real-time scheduling: Framework, modeling and algorithms," *J. Cont. Theoret. Approach. Real-Time Comput.*, vol. 23, pp. 85–126, 2002.

[32] P. Tabuada, "Cyber-physical systems," presented at the *National Science Foundation Workshop on Cyber-Physical Systems*, Austin, TX, October 16–17, 2006.

[33] J. Kleissl and Y. Agarwal, "Cyber-physical energy systems: focus on smart buildings," presented at the *Design Automation Conf.*, Anaheim, June 2010.

[34] R. Rajkumar, I. Lee, L. Sha, and J. Stankovic, "Cyber physical systems: The next computing revolution," presented at the *ACM Design Automation Conf.*, Anaheim, June 2010.

[35] S. I. Denisov, W. Horsthemke, and P. Hänggi, "Generalized Fokker–Planck equation: Derivation and exact solutions," *Eur. Phys. J. B*, vol. 68, no. 4, pp. 567–575, 2009.

[36] G. M. Zaslavsky, *The Physics of Chaos in Hamiltonian Systems*. London: Imperial College Press, 2007.

[37] M. S. Alencar and V. C. da Rocha, *Communication Systems*. New York: Springer, 2005.

[38] A. Krause, E. Horvitz, A. Kansal, and F. Zhao, "Toward Community sensing," presented at the *IEEE Computer Society Intl. Conf. on Information Processing in Sensor Networks*, St. Louis, MO, April 22–24, 2008.

[39] A. A. Kilbas, H. M. Srivastava, and J. J. Trujillo, *Theory and Applications of Fractional Differential Equations*. Amsterdam: Elsevier, 2006.

[40] P. Bogdan, S. Jain, and R. Marculescu, "Pacemaker control of heart rate variability: A cyber physical system perspective," *ACM Trans. Embed. Comput. Syst.*, vol. 12, article 50, Mar. 2013.

2 Effective Uncertainty Evaluation in Large-Scale Systems

Junfei Xie, Yan Wan, Yi Zhou, Kevin Mills, James J. Filliben, and Yu Lei

2.1 Introduction

One of the main challenges of managing large-scale cyber-physical systems (CPSs) is coping with uncertainty. In particular, the physical parts of CPSs typically operate in uncertain environments, with their dynamics significantly modulated by these environmental uncertainties (Figure 2.1). As such, the management of physical dynamics must consider uncertainties, and be designed in such a way to be robust to the uncertainties. For instance, strategic air traffic management is concerned with designing traffic management initiatives robust to weather uncertainty at a 2–15-hour look-ahead time [1, 2]. Similarly, the management of complex information systems seeks to avoid performance degradation subject to uncertain user behaviors [3].

Uncertainties may also inject into the cyber components of CPSs, including sensing, communication, computation, and control. For instance, the quality of communication channels can be easily affected by environmental uncertainties [4, 5]. In this chapter, the sources of uncertainties are not emphasized. Instead, the focus is on 2–15-hour systematic methods used for uncertainty evaluation, with an emphasis on an effective simulation-based uncertainty evaluation method called the probabilistic collocation method (PCM).

We note that the study of uncertainty evaluation contributes significantly to failure control and the design of countermeasures for cyberthreats to CPSs. As threats are typically unknown to CPSs in the operation stage, they can be treated as uncertain disturbances to the system. Uncertainty evaluation, which predicts system performance subject to uncertain disturbances, becomes critical toward failure control and the design of countermeasures.

Uncertainties are typically studied using either theoretical analysis or numerical simulation. Analytical approaches are developed based on models of dynamical CPSs integrated with uncertain environmental impact. For instance, in [6, 7], the master Markov chain and jump-linear approaches are adopted to analyze the performance of a simple air-traffic system under weather uncertainties. Simulation-based methods, on the other hand, do not require prior knowledge of CPS models, but instead use input–output simulation data to statistically analyze the impact of uncertainties.

Whether to use analytical or simulation-based approaches is dependent on the complexity of a specific application of study. Analytical approaches are more efficient if a tractable model is available. For large-scale CPSs whose models

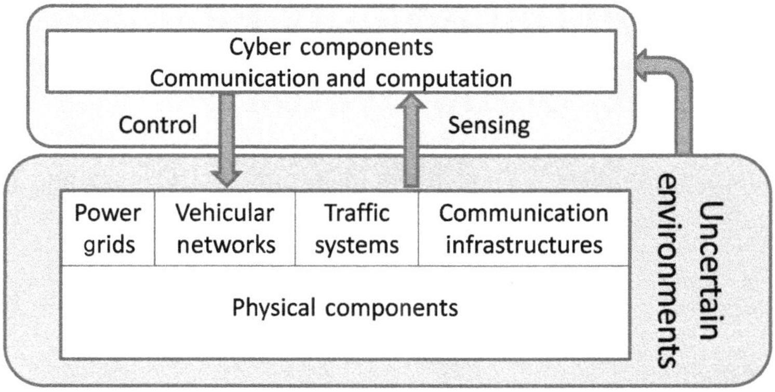

Figure 2.1 Uncertain environments modulate the dynamics of CPSs.

are too complicated to analyze, approximated simpler models may be sought to achieve a tractable analysis. When such an approximated model is difficult to achieve, simulation-based methods are utilized to obtain the statistics of uncertainty impacts. The two methods may also be integrated: Simulations are first used to produce approximate system dynamics, which are then used for tractable analysis. In this chapter, we focus on simulation-based uncertainty evaluation approaches when a tractable analysis is infeasible.

This chapter is organized as follows. In Section 2.2, the uncertainty evaluation problem is formulated. Several simulation-based evaluation methods are reviewed. In Section 2.3, the properties and procedures for the single-variable PCM approach are presented. Section 2.4 discusses an extension of the single-variable PCM to the multivariate case. Section 2.5 motivates the concerns of the multivariate PCM in terms of scalability and robustness to numerical errors, and then presents a design to address these issues. The design integrates statistical experimental design methods to make the PCM method applicable for large-scale system applications. Section 2.6 includes an example to demonstrate the use of simulation-based uncertainty evaluation methods in an exemplary CPS application – air-traffic flow management. Finally, Section 2.7 includes a brief conclusion and discussion of future work.

2.2 The Background of Simulation-Based Uncertainty Evaluation

This section provides a background study of simulation-based uncertainty evaluation. The uncertainty evaluation problem is first mathematically formulated in Section 2.2.1. The traditional Monte Carlo simulation method is then reviewed in Section 2.2.2. After that, several sampling-based simulation methods are reviewed in Section 2.2.3.

2.2.1 Problem Formulation

Many large-scale systems have complicated dynamics that require considerable time to simulate. In order to manage these systems in real time under uncertainty, a practical

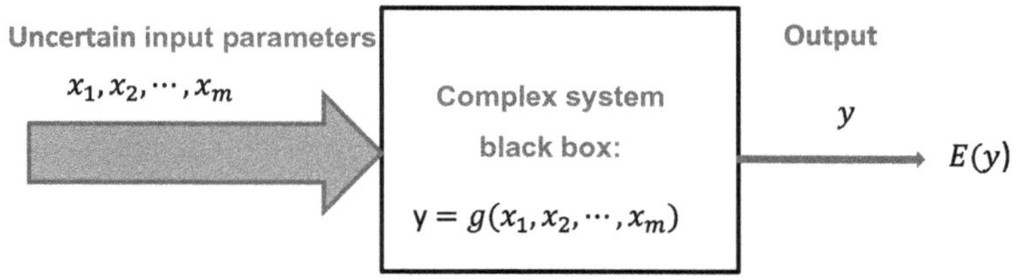

Figure 2.2 Estimation of mean output for a complex system subject to multiple uncertain input parameters. Copyright © 2015 AIAA. Reprinted, with permission, from [8].

need is to quickly evaluate system performance subject to multiple uncertain parameters. This problem can be formulated as a black-box estimation problem, shown in Figure 2.2 [8]. In particular, a system has m uncertain parameters (x_1, x_2, ..., x_m) as inputs and an output y that captures the system performance of interest. These input parameters modulate the dynamics of the system and shape the output y. The problem is to correctly estimate the mean output $E(y)$ using simulations. Examples 2.1 and 2.2 illustrate the problem formulation in two CPS applications.

EXAMPLE 2.1 *Air-traffic flow management.* Convective weather events reduce the capacity of affected airspace and cause en-route delays of aircraft [1, 2]. Traffic management initiatives are designed several hours ahead of possible weather events, based on predictions of traffic system performance, evaluated by, e.g., the total delay of aircraft over a given time frame. The start times, durations, and intensities of convective weather events at geographically distributed regions can be considered as uncertain input parameters, and the total delay can be captured by an output [9, 10].

EXAMPLE 2.2 *Energy management of smart homes.* Prediction of energy consumption is a step toward the closed-loop energy management of smart homes. Despite the detailed modeling of energy consumption for buildings using heat-transfer principles, large variations of predicted consumption values may occur due to uncertainties in an energy consumption system [11, 12]. To predict the mean energy consumption during a time span, key contributors to consumption dynamics may serve as uncertain input parameters, including outdoor temperatures, set-points of air conditioners, and parameters related to random indoor human activities [13].

2.2.2 Monte Carlo Methods

Monte Carlo methods are widely used for uncertainty evaluation. As shown in Figure 2.3, Monte Carlo methods typically involve three steps. First, a large number of simulations (N) are randomly selected. Each simulation point is a combination of m input parameters. Second, a simulation is conducted at each simulation point $(x_{1i}, x_{2i}, \ldots, x_{mi})$, and the corresponding output y_i is obtained. Finally, the statistic of the output is derived based on all sample outputs y_i according to $E(y) = \frac{1}{N}\sum_{j}^{N\to\infty} y_j$.

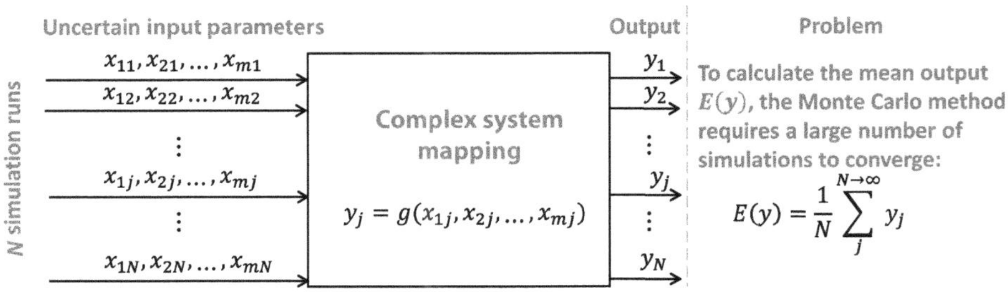

Figure 2.3 An illustration of the Monte Carlo method.

Monte Carlo methods can predict output statistics correctly, provided that the number of simulations $N \to \infty$. As each simulation for a large-scale CPS consumes considerable time, the large number of simulations required by Monte Carlo methods easily leads to heavy computational cost. As such, Monte Carlo methods are typically not suitable for real-time evaluation and management of large-scale systems under uncertainties.

2.2.3 Sampling-Based Methods

Sampling techniques have been developed to improve the convergence speed of Monte Carlo methods. Some widely used techniques are listed here. The Chevyshev nonuniform sampling method outperforms uniform sampling in overcoming Runge's phenomenon [14]. The stratified sampling method increases coverage through a combination of uniform and random sampling techniques [15–17]. The importance sampling method uses simple distributions to sample simulations and reweight the significance of simulations [18, 19]. Greedy and adaptive sampling techniques have also been developed [20, 21]. These sampling methods, however, may still not meet the stringent time-constraints for real time management.

An alternative sparse sampling method, PCM [22–27], uses the statistical knowledge of input parameters to smartly choose a very limited number of simulation points. These simulation points produce a low-order mapping that retains the mean statistics of the original higher-order system mapping. The method is based on the orthogonal polynomial expansions and their properties, and thus shares common merits with methods in the literature under different names, such as the stochastic collocation method (SCM), polynomial chaos expansion, and others [23, 28–32].

2.3 Single-Variable PCM

For a system mapping with only one input parameter, the single-variable PCM smartly selects a few simulations for uncertainty evaluation, using the properties of Gaussian quadrature integration. The method's key properties and design procedures are presented in this section.

2.3.1 Key Properties

Consider a system mapping $y = g(x)$ with a single input uncertain parameter x following a probabilistic distribution $f_X(x)$. Assume that $g(x)$ is a polynomial of degree at most $2n - 1$, i.e.,

$$g(x) = A_0 + A_1 x + \cdots + A_i x^i + \cdots + A_{2n-1} x^{2n-1} \tag{2.1}$$

where $A_i \in R$ are coefficients. A fundamental property of the single-variable PCM is that $g(x)$ can be approximated by a low-order polynomial of degree at most $n - 1$, i.e.,

$$g^*(x) = B_0 + \cdots + B_i x^i + \cdots + B_{n-1} x^{n-1}, B_i \in R \tag{2.2}$$

which predicts the same mean output. The low-order mapping can be found by n simulations at the roots of the nth normalized orthogonal polynomial $h^n(x)$, where $h^n(x)$ satisfies the orthogonality property

$$\int h^i(x) h^j(x) f_X(x) = 0 \, for \; i \neq j \tag{2.3}$$

$$\int h^i(x) h^j(x) f_X(x) = 1 \, for \; i = j, \; i, j \in \{0, 1, \ldots, n\}$$

$$h^0(x) = 0$$

This equality of $g(x)$ and $g^*(x)$ in mean output prediction is shown in Theorem 2.1 [25].

THEOREM 2.1 The original mapping $g(x)$ of degree $2n - 1$ and the low-order mapping $g^*(x)$ of degree $n - 1$ have the same mean output, i.e., $E[g(x)] = E[g^*(x)]$.

Theorem 2.1 suggests that the single-variable PCM can reduce the number of simulations from $2n$ to n with no error in the mean output prediction. The proof is omitted here due to space limitations. We only provide a conceptual illustration and sketch the proof mathematically. Interested readers can see [25] for the detailed proof.

Conceptually, this reduction is made possible by utilizing the probability density function (PDF) $f_X(x)$ of the input uncertain parameter x to produce a low-order mapping which approximates the original mapping well over the likely range of x. Mathematically, this property can be proved through expressing $g(x)$ as the expansion of orthogonal polynomials [24, 25]. In particular, $g(x)$ of degree $2n - 1$ can be represented as

$$g(x) = b_0 h^0(x) + \cdots + b_{n-1} h^{n-1}(x) + h^n(x)(b_0 h^0(x) + \cdots + b_{n-1} h^{n-1}(x)) \tag{2.4}$$

where $b_i \in R$ are coefficients. Similarly,

$$g^*(x) = b_0 h^0(x) + \cdots + b_{n-1} h^{n-1}(x) \tag{2.5}$$

The orthogonality of $h^i(x)$ suggests that $E[g(x)] = E[g^*(x)] = b_0$.

Note that besides predicting the correct mean, the low-order mapping also has several other important properties suggesting that it is a good approximation of the original mapping [25]. These properties include the correct prediction of cross-statistics up to certain degree (Theorem 2.2) and the minimum mean square estimation (MMSE) property of the low-order PCM mapping up to certain degree (Theorem 2.3). As shown elsewhere [25], these proofs can be developed using approaches similar to that for Theorem 2.1.

THEOREM 2.2 An original mapping $g(x)$ of degree $n + m, m \in \{0, \ldots, n-1\}$, and its corresponding low-order PCM mapping $g^*(x)$ of degree $n - 1$ have the same cross-statistics for all $i \in \{0, \ldots, n-1\}$, i.e., $E[x^i g(x)] = E[x^i g^*(x)]$.

THEOREM 2.3 The mean square error performance cannot be improved by adding polynomials of degree up to $n - 1 - m$ to the PCM mapping $g^*(x)$ of degree $n - 1$, when approximating an original mapping $g(x)$ of degree $n + m, m \in \{0, \ldots, n-1\}$. Furthermore, the MMSE mapping $g_r^*(x)$ for all polynomials of degree up to $n - m - 1$ is the low-order PCM mapping,

$$g_r^*(x) = b_0 h^0(x) + \cdots + b_{n-1-m} h^{n-1-m}(x) \tag{2.6}$$

2.3.2 Design Procedures

The design procedures for the single-variable PCM are summarized as follows [24].

Step 1: Selection of simulation points. The selection of simulation points is based on statistical information about the uncertain input parameter x. Four forms of statistical information are typically expected: (1) a PDF, (2) a large data set collected for x, (3) a small data set collected for x, and (4) low-order moments of x. The procedure for the first form is presented first, and then discussions follow for the other three forms (see [25] for a detailed development).

If the PDF $f_X(x)$ is known, Algorithm 2.1 can be used to find the normalized orthogonal polynomial $h^n(x)$ and its roots as the simulation points [33]. $H^j(x)$ represents the orthogonal polynomial of degree j for the uncertain parameter x, and $h^j(x)$ is its normalized form. $< p(x), q(x) >$ denotes the integration operation $\int p(x)q(x)f_X(x)dx$.

Algorithm 2.1: Find normalized orthogonal polynomial $h^n(x)$

1: **procedure** ORTHOPOLY(n, $f_X(x)$)
2: $H^{-1}(x) \leftarrow 0$; $h^{-1}(x) \leftarrow 0$
3: $H^0(x) \leftarrow 1$; $h^0(x) \leftarrow 1$
4: **for** $i \leftarrow 1, n$ **do**
5: $H^i(x) \leftarrow xh^{i-1}(x) - \langle xh^{i-1}(x), h^{i-1}(x)\rangle h^{i-1}(x) - \langle H^{i-1}(x), H^{i-1}(x)\rangle^{\frac{1}{2}} h^{i-2}(x)$
6: $h^i(x) \leftarrow H^i(x) / \langle H^i(x), H^i(x)\rangle^{\frac{1}{2}}$
7: Find the roots of $h^n(x) = 0$ as the n PCM simulation points for x, denoted as $x_{(1)}, x_{(2)}, \ldots, x_{(n)}$

In the case that a large data set of x is available, we can calculate its sample moments and replace the corresponding moment terms in the integrals in Steps 5 and 6 of Algorithm 2.1. If only a small data set or low-order moments are available, a feasible approach is to fit them with canonical distributions, such as the normal or beta distributions, and then use these distributions to select simulation points.

Step 2: Simulations. Evaluate $g(x_{(i)})$ at each of the n simulation points, where $i \in \{1,\ldots,n\}$.

Step 3: Low-order mapping construction and mean output calculation. Find coefficients $b_0, b_1, \ldots, b_{n-1}$ in the low-order mapping $g^*(x) = b_0 h^0(x) + \cdots + b_{n-1} h^{n-1}(x)$ using

$$\begin{bmatrix} b_0 \\ b_1 \\ \vdots \\ b_{n-1} \end{bmatrix} = \begin{bmatrix} h^0(x_{(1)}) & h^1(x_{(1)}) & \cdots & h^{n-1}(x_{(1)}) \\ h^0(x_{(2)}) & h^1(x_{(2)}) & \cdots & h^{n-1}(x_{(2)}) \\ \vdots & \vdots & \ddots & \vdots \\ h^0(x_{(n)}) & h^1(x_{(n)}) & \cdots & h^{n-1}(x_{(n)}) \end{bmatrix}^{-1} \begin{bmatrix} g(x_{(1)}) \\ g(x_{(2)}) \\ \vdots \\ g(x_{(n)}) \end{bmatrix} \tag{2.7}$$

where $h^i(x_{(j)})$ is the normalized orthogonal polynomial h^i evaluated at point $x_{(j)}$. $B_0, \ldots, B_{n-1}$ in the low-order mapping $g^*(x) = B_0 + \cdots + B_i x^i + \cdots + B_{n-1}x^{n-1}$ can be obtained accordingly. The mean output $E[g(x)] = b_0$.

Alternatively, we can directly calculate $B_0, \ldots, B_{n-1}$ in the original polynomial basis instead of the orthonormal polynomial basis, as (2.8) shows:

$$\begin{bmatrix} B_0 \\ B_1 \\ \vdots \\ B_{n-1} \end{bmatrix} = \begin{bmatrix} x^0(x_{(1)}) & x^1(x_{(1)}) & \cdots & x^{n-1}(x_{(1)}) \\ x^0(x_{(2)}) & x^1(x_{(2)}) & \cdots & x^{n-1}(x_{(2)}) \\ \vdots & \vdots & \ddots & \vdots \\ x^0(x_{(n)}) & x^1(x_{(n)}) & \cdots & x^{n-1}(x_{(n)}) \end{bmatrix}^{-1} \begin{bmatrix} g(x_{(1)}) \\ g(x_{(2)}) \\ \vdots \\ g(x_{(n)}) \end{bmatrix} \tag{2.8}$$

where $x^i(x_{(j)})$ is the ith power of x evaluated at point $x_{(j)}$. Example 2.3 illustrates the design procedures and performance of the single-variable PCM.

EXAMPLE 2.3 Consider a mapping of the form

$$g(x) = 1 + x^1 + x^2 + x^3 + x^4 + x^5$$

where x is normally distributed with mean of 10 and standard deviation of 1.8. Step 1 of the design procedure leads to the simulation points 6.8823, 10, and 13.1177. Step 2 results in the corresponding outputs 18066, 111112, and 420462. Step 3 then produces $b_0 = 147160$, $b_1 = 116160$, $b_2 = 50980$, and $B_0 = 578450$, $B_1 = -158001.82$, and $B_2 = 11126.79$. The low-order mapping is then

$$g^*(x) = 578450 - 158001.82x + 11126.79x^2$$

with mean 147160. Compared to using the original mapping $g(x)$, the number of simulations to calculate the correct output mean is halved with the low-order mapping $g^*(x)$. Figure 2.4(a) compares the single-variable PCM method with the Monte Carlo method in terms of the number of simulations required to correctly calculate the output

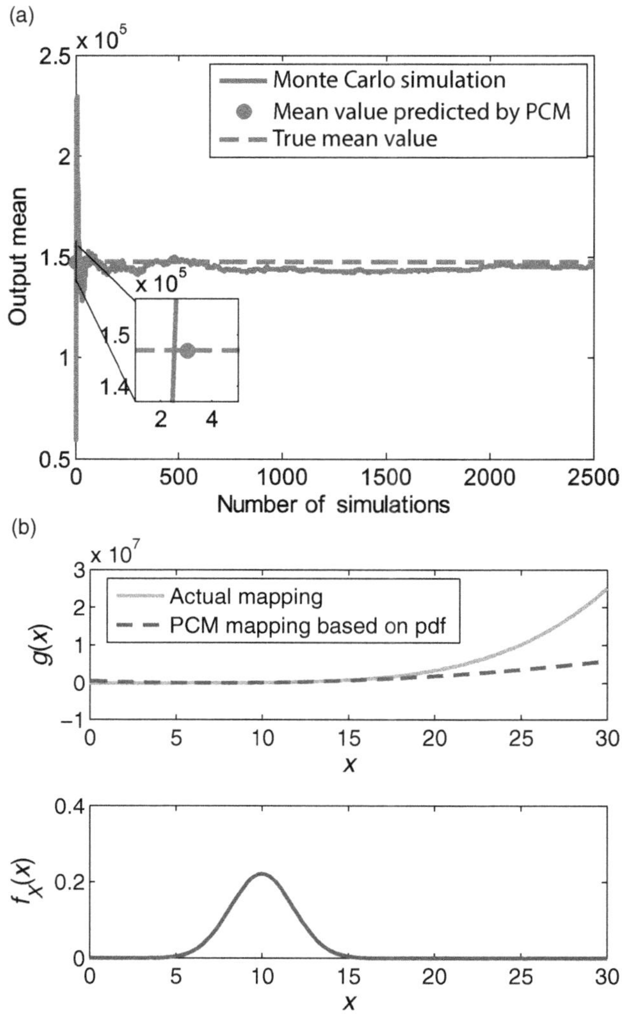

Figure 2.4 (a) A comparison of simulation time between the Monte Carlo and single-variable PCM approaches. (b) The upper figure shows a comparison between the original system mapping and the low-order mapping constructed by the single-variable PCM approach. The bottom figure shows the PDF $f_X(x)$ used to construct the low-order mapping.

mean. Figure 2.4(b) suggests that the low-order PCM mapping matches well with the original mapping over the likely range of parameter x.

2.4 Multivariate PCM

Large-scale system applications typically involve multiple uncertain input parameters, and require an extension of the single-variable PCM to the multivariate case. The properties and design procedures of the multivariate PCM (M-PCM) are presented in this section [9, 26, 34].

2.4.1 Independent M-PCM

When the multiple uncertain input parameters $x_1, \ldots, x_m$ are independent of each other, the properties and design procedures of M-PCM can be straightforwardly generalized from the single-variable PCM. In particular, Theorems 2.4–2.6 respectively illustrate properties of the independent M-PCM in mean output prediction, cross-statistics prediction, and mean square error estimation [9].

THEOREM 2.4 The original mapping

$$g(x_1, \ldots, x_m) = \sum_{k_1=0}^{2n_1-1} \sum_{k_2=0}^{2n_2-1} \cdots \sum_{k_m=0}^{2n_m-1} A_{k_1,\ldots,k_m} \prod_{i=1}^{m} x_i^{k_i} \tag{2.9}$$

of each parameter x_i up to degree $2n_i - 1$ and the low-order mapping

$$g^*(x_1, \ldots, x_m) = \sum_{k_1=0}^{n_1-1} \sum_{k_2=0}^{n_2-1} \cdots \sum_{k_m=0}^{n_m-1} B_{k_1,\ldots,k_m} \prod_{i=1}^{m} x_i^{k_i} \tag{2.10}$$

of each parameter x_i up to degree $n_i - 1$ have the same mean output, i.e., $E[g(x_1, \ldots, x_m)] = E[g^*(x_1, \ldots, x_m)]$.

Theorem 2.4 suggests that the M-PCM approach is capable of reducing the number of simulations from $2^m \Pi_{i=1}^m n_i$ to $\Pi_{i=1}^m n_i$ with the correct output mean prediction. The proof of this theorem is generalized from the proof of Theorem 2.1. As the joint PDF of m input parameters is a multiplication of individual PDFs for each parameter, the proof is constructed through recursively analyzing each parameter using the approach developed for the single-variable PCM. The proofs for Theorems 2.5 and 2.6 are constructed in a similar recursive fashion. Please refer to [9] for the detailed proofs.

THEOREM 2.5 The original mapping

$$g(x_1, x_2, \ldots, x_m) = \sum_{k_1=0}^{n_1+p_1} \sum_{k_2=0}^{n_2+p_2} \cdots \sum_{k_m=0}^{n_m+p_m} A_{k_1,\ldots,k_m} \prod_{i=1}^{m} x_i{}^{k_i} \tag{2.11}$$

of each parameter x_i up to degree $n_i + p_i$, $p_i \in \{0, \ldots, n_i - 1\}$, and the low-order PCM mapping $g^*(x_1, \ldots, x_m)$ of each parameter x_i up to degree $n_i - 1$ have the same cross-statistics for all $q_i \in \{0, \ldots, n_i - 1 - p_i\}$, i.e.,

$$E\left[\left(\prod_{i=1}^{m} x_i{}^{q_i}\right) g^*(x_1, x_2, \ldots, x_m)\right] = E\left[\left(\prod_{i=1}^{m} x_i{}^{q_i}\right) g(x_1, x_2, \ldots, x_m)\right]$$

THEOREM 2.6 The mean square error performance cannot be improved by adding polynomials of each parameter x_i up to degree $n_i - 1 - p_1$ to the PCM mapping $g^*(x_1, \ldots, x_m)$ of each parameter x_i up to degree $n_i - 1$ when approximating the original mapping

$$g(x_1, x_2, \ldots, x_m) = \sum_{k_1=0}^{n_1+p_1} \sum_{k_2=0}^{n_2+p_2} \cdots \sum_{k_m=0}^{n_m+p_m} A_{k_1,k_2,\ldots,k_m} \prod_{i=1}^{m} x_i{}^{k_i} \tag{2.12}$$

of each parameter x_i up to degree $n_i + p_i$, $p_i \in \{0, \ldots, n_i - 1\}$. Furthermore, the MMSE mapping $g_r^*(x_1, x_2, \ldots, x_m)$ for all polynomials of each parameter x_i up to degree $n_i - 1 - p_i$ is the low-order PCM mapping,

$$g_r^*(x_1, x_2, \ldots, x_m) = \sum_{k_1=0}^{n_1-1-p_1} \cdots \sum_{k_m=0}^{n_m-1-p_m} b'_{k_1,\ldots,k_m} \prod_{i=1}^{m} h_i^{k_i}(x_i) \tag{2.13}$$

The design procedures for independent M-PCM are summarized as follows. They are generalized from the procedures for single-variable PCM described in Section 2.3.2.

Step 1: Selection of simulation points. For each uncertain input parameter x_i, apply Step 1 of the single-variable PCM in Section 2.3.2 to find its PCM points $x_{i(1)}, \ldots, x_{i(n_i)}$, where n_i is the total number of PCM points selected for the input parameter x_i.

Step 2: Simulations. For each simulation point $(x_{1(j_1)}, \ldots, x_{m(j_m)})$, where $j_i \in \{1, \ldots, n_i\}$, run a simulation to obtain the corresponding output $g(x_{1(j_1)}, \ldots, x_{m(j_m)})$.

Step 3: Low-order mapping construction and mean output calculation. Find coefficients $b_{k_1,\ldots,k_m}$ in the low-order system mapping $g^*(x_1, \ldots, x_m) = \sum_{k_1=0}^{n_1-1} \sum_{k_2=0}^{n_2-1} \cdots \sum_{k_m=0}^{n_m-1} b_{k_1,\ldots,k_m} \prod_{i=1}^{m} h_i^{k_i}(x_i)$ using

$$\begin{bmatrix} b_{0,\ldots,0} \\ \vdots \\ b_{0,\ldots,n_m-1} \\ \vdots \\ b_{n_1-1,\ldots,n_m-1} \end{bmatrix} = \Gamma^{-1} \begin{bmatrix} g(x_{1(1)}, \ldots, x_{m(1)}) \\ g(x_{1(1)}, \ldots, x_{m(2)}) \\ \vdots \\ g(x_{1(n_1)}, \ldots, x_{m(n_m)}) \end{bmatrix} \tag{2.14}$$

$$\Gamma = \begin{bmatrix} h_1^0(x_{1(1)}) \ldots h_m^0(x_{m(1)}) & \cdots & h_1^0(x_{1(1)}) \ldots h_m^{n_m-1}(x_{m(1)}) & \cdots & h_1^{n_1-1}(x_{1(1)}) \ldots h_m^{n_m-1}(x_{m(1)}) \\ h_1^0(x_{1(1)}) \ldots h_m^0(x_{m(2)}) & \cdots & h_1^0(x_{1(1)}) \ldots h_m^{n_m-1}(x_{m(2)}) & \cdots & h_1^{n_1-1}(x_{1(1)}) \ldots h_m^{n_m-1}(x_{m(2)}) \\ \vdots & \ddots & \vdots & \ddots & \vdots \\ h_1^0(x_{1(n_1)}) \ldots h_m^0(x_{m(n_m)}) & \cdots & h_1^0(x_{1(n_1)}) \ldots h_m^{n_m-1}(x_{m(n_m)}) & \cdots & h_1^{n_1-1}(x_{1(n_1)}) \ldots h_m^{n_m-1}(x_{m(n_m)}) \end{bmatrix}$$

where $h_i^{k_i}(x_{i(j)})$ is the k_ith normalized orthogonal polynomial of x_i, $h_i^{k_i}$, evaluated at point $x_{i(j)}$. The subscripts of coefficients b in the coefficient matrix increase from $0, 0, \ldots, 0$ to $n_1, n_2, \ldots, n_m$, and the indices of roots in the output matrix increase from $1, 1, \ldots, 1$ to $n_1, n_2, \ldots, n_m$.

Then the coefficients $B_{k_1,\ldots,k_m}$ in the low-order mapping $g^*(x_1, \ldots, x_m) = \sum_{k_1=0}^{n_1-1} \sum_{k_2=0}^{n_2-1} \cdots \sum_{k_m=0}^{n_m-1} B_{k_1,\ldots,k_m} \prod_{i=1}^{m} x_i^{k_i}$ can be obtained accordingly. The mean output $E[g(x_1, \ldots, x_m)] = b_{\underbrace{0, \ldots, 0}_{m}}$.

Alternatively, we can calculate $B_{k_1,\ldots,k_m}$ directly using (2.15):

$$\begin{bmatrix} B_{0,\ldots,0} \\ \vdots \\ B_{0,\ldots,n_m-1} \\ \vdots \\ B_{n_1-1,\ldots,n_m-1} \end{bmatrix} = L^{-1} \begin{bmatrix} g(x_{1(1)}, \ldots, x_{m(1)}) \\ g(x_{1(1)}, \ldots, x_{m(2)}) \\ \vdots \\ g(x_{1(n_1)}, \ldots, x_{m(n_m)}) \end{bmatrix} \tag{2.15}$$

$$L = \begin{bmatrix} x_1^0(x_{1(1)})\ldots x_m^0(x_{m(1)}) & \cdots & x_1^0(x_{1(1)})\ldots x_m^{n_m-1}(x_{m(1)}) & \cdots & x_1^{n_1-1}(x_{1(1)})\ldots x_m^{n_m-1}(x_{m(1)}) \\ x_1^0(x_{1(1)})\ldots x_m^0(x_{m(2)}) & \cdots & x_1^0(x_{1(1)})\ldots x_m^{n_m-1}(x_{m(2)}) & \cdots & x_1^{n_1-1}(x_{1(1)})\ldots x_m^{n_m-1}(x_{m(2)}) \\ \vdots & \ddots & \vdots & \ddots & \vdots \\ x_1^0(x_{1(n_1)})\ldots x_m^0(x_{m(n_m)}) & \cdots & x_1^0(x_{1(n_1)})\ldots x_m^{n_m-1}(x_{m(n_m)}) & \cdots & x_1^{n_1-1}(x_{1(n_1)})\ldots x_m^{n_m-1}(x_{m(n_m)}) \end{bmatrix}$$

where $x_i^{k_i}(x_{i(j)})$ is the k_ith power of x_i evaluated at point $x_{i(j)}$. With $B_{k_1,\ldots,k_m}$, the output mean can be calculated through either an integration or a simple transformation: $E[g(x_1,\ldots,x_m)] = (\Gamma^{-1}L)_{(1,:)}[B_{0,\ldots,0}\cdots B_{0,\ldots,n_m-1}\cdots B_{n_1-1,\ldots,n_m-1}]^T$, where $(\Gamma^{-1}L)_{(1,:)}$ represents the first row of $\Gamma^{-1}L$.

Note that the calculation of coefficients requires inverting the matrix Γ or L. This inversion operation may fail or lead to inaccurate results if matrix Γ or L is ill-conditioned, as indicated by a very large condition number [35]. This often occurs when the dimension of the matrix is large. It is typically better to use Γ than L to avoid this numerical difficulty, because of the special structure of orthogonal polynomials used in Γ. For detailed discussion, please refer to [9].

2.4.2 Correlated M-PCM

In the case that uncertain input parameters are correlated, the correct output mean prediction property shown in Theorem 2.4 does not generally hold. The procedures in Section 2.4.1 may not even always lead to feasible solutions. As the example in Figure 2.5 shows, if two input parameters x_1 and x_2 follow a joint uniform distribution shown in the parallelogram region, the procedures for independent M-PCM in Section 2.4.1 lead to four simulation points marked by dark points, two of which are outside the feasible simulation region.

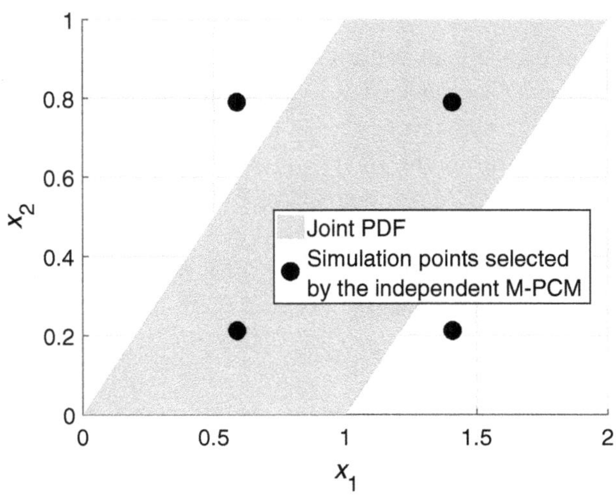

Figure 2.5 The parallelogram region shows the joint uniform distribution of the input parameters x_1 and x_2. Four simulation points selected using the independent M-PCM approach are marked by dark points.

Theorem 2.7 shows a special case in which the output mean of the original mapping can be correctly predicted using a low-order correlated M-PCM mapping. This case assumes certain special properties of conditional moments. Please refer to [9] for the proof.

THEOREM 2.7 Consider an original mapping

$$g_c(x_1, x_2, \ldots, x_m) = \sum_{k_1=0}^{2n-1} \sum_{k_2=0}^{2n-1-k_1} \sum_{k_3=0}^{2n-1-(k_1+k_2)} \cdots \sum_{k_m=0}^{2n-1-(k_1+k_2+\cdots+k_{m-1})} A_{k_1,\ldots,k_m} \prod_{i=1}^{m} x_i^{k_i} \quad (2.16)$$

If input parameters $x_1, \ldots, x_m$ follow a joint distribution $f_{X_1,X_2,\ldots,X_m}(x_1, x_2, \ldots, x_m)$, and the rth order conditional moment of x_i on $x_{i+1}, \ldots, x_m$ is a polynomial of $x_{i+1}, \ldots, x_m$ with the total degree not greater than r, then the low-order mapping $g_c^*(x_1, x_2, \ldots, x_m)$ of each parameter x_i up to the degree $n-1$ can predict the correct mean output, i.e., $E[g_c(x_1, \ldots, x_m)] = E[g_c^*(x_1, \ldots, x_m)]$.

To construct the low-order mapping $g_c^*(x_1, \ldots, x_m)$, similar procedures to those in Section 2.4.1 can be used, except for Steps 1 and 4. In Step 1, the simulation points are instead selected using the following recursive procedures: (1) select PCM points for the parameter x_m using the marginal probability of x_m; (2) select PCM points for the parameter x_i using the conditional probability $f(x_i|x_{i+1}, \ldots, x_m)$ for each selected point of the parameters $x_{i+1}, \ldots, x_m$; (3) repeat (2) for each $i = m-1, m-2, \ldots, 1$. In Step 4, (2.15) should be used to calculate the coefficients in the low-order mapping. The following example illustrates the design procedures for correlated M-PCM.

EXAMPLE 2.4 Consider an original mapping of the form

$$g(x_1, x_2) = 1 + x_2 + x_1 x_2 + x_1^2 + x_2^2 + x_1^2 x_2 + x_1 x_2^2 + x_1^3 + x_2^3$$

where x_1 and x_2 follow the joint uniform distribution shown in Figure 2.5. The four PCM points in Figure 2.5 are selected as follows. First, two PCM points are selected based on the marginal distribution of x_2, $x_{2(1)} = 0.2113$ and $x_{2(2)} = 0.7887$. After that, based on the conditional distribution $f(x_1|x_{2(1)}) = 1$ for $x_{2(1)}$, where $0.2113 \le x_1 \le 1.2113$, the corresponding PCM points are $(0.4226, 0.2113)$ and $(1, 0.2113)$. Similarly, based on the conditional distribution $f(x_1|x_{2(2)}) = 1$ for $x_{2(2)}$, where $0.7887 \le x_1 \le 1.7887$, the corresponding PCM points are $(1, 0.7887)$ and $(1.5774, 0.7887)$. The low-order mapping constructed from the four points is

$$g^*(x_1, x_2) = -1.67 + 0.83x_1 - 7.17x_2 + 13x_1x_2$$

The original mapping and the PCM mapping are shown in Figure 2.6(a) and 2.6(b) respectively, with the same mean $E[g(x_1, x_2)] = 6.5$. Figure 2.6(c) compares the number of simulations between the correlated PCM and the Monte Carlo method. Clearly, the Monte Carlo method requires a large number of simulations to converge, while the correlated M-PCM approach needs only four simulations to obtain the correct output mean.

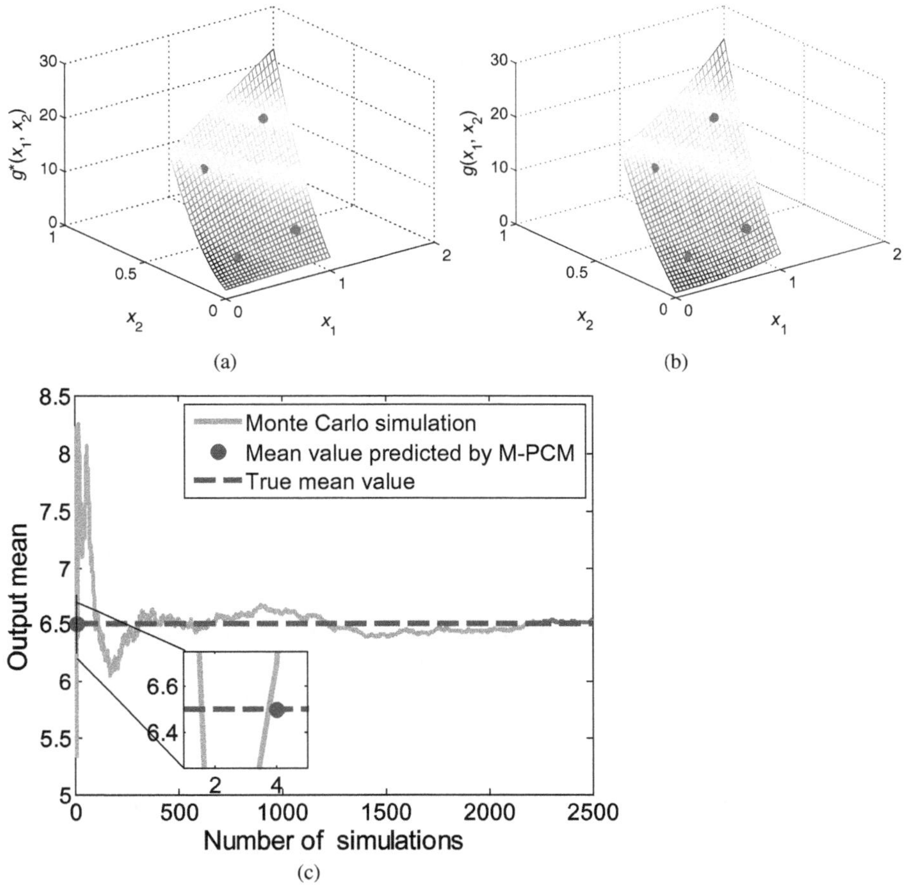

Figure 2.6 (a) The original mapping and four M-PCM points generated using the correlated M-PCM approach. (b) The four M-PCM points and low-order mapping generated using the correlated M-PCM approach. (c) Comparison of the number of simulations between the correlated M-PCM and Monte Carlo simulation methods.

2.5 Scalable M-PCM Design

2.5.1 Introduction

Despite the capability of M-PCM to reduce the number of simulations from $2^m \prod_{i=1}^{m} n_i$ to $\prod_{i=1}^{m} n_i$, such reduction may not yet be sufficient to meet the real-time requirement of evaluating and managing large-scale CPSs. Specifically, $\prod_{i=1}^{m} n_i$ grows exponentially with the number of uncertain input parameters m, and thus may cause potential computational load issues for systems with high dimensions of uncertainty.

EXAMPLE 2.5 Consider the air-traffic flow management problem described in Example 2.1. We evaluate total delay in an airspace of 50 regions (e.g., sectors) subject to convective weather events. The capacity of a region (e.g., number of aircraft allowed to pass a region) is halved if convective weather is present. We can use two uncertain variables – start time and duration – to describe uncertain weather at each region. A total number of $m = 100$ uncertain input parameters are thus included in this example. Assume that weather events are sporadic without much correlation across regions; these parameters follow independent distributions. Assume each parameter in the original input–output mapping has a degree of up to 3, i.e., $n_i = 2, i = 1,2,\ldots,m$, the M-PCM reduces the number of simulations from 4^{100} to 2^{100} to predict the correct output mean. If each simulation run takes 0.001 seconds, all 2^{100} runs take $2^{100} \times 0.001$ seconds $\approx 3.5 \times 10^{23}$ hours, which is still too expensive to execute.

There is a critical need, then, to further reduce the number of simulations from M-PCM. Note that M-PCM assumes the existence of all cross-terms in the original system mappings. However, realistic system mappings typically do not involve cross-terms of a large number of parameters. By removing these nonexistent/negligible cross-terms, a significant number of mapping coefficients to estimate can be reduced, resulting in a reduced number of simulations.

In this section, a scalable M-PCM design method is described to further improve computational efficiency. The method selects a subset from the M-PCM simulation points to predict the original system's output mean. Numerical issues should be considered in determining which subset to select. Simulation software for realistic systems typically has resolution constraints on input parameters. The truncation of parameter values due to resolution constraints introduces numerical errors to the estimation process, which may lead to the loss of precision or even the failure of calculating mapping coefficients. Example 2.6 illustrates such a case study.

EXAMPLE 2.6 Consider an original system of the input–output mapping

$$g(x_1,x_2,x_3) = \sum_{k_1=0}^{3}\sum_{k_2=0}^{3}\sum_{k_3=0}^{3} A_{k_1,k_2,k_3} x_1^{k_1} x_2^{k_2} x_3^{k_3}$$

where $x_i, i = 1,2,3$ are independent input parameters. $A_{k_1,k_2,k_3} \in R$ are coefficients. Using the procedures described in Section 2.4.1, M-PCM constructs a reduced-order mapping of the form

$$g^*(x_1,x_2,x_3) = \sum_{k_1=0}^{1}\sum_{k_2=0}^{1}\sum_{k_3=0}^{1} B_{k_1,k_2,k_3} x_1^{k_1} x_2^{k_2} x_3^{k_3}$$

where $B_{k_1,k_2,k_3} \in R$ are coefficients. Eight simulation points are selected by M-PCM, with each input parameter x_i evaluated at two levels (values), $x_{i(1)}$ and $x_{i(2)}, i = 1,2,3$. These eight simulation points are denoted as $p_1(x_{1(1)},x_{2(1)},x_{3(1)})$, $p_2(x_{1(1)},x_{2(1)},x_{3(2)})$, $p_3(x_{1(1)},x_{2(2)},x_{3(1)})$, $p_4(x_{1(1)},x_{2(2)},x_{3(2)})$, $p_5(x_{1(2)},x_{2(1)},x_{3(1)})$, $p_6(x_{1(2)},x_{2(1)},x_{3(2)})$, $p_7(x_{1(2)},x_{2(2)},x_{3(1)})$, and $p_8(x_{1(2)},x_{2(2)},x_{3(2)})$. With the assumption that all terms in the original system mapping $g(x_1,x_2,x_3)$ contain no more than one parameter, the low-order PCM mapping becomes

$$g^*(x_1,x_2,x_3) = B_{0,0,0} + B_{1,0,0}x_1 + B_{0,1,0}x_2 + B_{0,0,1}x_3 \\ = b_{0,0,0} + b_{1,0,0}h_1(x_1) + b_{0,1,0}h_2(x_2) + b_{0,0,1}h_3(x_3) \tag{2.17}$$

The number of coefficients required to estimate the output mean reduces from eight to four, and thus only four simulations are needed to uniquely determine these coefficients. If we select $\{p_1, p_2, p_3, p_4\}$ to calculate the coefficients, running simulations at these points leads to the following linear equations (similar to Step 3 of the M-PCM algorithm):

$$\Gamma' \begin{bmatrix} b_{0,0,0} \\ b_{1,0,0} \\ b_{0,1,0} \\ b_{0,0,1} \end{bmatrix} = \begin{bmatrix} g(x_{1(1)}, x_{2(1)}, x_{3(1)}) \\ g(x_{1(1)}, x_{2(1)}, x_{3(2)}) \\ g(x_{1(1)}, x_{2(2)}, x_{3(1)}) \\ g(x_{1(1)}, x_{2(2)}, x_{3(2)}) \end{bmatrix}, \quad \Gamma' = \begin{bmatrix} 1 & h_1(x_{1(1)}) & h_2(x_{2(1)}) & h_3(x_{3(1)}) \\ 1 & h_1(x_{1(1)}) & h_2(x_{2(1)}) & h_3(x_{3(2)}) \\ 1 & h_1(x_{1(1)}) & h_2(x_{2(2)}) & h_3(x_{3(1)}) \\ 1 & h_1(x_{1(1)}) & h_2(x_{2(2)}) & h_3(x_{3(2)}) \end{bmatrix} \tag{2.18}$$

or

$$L' \begin{bmatrix} B_{0,0,0} \\ B_{1,0,0} \\ B_{0,1,0} \\ B_{0,0,1} \end{bmatrix} = \begin{bmatrix} g(x_{1(1)}, x_{2(1)}, x_{3(1)}) \\ g(x_{1(1)}, x_{2(1)}, x_{3(2)}) \\ g(x_{1(1)}, x_{2(2)}, x_{3(1)}) \\ g(x_{1(1)}, x_{2(2)}, x_{3(2)}) \end{bmatrix}, \quad L' = \begin{bmatrix} 1 & x_{1(1)} & x_{2(1)} & x_{3(1)} \\ 1 & x_{1(1)} & x_{2(1)} & x_{3(2)} \\ 1 & x_{1(1)} & x_{2(2)} & x_{3(1)} \\ 1 & x_{1(1)} & x_{2(2)} & x_{3(2)} \end{bmatrix} \tag{2.19}$$

Clearly both equations are unsolvable, as both Γ' and L' are singular matrices (not full-rank). In cases that a matrix Γ' or L' constructed by a subset is full-rank but close to rank loss, potential numerical problems may still occur. In particular, a small perturbation may push Γ' or L' to lose rank, or introduce large errors to the calculation of coefficients $B_{0,0,0}$, $B_{1,0,0}$, $B_{0,1,0}$, and $B_{0,0,1}$.

Example 2.6 suggests the need to select a subset that is the most robust to numerical errors. The scalable M-PCM design method, called M-PCM-OFFD, achieves this goal by integrating M-PCM and orthogonal fractional factorial design (M-PCM-OFFD) [36, 37]. In particular, provided that all cross-terms have no more than a certain number of parameters, M-PCM is first used to find the full set of M-PCM points, and then a subset of M-PCM points are selected using the OFFD procedures [38–41]. Finally, mapping coefficients are calculated by running simulations at these points. Currently, M-PCM-OFFD has been developed for the case in which each input parameter in the original mapping has a degree up to three and follows an independent distribution. In this case, M-PCM-OFFD is able to maximally reduce the number of simulations from 2^{2m} to the range of $\left[2^{\lceil \log_2(m+1) \rceil}, 2^{m-1}\right]$ for a system of m input parameters. $\lceil x \rceil$ denotes the nearest integer larger than x. The very few simulation runs selected by M-PCM-OFFD can precisely predict the original system's output mean and are the most robust to numerical errors compared with all other designs of the same number of M-PCM points [36].

EXAMPLE 2.7 Consider the system in Example 2.5. M-PCM-OFFD can reduce the number of simulations to the range of $[2^7, 2^{99}]$, depending on the maximum number of parameters in cross-terms. This corresponds to $[3.6 \times 10^{-5}, 1.8 \times 10^{23}]$ hours of simulation time. Also of interest, simulations for air-traffic systems are typically executed at discrete time steps (e.g., every 15 minutes in [42]). Possible error led by the truncation of weather time data to meet the 15-minute temporal resolution constraint needs to be considered in selecting simulation points.

2.5.2 Design Procedures

Consider a system with m independent input parameters, each having a degree up to three. Provided that cross-terms have no more than τ parameters, the original system mapping has the form

$$g(x_1, x_2, \dots x_m) = \sum_{k_1=0}^{3} \sum_{k_2=0}^{3} \cdots \sum_{k_m=0}^{3} A_{k_1,\dots,k_m} \prod_{i=1}^{m} x_i^{k_i} \tag{2.20}$$

where the coefficients $A_{k_1,\dots,k_m} \in R$, and $A_{k_1,\dots,k_m} = 0$ if more than τ number of $k_1, \dots, k_m$ are nonzero in a term.

The following procedures for M-PCM-OFFD construct a low-order mapping

$$g^*(x_1, x_2, \dots, x_m) = \sum_{k_1=0}^{1} \sum_{k_2=0}^{1} \cdots \sum_{k_m=0}^{1} B_{k_1,\dots,k_m} \prod_{i=1}^{m} x_i^{k_i} \tag{2.21}$$

where the coefficients $B_{k_1,\dots,k_m} \in R$, and $B_{k_1,\dots,k_m} = 0$ if more than τ number of $k_1, \dots, k_m$ are nonzero in a term [36]. In another words, $B_{k_1,\dots,k_m} = 0$ if $\sum_{i=1}^{m} k_m > \tau$.

Step 1: Use M-PCM to choose 2^m simulation points. Follow the first step of the M-PCM discussed in Section 2.3.2 to select 2^m simulation points ($n_i = 2$ for all i).

Step 2: Check whether the number of simulations can be further reduced.

- *Step 2.1:* Compute the minimum number of simulations required to uniquely determine all coefficients, which is equal to the total number of coefficients, i.e., $l = \sum_{i=0}^{\tau} \binom{i}{m}$.
- *Step 2.2:* As the number of simulation runs chosen by the two-level OFFDs is a power of 2, at least $2^{\lceil \log_2(l) \rceil}$ runs are needed. If $2^{\lceil \log_2(l) \rceil} < 2^m$, then go to *Step 2.3*; otherwise directly go to *Step 4* as no further reduction is possible.
- *Step 2.3:* Check whether $2_R^{m-\gamma}$ OFFDs exist, where $0 < \gamma \leq m - \lceil \log_2(l) \rceil$ and $R \geq 2\tau + 1$. R is the resolution of OFFD, which suggests how significant effects of parameters are confounded with each other [38] and should be large enough and at least satisfy $R \geq 2\tau + 1$ to ensure the feasibility of calculation. If such OFFDs exist, find the one with the largest γ, i.e., $2_R^{m-\gamma_{max}}$ OFFD, which can maximally reduce the number of simulations to $l_{offd} = 2^{m-\gamma_{max}}$, and then go to *Step 3*. Otherwise, go to *Step 4* as no further reduction is possible.

Step 3: Use OFFD to further reduce the number of simulations. As discussed in the previous step, the number of simulations can be maximally reduced to $l_{offd} = 2^{m-\gamma_{max}}$, using the $2_R^{m-\gamma_{max}}$ OFFD. With γ_{max}, m, and R, the procedures to find the subset of simulations are shown below. Please refer to [36, 40, 41] for the detailed procedures for OFFDs, or directly apply standard designs [43].

- *Step 3.1:* Pick $m - \gamma_{max}$ parameters, and apply the $2^{m-\gamma_{max}}$ full factorial design, by listing all $2^{m-\gamma_{max}}$ combinations for these parameters. In listing the combinations, signs are used instead of parameter values. In particular, $x_{i(1)}$ is denoted as "-" and $x_{i(2)}$ is denoted as "+".
- *Step 3.2:* Choose γ_{max} generators that lead to the resolution R. Generators are the key of OFFDs, which determine the levels of the other γ_{max} parameters. The resolution R is the length of the shortest generator.
- *Step 3.3:* Assign values for the other γ_{max} parameters using the selected generators.

Step 4: Construct the low-order mapping. Run simulations at the selected $2^{m-\gamma_{max}}$ simulation points and then calculate the coefficients $B_{k_1,\ldots,k_m}$ in the low-order mapping $g^*(x_1, x_2, \ldots, x_m)$. In particular, if the number of simulations equals the number of coefficients, i.e., $l_{offd} = l$, follow a similar procedure to Step 3 of the M-PCM to find the coefficients, using the following equations:

$$H \begin{bmatrix} b_{0,\ldots,0} \\ \vdots \\ b_{0,\ldots,1} \\ \vdots \\ b_{1,\ldots,1} \end{bmatrix} = \Gamma'^{-1} P \begin{bmatrix} g(x_{1(1)}, \ldots, x_{m(1)}) \\ g(x_{1(1)}, \ldots, x_{m(2)}) \\ \vdots \\ g(x_{1(2)}, \ldots, x_{m(2)}) \end{bmatrix} \tag{2.22}$$

or

$$H \begin{bmatrix} B_{0,\ldots,0} \\ \vdots \\ B_{0,\ldots,1} \\ \vdots \\ B_{1,\ldots,1} \end{bmatrix} = L'^{-1} P \begin{bmatrix} g(x_{1(1)}, \ldots, x_{m(1)}) \\ g(x_{1(1)}, \ldots, x_{m(2)}) \\ \vdots \\ g(x_{1(2)}, \ldots, x_{m(2)}) \end{bmatrix} \tag{2.23}$$

where $H \in R^{l \times 2^m}$ and $P \in R^{l_{offd} \times 2^m}$ are transformation matrices with entries of 0s and 1s. Matrix H picks coefficients $b_{k_1,\ldots,k_m}$ or $B_{k_1,\ldots,k_m}$ with the subscripts satisfying $\sum_{i=1}^{m} k_m \leq \tau$, and matrix P selects entries related to the selected M-PCM-OFFD points. In each row of H or P, there is only one entry of 1, which corresponds to a coefficient or a selected M-PCM-OFFD point. Γ' is a submatrix of Γ, i.e., $\Gamma' = P\Gamma H^T$, which contains only rows corresponding to the selected M-PCM-OFFD points and columns corresponding to the cross-terms with no more than τ parameters. Similarly, $L' = PLH^T$ is a submatrix of L.

The coefficients $B_{k_1,\ldots,k_m}$ can then be directly computed using (2.23), or indirectly by first computing $b_{k_1,\ldots,k_m}$ using (2.22) and then following the equality

$$\begin{aligned} g^*(x_1, \ldots, x_m) &= \sum_{k_1=0}^{1} \sum_{k_2=0}^{1} \cdots \sum_{k_m=0}^{1} b_{k_1,\ldots,k_m} \prod_{i=1}^{m} h_i^{k_i}(x_i) \\ &= \sum_{k_1=0}^{1} \sum_{k_2=0}^{1} \cdots \sum_{k_m=0}^{1} B_{k_1,\ldots,k_m} \prod_{i=1}^{m} x_i^{k_i} \end{aligned} \tag{2.24}$$

If the number of simulations is larger than the number of coefficients, i.e., $l_{offd} > l$, we can use the least square estimation [44] to find the coefficients, by simply replacing Γ'^{-1} with $(\Gamma'^T\Gamma')^{-1}\Gamma'^T$, and L'^{-1} with $(L'^T L')^{-1}L'^T$.

The mean output $E[g(x_1,\ldots,x_m)] = b_{\underbrace{0,\ldots,0}_{m}}$. It can also be calculated through integrating the low-order mapping: $E[g(x_1,\ldots,x_m)] = \underbrace{\int\cdots\int}_{m} g^*(x_1,\ldots,x_m)$ $f_{X_1}(x_1)\cdots f_{X_m}(x_m)dx_1\ldots dx_m$. Another approach to calculate the output mean is through matrix manipulation. In particular, $E[g(x_1,\ldots,x_m)] = \left[\left(\Gamma'^{-1}L'\right)_{(1,:)}\right]B'$ in the case of $l_{offd} = l$ and $E[g(x_1,\ldots,x_m)] = \left[\left((\Gamma'^T\Gamma')^{-1}\Gamma'^T L'\right)_{(1,:)}\right]B'$ in the case of $l_{offd} > l$, where $B' = H[B_{0,\ldots,0}\ldots B_{0,\ldots,1}\ldots B_{1,\ldots,1}]^T$.

Example 2.8 illustrates the design procedures for M-PCM-OFFD. Example 2.9 suggests how to choose τ in realistic applications.

EXAMPLE 2.8 Continue the problem discussed in Example 2.6 and use M-PCM-OFFD to further reduce the number of simulations. In this example, $m = 3$ and $\tau = 1$. Step 1 of M-PCM-OFFD has been completed in Example 2.6. The following three steps are described here. Step 2 of M-PCM-OFFD leads to $l = \sum_{i=0}^{1}\binom{i}{3} = 4$. Since $4 < 2^3$, then check whether $2_R^{3-\gamma}$ OFFDs exist, with $0 < \gamma \leq 4 - \lceil\log_2(4)\rceil = 2$ and $R \geq 2 \times 1 + 1$. Since $2_{R=3}^{3-1}$ OFFD exists, $\gamma_{max} = 1$ and $l_{offd} = 2^{3-1} = 4$, suggesting that further reduction is possible. In Step 3 of M-PCM-OFFD, we follow the procedures for $2_R^{m-\gamma_{max}}$ OFFD to find the four simulation points as shown in Figure 2.7(a). "-" represents $x_{i(1)}$ and "+" represents $x_{i(2)}$ for all $i = 1,2,3$. The generator is $I = x_1x_2x_3$. This suggests that $x_3 = x_1x_2$, following the rule of signs for the multiplication of x_1 and x_2 (i.e., "$--=+$" and "$-+=-$," etc.). The third column of the second table in Figure 2.7(a) is filled accordingly. Figure 2.7(b) lists the four simulation points chosen by 2_{III}^{3-1} OFFD, translated from the second table in Figure 2.7(a). Note that the resolution R is typically written in roman numerals [38]. Step 4 of M-PCM-OFFD produces coefficients of low-order mapping $g^*(x_1,x_2,x_3)$, specifically, by using the following equation:

x_1	x_2	x_3
−	−	
−	+	
+	−	
+	+	

$x_1x_2 = x_3$ →

x_1	x_2	x_3
−	−	−
−	+	+
+	−	+
+	+	−

(a)

x_1	x_2	x_3
$x_{1(1)}$	$x_{2(1)}$	$x_{3(1)}$
$x_{1(1)}$	$x_{2(2)}$	$x_{3(2)}$
$x_{1(2)}$	$x_{2(1)}$	$x_{3(2)}$
$x_{1(2)}$	$x_{2(2)}$	$x_{3(1)}$

(b)

Figure 2.7 (a) Illustration of the 2_{III}^{3-1} OFFD design procedures. (b) Four simulation points selected by the 2_{III}^{3-1} OFFD.

$$\begin{bmatrix} b_{0,0,0} \\ b_{1,0,0} \\ b_{0,1,0} \\ b_{0,0,1} \end{bmatrix} = \begin{bmatrix} 1 & h_1(x_{1(1)}) & h_2(x_{2(1)}) & h_3(x_{3(1)}) \\ 1 & h_1(x_{1(1)}) & h_2(x_{2(2)}) & h_3(x_{3(2)}) \\ 1 & h_1(x_{1(2)}) & h_2(x_{2(1)}) & h_3(x_{3(2)}) \\ 1 & h_1(x_{1(2)}) & h_2(x_{2(2)}) & h_3(x_{3(1)}) \end{bmatrix}^{-1} \begin{bmatrix} g(x_{1(1)}, x_{2(1)}, x_{3(1)}) \\ g(x_{1(1)}, x_{2(2)}, x_{3(2)}) \\ g(x_{1(2)}, x_{2(1)}, x_{3(2)}) \\ g(x_{1(2)}, x_{2(2)}, x_{3(1)}) \end{bmatrix} \tag{2.25}$$

or

$$\begin{bmatrix} B_{0,0,0} \\ B_{1,0,0} \\ B_{0,1,0} \\ B_{0,0,1} \end{bmatrix} = \begin{bmatrix} 1 & x_{1(1)} & x_{2(1)} & x_{3(1)} \\ 1 & x_{1(1)} & x_{2(2)} & x_{3(2)} \\ 1 & x_{1(2)} & x_{2(1)} & x_{3(2)} \\ 1 & x_{1(2)} & x_{2(2)} & x_{3(1)} \end{bmatrix}^{-1} \begin{bmatrix} g(x_{1(1)}, x_{2(1)}, x_{3(1)}) \\ g(x_{1(1)}, x_{2(2)}, x_{3(2)}) \\ g(x_{1(2)}, x_{2(1)}, x_{3(2)}) \\ g(x_{1(2)}, x_{2(2)}, x_{3(1)}) \end{bmatrix} \tag{2.26}$$

EXAMPLE 2.9 Continue the problem discussed in Examples 2.5 and 2.7 to find τ in a realistic application. By noticing that the total delay in the 50 regions is a summation of delays in each region, each cross-term in the mapping between 100 uncertain input parameters and total delay output may involve only two parameters. With $\tau = 2$, a 2_V^{100-86} OFFD can be used to further reduce the number of simulations from 2^{100} to 2^{14}, saving a large number of simulations. The simulation time is 4.6×10^{-3} hours, which is executable.

2.5.3 Properties of the M-PCM-OFFD

This section introduces two key properties of M-PCM-OFFD: (1) correct mean output prediction (Theorem 2.8) and (2) robustness to numerical errors (Theorem 2.9) [36].

THEOREM 2.8 Provided that cross-terms have no more than $\tau(\tau \geq 1)$ number of parameters, the original system mapping of each parameter x_i up to degree three and the low-order M-PCM-OFFD mapping of each parameter up to degree one have the same output mean, i.e., $E[g(x_1, x_2, \ldots, x_m)] = E[g^*(x_1, x_2, \ldots, x_m)]$. Furthermore, the number of simulations reduces from 2^{2m} to $2^{m-\gamma_{max}}$, where $\gamma_{max} = \max\{\gamma \mid \gamma \leq m - \lceil \log_2(\sum_{i=0}^{\tau} \binom{i}{m}) \rceil,\ R \geq 2\tau + 1,\ and\ 2_R^{m-\gamma}\ OFFD\ exists\}$.

Before presenting Theorem 2.9 on the robustness of M-PCM-OFFD, we first introduce a novel metric to quantitatively measure robustness [36]. Recall Step 4 of M-PCM-OFFD. The calculation of coefficients in the low-order mapping involves the inversion of Γ', or $\Gamma'^T\Gamma'$, or L', or $L'^T L'$, which is feasible only when Γ' or L' is full column rank. Note that Γ' and L' have the same structure; therefore, we here only discuss the case for L'. Similar conclusions can be derived for Γ'. Due to the existence of numerical errors, satisfying the full column rank condition does not necessarily guarantee the success of the numerical calculation. Specifically, for a matrix L' that is close to rank loss, small perturbations may easily cause it to lose rank, and thus fail the calculation of coefficients. To measure the robustness of L' (or how close L' is to rank loss), a metric called the full-column-rank margin $D(L')$ is used. Mathematically,

$$D(L') = \min\{\|e\|_F \mid \text{rank}(L' + e) < l\} \tag{2.27}$$

where $e \in R^{l_{offd} \times l}$ is a perturbation matrix, and $||X||_F$ is the Frobenius norm calculated by summing the squares of all elements in matrix X, and then taking a square root of the sum [45]. Using this metric, Theorem 2.9 shows that L' constructed by OFFD has the largest full-column-rank margin, and thus is the most robust among all designs of the same number of simulations.

THEOREM 2.9 Among the 2^m M-PCM simulation points, the subset selected by M-PCM-OFFD has the largest full-column-rank margin among all subsets of $2^{m-\gamma_{max}}$ simulations points, where $\gamma_{max} = \max\{\gamma \mid \gamma \leq m - \lceil \log_2(\sum_{i=0}^{\tau} \binom{i}{m}) \rceil,\ R \geq 2\tau + 1,$ *and* $2_R^{m-\gamma}$ *OFFD exists*$\}$. Mathematically,

$$\max(D(L')) = D(L'_{offd}) \tag{2.28}$$

where L'_{offd} denotes the L' matrix of OFFD design.

2.6 Application to Air Traffic Flow Management

Examples 2.1, 2.5, 2.7, and 2.9 introduced an uncertainty evaluation problem in the application of air-traffic flow management – the evaluation of traffic delay in the presence of weather uncertainties. Convective weather reduces the capacities of these regions, and results in traffic delay. To reduce traffic delay, strategic air-traffic flow management is concerned with planning flows at a long look-ahead time by evaluating the impact of predicted, but uncertain, weather events on traffic delay and then designing strategies accordingly.

In this section, we use M-PCM and M-PCM-OFFD to complete the analysis of this problem and evaluate their performances. More studies along this direction can be found in [7–10, 34, 46, 47]. Here we consider the scenario that air-traffic flows pass through three regions which are subject to independent occasional weather events (Figure 2.8). The uncertain weather event at each region $i, i = 1,2,3$ is described by two uncertain parameters: start time S_i and duration D_i. Therefore, a total number of six parameters is needed. In addition, we assume that the start time S_i and duration D_i each follows a truncated Gaussian distribution with mean u_{si}, u_{di}, and variance σ_{si}^2, σ_{di}^2, in the range of $(0, \infty)$. In particular, for region $i = 1,2,3$, $u_{si} =$ 2h, 7h, 10h, $u_{di} =$ 1.5h, 0.8h, 1.2h, $\sigma_{si} =$ 0.5h, 0.3h, 0.4h, and $\sigma_{di} =$ 0.5h, 0.2h, 0.4h.

Now let us describe the simulator for evaluating the delay of this air-traffic system under weather uncertainty [6, 42, 48]. The simulator is built on an advanced discrete-time queuing network model that captures the effect of weather and management

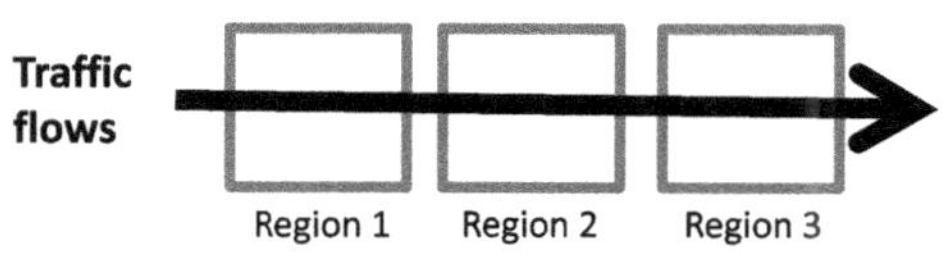

Figure 2.8 Illustration of a traffic flow passing through three regions in an airspace.

initiatives using service rate reduction. The capacity of a region (modulated by convective weather) at time k is captured by the service rate $N_i[k]$, which constrains the number of aircraft allowed to pass the queue, i.e., pass region i. In this simple example, $N_i[k]$ decreases by half in the presence of convective weather. The number of aircraft delayed at each time step k (sampled at $\Delta t = 15$ minutes) is denoted as backlog $B_i[k]$, which naturally captures traffic delay. In this study, we use the total backlog over a time span, $T_B = \sum_{i=1}^{3} \sum_{k=1}^{k_p} B_i[k]$ to measure traffic delay, where k_p is the length of the time span, equal to 96 (i.e., 24h) in this simulation study. The dynamics of the simulation model is captured by the following equations:

$$\begin{aligned} e_i[k] &= \min\{b_i[k-1], N_i[k]\} \\ b_i[k] &= b_i[k-1] + x_i[k] - e_i[k] \\ B_i[k] &= b_i[k-1] - e_i[k] \\ x_i[k] &= \begin{cases} x[k], & i = 1 \\ e_{i-1}[k-t], & i = 2,3 \end{cases} \end{aligned} \tag{2.29}$$

where $b_i[k]$ is the length of the queue for region i at time k; $e_i[k]$ is the number of aircraft passing the queue; and $x_i[k]$ is the number of aircraft entering the queue. $x[k]$, the incoming flow, is assumed to be a deterministic sequence sampled from a Poisson process of mean 5 per 15 minutes. The flight time between one region and the next, denoted as t, is assumed to be equal to one time step (15 minutes).

Given the probability density distributions of the six uncertain input parameters and the air-traffic simulation model, we apply M-PCM and M-PCM-OFFD to approximate the mean total backlog $E[T_B]$. Here, we approximate the original system mapping by assuming that each input parameter is of a degree of at most three. Therefore, we select two M-PCM points for each uncertain input parameter. In particular, we select 1.5h and 2.5h for S_1; 1h and 2h for D_1; 6.75h and 7.25h for S_2; 0.5h and 1h for D_2; 9.5h and 10.4h for S_3; and 0.75h and 1.5h for D_3. The total number of simulation points selected by M-PCM is $2^6 = 64$. Note that these M-PCM points are original values truncated to meet the 15-minute temporal resolution. By running simulations at these M-PCM points, we can calculate coefficients of the low-order mapping and predict the mean total backlog to be 43.3912.

We may further reduce the number of simulations using M-PCM-OFFD. Here, we eliminate all cross-terms that contain more than two parameters, and apply 2_V^{6-1} OFFD to select a subset of 32 simulation points. The mean total backlog predicted by M-PCM-OFFD is 43.3911. Figure 2.9 compares the number of simulations, using M-PCM, M-PCM-OFFD, and the Monte Carlo simulation method.

2.7 Concluding Remarks and Future Works

This chapter provides a detailed tutorial on the use of a family of probabilistic collocation methods to evaluate uncertainties for large-scale systems. The procedures

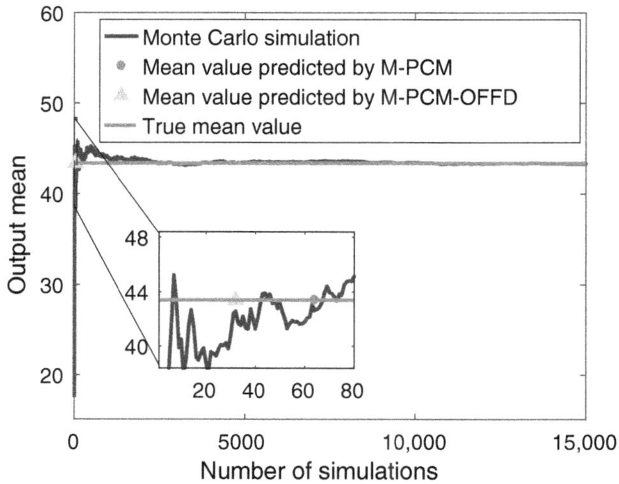

Figure 2.9 Comparison of number of simulations between Monte Carlo and M-PCM approaches.

and properties for both single-variable PCM and multivariate PCM are discussed in detail. The PCM methods construct low-order mappings that retain the correct means of higher-order mappings through smartly choosing a few simulations. The integrated multivariate PCM and orthogonal fractional factorial design method (M-PCM-OFFD) is also presented, which can significantly further reduce the number of simulations with the correct mean. The integrated design is also the most robust to numerical errors, making the method suitable for applications with input parameter resolution constraints. The family of PCM methods can also be used for optimal decision-making under uncertainties. Please refer to [6, 7, 49] for some preliminary studies.

Appendix

x_i	ith uncertain input parameter i eliminated if a single uncertain parameter is involved
m	number of uncertain input parameters
y	system output
$g(x_1, \ldots, x_m)$	original system mapping
$g^*(x_1, \ldots, x_m)$	low-order system mapping constructed using the PCM approach
n_i	highest degree of x_i in $g^*(x_1, \ldots, x_m)$
$A_{k_1,\ldots,k_m}$	coefficient of the term $x_1^{k_1}, \ldots, x_m^{k_m}$ in $g(x_1, \ldots, x_m)$
$B_{k_1,\ldots,k_m}$	coefficient of the term $x_1^{k_1}, \ldots, x_m^{k_m}$ in $g^*(x_1, \ldots, x_m)$
$f_{X_i}(x_i)$	PDF of x_i i eliminated if a single uncertain parameter x is involved
$f_{X_1 \ldots X_m}(x_1 \ldots x_m)$	joint probability distribution of input parameters $x_1, \ldots, x_m$
$x_{i(j)}$	jth PCM point for x_i

$h_i^{k_i}$	k_ith normalized orthogonal polynomial for x_i
	i eliminated if a single uncertain parameter x is involved
$h_i^{k_i}(x_{i(j)})$	$h_i^{k_i}$ evaluated at point $x_{i(j)}$
$x_i^{k_i}(x_{i(j)})$	$x_i^{k_i}$ evaluated at point $x_{i(j)}$
$D(\cdot)$	full-column-rank margin of the matrix in parentheses
e	perturbation matrix
L	matrix constructed by the full set of M-PCM points
L'	matrix constructed by a subset of M-PCM points
Γ	matrix constructed by normalized orthogonal polynomials evaluated at selected M-PCM points
Γ'	matrix constructed by normalized orthogonal polynomials evaluated at a subset of M-PCM points
l	number of coefficients in $g^*(x_1, \ldots, x_m)$
l_{offd}	number of simulation points selected by OFFD
γ	fraction constant of OFFD
R	resolution of OFFD
τ	maximal number of parameters in cross-terms in $g(x_1, \ldots, x_m)$
S_i	start time of a weather event in region i
D_i	duration of a weather event in region i
$N_i[k]$	capacity of region i at time k
$B_i[k]$	backlog of region i at time k
T_B	total backlog over a time span
$b_i[k]$	length of queue for region i at time k
$e_i[k]$	number of aircraft passing region i at time k
$x_i[k]$	number of aircraft entering region i at time k
t	flight time between two regions

Acknowledgments

Wan would like to thank NIST, MITRE, and NSF under grant numbers EAGER-1522458, EAGER-1839804/1839707, CAREER-1714519, and CPS-1714826 for supports germane to this work. The authors would also like to thank collaborators Dr. S. Roy, Dr. C. Taylor, and Dr. C. Wanke for prior results that contributed to this book chapter.

References

[1] C. Taylor, C. Wanke, Y. Wan, and S. Roy, "A decision support tool for flow contingency management," in *Proc. AIAA Guidance Navigation and Control Conference*, Minneapolis, MN, Aug. 2012.

[2] C. Taylor, C. Wanke, Y. Wan, and S. Roy, "A framework for flow contingency management," in *Proc. AIAA Aviation Technology, Integration, and Operations Conf.*, Virginia Beach, VA, Sept. 2011.

[3] B. Urgaonkar, P. Shenoy, A. Chandra, and P. Goyal, "Dynamic provisioning of multi-tier internet applications," in *Proc. Second Int. Conf. Autonomic Computing*, Seattle, WA, Jun. 2005.

[4] Y. Li, J. Ge, and J. Liu, "Wireless communication system in rainy and snowy environment using LED traffic light," *J. Inf. Comput. Sci.*, vol. 11, no. 1, pp. 268–287, Sept. 2005.

[5] P. Sadeghi, R. A. Kennedy, P. B. Rapajic, and R. Shams, "Finite-state Markov modeling of fading channels: A survey of principles and applications," *IEEE Sign. Proc. Mag.*, vol. 25, no. 5, pp. 57–80, Sept. 2008.

[6] Y. Zhou, Y. Wan, S. Roy, C. Taylor, and C. Wanke, "A stochastic modeling and analysis approach to strategic traffic flow management under weather uncertainty," in *Proc. AIAA Guidance, Navigation and Control Conf. and Exhibit*, Portland, OR, Aug. 2011.

[7] Y. Zhou, Y. Wan, C. Taylor, S. Roy, and C. Wanke, "Performance evaluation and optimal decision-making for strategic air traffic management under weather uncertainty," in *Proc. AIAA Infotech @ Aerospace Conf.*, Garden Grove, CA, Jun. 2012.

[8] J. Xie and Y. Wan, "Scalable multidimensional uncertainty evaluation approach to strategic air traffic flow management," in *Proc. AIAA Modeling and Simulation Technologies Conf.*, Dallas, TX, Jun. 2015.

[9] Y. Zhou, Y. Wan, S. Roy, et al., "Multivariate probabilistic collocation method for effective uncertainty evaluation with application to air traffic management," *IEEE Trans. Syst. Man Cybernet.: Syst.*, vol. 44, no. 10, pp. 1347–1363, 2014.

[10] Y. Zhou, Y. Wan, C. Wanke, C. Taylor, and S. Roy, "A probabilistic collocation method-based approach for optimal strategic air traffic flow management under weather uncertainties," in *Proc. AIAA ATIO Conf.*, Los Angeles, CA, Aug. 2013.

[11] N. C. Truong, J. McInerney, L. Tran-Thanh, E. Costanza, and S. D. Ramchurn, "Forecasting multi-appliance usage for smart home energy management," in *Proc. Twenty-Third Int. Joint Conf. Artificial Intelligence*, Beijing, China, Aug. 2013.

[12] D. Kim, W. Zuo, J. E. Braun, and M. Wetter, "Comparisons of building system modeling approaches for control system design," in *Proc. 13th Int. Conf. International Building Performance Simulation Association*, Chambery, France, Aug. 2013.

[13] G. Xiong, "Sensitivity and uncertainty analysis of occupancy-related parameters in energy modeling of UNT zero energy lab," Thesis, University of North Texas, June 2013.

[14] K. Atkinson, *An Introduction to Numerical Analysis*. Hoboken, NJ: Wiley, 2008.

[15] S. Hosder, R. W. Walters, and M. Balch, "Efficient samping for non-intrusive polynomial chaos applications with multiple uncertain input variables," in *Proc. AIAA/ASME/ASCE/AHS/ASC Structures, Structural Dynamics, and Materials Conf.*, Honolulu, Hawaii, Apr. 2007.

[16] R. Jin, W. Chen, and A. Sudjianto, "An efficient algorithm for constructing optimal design of computer experiments," *J. Statistical Planning Inference*, vol. 134, no. 1, pp. 268–287, Sept. 2005.

[17] M. D. Mckay, R. J. Beckman, and W. J. Conover, "A comparison of three methods for selecting values of input variables in the analysis of output from a computer code," *Technometrics*, vol. 21, no. 2, pp. 239–245, 1979.

[18] P. W. Glynn and D. L. Iglehart, "Importance sampling for stochastic simulations," *Manage. Sci.*, vol. 35, no. 11, pp. 1367–1392, 1989.

[19] D. Schuurmans and F. Southey, "Monte Carlo inference via greedy importance sampling," in *Proc. Sixteenth Conf. on Uncertainty in Artificial Intelligence*, San Francisco, CA, Jun. 2000.

[20] T. Hachisuka, W. Jarosz, R. P. Weistroffer, et al., "Multidimensional adaptive sampling and reconstruction for ray tracing," *ACM Trans. Graphics*, vol. 27, no. 3, pp. 1–10, 2008.

[21] J. Hesthaven, B. Stamm, and S. Zhang, "Efficient greedy algorithms for high-dimensional parameter spaces with applications to empirical interpolation and reduced basis methods," *ESAIM*, vol. 48, no. 1, pp. 259–283, 2011.

[22] M. A. Tatang, "Direct incorporation of uncertainty in chemical and environmental engineering systems," Ph.D. dissertation, Massachusetts Institute Technology, Cambridge, MA, 1995.

[23] M. Webster, M. A. Tang, and G. J. McRae, "Application of the probabilistic collocation method for an uncertainty analysis of a simple ocean model," Technical Report, MIT Joint Program on the Science and Policy of Global Change, no. 4, Jan. 1996.

[24] J. R. Hockenberry and B. C. Lesieutre, "Evaluation of uncertainty in dynamic simulations of power system models: The probabilistic collocation method," *IEEE Trans. Power Syst.*, vol. 19, no. 3, pp. 1483–1491, Aug. 2004.

[25] Y. Wan and S. Roy, "Uncertainty evaluation through mapping identification in intensive dynamic simulations," *IEEE Trans. Syst. Man Cybernet Syst. Hum.*, vol. 40, no. 5, pp. 1094–1104, Sept. 2010.

[26] D. Ramamurthy, "Smart simulation techniques for the evaluation of parametric uncertainties in black box systems," Master's thesis, Washington State University, Pullman, WA, 2005.

[27] S. Roy, D. Ramamurthy, and B. Lesieutre, "Studies on the probabilistic collocation method and its application to power system analysis," in *Proc. 36th North America Power Symposium*, Moscow, Idaho, Aug. 2004.

[28] I. Babuška, F. Nobile, and R. Tempone, "A stochastic collocation method for elliptic partial differential equations with random input data," *SIAM Rev.*, vol. 52, no. 2, pp. 317–355, 2010.

[29] T. Crestaux, O. Maître, and J.-M. Martinez, "Polynomial chaos expansion for sensitivity analysis," *Reliab. Eng. Syst. Safe.*, vol. 94, no. 7, pp. 1161–1172, 2009.

[30] S. S. Isukapalli, "Uncertainty analysis of transport-transformation models," Ph.D. dissertation, State University of New Jersey, New Brunswick, NJ, 1999.

[31] D. L. Wei, Z. S. Cui, and J. Chen, "Uncertainty quantification using polynomial chaos expansion with points of monomial cubature rules," *Comput. Struct.*, vol. 86, no. 23–24, pp. 2102–2108, Dec. 2008.

[32] F. Xiong, W. Chen, Y. Xiong, and S. Yang, "A new weighted stochastic response surface method for uncertainty propagation," in *Proc. AIAA/ISSMO Multidisciplinary Analysis Optimization Conf.*, Fort Worth, Sept. 2010.

[33] K. E. Atkinson, *An Introduction to Numerical Analysis*. Hoboken, NJ: Wiley, 1988.

[34] Y. Zhou, D. Ramamurthy, Y. Wan, C. Taylor, and C. Wanke, "Multivariate probabilistic collocation method for effective uncertainty evaluation with application to air traffic management," in *Proc. American Control Conference*, Washington, DC, Jun. 2013.

[35] E. W. Cheney and D. R. Kaincaid, *Numerical Mathematics and Computing*. Boston, MA: Cengage Learning, 2007.

[36] J. Xie, Y. Wan, Y. Zhou, et al., "Effective and scalable uncertainty evaluation for large-scale complex system applications," in *Proc. Winter Simulation Conference*, Savannah, GA, Dec. 2014.

[37] J. Xie, Y. Wan, K. Mills, J. J. Filliben, Y. Lei, and Z. Lin, "M-PCM-OFFD: An effective output statistics estimation method for systems of high dimensional uncertainties subject

to low-order parameter interactions," *Math. Comput. Simul.*, vol. 159, pp. 93–118, May 2019.

[38] J. Prinz, P. Tobias, W. F. Guthrie, et al., *NIST/Sematech Handbook of Statistical Methods.* Gaithersburg, MD: NIST, 2013.

[39] A. Dey, *Orthogonal Fractional Factorial Designs.* New York: Wiley, 1985

[40] G. E. P. Box, J. S. Hunter, and W. G. Hunter, *Statistics for Experimenters: Design, Innovation, and Discovery.* New York: Wiley, 1978.

[41] M. Robert, *A Comprehensive Guide to Factorial Two-Level Experimentation.* New York: Springer, 2009.

[42] Y. Wan, C. Taylor, S. Roy, C. Wanke, and Y. Zhou, "Dynamic queuing network model for flow contingency management," *IEEE Trans. Intell. Transport. Syst.*, vol. 14, no. 3, pp. 1380–1392, 2013.

[43] "Summary tables of useful fractional factorial designs," http://www.itl.nist.gov/div898/handbook/pri/section3/pri3347.htm.

[44] D. C. Montgomery, *Design and Analysis of Experiments.* New York: Wiley, 1997.

[45] P. Knupp, "Matrix norms and the condition number," in *Proc. 8th International Meshing Roundtable*, South Lake Tahoe, CA, Oct. 1999.

[46] Q. Ma, Y. Wan, and D. Sun, "Probabilistic collocation method in solving Eulerian model for air traffic flow management," in *Proc. Infotech@ Aerospace*, Garden Grove, CA, Jun. 2012.

[47] Q. Ma, Y. Wan, and D. Sun, "Stochastic analysis of air-traffic system and its corresponding application in parameters prediction," in *Proc. American Control Conf.*, Montreal, QC, Jun. 2012.

[48] Y. Wan and S. Roy, "A scalable methodology for evaluating and designing coordinated air-traffic flow management strategies under uncertainty," *IEEE Trans. Intell. Transport. Syst.*, vol. 9, no. 4, pp. 644–656, 2008.

[49] J. Xie, Y. Wan, K. Mills, J. J. Filliben, and F. L. Lewis, "A scalable sampling method to high-dimensional uncertainties for optimal and reinforcement learning-based controls," *IEEE Cont. Syst. Lett.*, vol. 1, no. 1, pp. 98–103, Jul. 2017.

3 A Flexible Graph Partitioning Algorithm for Cyber-Physical Systems

Rahul Dhal and Sandip Roy

3.1 Introduction

Graph-partitioning problems have been exhaustively studied, and a wide range of algorithms has been developed for fast partitioning of graphs with respect to diverse metrics. These partitioning algorithms have found applications in fields ranging from power-system operations to VLSI (very large-scale integration) design and thermal-process design, as well as in numerous computational domains (initially, parallel processing; more recently, computational biology, Internet, meteorology, etc.). In today's world, however, computational and physical-world engineered systems are becoming intrinsically linked to form cyber-physical systems (CPSs). These often constitute high-dimensional, highly networked amalgamations of computational and physical processes defined on a graph. Therefore, not surprisingly, there is a growing need for graph partitioning algorithms in CPS applications. The complex and hybrid nature of CPSs brings forth new challenges in partitioning, and indeed several new needs are coming to the forefront. First, the partitioning metrics are becoming increasingly complicated and multifaceted, since they must account for complex processes that span cyber and physical components, and tools for optimal or fast partitioning are needed for these new metrics. Additionally, CPSs require fast algorithms for partitioning in settings where real-time decisions are needed. In such settings, good approximation approaches may be needed for partitioning problems that are computationally difficult (e.g., NP-hard). Third, CPSs sometimes require distributed or self-partitioning algorithms, wherein network nodes achieve partitioning through distributed communication and/or local knowledge of the network topology; this is because large-scale CPSs are often spatially distributed and governed by multiple stakeholders, who coordinate only in a distributed way. Finally, because CPSs often serve safety-critical applications, there is a desire for algorithms with desired performance in terms of accuracy and speed. In short, CPSs require algorithms for graph partitioning that are extremely flexible, yet also have performance guarantees.

Recently, several stochastic algorithms have been put forth as tools for graph partitioning. These methods may be promising for CPS applications in that they yield rapid solutions for certain graph classes and partitioning metrics, and yet are potentially flexible enough to approximate optimal solutions for a range of problems, including some NP-hard ones. Many of the recent stochastic algorithms originate from Karger's classical work [1]. Using stochastic recursive contraction, [1] develops an algorithm to solve the problem of finding the minimal cut-set for partitioning an unweighted multigraph into two disjoint subgraphs. The recursive contraction algorithm is able to identify the minimal cuts with a very high probability (exceeding $1 - \frac{1}{n}$). The algorithm's runtime complexity is $\mathcal{O}(n^2 \log^3 n)$. Computational cost and storage needs compare favorably with other deterministic algorithms for the min-cut problem.

Recently our group proposed a class of stochastic algorithms for finding optimal partitions for a broad class of graphs as well as partitioning metrics [2–4]. The proposed algorithm utilizes a discrete-valued discrete-time stochastic network model called the influence model. The influence model was initially described in the controls engineering literature as a tractable modeling framework for stochastic interactions among components of engineered networks [5, 6]. The model has found applications in a variety of algorithmic tasks. In addition, the influence model has also been used for distributed-agreement and social-network decision-making tasks [7, 8]. Of particular interest to us, the authors in [9] develop a partitioning algorithm that is motivated by the influence model.

As shown in [2, 3], the influence-model-based partitioning algorithm is able to find optimal partitions for a variety of partitioning metrics, and for broad classes of weighted and directed graphs. It is especially suited for graph partitioning problems in which certain nodes are constrained to be in specified partitions (regardless of the specific cost metric). We refer to this particular problem as *partitioning with reference vertices* (cf. multiterminal cuts [10], and minimum multiway cut [11]). Examples in [2] and [3] suggest that the algorithm can achieve quick approximate solutions to these problems (many of which are NP-hard) for some graph classes. These previous results suggest that the influence-model-based algorithm provides a flexibility and accuracy guarantee that is particularly appealing for CPS applications. However, the algorithm's convergence rate was not formally analyzed in [2, 3], which is of critical interest for real-time operations in CPSs. Here, we characterize the convergence rate of the algorithm while focusing on a particular class of graphs, namely those whose optimal cut is sufficiently weak compared to other cuts in the graph. That is, we consider graphs with strongly connected sets of vertices, which are themselves weakly interconnected. Such topological structures are sometimes referred to as slow-coherency structures in the control engineering community, and are commonly encountered in infrastructure systems that are developed piecemeal in response to customer needs (e.g., power and transportation networks). As such, our analysis indicates that influence-model-based partitioning algorithms can solve hard problems for general graphs, and can be guaranteed to do so quickly for certain limited subclasses.

The performance analysis of the influence-model-based partitioner that we pursue here draws on common matrix-theoretic techniques (e.g., eigenvalue sensitivity notions), as well as graph-theoretic notions. In this sense, our study relates to a growing body of work on discrete graph algorithms (e.g., clustering algorithms) in the controls community (e.g., [12–14]). Although our work differs from these in its specifics, our work is similarly motivated by the need for flexible and distributed computational tools, and takes a similar linear algebraic perspective on performance analysis. The research presented here is also part of a broader effort to understand discrete-valued stochastic evolutions on networks, including Markov chains, stochastic automata, and stochastically time-varying network structures [15], from a graph-theoretic perspective. Most prominently, these graphical characterizations of stochastic network models are useful for evaluating uncertainties and noise in network dynamics [16]. However, as stochastic automata are increasingly used to define algorithms, the graph-theoretic analyses are also becoming important as tools for understanding algorithm performance [17].

In addition to contributing to the research on networked stochastic systems, our study also connects to efforts to provide quick solutions or approximations for subclasses of NP-complete problems (and hard graph-theory problems in particular) [18, 19]. The graph-partitioning problem considered in [19] is similar to ours. Pellegrini, in [19], also discusses a similar diffusion approach to graph partitioning problems. The article [20] provides a survey of recent developments in the theory of NP-completeness, and identifies several such results. Of interest to us, the authors in [21] propose a linear-time approximation algorithm for finding weighted matchings in graphs. Similarly, a faster algorithm for finding the girth of planar and bounded-genus graphs is reported in [22]. The article [23] also introduces a problem class called the polynomial constraint satisfaction problem, and presents algorithms for the maximum bisection problem for graphs from this viewpoint. Much like the mentioned examples, the broader theme of our research efforts is to propose fast solutions for subclasses of NP-hard and NP-complete graph-theoretic problems, in particular for partitioning with reference vertices in certain graph classes. One of the contributions of this chapter is to demonstrate how matrix algebra techniques can be leveraged to provide performance guarantees for graph algorithms. While our focus here is on graph partitioning, we believe that the techniques introduced here more generally may prove effective for addressing the difficult computational problems that arise in CPS applications. In using the influence model, they also identify an interesting connection between the discrete-valued processes that arise in computational settings and the linear-systems models that are commonly used for the physical aspects of CPS dynamics. We note that the merging of such discrete and continuous models has been explored in [24, 25], and the effort here makes concrete how such influence-type models (whether discrete or continuous) can be used for hard computational tasks in CPS and other complex networks.

The remainder of the chapter in organized as follows. Section 2 provides a brief review of the influence model. Next, Section 3 briefly reviews the influence-model-based partitioning algorithm. The remainder of the chapter is devoted to performance analysis of the algorithm for an m-way partitioning problem with reference vertices, for graphs with weak links (Sections 4 and 5). We note here that some preliminary work in this direction was presented at the 50th Control and Decision conference in 2011 [26]. Relative to the preliminary work, this chapter achieves the following contributions: (1) a comprehensive motivation for the work from a CPS standpoint; (2) tutorial content for readers in the form of illustrative figures and numerical examples, which make implementation of the algorithms in real-world CPSs easy; (3) complete proofs for the formal results; (4) extension of the initial analysis to a more general class of problems (m-way rather than two-way partitioning), which are indeed NP-hard problems that arise in diverse CPS applications; and (5) significant consideration of algorithm implementation. In keeping with the tutorial flavor of the book, we review the influence model and the partitioning algorithm by summarizing and illustrating our group's previous works in those directions [2–6, 26], before developing the main new results.

3.2 Influence Model: Review

An influence model is a network of discrete-time finite state Markov chains that interact with each other. Specifically, an influence model has n connected nodes (indexed $\{1, \ldots, n\}$, each of which can adopt one of a finite set of discrete-value statuses as they evolve in discrete time. The evolution of each node's status is based on some probabilistic update rule that depends not only on the status of the node, but also on those of its neighboring nodes. The update rules of the influence model is as follows:

1. At time step k, node $i \in \{1, \ldots, n\}$ picks another node $j \in \{1, \ldots, n\}$ (possibly itself) with probability d_{ij}. Node j is referred to as the *determining node* for node i for time step k.
2. The status of node i at time step $k + 1$ probabilistically depends on the current status of node j. That is the probability mass function (PMF) that the node i's next status is specified by the current status of the determining node j.

The flexible structure of the influence model allows it to be a modeling framework for a wide array of network stochastic systems. Our graph-partitioning algorithm utilizes a special class of influence model called the **copying influence model**, wherein each node simply copies the status of its determining node at time step k as its status at time step $k + 1$. In other words, node i copies the status of node j with probability d_{ij}. We refer the reader to [5, 6] for a detailed analysis of the general influence model.

3.2.1 Notations

To facilitate our analysis of the copying influence-model-based partitioning algorithm, let us introduce some notations and terminology. Each node in the copying influence model can adopt one of m possible values. The status of node i at time step k is represented by an m-length indicator vector $s_i(k)$, which has a single unity entry corresponding to the status of the node. The stochastic matrix $D(i, j) = d_{ij}$ is referred to as the network influence matrix. The network connecting the nodes is represented as a graph G with n vertices, corresponding to each of the n nodes. Graph G has a directed edge from vertex j to vertex i when $d_{ij} > 0$. The statuses of all nodes at time step k is represented in an $n \times m$ column vector $s[k] = \begin{bmatrix} s_1(k) \\ s_1(k) \\ \vdots \\ s_1(k) \end{bmatrix}$. For vector and matrix valued quantities we use the superscript ^{T} to denote the transpose quantity. I_n denotes an identity matrix of size $n \times n$. We use $\|.\|_p$ to denote the p-norm of the vector and matrix quantities.

At time step k, network status vector $s[k]$ has a total of m^n possible values. Characterizing probabilities of the network-wide status would require analysis of a *master Markov chain* of very high dimension. However, due to the special structure of the update rule, the status probabilities of individual nodes can be easily computed using a low-order linear recursion as follows:

$$\mathbb{E}(s^T[k+1]|s^T[k]) = s^T[0]H^{(k+1)} = H\,\mathbb{E}(s^T[k]|s^T[0]) \tag{3.1}$$

where the recursion matrix $H = D \otimes I_n$.The influence model admits numerous analyses beyond computing individual nodes' status probabilities. We refer our readers to [5, 6] for detailed development of (3.1) and further analysis.

The influence model is proving to be a compelling construct for CPS applications, for two reasons. First, it allows the simulation and analysis of discrete network dynamics, within a framework that shares the same tractabilities as continuous-valued linear dynamical systems. Thus, it naturally provides a pathway for capturing stochastic discrete events that arise in CPSs, as well as mixed continuous- and discrete-valued dynamics that are invariably present in CPSs. As one example, the influence model has been used to capture layered traffic, weather, and cyber dynamics in the air transportation infrastructure [27, 28].

3.3 Influence Model-Based Partitioning Algorithm

We use the *partitioning with reference vertices* problem to discuss the influence-model-based partitioning algorithm. Let us begin with a brief discussion of the partitioning with reference vertices problem.

Consider a graph $G(V, E : W)$, in which V is the set of vertices in the graph (which we label $1, \ldots, n$) and E is the set of directed weight edges in the graph. Weights of the edge from node i to node j are denoted as $w_{ij} > 0$. Consider a set of m distinct vertices labeled $(V_1, V_2, \ldots, V_m)$. The vertices in V need to be optimally divided into m groups (partitions); no two vertices in the set $(V_1, V_2, \ldots, V_m)$ are in the same partition. The partition is optimal if it minimizes a given cost function given the weights of edges between vertices in two different partitions. We do not need to choose a specific cost function for partitioning, but in general assume that the cost function increases with respect to the edge weights, i.e., that we are seeking partitions that are weakly connected. Even for the special case that the cost function is the sum of the edge weights between partitions (i.e., min-cut partitioning), this problem is known to be NP-hard [10, 11], and good approximations still need to be developed. It is worth noting that the m-way partitioning problem with reference vertices is particularly relevant to CPS and cyber-physical infrastructure applications, where decentralization in system operations often dictates that groupings of the system components must be based around existing federations of system resources or specified leader components.

For instance, resource partitioning schemes in transportation and power networks are often constrained to associate resources with particular fixed facilities like regional control centers or distribution facilities.

We proposed an algorithm for solving the m-way partitioning problem with reference vertices in [2, 3]. We now review the algorithm briefly (see [2, 3] for details).

The algorithm consists of three phases:

- **Mapping:** Each vertex in graph G is mapped to a node in the copying influence model, which can take on m possible statuses. The probability d_{ij} of node i copying the status of node j is chosen proportional to the edge-weight w_{ij}:

$$d_{ij} = \begin{cases} 0 & : \text{if there is no edge from } j \text{ to } i \text{ and } j \neq i \\ \Delta w_{ij} & : \text{if there is an edge from } j \text{ to } i \text{ and } j \neq i \\ 1 - \sum_{l \neq i} d_{il} & : \text{if } j = i \end{cases} \tag{3.2}$$

  where the scale factor Δ can be chosen as any value such that $\Delta < \frac{1}{\max_i \sum_j w_{ij}}$ to ensure d_{ij} represents viable probabilities.
- **Initialization:** Each vertex in the reference vertex set $(V_1, V_2, \ldots, V_m)$ is initialized to a different one of the m possible statuses (say V_1 to status 1, V_2 to status 2, and so forth without loss of generality). Also, the copying probabilities for the influence model node associated with reference vertex V_i are then modified as:

$$\forall_i d_{ij} = \begin{cases} 0 : j \neq i \\ 1 : j = i \end{cases} \tag{3.3}$$

  With the above modification to its copying probabilities, a reference will always choose itself as its determining node, and therefore its status will remain unchanged through time.

  The remaining nodes are initialized to any of the m possible statuses at random. The influence model is run as per the above-described update rules. At each time step, the model identifies an m-way partition of the network, i.e., nodes with the same status specify a partition of the original graph, with each reference vertex guaranteed to be in a different partition. The influence model thus stochastically scans through many potential partitions of the graph, which have varying costs according to a specified metric. The algorithm favors low-cost partitions (i.e., such partitions are visited more frequently), because tightly connected nodes are likely to influence each other and come to a common status, while weakly connected nodes remain in distinct statuses.
- **Stopping:** Stopping rules for the algorithm, including centralized and decentralized ones, have been developed, see [2, 3]. Because we are interested in the speed at which the algorithm achieves the optimal partition as it scans through potential partitions (for m-way partitioning with reference vertices assuming a min-cut cost), the stopping rule is not of significant interest.

Let us briefly illustrate the influence model partitioning algorithm. Figure 3.1 shows a graph that is to be mapped into an influence model. Figure 3.2 shows the mapped

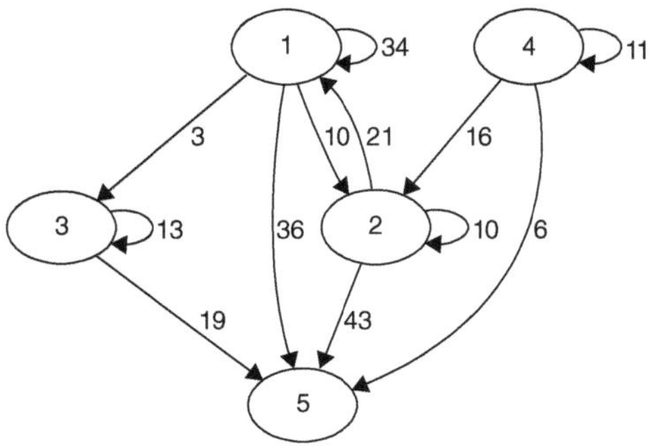

Figure 3.1 Graph to be mapped into influence model for two-way partitioning. Vertices 3 and 4 are chosen as reference vertices.

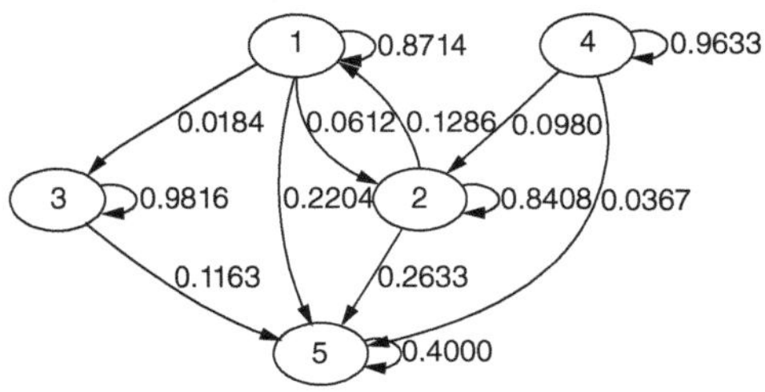

Figure 3.2 Mapped influence model.

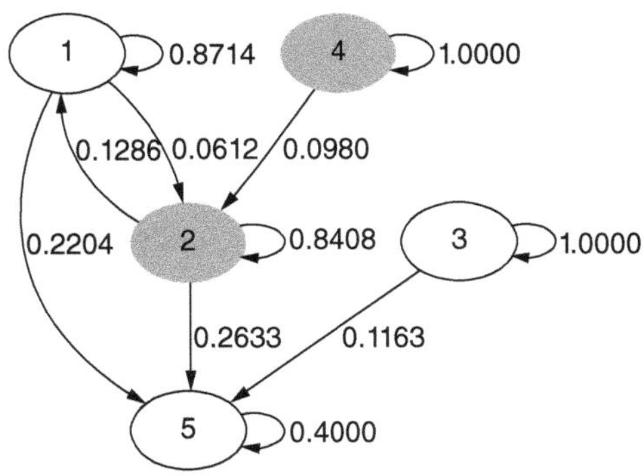

Figure 3.3 Modified influence model with nodes 3 and 4 as reference vertices.

influence model as per the mapping rules described above. Nodes 3 and 4 are chosen as the reference vertices. Figure 3.3 shows the modified copying probabilities with a sample initialization of the nodes. Node 3 is initialized to status 1 and node 4 to status 2. The other nodes are initialized to statuses randomly. Shaded nodes represent status 2. Figures 3.4 and 3.5 shows statuses at different time steps.

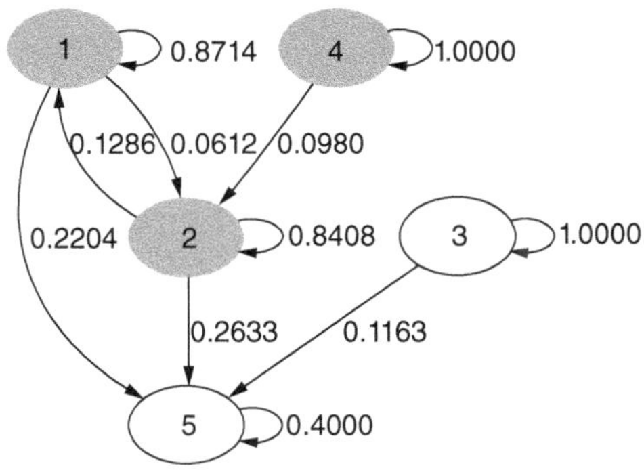

Figure 3.4 Statuses of the nodes at the 10th time step.

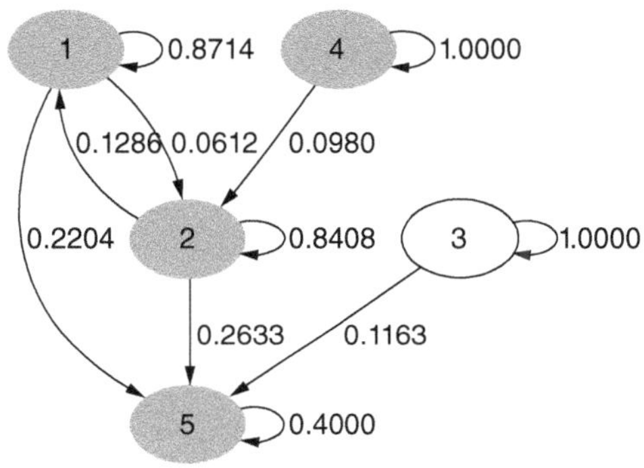

Figure 3.5 Statuses of nodes after the 50th time step (steady state).

3.4 Performance Analysis

Since the m-way partitioning problem with reference vertices is NP-hard, an algorithm that finds the optimal partition quickly for any graph cannot be expected. However, interestingly, we are able to show that the optimal partition can indeed be found quickly when the optimal cut is sufficiently weak compared to other cuts (while it is still guaranteed that the minimum cut will be found asymptotically no matter what, as proved before). This dual guarantee of an eventual and a fast solution for particular graph structures, for a flexible-cost function, is appealing for CPS applications as a practical solution for a range of essentially difficult partitioning problems.

We analyze the performance of the partitioning algorithm for the m-way partitioning problem in the following steps. First, we demonstrate for graphs where the cuts between the optimal partitions are sufficiently weak; asymptotically, the nodes in each optimal partition of the mapped copying influence model are equal to that reference vertex with high probability, given that there are weak links between the partitions. Next, we develop analytical bounds on time steps k, after which the nodes' status probabilities are close to their asymptotic value. We then show that the above characterizations imply that the partitions identified are optimal with a high probability. The results in this section focus on the limiting case of arbitrarily weak links between the optimal partitions. In Section 3.5, we present bounds on how weak these links need to be for fast partitioning, in terms of subgraph eigenvalues and the number of nodes in the graph.

We develop and prove results for the two-way partitioning problem with reference vertices for notational simplicity. The m-way partitioning with reference vertices is then studied, without proof since they follow a similar logic to the two-way partitioning case. Differences between the two-way and m-way partitioning case are briefly mentioned as needed. We stress that the m-way partitioning results are of particular importance because the m-way problem is NP-hard; our analysis identifies a set of networks for which a fast solution can be found.

For two-way partitioning, the graph is mapped to a copying influence model with two reference vertices, which we refer to as the binary copying influence model. Let the reference vertices be indexed as 1 and n, where n is the number of vertices in the network. We recall that each node's asymptotic status probabilities are specified by the eigenvector(s) of the influence matrix associated with eigenvalue(s) at one [5, 6]. These eigenvectors, and hence the individual nodes' status probabilities, are characterized in Theorem 3.1.

THEOREM 3.1 Consider two-way partitioning with reference vertices for a connected graph, comprising two strongly connected subgraphs (each of which contains one of the reference vertices) and consider that the edges between the subgraphs (or optimal partitions) can be made arbitrarily weak. The graph is mapped into a binary copying influence model so as to apply the partitioning algorithm. Then the asymptotic probability that the status of a node in a given optimal partition is equal to the status of the reference vertex of that partition approaches unity, as the edges between the partitions are made arbitrarily weak.

Proof. The organization of the proof is as follows. The proof draws on the observation that the asymptotic probabilities of nodes being in some given status (we will refer to these probabilities as asymptotic status occupancy probabilities henceforth) of a binary influence model can be determined from the eigenvector of the network influence matrix D associated with the eigenvalues at unity [5, 6]. Using this observation, we proceed as follows. As a first step, the eigenvectors of the D matrix associated with the eigenvalue at unity (and hence the steady-state occupancy probabilities) upon disconnection of the weak links between the optimal partitions are determined. Then, using matrix perturbation results, we determine the structure of the eigenvector of the matrix D when weak links are present in the graph, and hence prove the result.

Before we begin the eigen-analysis of the network influence matrix, we formalize the way in which the eigenvector(s) of D can be used to find the occupancy probabilities. Equations (3.4) and (3.5) give the status vector $s_i[k]$ at time step k for node i, and the occupancy probabilities $p_i[k]$ for node i at time step k:

$$s_i[k] = \begin{cases} \begin{bmatrix} 1 \\ 0 \end{bmatrix} : \text{for status} = 1 \\ \\ \begin{bmatrix} 0 \\ 1 \end{bmatrix} : \text{for status} = 2 \end{cases} \tag{3.4}$$

$$p_i[k] = \begin{bmatrix} \text{prob.(status} = 1) \\ \text{prob.(status} = 2) \end{bmatrix} \tag{3.5}$$

Accumulating the status vectors and occupancy probabilities for all nodes, we get the state vector and occupancy probability vector of the network as defined in [5, 6] as follows:

$$s_i[k] = \begin{bmatrix} s_1[k] \\ \vdots \\ s_2[k] \\ \vdots \\ s_n[k] \end{bmatrix} \in \Re^{2*n\times 1} \tag{3.6}$$

$$p_i[k] = \begin{bmatrix} p_1[k] \\ \vdots \\ p_2[k] \\ \vdots \\ p_n[k] \end{bmatrix} \in \Re^{2*n\times 1} \tag{3.7}$$

In [6], it was shown that the occupancy probability vector of any influence model is given by:

$$p^T[k+1] = p^T[0]H^{k+1} \tag{3.8}$$

where the matrix H is defined in generality in [5, 6] and is equal to $D^T \otimes I_2$ for the binary influence model. Due to the structure of H for the copying influence model, the entries in the occupancy probability vector evolve in a decoupled fashion. In particular, let us denote the vector of probabilities of each node being in status 1 (respectively status 2) at time step $[k+1]$ as $P_1[k+1]$ (respectively $P_2[k+1]$). It can be seen that $P_1^T[k=1] = P_1^T[0]\left(D^T\right)^{k+1}$ (respectively $P_2^T[k+1] = P_2^T[0]\left(D^T\right)^{k+1}$). Noticing that the matrix D is a stochastic one, we see immediately that the eigenvector(s) associated with its unity eigenvalue(s) determine the asymptotic status-occupancy probabilities. We now present the proof of the theorem. We assume that the reference vertex "1" is kept at status 1 and reference vertex "n" is kept at status 2. The first and last row of D always has the following form due to the assumed indexing of the reference vertices:

$$D(1, :) = \begin{bmatrix} 1 & 0 & 0\ldots & 0 \end{bmatrix}$$
$$D(n, :) = \begin{bmatrix} 0 & 0 & 0\ldots & 1 \end{bmatrix}$$

Given this structure of D and given the fact it is a stochastic matrix, D can be viewed as the state transition matrix of a Markov chain with two recurrent classes corresponding to the two reference vertices. Hence, D has two eigenvalues at unity while the rest of the eigenvalues are strictly less than 1 in magnitude. Therefore, the steady-state probability vector, $\lim_{k\to\inf} P_1^T[k] = \lim_{k\to\inf} P_1^T[0](D^T)^k$ is determined by the eigenvectors associated with the eigenvalues at unity. Furthermore, since the reference vertex "n" is always in status 2, the last entry of the probability vector is zero at each time step. Hence, the projection of the probability vector on one eigenvector direction associated with the unity eigenvalue is always zero. In other words, a unique eigenvector direction associated with the eigenvalues at unity gives the steady-state probability of being in status 1. Let us characterize these steady-state probabilities in the disconnected and then the weak link cases.

Consider the influence model defined on the graph upon removal of the weak links. The mapped influence model for this case would have two partitions. The network influence matrix (denoted as D_0) would have the structure:

$$D_0 = \begin{bmatrix} D_1 & \mathbf{0} \\ \mathbf{0} & D_2 \end{bmatrix} \Longrightarrow D_0^k = \begin{bmatrix} D_1^k & \mathbf{0} \\ \mathbf{0} & D_2^k \end{bmatrix}$$

where $\mathbf{0}$ represent zero matrices of appropriate dimension, while submatrices D_1 and D_2 represent the subgraphs defined by the partitions. Since D_0 is a stochastic matrix with unit row-sums, it follows that D_1 and D_2 are stochastic matrices as well. The probability vector P_1 at steady state is given by one right eigenvectors direction of D_0 at unity; specifically, we find that the asymptotic probability vector is $P_1 = \begin{bmatrix} \vec{\mathbf{1}} & \vec{\mathbf{0}} \end{bmatrix}^T$. That is asymptotically, all the nodes in partition "1" are in status 1 and those in partition "2" are in status 2.

Now we consider the graph with the weak links between the two optimal partitions. The network influence matrix (denoted as D) has the structure:

$$D = \begin{bmatrix} D_1 & \varepsilon_1 \\ \varepsilon_2 & D_2 \end{bmatrix}$$

The matrices D_1 and D_2 are no longer stochastic, but the matrix D is stochastic. Since the weak links are assumed as arbitrarily weak, the matrix can be viewed as a perturbation of D_0. Therefore, the steady-state probability vector for the connected graph can be viewed as a perturbation of an eigenvector of D_0. However, the multiplicity of the eigenvalue at unity poses a problem in applying eigenvector perturbation results directly. Luckily, the probability vector and the D matrix itself have some special structure that can be exploited. Since the last entry in the probability vector $P_1[k]$ is always 0[1] and the last row of both D_0 and D is $[0,\ldots,0,1]$, the last row and column do not affect the probability vector entries, other than that the last entry is always 0. Therefore, we can remove the last column and row from both D_0 and D and still be able to determine the eigenvector. The advantage of this reformulation

[1] The nth node is a reference vertex and is kept at status 2 always. Hence, its probability of being in status 1 is 0.

is that the multiplicity of the eigenvalue at unity of the resulting matrix (we refer to this matrix as the reduced network influence matrix of the graph) is reduced to 1 and hence becomes amenable to direct application of the matrix perturbation results for eigenvectors.

Let the reduced matrix of D_0 be denoted as $\widetilde{D}$ and that of D be $\widehat{D}$. It is easy to see that $\widehat{D} = \widetilde{D} + \mathcal{O}(\varepsilon)$ where ε is an arbitrarily small quantity as the edge weights between the optimal partitions are made small. From the matrix perturbation results, the appropriate eigenvector of $\widehat{D}$ is the eigenvector of $\widetilde{D}$ with a small perturbation [29]. Therefore, the steady-state probability vector of the nodes being in status 1 has a structure $\begin{bmatrix} \overrightarrow{\eta} \\ \overrightarrow{\kappa} \end{bmatrix}$ where entries in $\overrightarrow{\eta}$ are close to 1 and those in $\overrightarrow{\kappa}$ are close to 0.[2] Also, the steady-state probability of being in status 2 is given by the complementary probability vector .Thus we have proven the result.

The above theorem shows that the nodes in partition 1 have a high probability of being in status 1 and nodes in partition 2 have a high probability of being in status 2 in steady state when each optimal partition is internally strongly connected, but the two partitions are weakly linked. Now let us present the result for the m-way case. □

COROLLARY 3.1 Consider the m-way partitioning problem with reference vertices for a connected graph comprising m strongly connected subgraphs (where each subgraph has one of the reference vertices). Consider scaling down the weights of the edges between the subgraphs (or optimal partitions). The graph is mapped to an m-state copying influence model, and the partitioning algorithm is applied. Then the asymptotic probability that the status of a node in a given optimal partition is equal to the status of the reference vertex of that partition approaches unity, in the limit as the scaled edge weights become arbitrarily weak.

Proof. The proof for the m-way partitioning follows the same logic as for the two-way partition, except that the reduced influence matrix for the case of the m-way partitioning with reference vertices is obtained by deleting the ith row and column from the network influence matrix for all i such that vertex V_i is a reference vertex and $i \neq 1$. □

With the asymptotic behavior characterized in the above theorems, we investigate the transients of the model. Understanding the transient behavior can help us understand how quickly the model is able to identify the optimal partitions. We organize the analysis as follows. For a graph with two strongly connected partitions, which are connected via weak links, we show that after a specified number of time steps, nodes in each partition have a very high probability of being in the same status as that of the reference node for the partition. The m-way partitioning on a graph with m weakly connected partitions follows the same logic as the two-way partitioning.

We begin our analysis with a conceptual overview before presenting a formal result. As already described in Theorem 3.1, the probability of a node having status 1 at time step k is given by

[2] Note that the first entry in $\overrightarrow{\eta}$ is 1 and the last entry in $\overrightarrow{\kappa}$ is 0.

$$P_1^T[k] = P_1^T\left(D^T\right)^k$$

The eigen-analysis of the matrix D governs the evolution of the probability vector P_1 through the time steps. From the Frobenius theorem we see that all eigenvalues of D lie on or within the unit circle. As a result, for sufficiently large values of k, the modal contribution of eigenvalues with magnitude strictly less than unity in the product $P_1^T\left(D^T\right)^k$ will be arbitrarily small, and hence the probability vector P_1 will be arbitrarily close to their asymptotic values characterized in Theorem 3.1. This idea is the basis of the next theorem.

THEOREM 3.2 Consider a graph made up of two strongly connected subgraphs. Consider scaling down the weights of the edges between the subgraphs by a factor $\nu > 0$. Then the nodes in the mapped influence model in each partition (subgraph) have statuses equal to that of the reference vertex with a probability $> 1 - 0.01^q - \nu$, where ν is made arbitrarily small and q may be any positive number, if the number of time steps k satisfies

$$k \geq \frac{-q - \log_{10}(n-2)}{\log_{10}(|\lambda_{\max}|)}$$

Here, $\lambda_{\max}$ is the largest eigenvalue of the influence matrix whose magnitude is strictly less than 1.

Proof. As shown before, for any influence model defined on a connected graph with two reference vertices, the occupancy probability for all the nodes being in status 1 at any time step k can be written as:

$$P_1^T[k] = P_1^T[0]\left(D^T\right)^k = \sum_i \lambda_i^k\left(\vec{V}_i\vec{W}_i^T P_1[0]\right)$$

where $\vec{W}_i^T$ and $\vec{V}_i^T$ are the left and right eigenvectors associated with the eigenvalue λ_i. Since D is a stochastic matrix, all the eigenvalues are on or within the unit circle. If $k \geq \frac{-q-\log_{10}(n-2)}{\log_{10}(|\lambda_{\max}|)}$ then,

$$\left\|\lambda_i^k\left(\vec{V}_i\vec{W}_i^T P_1[0]\right)\right\|_\infty \leq \frac{10^{-q}}{n-2}$$

for every mode i with eigenvalue less than unity in magnitude. Therefore, if $k \geq \frac{-q-\log_{10}(n-2)}{\log_{10}(|\lambda_{\max}|)}$, then each term in the following expression is bounded as follows:

$$\left\|\lambda_i^k\left(\vec{V}_i\vec{W}_i^T P_1[0]\right)\right\|_\infty \leq (n-2)|\lambda_{\max}|^k \left\|\sum_{i,|\lambda_i|<1}\left(\vec{V}_i\vec{W}_i^T P_1[0]\right)\right\|_\infty \leq 10^{-q}$$

where, $\lambda_{\max}$ is the eigenvalue with the largest magnitude that is strictly less than unity. Please note that we have employed the fact that D has two eigenvalues at unity and $(n-2)$ nonunity eigenvalues. This implies that the occupancy probability of a node being in status 1 after the specified number of time steps is close (within 0.1^q) to the steady-state probability of the node being in status 1. In Theorem 3.1 we have shown that the steady-state probabilities of nodes in a mapped influence model of a connected graph with two weakly connected partitions can be viewed as a perturbation of the steady-state probabilities of the mapped influence model of the graph with the weak links removed. Let us denote the largest magnitude perturbation by ν. Thus, we can see that the nodes have the same status as the reference vertex in the partition with a probability of at least $(1 - 0.1^q - \nu)$ where q is an arbitrarily chosen positive quantity. □

COROLLARY 3.2 Consider a graph made up of m strongly connected subgraphs. Consider scaling down the weights of the edges between the subgraphs by a factor $\nu > 0$. Then the nodes in the mapped influence model in each partition (subgraph) have statuses equal to that of the reference vertex with a probability $(1 - 0.1^q - \nu)$ for sufficiently small ν and for any $q > 0$ after k time steps, if

$$k \geq \frac{-q - \log_{10}(n - m)}{\log_{10}(|\lambda_{\max}|)}$$

Here, $\lambda_{\max}$ is the largest eigenvalue of the model's influence matrix, which has magnitude strictly less than unity.

Our analyses in Theorems 3.1 and 3.2 depend on the graph having "sufficiently weak" links among the partitions. Specifically, the parameter ν can be viewed as a proportional scalar metric of the *weakness* of the links between partitions. We will revisit the perturbation of the eigenvector and present explicit bounds in terms of graph structure in Section 3.5.

Since each node has been shown to adopt the correct partition status with high probability quickly, it is reasonable to assume that the algorithm can identify partitions correctly with a high probability after a specified number of time steps. To verify this assumption, we should prove that all the nodes in a partition have the same status with high joint probability. Joint probability computations are developed and discussed in [5]. However, explicitly finding joint probabilities is taxing, and may not allow a formal analysis of performance. Instead of doing so, here we have bound the joint status probabilities in terms of the individual node status probabilities. Specifically, we have already shown that nodes in different partitions have high probability (nearly equal to 1) of being in the same status as the reference vertex of the partition. Thus, we can expect that the joint probability of the node in the same partition having the same status, and of nodes in different partitions having different statuses, is high. This notion is formalized in Theorem 3.3.

THEOREM 3.3 Consider solving the problem of m-way partitioning with reference vertices using the influence model. The optimal (min-cut) partitions are designated as Partition 1, ..., Partition m for notational convenience. Let us also define the individual node status occupancy probabilities in the influence-model algorithm at time step k as $\eta_i = p\,(s_i[k] = r_i)$, where r_i is the correct partition of node i in the optimal solution. Then the joint probability $P_J = p(\cap s_i[k] = r_i)$ is bounded by:

$$\max\left(0, 1 - \sum_{i=1}^{n}(1 - \eta_i)\right) \leq P_j \leq \min_i(\eta_i)$$

Proof. The results presented in the theorem can be deduced from simple probability theory constructs. We look at the lower bound first:

$$\begin{aligned} P_j &= p(\cap s_i[k] = r_i) \\ &= 1 - p(\cup s_i[k] \neq r_i) \\ &\geq 1 - \sum_{i=1}^{n} p(s_i[k] \neq r_i) \\ &\geq \sum_{i=1}^{n} \eta_i + 1 - n \\ &\geq 1 - \sum_{i=1}^{n}(1 - \eta_i) \end{aligned}$$

Meanwhile, the maximum value of the joint probability can be bounded by noting that the probability of the intersection events is less than or equal to each event's probability. Therefore, the joint probability is upper-bounded by $\min_i(\eta_i)$. □

From Theorems 3.1 and 3.2 and the associated corollaries, if the number of time steps k is greater than the specified value, then individual sites have a high probability of sharing the status of their corresponding reference vertices. Theorem 3.3 shows that the joint probability of all sites in a partition having the same status as that of the reference vertex is also high. Thus, the algorithm finds (with high probability) the correct partitions after the specified number of time steps, for graphs that have a weak cut.

While the formal results on partitioning are the main contributions of the work to CPS analysis, the proof methods draw on some concepts that may have more general implications in CPSs. Of particular note, the proofs: (1) demonstrate that certain discrete processes can be characterized using linear-system analysis techniques; and (2) demonstrate how perturbation arguments (which traditionally have been used to understand continuous-valued processes) have a bearing on analysis of discrete computations/algorithms.

3.5 Characterizing Weak Cuts

In Section 3.4 we showed that fast solutions to the m-way partitioning problem with reference vertices can be found for graphs with weak cuts. However, we have not yet given any idea of how weak the links between optimal partitions need to be for fast partitioning, since a perturbation argument has been used. Specifically, from the perturbation argument, the following qualitative results are apparent:

1. The network influence matrix's eigenvalues only change slightly with perturbation.
2. The eigenvectors of the network influence matrix corresponding to the two unity eigenvalues remain nearly the same after perturbation.

We next seek to give explicit bounds on the eigenvalue and eigenvector perturbation resulting from the graph perturbations, so as to give an idea of how weak the links need to be for the influence-model-based partitioner to be guaranteed to find the optimal partitions quickly. Specifically, we study the size of the weak cuts for which the eigenvectors and eigenvalues of the network influence matrix are not much perturbed. In presenting these results, it should be noted that the perturbation theory is typically conservative. This is mainly because there are no convenient tight characterizations of eigenvalue perturbations for Jordan blocks of sizes greater than 1. Nevertheless, we believe that the development of explicit graph theoretic bounds on eigenvalue and eigenvector perturbation (for both general and symmetric cases) is significant in characterizing the influence model partitioner, in that they make explicit the class of graphs for which fast partitioning is possible.

3.5.1 Perturbation of Eigenvalues

First, we give a bound on the cut size (formally, the norm of the influence-matrix partition effected by the weak cut) such that the eigenvalues of the network influence matrix remain close (in a relative sense).

THEOREM 3.4 We denote the influence matrix associated with a graph as $\bar{D}$, and the modified influence model when the cuts between the optimal partitions are removed as D. Further, we define a perturbation matrix $E = \bar{D} - D$. We also define as $\lambda_{\max}$ the largest eigenvalue of the influence matrix with magnitude strictly less than unity. We seek to determine conditions under which the change in $\lambda_{\max}$ due to the perturbation is bounded by a spectral gap, or more generally by an amount $G(1-\lambda_{\max})$, where $G(x)$ is any function such that $G(x) \to 0$ as $x \to 0$. If the perturbation matrix E satisfies the following inequality, then such a bound on the eigenvalue change is met:

$$\log\left(\|E\|_2\right) \leq n\left[\log\left(G(1-\lambda_{\max}) - \log(2n-1) - \log(4n)\frac{n-1}{2n}\right)\right]$$

Proof. The proof is a direct consequence of the Ostrowski–Elsner theorem [29]. The theorem defines a metric termed the “matching distance.” The matching distance of two matrices A and B with eigenvalues $\lambda(A) = \{\lambda_1, \lambda_2, \ldots, \lambda_n\}$ and $\lambda(B) = \{\mu_1, \mu_2, \ldots, \mu_n\}$ is defined as $md(A,B) = \min_{\pi \in S}\left(\max_i |\mu_{\pi(i)} - \lambda_i|\right)$, where S is the set of all permutations $\{1, 2, \ldots, n\}$. If the matching distance of a matrix A and its perturbation $\widetilde{A} = A + E$ is small, then the eigenvalues of the original matrix and those of the perturbed one can be grouped in nearby pairs (with the nearness specified by the matching distance). More specifically, the matching distance $md(A, \widetilde{A})$ is a bound on the amount that the eigenvalues move due to perturbation.

The Ostrowski–Elsner theorem shows that $md(A, \widetilde{A})$ can be bounded as:

$$md(A,\widetilde{A}) \leq (2n-1)\left(\|A\|_2 + \|\widetilde{A}\|_2\right)^{1-\frac{1}{n}} \|E\|_2^{\frac{1}{n}}$$

for any $A, \widetilde{A}$.

However, for the network influence matrix D, which is a stochastic matrix, the following then always holds:

$$\begin{aligned}
&\|D\|_2^2 \leq \|D\|_1 \|D\|_\infty \\
&\quad \leq \|D\|_1 \quad \text{since, for stochastic matrices } \|D\|_\infty = 1 \\
&\quad \leq \max_{1 \leq j \leq n} \sum_{i=1}^{n} |d_{ij}| \quad \text{(maximum column sum)} \\
&\quad \leq n \\
\Longrightarrow\ &\|D\|_2 \leq \sqrt{n}
\end{aligned}$$

Similarly, it can be shown that $\|\bar{D}\|_2 \leq \sqrt{n}$ since the matrix remains a stochastic one after the perturbation. Using these results, the Ostrowski–Elsner inequality can be restated for our purposes as:

$$md(D, \bar{D}) \leq (2n-1)\left(\|D\|_2 + \|\bar{D}\|_2\right)^{1-\frac{1}{n}} \|E\|_2^{\frac{1}{n}} \leq (2n-1)\left(2\sqrt{n}\right)^{1-\frac{1}{n}} \|E\|_2^{\frac{1}{n}}$$

We now define a function $G(x)$ as any function such that $G(x) \to 0$ as $x \to 0$. $G(1 - \lambda_{\max})$ gives the maximum allowed perturbation of the eigenvalue $\lambda_{\max}$ and so we need to ensure that the matching distance is smaller than this amount. We can ensure this by enforcing the following inequality:

$$\begin{aligned}
&(2n-1)\left(2\sqrt{n}\right)^{1-\frac{1}{n}} \|E\|_2^{\frac{1}{n}} \leq G(1 - \lambda_{\max}) \\
&\quad \Longrightarrow \log\left(\|E\|_2\right) \leq n\left[\log\left(G(1-\lambda_{\max}) - \log(2n-1) - \log(4n)\frac{n-1}{2n}\right)\right] \quad \square
\end{aligned}$$

This bound on the perturbation matrix is weak. A stronger bound cannot be developed in general because there is no guaranteed characterization on the size of the Jordan blocks in the network influence matrix. The bound can be improved for the case where the eigenvalues are simple, i.e., the size of the largest Jordan block is 1, through the application of Henrici's Theorem [29]. Unfortunately, it is difficult to provide general graph conditions such that the eigenvalues are simple for directed graphs, and so it is difficult to obtain a stronger bound on the perturbation of the eigenvalues in this way. For the special case of undirected graphs with weak links (which is a subset of the case that the network influence matrix has real eigenvalues), the Courant–Fischer theorem can provide strong bounds. Let us give the result for this case.

THEOREM 3.5 Consider two-way partitioning with a reference vertex for an undirected graph. Let the weights of the edges between the optimal partitions in the network influence matrix be bounded by ε. Let us denote the network influence matrix of the mapped influence model as $\bar{D}$, and that of the influence model with the weak cuts removed as D. Let $\lambda_{\max}$ be the largest eigenvalue of D with magnitude strictly less than unity. The change $\lambda_{\max}$ due to introduction of the weak cuts is bounded by $(n-2)\varepsilon$.

Proof. As shown in the proof for Theorem 3.1, the eigenvalues of any network influence matrix D are equal to the eigenvalues of the reduced matrix $\widetilde{D}$ where the reduced matrix has the structure:

$$\widetilde{D} = \begin{bmatrix} 1 & 0\ldots 0 \\ \vec{v}_1 & A \end{bmatrix}$$

We refer to the matrix block A as the *reference-free influence matrix*. Since the reduced matrix has a block-diagonal structure and A is a sub-stochastic matrix, the largest eigenvalue of D with magnitude less than unity is either the maximum or the minimum eigenvalue of matrix A.

If the graph is undirected, the matrix A is a symmetric matrix. The analysis of the undirected case is simpler because the symmetry of the matrix A preempts the occurrence of Jordan blocks of size greater than 1. For a disconnected graph, matrix A has the following structure:

$$A = \begin{bmatrix} A_1 & 0 \\ 0 & A_2 \end{bmatrix}$$

where the matrices A_1, and A_2 are symmetric square matrices. The size of the matrices is 1 less than the size of the corresponding partitions. When weak links are present, the reference-free influence matrix of the corresponding graph (denoted as $\widetilde{A}$) can be viewed as a perturbation of the matrix A, i.e.,

$$\widetilde{A} = A + E$$

where the matrix E denotes the perturbation and is a symmetric matrix with the following "Laplacian-like" structure:

$$E = \begin{bmatrix} Q & E_1 \\ E_1^T & W \end{bmatrix}$$

Matrices Q and W are diagonal matrices with negative entries on the diagonal (that are equal to the sum of the off-diagonal entries on that row). Since we are interested in bounding the change in the maximum and the minimum eigenvalues due to the perturbation, the Courant–Fischer theorem proves to be useful. We first focus on the maximum eigenvalue case.

The maximum eigenvalues $\lambda_{\max}$ of A and $\widetilde{A}$ are given by:

$$\lambda_{\max}(A) = \max_{x\in\Re^n:\|x\|_2=1} x^T Ax$$

$$\lambda_{\max}(\widetilde{A}) = \max_{x\in\Re^n:\|x\|_2=1} x^T \widetilde{A}x = \max_{x\in\Re^n:\|x\|_2=1} x^T(A+E)x$$

For any vector x,

$$x^T(A+E)x = x^T Ax + x^T Ex \geq x^T Ax - (n-2)\varepsilon$$

$$\Longrightarrow \max_{x\in\Re^n:\|x\|_2=1} x^T(A+E)x \geq \max_{x\in\Re^n:\|x\|_2=1} x^T Ax - (n-2)\varepsilon$$

$$\Longrightarrow \lambda_{\max}(\widetilde{A}) \geq \lambda_{\max}(A) - (n-2)\varepsilon$$

$$\Longrightarrow \lambda_{\max}(\widetilde{A}) - \lambda_{\max}(A) \leq (n-2)\varepsilon$$

Similarly,

$$x^T(A+E)x = x^TAx + x^TEx \leq x^TAx + (n-2)\varepsilon$$

$$\implies \max_{x\in\Re^n:\|x\|_2=1} x^T(A+E)x \leq \max_{x\in\Re^n:\|x\|_2=1} x^TAx + (n-2)\varepsilon$$

$$\implies \lambda_{\max}(\widetilde{A}) \leq \lambda_{\max}(A) + (n-2)\varepsilon$$

$$\implies \lambda_{\max}(\widetilde{A}) - \lambda_{\max}(A) \geq -(n-2)\varepsilon$$

Combining the two inequalities, we get that $|\lambda_{\max}(A) - \lambda_{\max}(\widetilde{A})| \leq (n-2)\varepsilon$. Therefore, if $(n-2)\varepsilon \leq \mu$, the change in the value of the maximum eigenvalue due to perturbation is upper bounded by μ. The exact same argument holds for the change in the value of the perturbation of the minimum eigenvalue.[3] The above result is a restatement of the well-known result in Hermitian perturbation of a Hermitian matrix [29]. □

Let us present the generalization of the above result to the m-way partitioning case.

COROLLARY 3.3 Consider m-way partitioning with reference vertices for an undirected graph. Let the weights of the edges between the optimal partitions in the network influence matrix be bounded by ε. Let us denote the network influence matrix of the mapped influence model as $\bar{D}$, and that of the influence model with the weak cuts removed as D. Let λ_{max} be the largest eigenvalue of D with magnitude strictly less than unity. The change in λ_{max} due to introduction of the weak cuts is bounded by $(n-m)\varepsilon$.

3.5.2 Eigenvector Sensitivity

In addition to the eigenvalue, sensitivity of eigenvectors to perturbation is key to our analysis in Section 3.4. While there are many well-known results that deal with sensitivity of eigenvectors to perturbation of general matrices, we find it worth our while to present a result that is applicable to our development. Exploiting the block lower-triangular structure of the network influence matrix, we present an explicit bound on the eigenvector change, in terms of the strength of the weak links between the partitions of the graph.

As mentioned before in Section 3.4, the reduced network influence matrix of the disconnected graph can be written in the following form:

$$\widetilde{D} = \begin{bmatrix} 1 & 0,\ldots,0 \\ \vec{v}_1 & A \end{bmatrix}$$

After the perturbation, the reduced matrix has the form:

$$\widehat{D} = \begin{bmatrix} 1 & 0,\ldots,0 \\ \vec{v}_1 & \widehat{A} \end{bmatrix}$$

[3] Instead of the max operator, we would use the min operator. The result for the change in the value of the minimum eigenvalue follows directly.

where $\widehat{A} = A + E$, $\mathcal{O}(E) = \varepsilon$ and $\overrightarrow{\widehat{v}}_1 = \overrightarrow{v}_1 + \mathcal{O}(\varepsilon)$. We will show that the relevant eigenvectors of these two matrices (which give the steady-state probabilities) can be found by solving a certain related system of linear equations, and bound the change in the eigenvector from this perspective. Here follows the formal result.

THEOREM 3.6 Let the reduced network influence matrix of the mapped binary copying influence model be $\widehat{D}$ and that of the influence model with the weak cuts between the optimal partitions removed be $\widetilde{D}$. Note that $\widehat{D}$ and $\widetilde{D}$ have the following structure:

$$\widetilde{D} = \begin{bmatrix} 1 & 0, \ldots, 0 \\ \overrightarrow{v}_1 & A \end{bmatrix}$$

$$\widehat{D} = \begin{bmatrix} 1 & 0, \ldots, 0 \\ \overrightarrow{\widehat{v}}_1 & \widehat{A} \end{bmatrix}$$

where $\widehat{A} = A + E$ and $\overrightarrow{\widehat{v}}_1 = \overrightarrow{v}_1 + \mathcal{O}(\varepsilon)$. If

$$E_{i,j} < \frac{0.1^q}{0.1^q(n-2) + n} \frac{1}{\|(I-A)^{-1}\|_\infty}$$

and

$$e_i < \frac{0.1^q}{0.1^q(n-2) + n} \frac{1}{\|(I-A)^{-1}\|_\infty}$$

Then the change in the entries of the eigenvector for the unity eigenvalue due to perturbation is at most 0.1^q, where q is an arbitrarily chosen positive quantity.

Proof. The result is a straightforward derivative of the Bauer–Skeel Theorem [29] for perturbation of systems of linear equations. Specifically, the eigenvector entries for the unity eigenvalue of the unperturbed reduced network influence matrix (i.e., for a graph of n nodes with weak cuts removed) can be obtained from the solution of the following system of linear equations:

$$(I - A)\overrightarrow{x} = \overrightarrow{v}_1$$

where the dimension of the vector $\overrightarrow{x}$ is $(n-2) \times 1$. Solving the equation yields that $\overrightarrow{x} = \begin{bmatrix} 1_{m_1 - 1} \\ 0_{m_2 - 1} \end{bmatrix}$ (where m_1 and m_2 are the sizes of the partitions), i.e., the solution is a vector of 1s and 0s. The eigenvector entries of the perturbed reduced network influence model (i.e., for the graph with the weak cuts) can be obtained from the solution of the following linear system.

$$(I - \widehat{A})\overrightarrow{y} = \overrightarrow{\widehat{v}}_1$$

where $\widehat{A} = A + E$ and $\overrightarrow{\widehat{v}}_1 = \overrightarrow{v}_1 + \overrightarrow{e}_1$. The conditions for the Bauer–Skeel Theorem when applied to the current problem are as follows:

If we can show that there exists a nonnegative matrix S, vector $\overrightarrow{s}$ and a scalar ε and an absolute vector norm and its consistent matrix norm such that the following holds true:

1. $|E| \leq \varepsilon S$ and $|e| \leq \varepsilon s$
2. $\varepsilon \| |(I - A)^{-1}| S \| < 1$

then the change in the solution of the linear system due to perturbation is bounded by:

$$\|(y - x)\| \leq \frac{\varepsilon \| |(I - A)^{-1}|(S|x| + s)}{1 - \varepsilon \| |(I - A)^{-1}| S \|}$$

To arrive at the result presented in Theorem 3.6, we choose S and s as matrix and vector of all 1s of appropriate dimensions, and choose the norm to be ∞-norm. This choice yields three benefits, namely:

1. The condition (1) mentioned above is directly converted into a condition on the entries of E and e, i.e., that the absolute values of the entries are less than ε.
2. The condition (2) is simplified into a condition on the ∞-norm of $(I - A)^{-1}$, as follows

$$\| |(I - A)^{-1}| \|_\infty = (n - 2)\|(I - A)^{-1}\|_\infty$$

3. The term $\varepsilon \| |(I - A)^{-1}|(S|x| + s)$ can be written as $n_1 \|(I - A)^{-1}\|_\infty$, where $n_1 < n$ is the size of the first partition as determined by the number of 1s in the eigenvector of the unperturbed network influence matrix.

With these simplifications, we arrive at the condition that

$$\varepsilon \| |(I - A)^{-1}| S \| = \varepsilon(n - 2)\|(I - A)^{-1}\|_\infty \leq \frac{0.1^q}{0.1^q + \frac{n}{n-2}} \leftrightarrow \|(y - x)\| 0.1^q$$

This inequality also implies that the second condition is satisfied automatically. Simplifying the first inequality, we arrive at the following condition which directly yields the conditions on the entries of E and $\overline{e}$ given in the theorem:

$$\varepsilon \leq \frac{0.1^q}{0.1^q(n - 2) + n} \frac{1}{\|(I - A)^{-1}\|_\infty}$$

Since we have used the ∞-norm to arrive at the condition, it ensures that $\|(y-x)\|_\infty \leq 0.1^q$, which means that the steady-state status probabilities of the influence model with the weak links as given by the entries of the eigenvector for perturbed matrix (which has the form $\begin{bmatrix} 1 \\ \vec{y} \\ 0 \end{bmatrix}$) would be close (within 0.1^q) to the entries of the vector $\begin{bmatrix} 1 \\ \vec{x} \\ 0 \end{bmatrix}$. This implies that the statuses of the nodes are equal to those of the reference nodes of the partitions they belong to with a high probability in the steady state. □

COROLLARY 3.4 Let the reduced network influence matrix of the mapped binary copying influence model be $\widehat{D}$ and that of the influence model with the weak cuts between the optimal partitions removed be $\widetilde{D}$. Note that $\widehat{D}$ and $\widetilde{D}$ have the following structure:

$$\widetilde{D} = \begin{bmatrix} 1 & 0,\ldots,0 \\ \vec{v}_1 & A \end{bmatrix}$$

$$\widehat{D} = \begin{bmatrix} 1 & 0,\ldots,0 \\ \hat{\vec{v}}_1 & \widehat{A} \end{bmatrix}$$

where $\widehat{A} = A + E$ and $\hat{\vec{v}}_1 = \vec{v}_1 + \mathcal{O}(\varepsilon)$. If

$$E_{i,j} < \frac{0.1^q}{0.1^q(n-m)+n} \frac{1}{\|(I-A)^{-1}\|_\infty}$$

and

$$e_i < \frac{0.1^q}{0.1^q(n-m)+n} \frac{1}{\|(I-A)^{-1}\|_\infty}$$

then the change in the entries of the eigenvector for the unity eigenvalue due to perturbation is at most 0.1^q, where q is an arbitrarily chosen positive quantity.

REMARK 3.1 The term $\frac{1}{\|(I-A)^{-1}\|_\infty}$ is undesirable in a sense. However, the algebraic graph theory and stochastic matrix literature provides means to bound the quantity in terms of the network graph; we omit the details.

As an aside, the linear-algebraic techniques used to develop bounds in this section may be of broader interest in the study of CPSs, and in particular this work shows that the techniques can be used to characterize the discrete and stochastic behaviors that are commonplace in CPSs.

3.6 Integrative Theorem and Discussion

We find it helpful to combine all the results in Sections 3.4 and 3.5 into a single result that informs implementation of a fast influence model partitioning algorithm. We present the combined result in Theorem 3.7. We directly present the result for the general m-way partitioning problem, since we envision the algorithm as being of particular interest in that case.

THEOREM 3.7 Consider m-way partitioning with reference vertices for a connected graph with m strongly connected subgraphs (each of which contains one reference vertex). The partitions found by the influence model-based partitioning algorithm after k time steps, where

$$k \geq \frac{-q - \log_{10}(n-m)}{\log_{10}(|\lambda_{max}|)}$$

and $\lambda_{\max}$ is the largest eigenvalue of the network influence matrix of the mapped influence model with magnitude strictly less than unity, are the optimal partitions with a probability of at least $1 - 2(0.1^q)$ (where q is an arbitrarily chosen positive quantity), if the strength of the connections w between sites i and j are such that they are in different identified partitions and satisfy the following inequality for all i and j:

$$w_{i,j} < \frac{0.1^q}{0.1^q(n-m)+n} \frac{1}{\|(I-A)^{-1}\|_\infty}$$

Here, matrix I is the identity matrix and matrix A is the reference-free influence matrix as defined in Section 3.5.

The result can be obtained by simply coalescing all the previously stated results. We omit the details.

However, we note that this result depends on the magnitude of the sub-dominant eigenvalue. Calculation of the sub-dominant eigenvalue is computationally taxing. There exists a large body of results in the field of algebraic graph theory that provide bounds on the same in terms of the strength of the connection. For example, an approach for estimating spectral properties of the graph-adjacency matrix from local measurements as well as the relation between spectral properties of graph-adjacency matrix and graph structures in discussed in [30]. Theorems 3.4 and 3.5 are also useful tools in terms of providing bounds on the magnitude of λ_{max}.

We suggest the following working algorithm to identify the partitions, as well as to verify if the solution generated by the influence model algorithm is a good approximation of the optimal solution for the case of m-way partitioning with reference vertices on graphs known to have weak cuts.

1. The graph to be partitioned is mapped to an influence model as described in Section 3.3. Please note that this can be done in a distributed manner, by assuming additional computation capability for each vertex to run separate consensus-like algorithms.
2. The influence-model-based algorithm is allowed to run with suitable stopping criteria. Various stopping methods are described in [2, 3].
3. Based on the weak links identified by the algorithm and a user-chosen value of q, if the various conditions in Theorem 3.7 are satisfied, the partitions identified by the algorithm are a good approximation of the optimal solution.
4. If the conditions are not satisfied, the algorithm is run again until the partitions identified are a good approximation of the solutions.

As a discussion, let us revisit the motivation for this work and explicitly mention the contribution made by this chapter. The problem of partitioning graphs with reference vertices is a known NP-hard problem when the number of reference vertices is greater than two. The results presented here show that the influence-model-based partitioning algorithm can quickly find the optimal solutions to the partitioning with the reference vertices problem for graphs with sufficiently weak cuts by various metrics. Such problems are of common interest in modern CPSs and infrastructures because these networks are so complex that they invariably require distributed partitioning (often based on sophisticated metrics). The research described here addresses this need, and presents a promising solution.

Acknowledgements

This project was partially supported by NSF grants ECS-0901137, CNS-1035369, CNS-1058124, and CNS-1545104.

References

[1] D. R. Karger and C. Stein, "A new approach to the minimum cut problem," *JACM*, vol. 43, no. 4, pp. 601–640, 1996.
[2] S. Roy, Y. Wan, and A. Saberi, "A flexible algorithm for sensor network partitioning and self-partitioning problems," in *Int. Symp. Algorithms and Experiments for Sensor Systems, Wireless Networks and Distributed Robotics*, 2006, pp. 152–163.
[3] Y. Wan, S. Roy, A. Saberi, and B. C. Lesieutre, "A stochastic automaton-based algorithm for flexible and distributed network partitioning," in *Proc. Swarm Intelligence Symp.*, 2005, pp. 273–280.
[4] Y. Wan, S. Roy, A. Saberi, and B. Lesieutre, "A flexible stochastic automaton-based algorithm for network self-partitioning," *Int. J. Distrib. Sensor Netw.*, vol. 4, no. 3, pp. 223–246, 2008.
[5] C. Asavathiratham, "The influence model: A tractable representation for the dynamics of networked Markov chains," Ph.D. dissertation, Massachusetts Institute of Technology, Cambridge, MA, 2001.
[6] C. Asavathiratham, S. Roy, B. Lesieutre, and G. Verghese, "The influence model," *IEEE Contr. Syst.*, vol. 21, no. 6, pp. 52–64, 2001.
[7] D. Kempe, J. Kleinberg, and É. Tardos, "Maximizing the spread of influence through a social network," in *Proc. Ninth ACM SIGKDD Int. Conf. Knowledge Discovery and Data Mining*, 2003, pp. 137–146.
[8] S. Roy, K. Herlugson, and A. Saberi, "A control-theoretic approach to distributed discrete-valued decision-making in networks of sensing agents," *IEEE Trans. Mobile Comput.*, vol. 5, no. 8, pp. 945–957, 2006.
[9] F. Pellegrini, "A parallelisable multi-level banded diffusion scheme for computing balanced partitions with smooth boundaries," in *European Conf. Parallel Processing*, 2007, pp. 195–204.
[10] E. Dahlhaus, D. S. Johnson, C. H. Papadimitriou, P. D. Seymour, and M. Yannakakis, "The complexity of multiterminal cuts," *SIAM J. Comput.*, vol. 23, no. 4, pp. 864–894, 1994.
[11] P. Crescenzi, V. Kann, M. Halldórsson, M. Karpinski, and G. Woeginger, "A compendium of NP optimization problems," http://www.nada.kth.se/~viggo/problemlist/compendium.html, 1997.
[12] S. Ghiasi, A. Srivastava, X. Yang, and M. Sarrafzadeh, "Optimal energy aware clustering in sensor networks," *Sensors*, vol. 2, no. 7, pp. 258–269, 2002.
[13] Y. Wang, C. Zhang, and Z. Liu, "A matrix approach to graph maximum stable set and coloring problems with application to multi-agent systems," *Automatica*, vol. 48, no. 7, pp. 1227–1236, 2012.
[14] T. Sahai, A. Speranzon, and A. Banaszuk, "Hearing the clusters of a graph: A distributed algorithm," *Automatica*, vol. 48, no. 1, pp. 15–24, 2012.
[15] J. Xu, "Average case analysis of the Markov chain small-world model," *J. Franklin Instit.*, vol. 349, no. 7, pp. 2431–2444, 2012.

[16] Y. Sun, "Average consensus in networks of dynamic agents with uncertain topologies and time-varying delays," *J. Franklin Instit.*, vol. 349, no. 3, pp. 1061–1073, 2012.
[17] M. Xue and S. Roy, "Spectral and graph-theoretic bounds on steady-state-probability estimation performance for an ergodic Markov chain," *J. Franklin Instit.*, vol. 348, no. 9, pp. 2448–2467, 2011.
[18] B. L. Chamberlain, "Graph partitioning algorithms for distributing workloads of parallel computations," University of Washington Technical Report UW-CSE-98-10, vol. 3, 1998.
[19] H. Meyerhenke, B. Monien, and T. Sauerwald, "A new diffusion-based multilevel algorithm for computing graph partitions," *J. Parallel Distrib. Comput.*, vol. 69, no. 9, pp. 750–761, 2009.
[20] D. S. Johnson, "The NP-completeness column: Finding needles in haystacks," *ACM Trans. Algorithms*, vol. 3, no. 2, p. 24, 2007.
[21] D. E. D. Vinkemeier and S. Hougardy, "A linear-time approximation algorithm for weighted matchings in graphs," *ACM Trans. Algorithms*, vol. 1, no. 1, pp. 107–122, 2005.
[22] H. N. Djidjev, "A faster algorithm for computing the girth of planar and bounded genus graphs," *ACM Trans. Algorithms*, vol. 7, no. 1, p. 3, 2010.
[23] A. D. Scott and G. B. Sorkin, "Polynomial constraint satisfaction problems, graph bisection, and the Ising partition function," *ACM Trans. Algorithms*, vol. 5, no. 4, p. 45, 2009.
[24] M. Xue and S. Roy, "Averager-copier–voter models for hybrid consensus," *IFAC-PapersOnLine*, vol. 49, no. 4, pp. 1–6, 2016.
[25] M. Xue and S. Roy, "A framework for modeling stochastic flow and synchronization networks," in *Proc. 2011 Winter Simulation Conference (WSC)*, 2011, pp. 3224–3235.
[26] R. Dhal and S. Roy, "Performance analysis of an influence-model-based graph partitioning algorithm," in *Decision and Control and European Control Conference (CDC-ECC)*, 2011, pp. 6578–6583.
[27] R. Dhal and S. Roy, "Layered moment-linear network models as tools for strategic air traffic flow management," in *Proc. Infotech@ Aerospace 2012*, 2012, p. 2514.
[28] R. Dhal and S. Roy, "Layers of interacting dynamic networks: Motivation and theory," in *2013 American Control Conf.*, 2013, pp. 6339–6344.
[29] G. Stewart and J. Sun, *Matrix Perturbation Theory*. New York: Academic Press, 1990.
[30] V. M. Preciado, J. Oh, and A. Jadbabaie, "Analysis of equilibria and strategic interaction in complex networks," in *Decision and Control and European Control Conf. (CDC-ECC)*, 2011, pp. 4498–4503.

Part I Exercises

1. Please list and discuss several factors that make uncertainty evaluation for cyber-physical systems difficult, as compared to traditional computing or engineering systems.
2. The three chapters in Part I introduce a number of different analytical and numerical tools for uncertainty analysis in cyber-physical systems. Please make a list of the tools introduced in the chapter, and the advantages and disadvantages of each. Can you think of other uncertainty evaluation tools that are applicable to cyber-physical systems, which were not discussed in the chapters?
3. Chapter 1 (Bogdan and Marculescu) is focused on modeling workload uncertainties in cyber-physical systems. What models have been traditionally used to capture workloads? What are the new innovations on workload modeling proposed in the chapter?
4. Please explain what the master equation for a stochastic process is. In what contexts have master equations traditionally been used? How can they be used for uncertainty evaluation of cyber-physical systems?
5. Refer to Chapter 2 and compare the advantages and disadvantages of analytical and simulation-based approaches to evaluating uncertainties. Provide examples of where one method is superior to another.
6. Using ground transportation cyber-physical systems as an example, describe the uncertainties that may occur in this system. How do they impact the system's dynamics? Precisely identify uncertain input parameters and outputs that may be used for ground transportation cyber-physical systems.
7. Chapter 2 introduces the probabilistic collocation method (PCM) as a tool for uncertainty analysis. What is the advantage of PCM compared to the Monte Carlo simulation method? Explain conceptually the reason for PCM to reduce the number of simulations in predicting output statistics.
8. Find five PCM points for an uncertain input parameter of uniform distribution between 0 and 1.
9. The M-PCM-OFFD technique introduced in Section 2.5.1 enables further reduction of the number of simulation points. Describe the assumptions that allow this further reduction. Are these assumptions reasonable for realistic CPS applications?

10. Consider a system of 20 uncertain input parameters, each of which has a degree of at most three. Answer the following questions.
    a. How many simulations are needed for M-PCM to correctly predict the system's output mean?
    b. If all cross-terms have fewer than three different parameters, can we further reduce the number of simulations with the correct output mean? How many simulations can be reduced?
11. Consider an original system mapping of the form $g(x_1, x_2, x_3) = 2 + 5x_1 + 3x_2 + x_1^2 + x_3^2 + x_2^3$, where x_1 follows a uniform distribution of $f_{X_1}(x_1) = \frac{1}{10}, 1 \leq x_1 \leq 10$, x_2 follows a normal distribution of $f_{X_2}(x_2) = \frac{1}{\sqrt{2\pi}} e^{-\frac{x_2^2}{2}}$, and x_3 follows an exponential distribution of $f_{X_3}(x_3) = e^{-x_3^2}$. What is the reduced-order mapping constructed by M-PCM? How is the mapping constructed by M-PCM-OFFD?
12. Please discuss another application area, besides those discussed in the book so far, where inverse modeling techniques to extract uncertain parameters from output data using low-order mappings may be used.
13. Chapter 3 introduces a method for exploiting uncertainty to address a difficult computational problem (in this case graph partitioning), rather than seeking to analyze the impact of environmental uncertainties. Develop a literature review on randomized algorithms/methods that exploit uncertainty to solve difficult computational or engineering problems.
14. The authors of Chapter 3 claim that the developed partitioning algorithm is particularly apt for CPS applications, because it allows distributed partitioning and partitioning based on multiple cost metrics. Can you think of cyber-physical systems applications where these characteristics would be useful?

Part II

Exploiting Structure for Control

Complex engineered systems invariably require automatic feedback control functions to regulate system states and achieve mission objectives in the face of disruptions and noise. The pervasive and tight integration of cyber technologies in physical-world systems is changing the feedback-control paradigm in competing directions. On one hand, the resulting cyber-physical system (CPS) allows unprecedented opportunity for information-rich control, because the cyber technologies permit wide-area sensing and communication, implementation of sophisticated controls, mission-level coordination of control functions, etc. Thus, while traditionally control systems were necessarily simple, specific, and local, CPSs enable much more complex controls that can use diverse information sources and actuate complex mission objectives. On the other hand, the tight interface of cyber and physical components also means that control functions must account for dynamics that cross into the cyber systems, as well as disruptions and noise that originate from the cyber world. Thus, traditional modeling paradigms for control design, which are based on continuous state-space differential equations of the physical world as well as engineering systems representations of mission objectives and disruptions, become obsolete. Control of CPSs is also complicated by issues of scale and complexity, since integrated cyber systems may network and interface components that were traditionally standalone. These differences in control for CPSs as compared to traditional engineered systems are especially pronounced in large-scale infrastructures, in which many new sensing, communication, and control functionalities are being rapidly integrated. For example, both the electric power network and the air transportation system are undergoing rapid evolution to cyber-pervasive systems, and control/management functions must both account for and exploit these new technologies.

The chapters in Part II describe the state-of-the-art in control of CPSs. Broadly, the four chapters are all concerned with how different structures of CPSs inform their control. Specifically, Chapter 4, by Vasconcelos and Martins, discusses state estimation (which is an integral part of control design) using remote sensor measurements; this is an intrinsically CPS problem, since the physical dynamics being estimated as well as the cyber communication channel must be accounted for. Chapter 5 addresses event-driven control for CPSs, wherein control functions are implemented in response to discrete events rather than via continual feedback. Event-driven control paradigms are often apt for CPSs, since controls must depend on communication channels with limited bandwidth and perhaps spotty connectivity. The chapter characterizes event-driven controls for systems with explicit models of meshed cyber and physical components, thus providing a comprehensive treatment of event-driven control for CPSs. Chapters 6 and 7 are concerned with exploiting network structure in the identification and control of CPSs. Many CPSs, including both infrastructures and the Internet of Things, involve networking of a large number of subsystems or components. These two chapters advance research into the control and identification of networked and multi-agent systems. The work of Chapman et al. pursues the online and adaptive design of controllers for multi-agent systems, so as to reject disruptions. Meanwhile, the work of Chetty and Warnick introduces a new notion of structure for networks, based on signals rather than subsystems, which is promising for control, identification, and model reduction. Together, the chapters in Part II address diverse problems that arise in the control of CPSs, and show how structures can be exploited to address these problems.

4 A Survey on Remote Estimation Problems

Marcos M. Vasconcelos and Nuno Martins

4.1 Introduction

Signal estimation is a fundamental component in any cyber-physical system (CPS). Algorithms that estimate or track the state (or a function of the state) of a physical system provide information on its dynamics that is essential for monitoring or decision-making. Modern technology allows the use of smart sensors that cooperate through a network to achieve a common task. Although these networks of sensors provide convenience, flexibility, and robustness to modern systems, design methods to achieve optimal performance are elusive as they must account for certain detrimental characteristics of the underlying network. These include constrained connectivity among agents, rate-limited communication links, physical noise at the antennas, packet-drops, and interference.

One concrete application in which remote signal estimation is paramount is the remote monitoring of wildlife being pursued as part of the NSF project *Remote Imaging of Community Ecology via Animal-Borne Wireless Networks*. Here, sensors are mounted on animals with the goal of understanding and preserving endangered species. The mobile smart sensors perform collaborative tasks such as tracking, making autonomous decisions such as when to turn video cameras on and off, and detecting events such as predatory attacks. The observation of the natural behavior and social interactions of these animals can only be done remotely with minimal, if any, human interference.

As is usually the case in such extreme scenarios, deployments are rare and costly. To ensure the success of each experiment, it is fundamental to maximize the operational life of each sensor. Therefore, battery power is a valuable resource that must be saved and used wisely. Each sensor comprises several modules and our experiments showed that turning on the communication module alone consumes more power than any other component in the system. If sensors were allowed to communicate constantly, the deployment would last no longer than a few hours, whereas using techniques similar to the ones described in this chapter, we were able to extend the deployment duration to about 10 days. This highlights the importance of understanding the fundamental issues and design guidelines that can be drawn from optimizing a remote estimation system with communication constraints.

This chapter provides some insights on essentially three ways in which communication is penalized in these systems: a limited number of transmissions, costly transmissions, and interfering communication links. In particular, we are interested in reviewing basic remote estimation problems that deal with each of these constraints. A common feature shared by the results shown here is that when communication is penalized, optimal solutions tend to be event-driven, i.e., transmission is done when the sensor observations exceed certain thresholds, in engineering terms, the design principle that packets should be generated and transmitted only when new information is observed. At all other times, the communication module of sensors should be idle.

4.1.1 Organization

The chapter is organized in three main parts. In the first part we present in some detail the results on four different problem formulations which incorporate limited communications in remote estimation. To avoid ambiguity and simplify the nomenclature, each problem is referred to by the names of its authors. Our approach is to present only the main results and solution methods in a unified notation. Due to the highly technical arguments involved in the original proofs, we attempt to guide the reader over the main ideas rather than presenting a detailed discussion. These problems are also diverse in their solution strategies, which we highlight at the end of each section. The second part of the chapter consists of a survey of results in sensor scheduling problems, event-driven remote estimation, and estimation over shared networks. This part aims at pointing the reader to only a few papers in those areas and is not, by any means, an exhaustive list. The third and last part of the chapter is a brief discussion on a new remote estimation problem over a simple collision network, which is also related to some of the works in the first part.

If the reader is interested in the cases of additive noise, rate-limited links, and packet-drop networks, we would like to point out that these have been extensively studied and surveyed in other papers, such as [1–3].

4.1.2 Notation

Throughout the chapter we adopt the following notation. Random variables are represented using upper case letters, such as X. Realizations of random variables are represented by the corresponding lower case letter, such as x. When two random variables X, Y are independent, we represent this relationship using the symbol $\perp\!\!\!\perp$, such as $X \perp\!\!\!\perp Y$. Finally, $\mathcal{B}(p)$ and $\mathcal{N}(m, \sigma^2)$ are used to represent the Bernoulli distribution of parameter p and the Gaussian distribution of mean m and variance σ^2, respectively. Time indices are represented by the subindex k (in discrete time) and t (continuous time), as in X_k or X_t. Sequences such as $\{x_k\}_{k=0}^{N}$ are sometimes represented by $\{x_k\}$ or x_0^N. The set of real numbers is denoted by $\mathbb{R}$ and the empty set is denoted by $\emptyset$; the complement with respect to $\mathbb{R}$ of a subset of real numbers $\mathcal{A}$ is denoted by $\mathcal{A}^c$.

4.2 Optimal Estimation with Limited Transmissions

In most applications, sensing devices have limited battery life. This limitation can be mathematically modeled by a finite number of transmissions (or channel uses) less than the total problem's horizon. This means that communication at every time step is simply not possible and so it should be intermittent. One of the first papers that addressed such problems is by Imer and Basar [4]. We shall refer to it as the Imer–Basar Problem.

4.2.1 The Imer–Basar Problem

Consider the system depicted in Figure 4.1, where a sensor makes measurements of a zero-mean independent and identically distributed (i.i.d.) Gaussian process $\{X_k\}_{k=1}^N$ with variance σ^2, and is constrained to make only $M < N$ transmissions over a noiseless channel to a remote estimator. It is assumed that the statistical model of the random process is available to the remote estimator. At each time k, the sensor decides whether to transmit or to remain silent based on all the information available to it up to the current time. When the sensor decides to transmit, a packet containing the most recent observation x_k is sent through the channel, and when the sensor decides not to transmit, the channel is idle. Every packet transmitted by the sensor is noiselessly received by the estimator. The channel input at time k is denoted by $s_k \in \mathcal{S}$, where

$$\mathcal{S} = \mathbb{R} \cup \{\varnothing\} \tag{4.1}$$

in which the special symbol $\varnothing$ denotes *no-transmission*. Only when $s_k = x_k$ does a transmission count toward the maximum allowed limit M. The remote estimator generates a sequence of estimates $\{\hat{x}_k\}_{k=1}^N$, where each sample $\hat{x}_k$ depends causally on the information received until time k, i.e., s_1^k. The goal is to minimize a cost function corresponding to the aggregate expected quadratic estimation error

$$J_{(M,N)}(\gamma, \eta) = \mathbb{E}\left[\sum_{k=1}^{N}\left(X_k - \hat{X}_k\right)^2\right] \tag{4.2}$$

where γ and η are the sequences of transmission and estimation policies implemented by the sensor and the remote estimator, respectively. Denoting the sensor's decision to transmit by $u_k = 1$ and the decision not to transmit by $u_k = 0$, the hard constraint on the optimization problem, which is the total number of packets the sensor is allowed to transmit, is given by

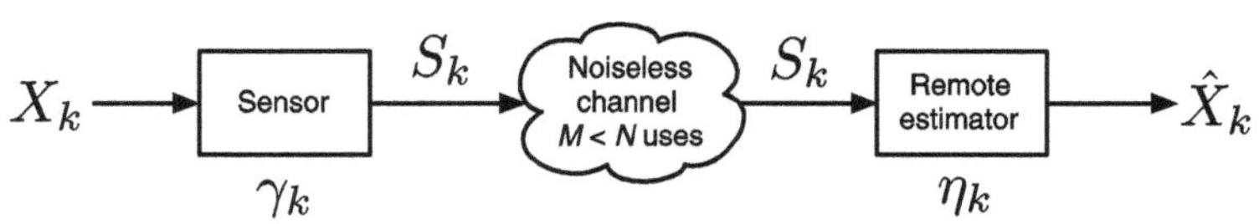

Figure 4.1 Estimation with limited transmissions.

$$\sum_{k=1}^{N} u_k \leq M \tag{4.3}$$

In order to specify the information available to each of the decision-makers involved in this problem, we must define two additional variables: t_k, which denotes the number of transmissions left at time k; and d_k, which denotes the number of decision slots left at time k. The channel input at time k is a measurable function of $\left((t_k, d_k), x_1^k, s_1^{k-1}\right)$, i.e.,

$$s_k = \gamma_k \left((t_k, d_k), x_1^k, s_1^{k-1}\right) \tag{4.4}$$

whereas the estimate at time k is a measurable function of $\left((t_k, d_k), s_1^k\right)$, i.e.,

$$\hat{x}_k = \eta_k \left((t_k, d_k), s_1^k\right) \tag{4.5}$$

As one would expect from the expected quadratic cost structure, the optimal estimator is the conditional expectation of X_k given the information available to the estimator at time k, i.e.,

$$\eta_k^* \left((t_k, d_k), s_1^k\right) = \mathbb{E}\left[X_k \mid (t_k, d_k), s_1^k\right] \tag{4.6}$$

Therefore, the cost we would like to optimize over the space of transmission policies is

$$J_{(M,N)}(\gamma, \eta^*) = \mathbb{E}\left[\sum_{k=1}^{N} \left(X_k - \mathbb{E}\left[X_k \mid (T_k, D_k), X_1^k, S_1^{k-1}\right]\right)^2\right] \tag{4.7}$$

Imer and Basar showed in [4] that $((t_k, d_k), x_k)$ is a *sufficient statistic* [5] for the optimal transmission policies. Therefore, optimal decisions can be made based on the current measurement x_k and its past actions through (t_k, d_k). Based on this structural result, one can constrain the transmission policies to be of the form

$$\gamma_k \left((t_k, d_k), x_k\right) = \begin{cases} x_k, & \text{if } x_k \in \mathcal{M}_{(t_k, d_k)} \\ \varnothing, & \text{if } x_k \in \mathcal{M}^c_{(t_k, d_k)} \end{cases} \tag{4.8}$$

where $\mathcal{M}_{(t_k, d_k)}$ is a measurable subset of $\mathbb{R}$ corresponding to the range of values of x_k that are to be transmitted at time k if there are t_k transmissions out of d_k decision slots left. The optimization can be recursively carried out using a *dynamic programming* (DP) equation [5]:

$$J^*_{(t,d)} \inf_{\mathcal{M}_{(t,d)}} \left\{ J^*_{(t-1,d-1)} \cdot \int_{\mathcal{M}_{(t,d)}} f_X(x)dx + J^*_{(t,d-1)} \cdot \int_{\mathcal{M}^c_{(t,d)}} f_X(x)dx \right.$$
$$\left. + \int_{\mathcal{M}^c_{(t,d)}} \left(x - \mathbb{E}\left[X \mid X \in \mathcal{M}^c_{(t,d)}\right]\right)^2 f_X(x)dx \right\} \tag{4.9}$$

together with the boundary conditions $J^*_{(d,d)} = 0$, $J^*_{(0,d)} = d \cdot \sigma_X^2$, $\mathcal{M}^*_{(d,d)} = \mathbb{R}$ and $\mathcal{M}^*_{(0,d)} = \emptyset$, $\forall d \geq 0$. The minimization step in the DP equation is over the space of measurable subsets of $\mathbb{R}$. This is addressed by further constraining the sets $\mathcal{M}^c_{(t,d)}$ to be symmetric intervals, i.e., $\mathcal{M}^c_{(t,d)} = \left[-\beta_{(t,d)}, \beta_{(t,d)}\right]$, where $0 \le \beta_{(t,d)} \le \infty$. Due to the symmetry of the Gaussian probability density function (PDF) $f_X(x)$, we have $\mathbb{E}\left[X \mid X \in \left[-\beta_{(t,d)}, \beta_{(t,d)}\right]\right] = 0$. The DP equation can be rewritten as

$$J^*_{(t,d)} = J^*_{(t-1,d-1)} + \inf_{\beta_{(t,d)}} \left\{ \left[J^*_{(t,d-1)} - J^*_{(t-1,d-1)} \right] \cdot \int_{-\beta_{(t,d)}}^{\beta_{(t,d)}} f_X(x)dx + \int_{-\beta_{(t,d)}}^{\beta_{(t,d)}} x^2 f_X(x)dx \right\} \tag{4.10}$$

Now the minimization step can be easily carried out by simply taking the derivative of the argument with respect to $\beta_{(t,d)}$ and setting it to zero, which yields the following optimal threshold:

$$\beta^*_{t,d} = \sqrt{J^*_{(t-1,d-1)} - J^*_{(t,d-1)}} \tag{4.11}$$

It is important to note that $J^*_{(t-1,d-1)} \ge J^*_{(t,d-1)}$. This is intuitively verified by arguing that if more transmissions are allowed over the same horizon, the optimal cost cannot increase. This implies that the threshold is always a real number.

4.2.2 Variations and Extensions

This problem was introduced in [4] and a more detailed version later appeared in [6]. The original paper reporting the results for the i.i.d. Gaussian source also solves the scalar Gauss–Markov case where

$$X_{k+1} = aX_k + W_k, \;\; k = 1, \ldots, N-1 \tag{4.12}$$

where $\{W_k\}_{k=1}^{N-1}$ is i.i.d. such that $W_k \sim \mathcal{N}(0, \sigma^2)$, and $a \in \mathbb{R}$. The sequence of steps involved in solving the problem is essentially the same, but in this case the DP recursion is in three dimensions instead of two. The reason why this is indeed the case is that the sensor and the remote estimator also need to keep track of the number of time steps since the last transmission in addition to t_k and d_k. Also, due to the presence of a homogeneous term in the state equation when a transmission is made, the thresholds become symmetric around a nonzero mean. Other variations addressed by [6] include the case when X_k is a discrete state-space random process and the distortion metric is the probability of error, i.e.,

$$J_{(M,N)}(\gamma, \eta) = \sum_{k=1}^{N} \Pr(X_k \ne \hat{X}_k) \tag{4.13}$$

When $X_k \in \mathbb{R}^n$, $n > 1$, MacKunis et al. [7] used essentially the same techniques as Imer and Basar, constraining the sets $\mathcal{M}^c_{(t,d)}$ to be hyper-ellipsoids which required a re-parametrization of the integrals in the argument of the minimization problem in the DP equation. In that case, the optimization variables are positive-definite matrices that define the hyper-ellipsoidal integration region and its complement.

4.2.3 Main Features of the Imer–Basar Problem

- There is a hard constraint on the number of transmissions.
- Transmitted packets are received error-free by the remote estimator.
- They provide a structural result which determines a sufficient statistic for the optimal decision rule. The dimension of the sufficient statistic does not grow with time.
- The solution can be recursively computed using a DP equation whose solution requires the offline computation and storage of a two-dimensional matrix.
- To solve the minimization step of the DP equation, the authors constrain the strategies to be of the threshold type. Therefore, the threshold structure does not arise as the optimal solution, although it is the most *natural* candidate.

4.3 Optimal Communication Logics

In [8] and [9], Xu and Hespanha explored the trade-off between communication rate and estimation error in networked estimation networks. The communication policies, which they denominated *logics*, were designed exclusively to stabilize the estimation error process both in continuous time and discrete time. In [10], Xu and Hespanha proposed the design of *optimal* communication policies based on a cost criterion that incorporated the estimation error as well as the communication rate.

4.3.1 The Xu–Hespanha Problem

Consider the system shown in Figure 4.2. The goal is to characterize the optimal trade-off between communication rate and estimation performance of the vector-valued linear stochastic process defined by

$$X_{k+1} = AX_k + W_k,\ k \geq 1 \tag{4.14}$$

where $X_k, W_k \in \mathbb{R}^n$ where $\{W_k\}$ is i.i.d. with $W_k \sim \mathcal{N}(0, \Sigma)$, and $A \in \mathbb{R}^{n \times n}$. As in the Imer–Basar problem, the sensor's decision to transmit the current observation x_k over the network at time k is binary and denoted by $u_k = 1$ in case of transmission and $u_k = 0$ in case of remaining silent. It is assumed that communication is noiseless but incurs a nonnegative cost $\rho \in \mathbb{R}$ to the sensor. It is assumed that there is a fixed unit delay on packet reception. Therefore, the information available to the remote estimator at time k is

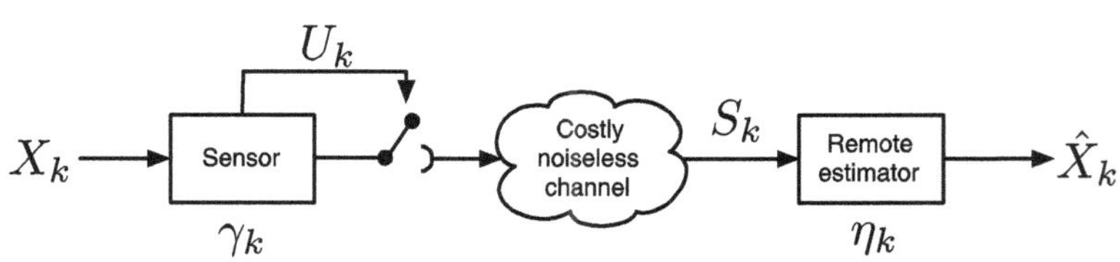

Figure 4.2 Optimal communication logics for remote estimation.

$$\mathcal{I}_k = \{(t, x_t) \mid u_t = 1, \, t \leq k-1\} \tag{4.15}$$

The estimator is constrained to be

$$\hat{x}_k = \mathbb{E}\left[X_k \mid \mathcal{I}_k\right] \tag{4.16}$$

which is the minimum mean squared error (MMSE) estimate of X_k given $\mathcal{I}_k$. The sequence $\{\hat{x}_k\}$ is assumed to satisfy the following recursion:

$$\hat{x}_{k+1} = \begin{cases} Ax_k, & \text{if } u_k = 1 \\ A\hat{x}_k, & \text{if } u_k = 0 \end{cases} \tag{4.17}$$

This recursive structure is called *Kalman-like* due to its similarity with the recursive structure of the Kalman filter. Finally, the transmission policies are randomized, i.e.,

$$U_k = \begin{cases} 1, & \text{with probability } p_k \\ 0, & \text{with probability } 1 - p_k \end{cases} \tag{4.18}$$

The goal is to design the sequence of transmission policies $\gamma = \{\gamma_1, \gamma_2, \ldots\}$, where $p_k = \gamma_k\left(\mathcal{I}_k, x_1^k, u_1^{k-1}\right) \in [0,1]$ that minimizes the infinite horizon average cost criterion:

$$J(\gamma) = \lim_{N\to\infty} \frac{1}{N}\mathbb{E}\left[\sum_{k=1}^{N}\left((X_k - \hat{X}_k)^{\mathsf{T}} Q (X_k - \hat{X}_k) + \rho \cdot U_k\right)\right] \tag{4.19}$$

This cost penalizes both the quadratic estimation error and the communication rate. In fact, the second term in the cost functional is equivalent to a soft constraint on the average number of transmissions over the infinite horizon.

Defining the estimation error as $Y_k = X_k - \hat{X}_k$, the error evolves as

$$Y_{k+1} = \begin{cases} AY_k + W_k, & \text{if } U_k = 0 \\ W_k, & \text{if } U_k = 1 \end{cases} \tag{4.20}$$

inducing a *stationary stochastic kernel*, P, which describes the transition from $Y_k = e$ to $Y_{k+1} = y$ given $U_k = u$ by

$$P(dy \mid e, u) = \left[u \cdot f_W(y) + (1-u) \cdot f_W(y - Ae)\right] dy \tag{4.21}$$

where f_W is the PDF of W_k.

The strategy is to formulate an auxiliary Markov decision process (MDP) [11] whose solution will also solve the original problem. This kernel is used to define a DP operator associated with the auxiliary Markov decision process whose cost is

$$J^{\text{aux}}(\pi) = \lim_{N\to\infty} \frac{1}{N}\mathbb{E}\left[\sum_{k=1}^{N} c(Y_k, U_k)\right] \tag{4.22}$$

where the per-stage cost is given by

$$c(e,u) = e^{\mathsf{T}} A^{\mathsf{T}} Q A e \cdot (1-u) + \rho \cdot u + \text{trace}\,(\Sigma Q) \tag{4.23}$$

and π is a feedback policy for the auxiliary MDP. The DP operator associated to this MDP is

$$(Th)(e) = \inf_{p \in [0,1]} \mathbb{E}\left[c(e,U) + \int_{\mathbb{R}^n} h(y) \cdot P(dy \mid e, U) \right] \tag{4.24}$$

where $h : \mathbb{R}^n \to \mathbb{R}$ is a bounded function and $p = \Pr(U = 1)$.

The main result in the Xu–Hespanha problem is that the optimal communication policy γ^* is deterministic, stationary, and given by

$$\gamma_k^*(\mathcal{I}_k) = \pi^*(y_k) \tag{4.25}$$

where π^* is the optimal feedback policy that solves the auxiliary MDP. The optimal policy is determined by the unique function h that satisfies the average cost optimality equation,

$$(Th)(e) = h(e) + \bar{J} \tag{4.26}$$

and is equal to

$$\pi^*(e) = \begin{cases} 0, & \text{if } e^\mathsf{T} A^\mathsf{T} Q A e < \rho + \mathbb{E}\left[h(Ae + W) - h(W)\right] \\ 1, & \text{otherwise} \end{cases} \tag{4.27}$$

The proofs of these results are very technical and can be found in detail in Xu's Ph.D. dissertation [12], but the idea is essentially as follows: first, it is shown that the DP operator is a contraction mapping,[1] which guarantees the existence and uniqueness of h. Then, it is shown that $\bar{J} = \text{trace}\,(\Sigma Q) + \mathbb{E}[h(W)]$ is a lower bound for the optimal cost. Finally, it is shown that the policy $\pi^*(e)$ above achieves the lower bound, and therefore it must be optimal. Because the DP operator is a contraction, the function h can be computed by value iteration [11].

4.3.2 Variations and Extensions

The task of computing the optimal policies for the Xu–Hespanha problem through value iteration is computationally intense. Cogill et al. [13] proposed a suboptimal scheme for the case when $A^\mathsf{T} Q A - Q \preceq 0$, i.e., the system is at least marginally stable. They devised a simple and easily computable suboptimal threshold policy μ such that

$$\mu(e) = \begin{cases} 0, & \text{if } e^\mathsf{T} M e \leq 1 \\ 1, & \text{otherwise} \end{cases} \tag{4.28}$$

where M is the unique solution to the following matrix equation:

$$\frac{1}{1 + \text{trace}(\Sigma M)} A^\mathsf{T} M A - M + \frac{1}{\rho} Q = 0 \tag{4.29}$$

This policy satisfies $J^* \leq J(\mu) \leq 6J^*$, where J^* is the cost of the optimal threshold policy.

[1] A special type of contraction in the total variation metric called *span-contraction*.

4.3.3 Main Features of the Xu–Hespanha Problem

- The solution strategy first allows randomized policies, but optimal policies are found to be deterministic.
- By constraining the estimator to have a Kalman-like structure, the problem can be cast as a centralized MDP.
- The optimal policies are of the threshold type defined by hyper-ellipsoids.
- An algorithm to compute the optimal policy via value iteration is provided.

4.4 Remote Estimation with Communication Costs

Lipsa and Martins proposed in [14] a scalar, finite-horizon version of the Xu–Hespanha problem when there is no communication delay. The basic framework is illustrated in Figure 4.3. The main goal was to establish the *joint* optimality of threshold transmission policies and a Kalman-like estimator structure, which was assumed in [10], using majorization theory [15]. One curious historical note about this paper is that it was developed concurrently with the work of Hajek et al. in [16], who used similar techniques to establish joint optimality of paging and registration policies in cellular networks.

4.4.1 The Lipsa–Martins Problem

Consider the scalar, linear time-invariant system

$$X_{k+1} = aX_k + W_k, \;\; k = 0, \ldots, N \tag{4.30}$$

where $\{W_k\}$ is i.i.d. with $W_k \sim \mathcal{N}(0, \sigma^2)$, $a \in \mathbb{R}$ and $X_0 = x_0$ with probability 1. The sequence of transmission policies is denoted by $\gamma = \{\gamma_1, \gamma_2, \ldots, \gamma_N\}$, and the sequence of estimation policies as $\eta = \{\eta_1, \eta_2, \ldots, \eta_N\}$. Furthermore, γ_k and η_k are constrained to be measurable functions of $\left(x_0^k, u_0^{k-1}\right)$ and s_0^k, respectively. The problem is to find the sequence of policies γ and η that jointly minimize the following cost:

$$J(\gamma, \eta) = \mathbb{E}\left[\sum_{k=1}^{N} d^{k-1} \cdot \left(X_k - \hat{X}_k\right)^2 + \rho \cdot U_k\right] \tag{4.31}$$

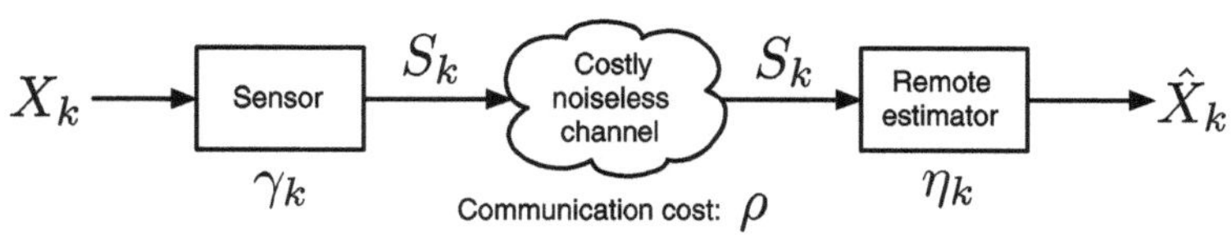

Figure 4.3 Estimation with communication costs.

where $d < 1$ and $\rho \in \mathbb{R}$ is the cost of transmitting a packet. The discount factor d does not play any role in the structural results described here and we will drop it for the remainder of this section.

As one might expect, for a fixed transmission policy γ, the optimal estimator is the MMSE estimate, i.e.,

$$\hat{x}_k = \eta_k^*(s_0^k) = \mathbb{E}\left[X_k \mid s_0^k\right] \tag{4.32}$$

Therefore, assuming the estimator is the MMSE, the designer wants to determine when the sensor should transmit its current observation x_k to the remote estimator. Letting $u_k = \gamma_k\left(x_0^k, u_0^{k-1}\right) \in \{0, 1\}$, the transmitted packet s_k is determined by

$$s_k = \begin{cases} x_k, & \text{if } u_k = 1 \\ \varnothing, & \text{if } u_k = 0 \end{cases} \tag{4.33}$$

The recursive arguments used throughout the paper require that the initial state x_0 must be transmitted, i.e., $s_0 = x_0$, but note that the cost only runs from $k = 1$ onwards. On the remote estimator end, a particularly important instance of causal policies is $\eta_k : s_0^k \mapsto z_k$, where

$$z_k = \begin{cases} a z_{k-1}, & \text{if } s_k = \varnothing \\ s_k, & \text{if } s_k = x_k \end{cases} \tag{4.34}$$

This is a Kalman-like estimation policy where the estimator propagates the previous estimate if the sensor does not transmit, and reproduces the received information when a packet is transmitted.

Before describing the results, we must first reparametrize the cost functional using

$$Y_k = X_k - aZ_{k-1} \tag{4.35}$$

The variable Y_k can be understood as the error incurred by the Kalman-like estimator at time k, given that $U_{k-1} = 0$. The following recursive relation holds

$$Y_{k+1} = \begin{cases} aY_k + W_k, & \text{if } U_k = 0 \\ W_k, & \text{if } U_k = 1 \end{cases} \tag{4.36}$$

Defining $\hat{Y}_k = \mathbb{E}\left[Y_k \mid S_0^k\right]$, the following holds:

$$\hat{Y}_k = \hat{X}_k - aZ_{k-1} \tag{4.37}$$

which in turn implies that

$$Y_k - \hat{Y}_k = X_k - \hat{X}_k \tag{4.38}$$

Therefore, the cost can be recast as

$$J(\gamma) = \mathbb{E}\left[\sum_{k=1}^{N} \left(Y_k - \hat{Y}_k\right)^2 + \rho \cdot U_k\right] \tag{4.39}$$

Note that y_k can be computed from $\left(x_0^k, u_0^{k-1}\right)$. To understand the sequence of results in this paper we need to define three classes of transmission policies:

1. $\mathcal{P}$ – the class of admissible policies, where $u_k = \gamma_k(x_1^k)$.
2. $\mathcal{D}$ – the class of path-dependent symmetric threshold policies, where

$$u_k = \begin{cases} 1, & \text{if } |y_k| > \tau_k(u_1^{k-1}) \\ 0, & \text{otherwise} \end{cases}$$

3. $\mathcal{S}$ – the class of path-independent symmetric threshold policies, where

$$u_k = \begin{cases} 1, & \text{if } |y_k| > \tau_k \\ 0, & \text{otherwise} \end{cases}$$

In the policy classes $\mathcal{D}$ and $\mathcal{S}$, each τ_k is a positive real number representing the error threshold. Note that in $\mathcal{D}$, the threshold τ_k is a function of previous actions. These sets of policies satisfy the inclusion relation $\mathcal{S} \subseteq \mathcal{D} \subseteq \mathcal{P}$.

Lipsa and Martins first show that when $\gamma \in \mathcal{D}$, the MMSE estimator $\hat{X}_k = \mathbb{E}\left[X_k \mid S_0^k\right] = Z_k$, i.e., the Kalman-like estimator is optimal. The proof for this fact essentially relies on the even symmetry of the PDF of $\widetilde{X}_k = X_k - Z_k$, given S_0^k under path-dependent symmetric threshold policies. Then, it is shown that for every admissible transmission policy $\gamma \in \mathcal{P}$, there exists a path-dependent symmetric threshold policy $\gamma \in \mathcal{D}$, such that

$$J(\gamma) \leq J(\gamma) \tag{4.40}$$

The fundamental mathematical tools in their proof come from majorization theory, which is used to establish a partial ordering of candidate solutions. Also note that due to the additive structure of the cost, the system essentially restarts when there is a transmission; this is key to understanding the recursive argument used here onward. The idea is to establish the following lower bound to the cost:

$$\begin{aligned} J(\gamma_1^N) \geq \sum_{k=1}^{N} \Big(& \mathbb{E}\left[(Y_k - \hat{Y}_k)^2 \mid U_1^k = \mathbf{0}\right] \cdot \Pr(U_1^k = \mathbf{0}) \\ & + \left(\rho + J_k^*(\gamma_k^N)\right) \cdot \Pr(U_1^k = \mathbf{0}) \cdot (1 - \Pr(U_k = 0 | U_1^k - 1 = \mathbf{0})\Big) \end{aligned} \tag{4.41}$$

where $J_k^*(\gamma_k^N)$ is the optimal cost-to-go at time k. Each choice of γ induces a sequence of conditional PDFs:

$$\mu_{k|k} = f_{Y_k | U_1^k = \mathbf{0}}(y) \tag{4.42}$$

and for every $\mu_{k|k}$ one can choose another $\mu_{k|k}$ induced by a path-dependent threshold policy such that $\mu_{k|k}$ majorizes $\mu_{k|k}$, i.e.,

$$\mathring{\mu}_{k|k} \succ \mu_{k|k} \tag{4.43}$$

Then, from majorization theory, we have

$$\mu_{k|k} \succ \mu_{k|k} \Rightarrow \mathbb{E}_{\mu_{k|k}}\left[(Y_k - \hat{Y}_k)^2\right] \leq \mathbb{E}_{\mu_{k|k}}\left[(Y_k - \hat{Y}_k)^2\right] \tag{4.44}$$

Therefore, using a recursive argument, it is shown that the optimal policies must lie in the path-dependent symmetric threshold class.

The final step is to show that the optimal path-dependent symmetric threshold policies are in fact path-independent. The proof of this fact is not trivial and is detailed in [14]. Once it is shown that the optimal policies are of the symmetric threshold type, the optimal threshold sequence $\{\tau_k\}$ can be computed using dynamic programming.

4.4.2 Variations and Extensions

The work of Lipsa and Martins can be modified, as in [17], to account for the possibility of random packet-drops. In that case, an acknowledgment bit must be sent from the estimator to the sensor so that the state of the estimator is known at the sensor. The same structural results described in this section hold.

The work of Nayyar et al. [18] considered an energy-harvesting sensor which generalizes the problem described in this section in two ways: the sensor has a random limited energy supply, and the dynamic system is multidimensional. The energy level at time k is denoted by $E_k \in \mathbb{Z}_+$, and at each time $T_k \in \mathbb{Z}_+$ i.i.d. random energy levels are added to the budget. If a single transmission consumes one energy level, the state evolves as

$$E_{k+1} = \min\{E_k + T_k - U_k, B\} \tag{4.45}$$

where B is the capacity of the sensor's battery. When a transmission is made, in order to allow a partially observed MDP (POMDP) formulation and the characterization of its solution, it is necessary to transmit the state e_k along with x_k. This creates enough common information between the sensor and estimator to use a technique from decentralized POMDPs called the *common information approach*. Their main result is to show that symmetric threshold policies are optimal for this problem; however, the optimal thresholds depend on the current energy level at the sensor. The steps in the proof also use results from majorization theory.

4.4.3 Main Features of the Lipsa–Martins Problem

- It establishes the joint optimality of symmetric threshold policies and a Kalman-like estimator for a cost combining quadratic estimation error and communication costs.
- It presents a constructive proof based on techniques from majorization theory.
- It is considered only a scalar system, but was generalized in [18] to the multidimensional case as well as to bounded random energy levels.
- Optimal thresholds can be computed using dynamic programming.

4.5 Remote Estimation in Continuous Time

The problem formulations described in previous sections are all in discrete time. However, it is important to include in this survey a class of optimal remote estimation problems in continuous time. Here we display the similarities, differences, and new techniques used to solve them. Our focus will be on a continuous-time version of the Imer–Basar problem introduced by Rabi and Baras in [19] and more extensively studied by Rabi et al. in [20].

4.5.1 The Rabi–Moustakides–Baras Problem

Consider a scalar linear diffusion continuously and fully observed by a sensor, which is limited to making at most a finite number N of transmissions over a noiseless channel to a remote estimator. The system state X_t is described by

$$\dot{X}_t = aX_t + b\dot{W}_t, \quad 0 \leq t \leq T \tag{4.46}$$

where W_t is a standard Brownian motion process, and $a, b \in \mathbb{R}$. The goal is to find the increasing sequence of sampling instants $\tau = \{\tau_1, \tau_2, \ldots, \tau_N\}$ that minimizes the cumulative mean squared error:

$$J_{(T,N)}(\tau) = \mathbb{E}\left[\int_0^T \left(X_s - \hat{X}_s\right)^2 ds\right] \tag{4.47}$$

The information available to the estimator at time $t \in [\tau_i, \tau_{i+1})$ is

$$\mathcal{I}_t = \bigcup_{j \leq i} \{(\tau_j, x_{\tau_j})\} \tag{4.48}$$

Note that $\mathcal{I}_t$ implicitly carries additional information about the process X_t in the time passed since the last transmission, i.e., the fact that $(\tau_{i+1}, x_{\tau_{i+1}})$ has not arrived. This is similar to the discrete-time case, where some information about the observation x_k is implicitly transmitted even when a sensor decides not to transmit a packet. The times at which packets are generated are *stopping times* with respect to the sensor's observation process. The problem of finding optimal stopping times is significantly different and more involved than optimizing over-deterministic sampling times.

A suboptimal estimate that does not consider this implicit additional information is

$$\hat{x}_t = \mathbb{E}\left[X_t \mid \{(\tau_j, x_{\tau_j}) : j \leq i\}\right] \tag{4.49}$$

$$= \mathbb{E}\left[X_t \mid (\tau_i, x_{\tau_i})\right] \tag{4.50}$$

which satisfies the following linear differential equation

$$\dot{\hat{x}}_t = a\hat{x}_t, \quad \tau_i \leq t < \tau_{i+1}, \quad \hat{x}_{\tau_i} = x_{\tau_i} \tag{4.51}$$

There is a remarkable similarity of the estimator above and the Kalman-like estimator of the previous section. In fact, the authors of [20] conjecture this is in fact optimal when the increments of the state process have symmetric PDFs, but there is no guarantee that this class of estimators is optimal in general.

The optimization of the transmission times can be carried out recursively using the following DP equation:

$$\inf_{\tau_i \geq 0} \mathbb{E}\left[\int_0^{\tau_i} \left(X_s - \hat{X}_s\right)^2 ds + J^*_{(T-\tau_i, N-i)}\right] \tag{4.52}$$

where $J^*_{T-\tau_i,N-i}$ is the cost-to-go for the subproblem of choosing the $\{\tau_{i+1},\dots,\tau_N\}$ that minimize J over the interval $t \in [\tau_i, T]$. Consider the optimal choice of the first transmission time τ_1^*, i.e.,

$$\tau_1^* = \arg \inf_{\tau\in[0,T]} \mathbb{E}\left[\int_0^\tau (X_s - \hat{X}_s)^2 ds + J^*_{(T-\tau,N-1)}\right] \tag{4.53}$$

where

$$J^*_{(T-\tau,N-1)} = \inf_{\{\tau_2,\cdots,\tau_N\}} \mathbb{E}\left[J_{(T-\tau,N-1)}(\tau_2,\cdots,\tau_N)\right] \tag{4.54}$$

This optimal stopping time problem is solved using the *Snell envelope* process, which in this case is

$$S_t = \inf_{\tau\geq t} \mathbb{E}\left[\int_0^\tau (X_s - \hat{X}_s)^2 ds + J^*_{(T-\tau,N-1)}\middle|\{x_s,\ 0 \leq s \leq t\}\right] \tag{4.55}$$

The earliest time at which the cost of stopping at time t meets the envelope S_t is an optimal stopping time, i.e.,

$$\tau_1^* = \inf\left\{t : \int_0^t (x_s - \hat{x}_s)^2 ds + J^*_{(T-t,N-1)} = S_t\right\} \tag{4.56}$$

Therefore, the optimal transmission policy for this continuous-time remote estimation problem with limited measurements is a special type of threshold function, also called a decision envelope. The results described here only provide a characterization of the optimal solutions. Analytic results can be explicitly computed in the case in which $a = 0$, i.e., the process X_t is a Brownian motion.

4.5.2 Variations and Extensions

Other continuous-time formulations of remote estimation problems exist. In [9], Xu and Hespanha studied the trade-off between estimation error and communication rate in a continuous-time vector linear time-invariant system driven by Gaussian noise. The transmissions occurred according to a Poisson process whose rate could be fixed (uncontrolled) or dependent on the estimation error (controlled). In their formulation there was no concept of optimality, and threshold policies did not play any role.

As an extension, Rabi et al. [20] mention that the hard constraint on the number of transmissions could be replaced by a constraint on the expected number of transmissions per unit time. Such variation would turn out to be a continuous-time formulation of the Lipsa–Martins problem, which seems to be an interesting open research problem.

4.5.3 Main Features of the Rabi–Moustakides–Baras Problem

- It is a finite-horizon continuous-time remote estimation problem with a hard constraint on the number of transmissions.
- It can be understood as the continuous-time version of the Imer–Basar problem.
- Fixing the structure of the estimator to a Kalman-like filter, the problem is reduced to one of optimal stopping times.

- The solution is characterized using a technique involving the Snell envelope.
- An analytical solution for the optimal decision envelopes can only be obtained in a few cases, e.g., when X_t is a Brownian motion process.

4.6 Sensor Scheduling versus Event-Driven Strategies for Remote Estimation

The goal in the remote estimation problems described in the previous sections is to optimally decide when to transmit a packet containing a sample of the measurements made by a sensor. This relates to the problem of sensor scheduling, which aims to determine how to optimally choose a sensor (or a set of sensors) at a given time step to transmit their measurements over a shared communication medium, e.g., a wireless network. Each sensor may have a different cost depending on its accuracy (how much measurement noise is added to the state process we are trying to estimate) or how much battery power is left. The case of intermittent communications can be addressed by introducing to the scheduling problem a cost-free dummy sensor that does not take any measurements. When this sensor is selected, it is as if a packet is not received by the remote estimator. This situation is illustrated in Figure 4.4. Note that in the scheduling problem the sensors are "passive" and the decisions are effectively taken by the estimator. However, remote estimation problems have a very distinctive feature: It is the sensors that make the decisions based on their local measurements at each time, which is not the case in scheduling problems.

4.6.1 Separation of Sensor Scheduling and Control

Work on sensor scheduling problems began in the 1960s with Meier et al. in [21]. This work considered a discrete-time stochastic control problem in which the control signal had two components $u_k = (u_k^M, u_k^P)$, where u_k^M controlled the measurement

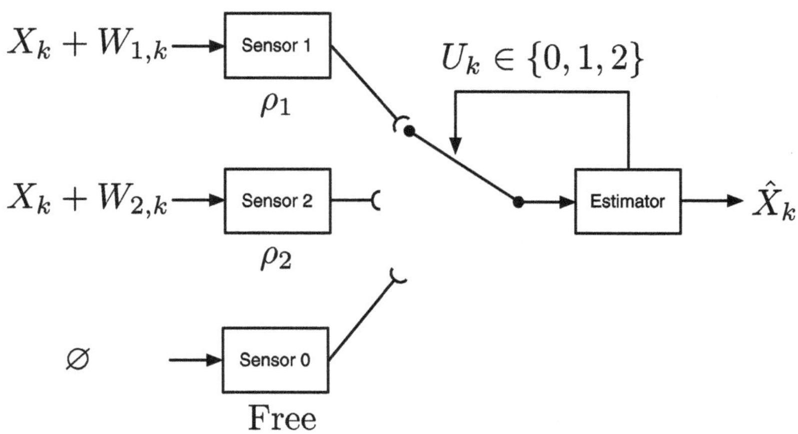

Figure 4.4 Sensor scheduling with measurement costs and a free dummy sensor.

equation and u_k^P controlled the plant. They showed that when the problem has linear dynamics, quadratic cost, and Gaussian noise (LQG), a separation result holds, i.e., the measurement control can be designed independently of the plant control by solving a deterministic optimal control problem, which is essentially a sensor scheduling problem with costly measurements. The optimal schedule can then be found with an exhaustive search of paths in a rooted tree, whose complexity is exponential in the problem's horizon, rendering the problem intractable.

4.6.2 Sensor Scheduling in Continuous Time

A few years later, Athans in [22] reported results on a finite-horizon continuous-time system observed by multiple costly sensors, but only one of them could be selected by the estimator at time t. The goal was to design the switching function among sensors to minimize a cost function combining the estimation error and the measurement cost. By rewriting the problem as one of deterministic optimal control, Athans showed that the solution to this problem could be obtained through the so-called *matrix minimum principle* and computed using numerical techniques. This work was later generalized by Herring and Melsa [23] to allow the selection of more than one sensor at a time.

4.6.3 Sensor Scheduling in Discrete Time

Due to the inherent complexity of finding the optimal scheduling policies in discrete time, researchers started to focus on formulating problems using different cost functions, proposing suboptimal solutions, or using heuristics to avoid a complete search on a decision tree.

A nonconventional formulation for the scheduling problem was proposed by Oshman in [24], where a linear system driven by Gaussian noise is observed by multiple sensors through linear measurement equations. At each time k, a single sensor must be selected to maximize the amount of "information"[2] along the state space direction associated with the maximal estimation error. This can be accomplished by a technique Oshman called *V-Lambda estimation*.

Savage and La Scala [25] studied the scheduling of multiple scalar Gauss–Markov systems with a cost that involves only the terminal state of the estimation error processes. By reordering the systems according to the variances of the plant (or observation) noise, the authors showed that the optimal schedules are simple functions of the system index in this new ordering. Sandberg et al. incorporated the effect of delays [26] by formulating a problem involving the scheduling of two types of sensors: one of high measurement noise, small delay, and cheap communication cost, and a second one with low measurement noise, large delay, and high communication cost. The cost function explicitly involved the cost of the sensor and the effect of delay is reflected indirectly on the estimation error.

[2] The quantification of information here is not in the Shannon-theoretic sense.

In the LQG setting, Gupta et al. in [27] avoided the search over a rooted tree by considering stochastic sensor schedules rather than deterministic ones. The idea is to design the probability distributions according to which sensors are chosen at each time in order to minimize the trace of the expected estimation error covariance matrix. The cost function does not include any measurement/communication costs and the optimal estimator in this case is a time-varying Kalman filter. Using a random selection of sensors it is possible to find an upper bound for the trace of the expected error covariance matrix in the steady-state regime, which is the quantity minimized over the space of probability distributions.

In [28], Vitus et al. proposed another way of dealing with the computational complexity in searching for the optimal sensor schedule by pruning the search tree. Again, this formulation does not incorporate the cost of selecting a particular sensor, but their proposed algorithm trades-off system performance with problem complexity, which is an interesting way of analyzing the problem. In [29], He and Chong proposed a POMDP formulation of the scheduling problem combining estimation error and measurement costs. The solution to this problem was approximately computed using particle filters.

4.6.4 Event-Driven Strategies for Remote Estimation

As we have seen in previous sections, event-driven strategies arise naturally as solutions to optimal estimation problems with communication constraints. Once these structural results are established, the system designer optimizes over the appropriate class of event triggers in order to improve the performance depending on his or her applications. In general, even if suboptimal, appropriately designed thresholding strategies will potentially decrease the traffic of information over the communication network, reducing the occurrence of collisions and packet loss due to long delays in the system.

As opposed to scheduling problems, event-driven (or event-triggered) remote estimation is based on the idea that sensors should transmit information only when measurements exceed a certain threshold. Aström and Bernhardson were among the first researchers to explore this idea in [30]. Their work analyzes the performance of state estimators when the act of accessing the system is dictated by a simple event-driven strategy at the sensor, which the authors called *Lebesgue sampling*. They showed that such sampling policies can perform better than strategies based on periodic sampling with the same average sampling rate. The application of Lebesgue sampling in estimation was extended to control as in, for example, the works of Tabuada [31] and Mazo and Tabuada [32, 33].

4.6.5 Estimation over Shared Networks

All the works cited so far assume that there is only one transmitter–receiver pair over the communication channel. Although certain networks may be modeled by nodes connected by such links, more complex network models, such as wireless networks, must accommodate multiple sensors communicating at the same time over the same channel. In this case, the system designer must address the case in which simultaneous packets collide, either by resolving the collision events or by avoiding them through scheduling or random access such as the ALOHA protocol [34].

The modeling of packets dropped due to collisions in a carrier sense multiple access (CSMA) protocol was done by Zhu et al. in [35] with the aid of the machinery of Kalman filtering with intermittent observations [36]. Another approach to avoid collisions by assigning priority to packets was considered by Xia et al. in [37]. Also assuming CSMA, Henningsson and Cervin in [38] proposed a model for interference when control systems share the same wireless network. Dealing specifically with the possibility of collisions, Hong et al. [39] allow the sensors to measure the quality of the channels in the physical layer prior to transmission. Sensors then transmit their measurements only if the quality of their channel is above a certain threshold. The work in [40] partially addresses collisions by restricting the estimation filter to be linear and optimizing over the transmission probabilities of each sensor.

4.7 Estimation over the Collision Channel

In the general setting of estimation over shared networks, the main challenge is dealing with multiple simultaneous transmissions, which may congest the network and incur packet losses. A simple way of modeling the effects of interference in multi-access communication is by the so-called *collision channel* model. At each time instant, the channel can only convey a single packet, and if two or more sensors transmit at the same time, a collision is declared and the estimator receives a garbled waveform, which is represented by a collision symbol $\mathfrak{C}$. If both sensors decide not to transmit, the channel is idle and the estimator receives a no-transmission symbol $\varnothing$. The basic framework is depicted in Figure 4.5.

The following remote estimation problem over the collision channel was formulated in [41]. Sensor i observes a random variable $X_i \sim \mathcal{N}(0, \sigma_i^2)$, $i = 1, 2$, such that

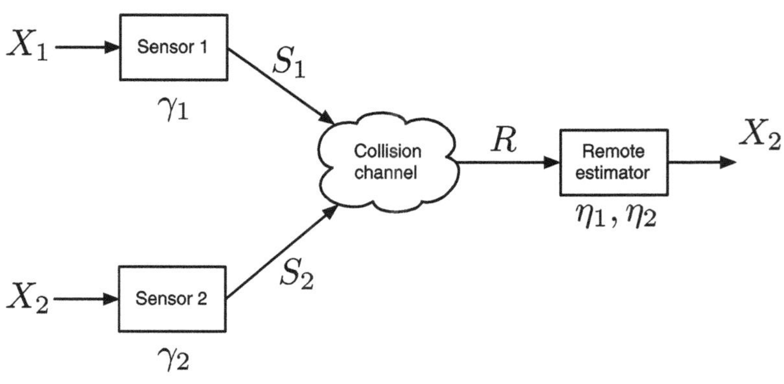

Figure 4.5 Estimation over the collision channel.

$X_1 \perp\!\!\!\perp X_2$. Then, sensor i must decide whether it will attempt to access the channel and send a packet containing its observation to the estimator or remain silent. The decision variable is denoted by $u_i = \gamma_i(x_i)$, where $\gamma_i : \mathbb{R} \to \{0, 1\}$ is called the communication policy of sensor i. When $u_i = 1$, sensor i attempts to send a packet containing its observation over the collision channel, i.e.,

$$s_i = \begin{cases} x_i, & \text{if } u_i = 1 \\ \varnothing, & \text{if } u_i = 0 \end{cases} \tag{4.57}$$

The channel output r is either $\varnothing$, when both sensors decide not to transmit; $\mathfrak{C}$, when both sensors decide to transmit; or the corresponding transmitted packet, when a single sensor transmits. The estimation policy is a pair of mappings $\eta(r) = \big(\eta_1(r), \eta_2(r)\big)$, where $\eta_i : \mathbb{R} \cup \{\mathfrak{C}, \varnothing\} \to \mathbb{R}$, $i = 1, 2$. The problem is to design γ_1, γ_2, and η that jointly minimizes the following cost:

$$J(\gamma_1, \gamma_2, \eta) = \mathbb{E}\left[(X_1 - \hat{X}_1)^2 + (X_2 - \hat{X}_2)^2\right] \tag{4.58}$$

In general, there is no systematic way of solving the problem as is. Instead, we can use the *person-by-person optimality* approach borrowed from team decision theory. By fixing the policies (γ_2, η_2), the problem from the perspective of sensor 1 becomes a remote estimation with communication costs. One can intuitively see this by arguing that with (γ_2, η_2) fixed, the channel is occupied with probability $\Pr(U_2 = 1)$. If sensor 1 attempts to access the channel more often with the intent of decreasing $\mathbb{E}[(X_1 - \hat{X}_1)^2]$, it will cause more collisions with packets sent from sensor 2, increasing $\mathbb{E}[(X_2 - \hat{X}_2)^2]$. It can be analytically shown that from the perspective of sensor 1, the problem is reduced to minimizing

$$J(\gamma_1, \eta_1) = \mathbb{E}\left[(X_1 - \hat{X}_1)^2\right] + \rho_{21} \cdot \Pr(U_1 = 1) \tag{4.59}$$

where $\rho_{21} = \mathbb{E}[(X_2 - \hat{X}_2)^2 | U_1 = 1, U_2 = 1] \cdot \Pr(U_2 = 1)$, which is exactly the same cost structure as for the Lipsa–Martins problem with horizon $N = 1$. In fact, one can readily use their results to establish the structure of the optimal γ_1^* in the case when $\mathfrak{C} = \varnothing$. If the estimator cannot distinguish between collision and no-transmission events, the problem is one of remote estimation with communication costs and packet-drops for which the optimality of symmetric threshold policies has already been established [17].

A more interesting problem is to determine the structure of γ_1^* in the case in which $\mathfrak{C} \neq \varnothing$. We analyze the equivalent problem in the setup of Figure 4.6. The random variable $D \sim \mathcal{B}(\beta)$, where $D \perp\!\!\!\perp X$, denotes the channel being occupied by another transmission at the same time. Using $\hat{x}_0 = \eta(\varnothing)$ and $\hat{x}_1 = \eta(\mathfrak{C})$, the cost can be rewritten as

$$\begin{aligned} J(\gamma, \hat{x}_0, \hat{x}_1) = {} & \mathbb{E}[(X - \hat{x}_0)^2 \mid U = 0] \Pr(U = 0) \\ & + \mathbb{E}\left[\beta \cdot (X - \hat{x}_1)^2 + \rho \mid U = 1\right] \Pr(U = 1). \end{aligned} \tag{4.60}$$

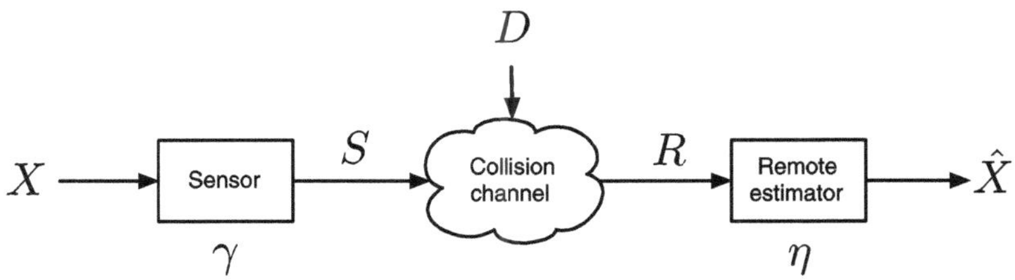

Figure 4.6 Single sensor collision channel.

Noticing that $\hat{x}_0^* = \mathbb{E}[X|U = 0]$ and $\hat{x}_1^* = \mathbb{E}[X|U = 1]$, it is possible to optimize γ over the class of randomized policies, $U \sim \mathcal{B}(\gamma(x))$, $\gamma : \mathbb{R} \to [0,1]$. Using standard techniques from [42] and fixing $\mathbb{E}[U] = \alpha \in (0,1)$, the resulting optimization problem can be cast as the following infinite dimensional linear program:

$$\begin{aligned} \underset{\gamma}{\text{minimize}} \quad & \mathbb{E}[X^2|U = 0] \\ \text{subject to} \quad & \mathbb{E}[U] = \alpha \\ & \mathbb{E}[X|U = 0] = \hat{x}_0 \\ & 0 \leq \gamma(x) \leq 1, \forall x \in \mathbb{R} \end{aligned} \tag{4.61}$$

whose solution is found using Lagrange duality theory [42]. It turns out that the optimal policy in this case is characterized as a deterministic threshold function, i.e.,

$$\gamma^*(x) = \begin{cases} 0, & \text{if } x \in [a,b] \\ 1, & \text{otherwise} \end{cases} \tag{4.62}$$

where $a \neq b$, in general. Based on the characterization of the optimality of threshold policies for this problem, the cost function can be rewritten as

$$\begin{aligned} J(a,b,\hat{x}_0,\hat{x}_1) = & \int_{x\in[a,b]} (x - \hat{x}_0)^2 f_X(x)dx \\ & + \int_{x\notin[a,b]} \left[\beta(x - \hat{x}_1)^2 + \rho\right] f_X(x)dx \end{aligned} \tag{4.63}$$

This reduces the design of the optimal transmission and estimation policies to a binary optimal quantization problem with an asymmetric distortion function in which $\hat{x}_0$ and $\hat{x}_1$ are the representation points of the interval $[a,b]$ and its complement $[a,b]^c$, respectively. The minimization of $J(a,b,\hat{x}_0,\hat{x}_1)$ can then be carried out through a nonlinear optimization solver, or by an iterative procedure known as the modified-lloyd-max algorithm proposed in [41]. It is important to note that the asymmetry of the optimal threshold policy allows the sensor to use the collision symbol to carry information about X even when the packet is not successfully received by the estimator. Therefore, through careful design, the estimator receives some level of useful information even when the communication fails due to a collision.

4.8 Conclusion

This chapter presents a review of remote estimation problems. We focused on different problems of optimal policy design for three classes of communication constraints: limited channel uses, transmission costs, and interference. The common feature in each of these cases is that the optimal transmission policies are event-driven, i.e., a measurement made by a sensor is transmitted to the estimator if it exceeds a certain threshold. A literature review on sensor scheduling, event-driven remote estimation, and networked estimation is also included. Finally, a problem of remote estimation over the collision channel was discussed. This problem displays some of the opportunities and challenges for future research in networked estimation, such as how to design optimal communication protocols for a system consisting of multiple sensors that attempt to communicate their measurements in a distributed manner over an interference-limited wireless network.

Acknowledgments

The authors would like to thank the anonymous reviewers who carefully read the initial draft of this chapter and suggested modifications that improved the clarity of the final text. A special thanks goes to Eduardo Arvelo for proofreading the final manuscript.

References

[1] J. P. Hespanha, P. Naghshtabrizi, and Y. Xu, "A survey of recent results in networked control systems," *Proc. IEEE*, vol. 95, no. 1, pp. 138–162, Jan. 2007.

[2] L. Schenato, B. Sinopoli, M. Franceschetti, K. Poolla, and S. Sastry, "Foundations of control and estimation over lossy networks," *Proc. IEEE*, vol. 95, no. 1, pp. 163–187, Jan. 2007.

[3] G. N. Nair, F. Fagnani, S. Zampieri, and R. J. Evans, "Feedback control under data rate constraints: An overview," *Proc. IEEE*, vol. 95, no. 1, pp. 108–137, Jan. 2007.

[4] O. C. Imer and T. Basar, "Optimal estimation with limited measurements," in *Proc. 44th Conf. Decision and Control*, 2005, pp. 1029–1034.

[5] D. P. Bertsekas, *Dynamic Programming and Optimal Control*, vol. 1. Belmont, MA: Athena Scientific, 2005.

[6] O. C. Imer and T. Basar, "Optimal estimation with limited measurements," *Int. J. Syst. Contr. Commun.*, vol. 2, no. 1–3, pp. 5–29, 2010.

[7] W. MacKunis, J. W. Curtis, and P. E. K. Berg-Yuen, "Optimal estimation of multidimensional data with limited measurements," in *Proc. American Control Conf.*, 2011, pp. 4257–4262.

[8] Y. Xu and J. P. Hespanha, "Communication logics for networked control systems," in *Proc. American Control Conf.*, 2004, pp. 572–577.

[9] Y. Xu and J. P. Hespanha, "Estimation under uncontrolled and controlled communications in networked control systems," in *Proc. 44th IEEE Conf. Decision and Control*, 2005, pp. 842–847.

[10] Y. Xu and J. P. Hespanha, "Optimal communication logics in networked control systems," in *Proc. 43rd IEEE Conf. Decision and Control*, 2004, pp. 3527–3552.

[11] P. R. Kumar and P. Varaiya, *Stochastic Systems: Estimation, Identification, and Adaptive Control*. Englewood Cliffs, NJ: Prentice Hall, 1986.

[12] Y. Xu, "Communication scheduling methods for estimation over networks," Ph.D. dissertation, University of California Santa Barbara, CA, 2006.

[13] R. Cogill, S. Lall, and J. P. Hespanha, "A constant factor approximation algorithm for event-based sampling," in *Proc. American Control Conf.*, 2007, pp. 305–311.

[14] G. M. Lipsa and N. C. Martins, "Remote state estimation with communication costs for first-order LTI systems," *IEEE Trans. Autom. Control*, vol. 56, no. 9, pp. 2013–2025, 2011.

[15] A. W. Marshall, I. Olkin, and B. C. Arnold, *Inequalities: Theory of Majorization and Its Applications*, 2nd ed. New York: Springer, 2011.

[16] B. Hajek, K. Mitzel, and S. Yang, "Paging and registration in cellular networks: Jointly optimal policies and an iterative algorithm," *IEEE Trans. Inf. Theor.*, vol. 54, no. 2, pp. 608–622, 2008.

[17] G. M. Lipsa and N. C. Martins, "Optimal state estimation in the presence of communication costs and packet drops," in *Proc. 47th Annual Allerton Conf. Communication, Control and Computing*, 2009, pp. 160–169.

[18] A. Nayyar, T. Basar, D. Teneketzis, and V. Veeravalli, "Optimal strategies for communication and remote estimation with an energy harvesting sensor," *IEEE Trans. Autom. Control*, vol. 58, no. 9, pp. 2246–2260, Sept. 2013.

[19] M. Rabi, G. V. Moustakides, and J. S. Baras, "Multiple sampling for estimation on a finite horizon," in *Proc. 45th IEEE Conf. Decision and Control*, 2006, pp. 1351–1357.

[20] M. Rabi, G. V. Moustakides, and J. S. Baras, "Adaptive sampling for linear state estimation," *SIAM J. Contr. Optim.*, vol. 50, no. 2, pp. 672–702, Mar. 2012.

[21] L. Meier, III, J. Peschon, and R. M. Dressler, "Optimal control of measurement subsystems," *IEEE Trans. Autom. Control*, vol. 12, no. 5, pp. 528–536, Oct. 1967.

[22] M. Athans, "On the determination of optimal costly measurement strategies for linear stochastic systems," *Automatica*, vol. 8, no. 4, pp. 397–412, Jul. 1972.

[23] K. Herring and J. L. Melsa, "Optimum measurements for estimation," *IEEE Trans. Autom. Control*, vol. 19, no. 3, pp. 264–266, Jun. 1974.

[24] Y. Oshman, "Optimal sensor selection strategy for discrete-time state estimator," *IEEE Trans. Aerospace Electron. Syst.*, vol. 30, no. 2, pp. 307–314, Apr. 1994.

[25] C. O. Savage and B. F. La Scala, "Optimal scheduling of scalar Gauss–Markov systems with a terminal cost," *IEEE Trans. Autom. Control*, vol. 54, no. 5, pp. 1100–1105, May 2009.

[26] H. Sandberg, M. Rabi, M. Skoglund, and K. Johansson, "Estimation over heterogeneous sensor networks," in *Proc. 47th IEEE Conf. Decision and Control*, 2008, pp. 4898–4903.

[27] V. Gupta, T. Chung, B. Hassibi, and R. Murray, "On a stochastic sensor selection algorithm with applications in sensor scheduling and dynamic sensor coverage," *Automatica*, vol. 42, no. 2, pp. 251–260, Feb. 2006.

[28] M. P. Vitus, W. Zhang, A. Abate, J. Hu, and C. J. Tomlin, "On efficient sensor scheduling for linear dynamical systems," *Automatica*, vol. 48, no. 10, pp. 2482–2493, Oct. 2012.

[29] Y. He and E. K. P. Chong, "Sensor scheduling for target tracking in sensor networks," in *Proc. 43rd IEEE Conf. Decision and Control*, 2004, pp. 743–748.

[30] K. J. Astrom and B. Bernhardson, "Comparison of Riemann and Lebesgue sampling for first order stochastic systems," in *Proc. 41st IEEE Conf. Decision and Control*, 2002, pp. 2011–2016.

[31] P. Tabuada, "Event-triggered real-time scheduling of stabilizing control tasks," *IEEE Trans. Autom. Control*, vol. 52, no. 9, pp. 1680–1685, Sept. 2007.

[32] M. Mazo and P. Tabuada, "On event-triggered and self-triggered control over sensor/actuator networks," in *Proc. 47th IEEE Conf. Decision and Control*, 2008, pp. 435–440.

[33] M. Mazo and P. Tabuada, "Decentralized event-triggered control over wireless sensor/actuator networks," *IEEE Trans. Autom. Control*, vol. 56, no. 10, pp. 2456–2461, Oct. 2011.

[34] N. Abramson, "The ALOHA system: Another alternative for computer communications," in *AFIPS Conf. Proc.*, vol. 37, Nov. 1970, pp. 281–285.

[35] B. Zhu, B. Sinopoli, K. Poolla, and S. Sastry, "Estimation over wireless sensor networks," in *Proc. American Control Conf.*, 2007, pp. 2732–2737.

[36] B. Sinopoli, L. Schenato, M. Franceschetti, et al. "Kalman filtering with intermittent observations," *IEEE Trans. Autom. Control*, vol. 49, no. 9, pp. 1453–1464, 2004.

[37] M. Xia, V. Gupta, and P. Antsaklis, "Networked state estimation over a shared communication medium," in *Proc. American Control Conf.*, 2013, pp. 4128–4133.

[38] T. Henningsson and A. Cervin, "A simple model for the interference between event-based control loops using a shared medium," in *Proc. 49th IEEE Conf. Decision and Control*, 2010, pp. 3240–3245.

[39] Y. W. P. Hong, K.-U. Lei, and C.-Y. Chi, "Channel-aware random access control for distributed estimation in sensor networks," *IEEE Trans. Sign. Process.*, vol. 56, no. 7, pp. 2967–2980, Jul. 2010.

[40] W.-A. Zhang, L. Yu, and G. Feng, "Optimal linear estimation for networked systems with communication constraints," *Automatica*, vol. 47, no. 9, pp. 1992–2000, Sept. 2011.

[41] M. M. Vasconcelos and N. C. Martins, "Estimation over the collision channel: Structural results," in *Proc. 51st Annual Allerton Conf. Communication, Control and Computing*, 2013, pp. 1114–1119.

[42] S. Boyd and L. Vandenbergue, *Convex Optimization*. New York, NY: Cambridge University Press, 2004.

5 Joint Scheduling and Optimal Adaptive Event-Triggered Control of Distributed Cyber-Physical Systems

Hao Xu, Avimanyu Sahoo, and S. Jagannathan

5.1 Introduction

Embedded intelligent control [1] and cross-layer network protocol designs [2] have been considered as two of the fastest-growing research areas in recent years. As mentioned in [3], combining these two areas can introduce significant advantages, such as reduction in installation costs, increased adaptability and reliability, and improved usability, for both modern control and networking communities. This novel class of systems is referred to as cyber-physical systems (CPSs) [4–7].

Since the control and communication aspects are tightly coupled in CPSs [8], one needs to consider these aspects together for efficient CPS design. Therefore, incorporating the fixed communication network effects, the authors in [4] proposed a control scheme to maintain the stability of the control system part of the CPS in the presence of a deterministic packet dropout rate. In [5], from the communication network protocol side, the authors implemented a stable control scheme and evaluated the performance of the widely used IEEE 802.11 protocol (e.g., CSMA/CS etc.) for distributed CPSs.

However, the research efforts in [4, 5] have not considered the real-time interaction between the control and communication subsystems. A revolutionary scheme for distributed CPSs should utilize the real-time interaction between the physical and cyber layers to optimize the performance of the CPS. From the well-known open system interconnection (OSI) architecture [8] perspective, the physical systems of a CPS are connected through a communication network protocol which resides in the network layers. Therefore, cross-layer design [9, 10], considering the interaction properly among different layers, is necessary.

In [9], the authors showed that cross-layer design can achieve performance gains by exploiting the dependence among protocol layers when compared to the traditional individual-layer protocol designs. However, most cross-layer designs are implemented for the network layer and the cyber layer [10] wherein the cyber layer is neglected. For distributed CPSs, the controller and the IEEE 802.11 network protocol design for communication have to be considered jointly.

Among the different protocols utilized for communication networks, the distributed scheduling [11] scheme is critical for the communication protocol design when there are multiple systems sharing a common network. Compared to traditional centralized scheduling [12], the main advantage of distributed scheduling is that it does not require a central processor to deliver the schedules after collecting information from all the communication links in the network. According to the IEEE 802.11 standard [11], the carrier sense multiple access (CSMA) protocol is introduced to schedule communication links in a distributed manner in which a communication link requests network resources to transmit the message only if it does not hear an ongoing transmission from the shared network.

Further, the authors of [13] derived a throughput-optimal distributed scheduling algorithm and proved that even a distributed scheduling scheme can still maximize throughput. Moreover, a random-access scheme is widely incorporated in most CSMA-based distributed scheduling schemes [13, 14]. However, these protocols [13, 14] are neither optimal nor suitable for CPSs since they focus on improving network layer performance alone, which in turn neglects the effects on the control system in the cyber layer. In particular, the feedback signals and the control inputs are packetized and transmitted to the controller and actuator respectively via the communication network. Neglecting the effects of the network layer on the control scheme can result in significant deterioration in performance.

At the cyber layer, the control of such CPSs has to be carefully dealt with since existing periodically sampled schemes [15] are unsuitable for handling the continuous dynamics of the physical system, the event-sampled state vector, and its transmission over the communication network. The implementation of the control system at the cyber layer coupled with the physical system will form a CPS. Such a CPS requires an event-sampled control system design framework wherein the state vector is transmitted to the controller in an aperiodic manner in order to save communication network resources. Specifically, the optimal event-triggered control design is necessary for CPSs in order to minimize use of network resources.

Controlling a linear CPS is quite challenging, especially in the presence of uncertain system dynamics [16]. It is important to note that the traditional optimal control schemes require a periodically sampled state and control vector, which requires significant network resources such as bandwidth for communication between the sensor and the controller, which could potentially cause congestion. In contrast, in the event-triggered design, the control input updates are nonperiodic, which creates challenges in terms of stability and convergence.

This chapter proposes a novel cross-layer scheme for each system in a distributed CPS that includes an optimal event-triggered controller design in the cyber layer and a distributed scheduling scheme for the communication network in the network layers. The cross-layer design in the event-sampled domain brings several challenges mentioned before. Significant benefits are expected from the cross-layer design.

The main contributions of this chapter are: (1) a novel adaptive optimal event-triggered control scheme designed in a forward-in-time manner without using knowledge of the linear dynamics of each system, whereas most existing designs are either backward-in-time or require system dynamics; and (2) the distributed scheduling via the cross-layer approach which improves the performance of CPSs by minimizing the cost function from both the network layer and physical systems. Compared with time-based periodic sampling, the proposed event-triggered scheme designed using the cross-layer approach utilizing event-based sampling saves significant computational and network resources.

5.2 Background and Distributed CPSs

During the past several decades, great advances have been made such that novel communication network and control designs provide benefits such as robustness, optimality, and flexibility. Here, we want to summarize these recent trends in the area of network and controller design. The three critical trends in the networking aspects are energy efficiency, spectrum management, and game theory.

Energy efficiency. Energy consumption has been considered as one of the critical challenges in advanced communication network design. Especially for wireless sensor networks (WSN), an energy-efficient protocol is necessary for saving energy and prolonging the network's lifetime. Wei et al. [17] proposed a novel medium-access control (MAC) for WSN which reduces energy consumption significantly while achieving scalability and avoiding the collisions with interfering nodes. Recently, researchers have demonstrated that clustering is an efficient technique to increase network lifetime and scalability [18, 19]. Kumar [20] proposed and evaluated the single-hop and multi-hop energy-efficient clustering protocol for WSN.

Spectrum management. Due to limited spectrum resource availability and with increasing numbers of users on wireless networks, it is necessary to manage spectrum resources efficiently. Recently, cognitive radio (CR) technology has been envisaged to solve spectrum management challenges. Salem et al. [21] developed an opportunistic spectrum access protocol for CR ad hoc networks that can improve spectrum usage significantly.

Game theory. Due to the distributed and autonomous nature of existing wireless networks, users might behave selfishly to maximize their own benefit, thus affecting the entire network performance. Therefore, a proper distributed network protocol is needed to manage network resources among different users. Meanwhile, game theory has been considered as a promising technique to address multi-user resource allocation problems. Engaging game theory with network protocol design is a future direction for wireless network research. Recently, Guan et al. [6] introduced a novel routing algorithm to solve the obstacle problem in WSN based on a game-theory model. Nie and Comaniciu [7] propose a game-theoretic framework to analyze the behavior of CRs for distributed adaptive channel allocation.

On the other hand, modern control theory has been developing quickly during the past few decades to address challenges in systems with constrained resources. Of all the methods, optimal control and event-triggered control are the two most important and relevant ones for discussion.

Optimal control. Optimality is more preferred than stability alone for a dynamic system. However, according to [22], optimal control of nonlinear systems is a very difficult, even impossible, feat when the system dynamics are unknown. Inspired by computational intelligence, novel adaptive dynamic programming (ADP) [22] has been introduced to overcome this deficiency. For solving optimal control, one has to find the solution to either the Bellman equation or the Hamilton–Jacobi–Bellman (HJB) equation. Unfortunately, no closed-form solution can be determined for these equations and one has to resort to iterative methods.

Recently, the authors of [23, 24] proposed policy and/or value iteration-based ADP schemes to attain near-optimal control without knowing the system dynamics. However, these approaches require a significantly large number of iterations within a sampling interval that is difficult to realize in practice. To overcome this deficiency, Dierks and Jagannathan [25] introduced a time-based ADP algorithm to solve for optimal control by using system historical information instead of iteration-based information. However, traditional methods are periodically time-driven.

Event-triggered control. Traditional periodic transmission of feedback control signals in a closed-loop environment requires higher communication bandwidth. Event-triggered control, on the other hand, has emerged as an alternative to reduce the network communication and controller execution. In an earlier work for a continuous-time system, the authors [26] ensured the system was input-to-state stable (ISS) with respect to an event-triggering condition that decided when to sample the state and input vector. Recently, a variety of event-triggered control techniques (e.g., zero-order-hold scheme [27] and model-based schemes [28]) have been developed for both linear and nonlinear systems. All these works have demonstrated the effectiveness of event-triggered control over periodic time-driven control scheme.

Due to the explosive growth in these areas, an interesting idea is to combine the network protocol design and control theory to harvest the benefits. This novel class of systems is referred to as CPSs [2, 3]. The basic structure of a distributed CPS is described in Figure 5.1, where numerous systems communicate with their corresponding controllers through a common communication network such as IEEE 802.1x. It is considered as a distributed CPS since multiple subsystems are distributed without any interconnection among them in the physical layer. It is clear that the shared communication network will affect the performance of the control system. For example, when the shared communication network gets congested due to incorrect scheduling, the CPS subsystem under consideration cannot maintain stability since the information from the physical system cannot be transmitted to their respective controllers. Therefore, a novel cross-layer design is necessary for this distributed CPS, which incorporates the control and distributed scheduling simultaneously. Without loss of generality, numerous coupled homogeneous systems are considered as part of a distributed CPS in this chapter.

Consider each system in the overall CPS described as a linear time-invariant (LTI) continuous-time system described by $\dot{x}_l(t) = A_l x(t) + B_l u_l(t)$, $\forall l = 1, 2, \ldots, N$ with A_l, B_l representing the dynamics of the lth system and T_s being the sampling interval. For optimal control design, the communication network delay for each system is assumed to be unknown but bounded as $\tau_l \leq \bar{d}T_s$, $\forall l = 1, \ldots, N$, and this needs to be ensured by selecting the proper cross-layer distributed network scheduling protocol. Then, incorporating the network delay and packet losses, the dynamics of the

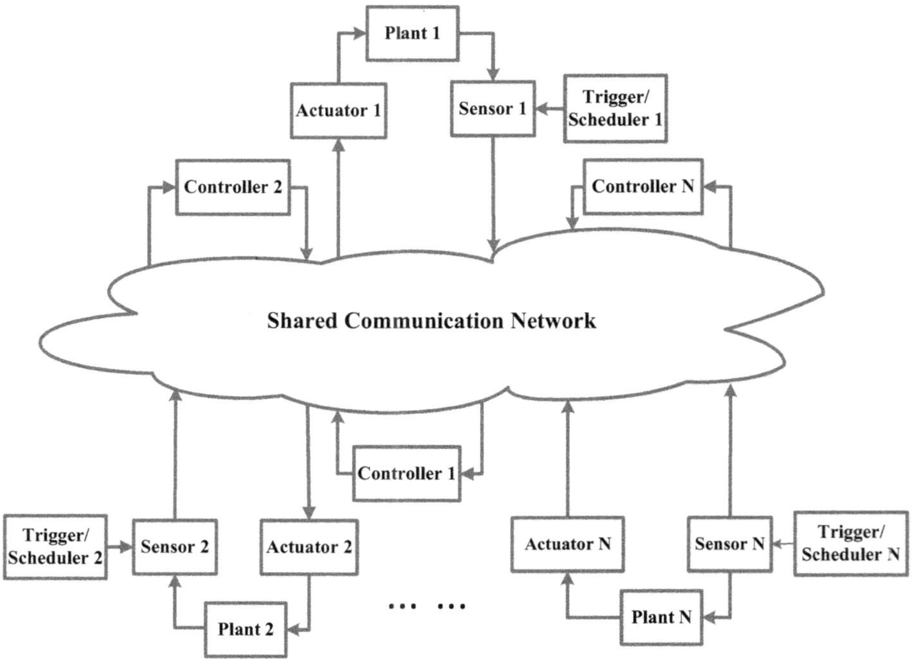

Figure 5.1 Distributed CPS.

lth system can be represented [17] as a stochastic time-varying discrete-time system described by

$$x_{l,k+1} = A_{l,s}x_{l,k} + B_{l1,k}u_{la,k-1} + B_{l2,k}u_{la,k-2} + \ldots B_{l\bar{d},k}u_{la,k-\bar{d}} + B_{l0,k}u_{la,k}$$
$$u_{la,k-i} = \gamma_{l,k-i}u_{l,k-i},$$
$$\forall l = 0,1,2,\ldots,\bar{d},\ \forall k = 0,1,2\ldots \qquad (5.1)$$

where $u_{la,k}$ is the actual control input received by the actuator of the lth system at the time instant kT_s; $u_{l,k}$ is the control input computed by the lth controller at the time instant kT_s; $\gamma_{l,k}$ models the stochastic variable representing the packet losses for the lth system at the time instant kT_s which follows a Bernoulli distribution with $P(\gamma_{l,k} = 1) = \bar{\gamma}_l$; and $A_{l,s}, B_{l1,k}, \ldots, B_{l\bar{d},k}$ denote the augmented system dynamics resulting from incorporating the communication network delay and packet losses for the lth system at the time instant kT_s. Defining the augmented state vector to be $z_{l,k} = [(x_{l,k})^T(u_{l,k-1})^T(u_{l,k-2})^T \cdots (u_{l,k-\bar{d}})^T]^T$ for the lth system at time instant kT_s, the system dynamics (5.1) can be rewritten as a linear stochastic time-varying system described by

$$z_{l,k+1} = A_{l,zk}z_{l,k} + B_{l,zk}u_{l,k}, \quad \forall l = 0,1,2,\ldots,\bar{d},\ \forall k = 0,1,2\ldots \qquad (5.2)$$

Note the detailed time-varying augmented system matrices are given in [16]. After incorporating the random delays and packet losses from the network, the linear continuous-time system becomes an uncertain stochastic linear discrete-time system, as shown in (5.2).

It is well-known that event-triggered control [29–31] can bring significant advantages to distributed CPSs in terms of reducing computational costs and network traffic enormously. The event-triggered control framework is incorporated into the distributed CPS design. However, optimality is preferred over stabilizing event-triggered control [14]. Therefore, a novel optimal adaptive event-triggered control is developed in the next section.

5.3 Optimal Adaptive Event-Triggered Control

In this section, a novel optimal adaptive zero-order-hold (ZOH)-based event-triggered control of distributed CPSs is developed. First, the general ZOH event-triggered control system is introduced. Subsequently, the optimal adaptive ZOH event-triggered control scheme is derived for a distributed CPS that can maintain stability even when the system dynamics are unknown and in the presence of event-based sampling of the system state and control input vector. Without loss of generality, the lth system is selected to describe the optimal adaptive event-triggered control as follows.

5.3.1 Zero-Order-Hold-Based Event-Triggered Control System

Recently, event-triggered control of linear systems has been a topic of significant interest to the CPS community due to its network benefits [32]. In Figure 5.2, the structure of a ZOH-based event-triggered control system is shown. Unlike time-driven control systems, a trigger mechanism is included at the sensor node of each system

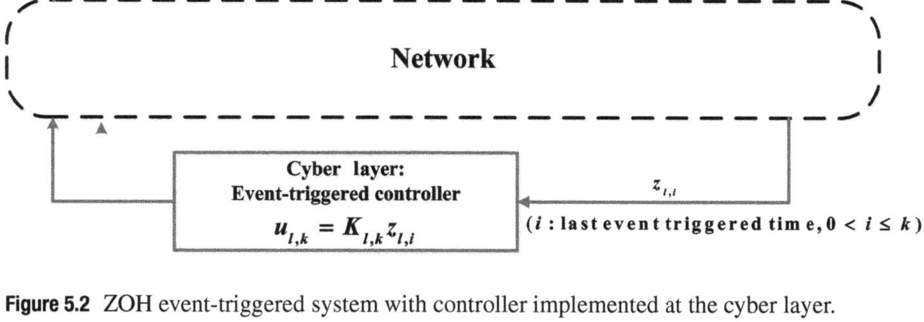

Figure 5.2 ZOH event-triggered system with controller implemented at the cyber layer.

to decide when to transmit the system state information z_k over the communication network to the controller.

According to (5.1) and (5.2), the lth system is described as $z_{l,k+1} = A_{l,zk}z_{l,k} + B_{l,zk}u_{l,k}$. Moreover, a ZOH event-triggered controller will hold the last received system state vector until a new state vector is received. In order to improve the performance of the ZOH event-triggered system, the optimal control theory [13] has to be incorporated into the event-triggered control design. However, since network imperfections can bring uncertainty into system dynamics, a novel ADP-based stochastic optimal event-triggered control [32] scheme will be introduced as follows.

5.3.2 Optimal Adaptive ZOH Event-Triggered Control

In this subsection, the optimal ZOH event-triggered control system design is given in a detailed manner. First, a value function is introduced and estimated adaptively. Then, the optimal ZOH event-triggered control scheme is developed by utilizing the estimated value function.

Value Function Setup

According to optimal control [13], the stochastic value function for the lth system can be represented as

$$V_{l,k} = \sum_{k=0}^{\infty}(z_{l,k}^T Q_{l,z}z_{l,k}+u_{l,k}^T S_{l,z}u_{l,k}) = \underset{\tau,\gamma}{E}\{z_{l,k}^T P_{l,k}z_{l,k}\} \tag{5.3}$$

where $Q_{l,z}, S_{l,z}$ are positive semidefinite and positive definite matrices respectively; $P_{l,k} \geq 0$ is the solution to the stochastic Riccati equation (SRE) for the lth system with $\underset{\tau,\gamma}{E}\{\bullet\}$ as the expectation operator (i.e., mean value) of $\{(z_{l,k})^T P_{l,k}z_{l,k}\}$. Then, the Hamiltonian for the lth system can be represented as

$$H(z_{l,k}, u_{l,k}) = r(z_{l,k}, u_{l,k}) + V(z_{l,k+1}, u_{l,k+1}) - V(z_{l,k}, u_{l,k}) \tag{5.4}$$

with $r(z_{l,k}, u_{l,k}) = z_{l,k}^T Q_{l,z}z_{l,k} + u_{l,k}^T S_{l,z}u_{l,k}$ as a one-step cost-to-go function. Utilizing the standard optimal control theory [13], the stochastic optimal control signal of the lth system at the time instant kT_s can be solved by minimizing this stochastic value function (5.3) (i.e., $(u_{l,k})^* = \arg\min V_{l,k}, \forall l = 1, 2, \ldots, N, \forall k = 1, 2, \ldots$), which yields

$$u_{l,k}^* = K_{l,k}^* z_{l,k} = -[R_{l,z} + \underset{\tau,\gamma}{E}(B_{l,zk}^T P_{l,k+1}B_{l,zk})]^{-1}\underset{\tau,\gamma}{E}(B_{l,zk}^T P_{l,k+1}A_{l,zk})z_{l,k} \tag{5.5}$$

REMARK 5.1 It is important to note that computing optimal control input in (5.5) requires the system dynamics, which is not usually known in practice.

On the other hand, the stochastic optimal action-dependent value function with event-sampled state vector can be expressed as

$$V_{l,k} = \underset{\tau,\gamma}{E}\{[(z_{l,k})^T \; u_{l,k}^T]H_{l,k}[(z_{l,k})^T \; u_{l,k}^T]^T\} = \bar{h}_{l,k}^T \bar{\chi}_{l,k} \tag{5.6}$$

where

$$\bar{H}_{l,k} = \underset{\tau,\gamma}{E}(H_{l,k}) = \begin{bmatrix} \bar{H}_{l,k}^{zz} & \bar{H}_{l,k}^{zu} \\ \bar{H}_{l,k}^{uz} & \bar{H}_{l,k}^{uu} \end{bmatrix}$$

$$= \begin{bmatrix} Q_{l,z} + \underset{\tau,\gamma}{E}[(A_{l,zk})^T P_{l,k+1} A_{l,zk}] & \underset{\tau,\gamma}{E}[(A_{l,zk})^T P_{l,k+1} B_{l,zk}] \\ \underset{\tau,\gamma}{E}[(B_{l,zk})^T P_{l,k+1} A_{zk}^l] & S_{l,z} + \underset{\tau,\gamma}{E}[(B_{l,zk})^T P_{l,k+1} B_{l,zk}] \end{bmatrix} \chi_{l,k}$$

$$= [(z_{l,k})^T \; u^T (z_{l,k})]^T$$

and $\bar{\chi}_{l,k}$ is the Kronecker product quadratic polynomial basis vector of the lth system and $\bar{h}_{l,k} = \text{vec}(\bar{H}_{l,k})$ with the vector function acting on a square matrix thus yielding a column vector. Note that the vec $(\bullet)$ function is constructed by stacking the columns of the matrix into one column vector with the off-diagonal elements which can be combined as $H_{l,mn} + H_{l,nm}$ [14].

Next, according to [13] and [32], the optimal control gain of the lth system can be represented by using the H_l matrix as

$$K_{l,k}^* = -[R_{l,z} + \underset{\tau,\gamma}{E}(B_{l,zk}^T P_{l,k+1} B_{l,zk})]^{-1} \underset{\tau,\gamma}{E}(B_{l,zk}^T P_{l,k+1} A_{l,zk})$$
$$= -(\bar{H}_{l,k}^{uu})^{-1} \bar{H}_{l,k}^{uz} \tag{5.7}$$

Therefore, if the H_l matrix is solved for the lth system, then the stochastic optimal control gain can be obtained by utilizing (5.5). However, since the system dynamics are unknown, H_l cannot be solved directly. Similar to [32], a novel adaptive estimator will be proposed to estimate the value function, H matrix (i.e., H_l), and the stochastic optimal control gain.

Model-Free Online Tuning of Value Function

From the ZOH event-triggered control scheme in Section 5.2.1 and the novel ADP technique in [27], the value function can be estimated when an event is triggered:

$$\hat{V}_{l,i} = \underset{\tau,\gamma}{E}\{[(z_{l,i})^T \; u_{l,i}^T]\hat{H}_{l,i}[(z_{l,i})^T \; u_{l,i}^T]^T\} = \hat{\bar{h}}_{l,i}^T \bar{\chi}_{l,i} \tag{5.8}$$

where iT_s is the triggering time instant with T_s denoting the sampling time. According to traditional optimal control theory [13], the Bellman equation is represented in terms of the value function as

$$V_{l,i+1} - V_{l,i} + r(z_{l,i}, u_{l,i}) = 0 \tag{5.9}$$

However, (5.9) cannot be guaranteed while the estimated $\hat{H}_{l,i}$ is utilized. After incorporating the estimated $\hat{H}_{l,i}$ into the Bellman equation, (5.9) can be presented as

$$\hat{V}_{l,i+1} - \hat{V}_{l,i} + r(z_{l,i}, u_{l,i}) = e_{l,i+1} \tag{5.10}$$

where $e_{l,i+1}$ denotes temporary difference (TD) error [14] in the Bellman equation. Moreover, substituting (5.8) into (5.10), (5.10) can be presented as

$$\begin{aligned} e_{l,i+1} &= r(z_{l,i},u_{l,i}) + \hat{V}_{l,i+1} - \hat{V}_{l,i} = r(z_{l,i},u_{l,i}) + \hat{\bar{h}}_{l,i+1}^T \bar{\chi}_{l,i+1} - \hat{\bar{h}}_{l,i+1}^T \bar{\chi}_{l,i} \\ &= r(z_{l,i},u_{l,i}) + \hat{\bar{h}}_{l,i+1}^T (\bar{\chi}_{l,i+1} - \bar{\chi}_{l,i}) = r(z_{l,i},u_{l,i}) + \hat{\bar{h}}_{l,i+1}^T \Delta\bar{\chi}_{l,i} \end{aligned} \tag{5.11}$$

where $\Delta\bar{\chi}_{l,i} = \bar{\chi}_{l,i+1} - \bar{\chi}_{l,i}$. Subsequently, the update law of parameter $\hat{H}_{l,i}$ for the lth system can be introduced as

$$\hat{\bar{h}}_{l,i+1} = \hat{\bar{h}}_{l,i} + \alpha_l \frac{\Delta\bar{\chi}_{l,i} e_{l,i+1}^T}{\Delta\bar{\chi}_{l,i}^T \Delta\bar{\chi}_{l,i} + 1} \tag{5.12}$$

where the TD error with event-sampled inputs drives the parameter update. Define the parameter estimation error as $\tilde{\bar{h}}_{l,i} = \bar{h}_l - \hat{\bar{h}}_{l,i}$, and the parameter estimation error dynamics can be expressed as

$$\tilde{\bar{h}}_{l,i+1} = \tilde{\bar{h}}_{l,i} - \alpha_l \frac{\Delta\bar{\chi}_{l,i} e_{l,i+1}^T}{\Delta\bar{\chi}_{l,i}^T \Delta\bar{\chi}_{l,i} + 1} \tag{5.13}$$

REMARK 5.2 It is observed that the estimated value function will no longer be updated while the system state vector converges to zero. It can be seen as a persistence of excitation (PE) [33] requirement for the input to the value function estimator, whereas the optimal adaptive ZOH event-triggered control system state vector must persistently exist long enough for the estimator to learn the value function. The PE condition is well known in adaptive control theory and can be ensured by adding exploration noise [13]. This PE condition is satisfied only when there are enough trigger events such that the adaptive estimator generates the H matrix.

Next, the estimation of optimal adaptive ZOH event-triggered control is derived. According to [27], the optimal ZOH event-triggered control can be attained by minimizing the value function. Recall from (5.7) that the optimal control gain can be designed by using the estimated parameter $\hat{H}_{l,i}$ as

$$\hat{u}_{l,i} = \hat{K}_{l,i} z_{l,i} = -(\hat{H}_{l,i}^{uu})^{-1} \hat{H}_{l,i}^{ux} z_{l,i} \tag{5.14}$$

Obviously, optimal adaptive ZOH event-triggered control gain can be obtained in terms of the $\hat{H}_{l,i}$ matrix, which is solved by estimating the value function with the event-sampled state vector. It is important to note that the proposed design (5.12) and (5.14) not only relaxes the requirement of the system dynamics, but also eliminates the value and policy iterations.

Next, the event-triggering condition that facilitates the events is derived. Since the controller will hold the latest received system state (i.e., $x_i, 0 < i \le k$) at the time instant kT_s, the measurement error e_k^{ZOH}, also referred to as event-trigger error, can be presented in terms of the state vector as

$$e_k^{ZOH} = x_k - x_i, \qquad 0 < i \le k \tag{5.15}$$

When the measurement error e_k^{ZOH} exceeds the threshold defined as the event-trigger condition, the sensor will transmit a new state vector to the controller. Upon receiving the state vector, the controller generates the control input and resets the event trigger error e_k^{ZOH} to zero. Moreover, the ZOH-based event-triggered closed-loop control system dynamics can be presented as

$$x_{k+1} = (A_{zk} + B_{zk}K_k)x_k - B_{zk}K_k e_k^{ZOH} \tag{5.16}$$

with

$$e_k^{ZOH} = \begin{cases} 0 & \text{event is initiated} \\ x_k - x_i & \text{event is not initiated} \end{cases}$$

Next, to maintain the stability of the ZOH event-triggered control system, a novel event-trigger condition [32] can be presented as

$$\left\| e_k^{ZOH} \right\| \leq \sigma(A_{zk}, B_{zk}, K_k) \left\| x_k \right\| \tag{5.17}$$

where $\sigma(A_{zk}, B_{zk}, K_k)$ is the state-dependent threshold of the event-trigger condition to be designed. The event will be triggered and the system state vector will be transmitted to the controller when the event trigger error, e_k^{ZOH}, fails to meet the event-triggering condition given by (5.8). On the other hand, an event will not be triggered and the controller will hold the previous received system state vector when the event-triggering condition (5.17) is satisfied. The design of the event-trigger condition is very challenging since the system matrices, A_{zk}, B_{zk}, are unknown. Next, the stability of the event-triggered control scheme is introduced.

THEOREM 5.1 (*ZOH based optimal event trigger control scheme*) Consider the linear discrete-time systems in a distributed CPS (5.2) along with the adaptive estimator (5.12) and adaptive estimator-based optimal control (5.14). Let $u_0\left(x_{l,k}\right)$ be an initial stabilizing control policy for the system (5.2). Let the adaptive estimator parameter $\hat{\bar{h}}_{l,0}$ be initialized in $D \subset \Re^n$. Further, there exists positive constant α_l such that the closed-loop event-triggered augmented system state vector $z_{l,k}$ and adaptive estimator parameter estimation errors $\hat{\bar{h}}_{l,k}$ are asymptotically stable, provided the system states are transmitted to the controller and the adaptive estimator parameters are updated using (5.12) through the violation of the event-trigger condition defined by

$$D\left(\left\| e_{l,k}^{ZOH} \right\|\right) \leq \sigma_{l,k} \left\| z_{l,k} \right\|$$

where $\sigma_{l,k} = \sqrt{(1-2l_o^2)\Gamma_l \Big/ 4B_{l,M}^2 \left\| \hat{K}_{l,k} \right\|^2}$ with, $0 < l_o < 1\big/\sqrt{2}, 0 < \Gamma_l < 1, B_{l,M}$ is the upper bound on the control coefficient matrix $B_{l,zk}$, and $\hat{K}_{l,k}$ is the estimated control gain defined in (5.14).

5.4 Novel Cross-Layer Distributed Scheduling Design

In this section, a novel cross-layer design [34, 35] is proposed for a distributed CPS. First, the main idea of cross-layer design is provided. Subsequently, novel distributed scheduling [36] is proposed to optimize the distributed CPS by minimizing the cost function which includes the information from both cyber and network layers. To the best knowledge of the authors, it is the first time that both the ZOH-based optimal adaptive event-triggered control design in the cyber layer and distributed scheduling scheme design in the network layer have been designed for a distributed CPS via a cross-layer approach.

5.4.1 Cross-Layer Design

In the proposed scheme, the novel event-triggered control design and distributed scheduling protocol are implemented into the cyber layer and network layer respectively for a CPS with a shared communication network. Since the CPS system dynamics and the scheduling scheme – as part of the communication network protocol – can affect each other, a cross-layer approach will be introduced to jointly optimize the CPS and communication network based on the information from both the cyber and the network layers.

Each CPS will tune its stochastic optimal controller in the presence of network imperfections such as network-induced delay and packet losses resulting from current distributed scheduling design, estimate its value function [14] based on tuned control design, and transmit that information (i.e., value function) to the network layer. The network layer tunes the distributed scheduling scheme based on throughput from the network layer and the value function information received from the physical systems. The cross-layer CPS design framework is shown in Figure 5.3.

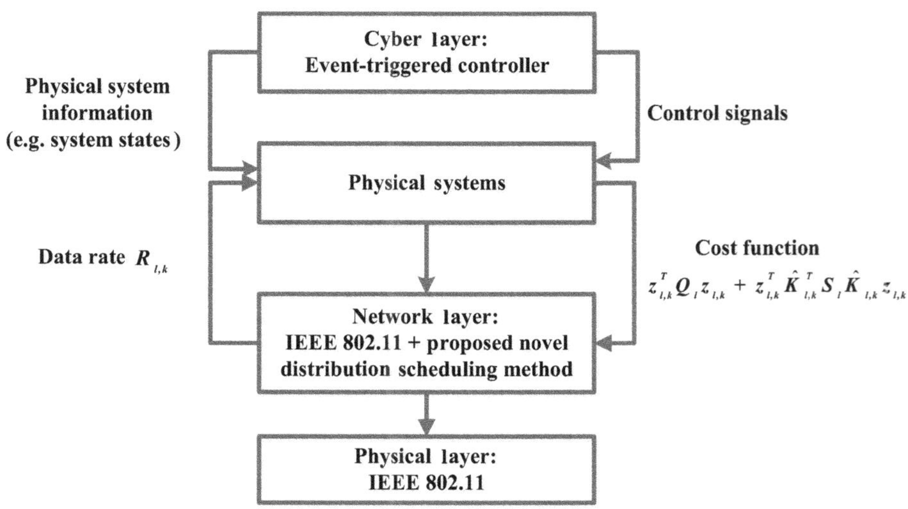

Figure 5.3 Framework for the cross-layer design.

5.4.2 Novel Distributed Scheduling

In this subsection, optimal distributed scheduling is developed at the network layer. Without loss of generality, the traditional wireless ad hoc network protocol [15] is implemented at the other protocol layers. In order to optimize the performance of the overall CPS, which includes the performance of both physical systems and network layers, a novel optimal cross-layer distributed scheduling algorithm is derived by incorporating control system information from physical systems. Similar to Section 5.2.1, the lth CPS system is considered here.

First, to design the distributed scheduling and optimal control jointly, the cost function for the lth system can be represented as

$$J_{l,k} = z_{l,k}^T Q_{l,z} z_{l,k} + u_{l,k}^T S_{l,z} u_{l,k} + \beta_l R_{l,k} \tag{5.18}$$

where $R_{l,k}$ denotes the lth system average traffic payload during $[kT_s, (k+1)T_s]$ and β_l is the weight parameter of average traffic payload. When β_l becomes large, it indicates the average traffic payload will affect the total cost. Subsequently, the entire cost function for distributed CPS can be expressed as

$$J_k = \sum_{l=1}^{M} J_{l,k} = \sum_{l=1}^{M} (z_{l,k}^T Q_{l,z} z_{l,k} + u_{l,k}^T S_{l,z} u_{l,k} + \beta_l R_{l,k}) \tag{5.19}$$

with M denoting the number of CPS systems. Then, the optimal design of the distributed CPS should minimize the cost function (5.19):

$$J_k^* = \min_{u,\pi} \sum_{l=1}^{M} (z_{l,k}^T Q_{l,z} z_{l,k} + u_{l,k}^T S_{l,z} u_{l,k} + \beta_l R_{l,k}) \tag{5.20}$$

where u is the control design and π is the scheduling policy.

Minimizing the cost function (5.20) requires: (1) a novel distributed scheduling scheme design, and (2) optimal control input design. For the control input design, the optimal adaptive ZOH event-triggered control scheme has been derived in Section 5.2. For the distributed scheduling part, a novel scheme will be developed in this subsection. Moreover, each CPS system has two options for scheduling its transmission of the event-sampled state vector, either to schedule or not to schedule. It is important to note that whether or not each CPS system is scheduled depends on which option will reduce the cost function. For example,

***Case 1:* lth system is scheduled**

$$\begin{aligned} J_{l,k}^{S,1} &= z_{l,k}^T Q_{l,z} z_{l,k} + z_{l,k}^T \hat{K}_{l,k}^T S_{l,z} \hat{K}_{l,k} z_{l,k} + \beta_l R_{l,k}^Y \\ &= z_{l,k}^T Q_{l,z} z_{l,k} + z_{l,k}^T \Lambda_{l,z} z_{l,k} + \beta_l R_{l,k}^Y \end{aligned} \tag{5.21}$$

where $\Lambda_{l,z} = \hat{K}_{l,k}^T S_{l,z} \hat{K}_{l,k}$ and $R_{l,k}^Y$ is the average traffic payload when the lth CPS system has been scheduled.

Case 2: lth **system is not scheduled**

$$\begin{aligned} J_{l,k}^{S,2} &= z_{l,k}^T Q_{l,z} z_{l,k} + z_{l,i}^T \hat{K}_{l,i}^T S_{l,z} \hat{K}_{l,i} z_{l,i} + \beta_l R_{l,k}^N \\ &= z_{l,k}^T Q_{l,z} z_{l,k} + z_{l,i}^T \Lambda_{l,i} z_{l,i} + \beta_l R_{l,k}^N \end{aligned} \quad (5.22)$$

where $\Lambda_{l,i} = \hat{K}_{l,i}^T S_{l,z} \hat{K}_{l,i}$, where $\hat{K}_{l,i}$ and $z_{l,i}$ are held control gain and system state vector as defined in Section 5.2, and $R_{l,k}^N$ is the average traffic payload when the lth CPS system has not been scheduled.

Next, the difference between these two cases can be considered in the utility function and represented as

$$\begin{aligned} \Delta J_{l,k}^S &= \pi_{l,k}\left(J_{l,k}^{S,2} - J_{l,k}^{S,1}\right) \\ &= \pi_{l,k}\left(z_{l,k}^T Q_l x_{l,k} + z_{l,i}^T \Lambda_{l,i} z_{l,i} + \beta_l R_{l,k}^N\right) \\ &\quad - \pi_{l,k}\left(z_{l,k}^T Q_l z_{l,k} + z_k^T \Lambda_{l,k} z_{l,k} + \beta_l R_{l,k}^Y\right) \\ &= \pi_{l,k}\varphi(e_{l,k}^{ZOH}) - \pi_{l,k}\beta_l D_{l,k} \end{aligned} \quad (5.23)$$

where $D_{l,k} = R_{l,k}^Y - R_{l,k}^N$ is the difference of average traffic payload for the lth CPS between the two scenarios, which is expressed as

$$D_{l,k} = R_{l,k}^Y - R_{l,k}^N = \left(\frac{N_k + 1}{kT_s} - \frac{N_k}{kT_s}\right) N_{l,bit} = \frac{1}{kT_s} N_{l,bit} \quad (5.24)$$

where $N_{l,bit}$ is the number of bits for packetizing the sensed event of the lth CPS system, $\pi_{l,k}$ is the schedule indicator of the lth CPS system (i.e., $\pi_{l,k} = \begin{cases} 1, & \text{Scheduled} \\ 0, & \text{Not Scheduled} \end{cases}$), $e_{l,k}^{ZOH} = z_{l,k} - z_{l,i}$, and $\varphi(e_{l,k}^{ZOH}) = (z_{l,i}^T \Lambda_{l,i} z_{l,i} - z_{l,k}^T \Lambda_{l,k} z_{l,k})$. Obviously, when $\Delta J_{l,k}^S > 0$, it indicates that scheduling the lth CPS system to transmit over the network yields additional advantage. Otherwise, scheduling the lth CPS system will degrade the performance. Therefore, when $\Delta J_{l,k}^S > 0$, this CPS system can be considered as scheduled. It is important to note that there are multiple CPS systems in a distributed CPS (i.e., M coupled CPS systems) eligible to transmit, and probably the utility functions of several subsystems are higher than zero, which indicates that all of these CPS systems have to be scheduled. However, according to the literature on networking [10], only one CPS system can access the network.

In order to optimize the performance of the network, the optimal scheduling policy should maximize the total utility function given by

$$(\Delta J_k^S)^* = \max_{\pi} \sum_{l \in \mathbf{G}_k} \Delta J_{l,k}^S \quad (5.25)$$

where $\mathbf{G}_k$ is the subsystem set with positive value of utility function at time instant kT_s i.e., $\Delta J_{l,k}^S > 0$ for $l \in \mathbf{G}_k$. Obviously, for centralized scheduler design, the optimal schedule has to select a subsystem that has the maximum value of $\Delta J_{l,k}^S$. However, in the centralized scheduling scheme, finding the maximum value $\Delta J_{l,k}^S$ requires significant information from every system, which might overload the network. Therefore, a novel distributed scheduling scheme is needed to solve this drawback.

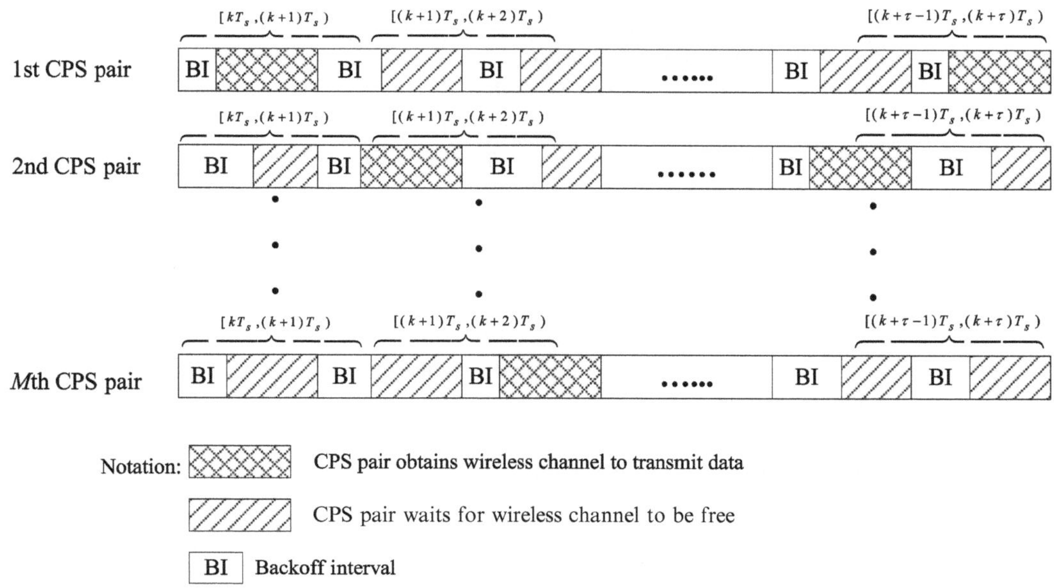

Figure 5.4 Framework of proposed cross-layer distributed scheduling scheme.

In this chapter, the main idea of the scheduling algorithm as shown in Figure 5.4 is to separate the transmission time of different systems by using a back-off interval (BI) [15], which is designed based on a related utility function in the distributed manner. To solve the optimal scheduling problem (5.25) for a distributed CPS, the BI can be designed as

$$BI_{l,k} = \varsigma \times (e^{-\Delta J_{l,k}} + n_{l,k}) \text{ for } l \in \mathbf{G}_k \tag{5.26}$$

where ς is the scaling factor, $n_{l,k}$ is a random variable satisfying the Gaussian distribution (i.e., $n_{l,k} \sim L * N(0, \sigma^2)$), and $L = \min_{l,j \in G_k} (e^{-\Delta J_{l,k}} - e^{-\Delta J_{j,k}})$ is the range of the random value $n_{l,k}$. Next, the proposed novel distributed scheduling algorithm is introduced.

REMARK 5.3 Since each CPS system decides its scheduling by only using local information from the physical systems and network layers, the proposed novel cross-layer scheduling scheme is distributed.

REMARK 5.4 Compared to other distributed scheduling schemes [11–13] the proposed algorithm generates the BI intelligently by optimizing a utility function instead of selecting it at random as in [11–13], which can be considered as the main contribution of this distributed scheduling algorithm. The optimality of proposed novel distributed scheduling is shown in Theorem 5.2.

THEOREM 5.2 (*Optimal distributed scheduler performance*) Given the multiple systems in a CPS and adaptive optimal event-triggered control scheme given by (5.14),

Algorithm 5.1: Novel optimal distributed scheduling scheme

1: **Initialize:** The utility function is initialized as $\Delta J_{l,0} = 0, \forall l = 1, 2, \ldots, M$
2: **While** $\{kT_s \leq t < (k+1)T_s\}$ **do**
3: **Calculate** BI for different subsystems of the overall CPS adaptive ZOH event-triggered control system $BI_{l,k} = \varsigma \times (e^{-\Delta J_{l,k}} + n_{l,k})$ for $l \in \mathbf{G}_k$.
4:     **Contend** shared communication network resource.
5:     **If** the lth system of the distributed CPS has the smallest BI **then**
6:         **Schedule** event-triggered pair l and transmit the lth system's data through the shared communication network.
7:             **If** transmission is over, **then**
8:                 **Update** the scheduled system's utility function $\Delta J_{l,k}$.
9:             **end if**
10:     **else**
11:         **Wait** for shared communication network channel to be free.
12:     **end if**
13:     **Update** time stamp: $t = t + BI_{l,k} + T_{l,k}$($BI_{l,k}$ is the BI of the scheduled distributed CPS system. $T_{l,k}$ is the transmission time of the scheduled system.)
14: **end while**
15: **Update and broadcast** utility function $\Delta J_{l,k}$ from each system of the overall CPS.
16: **Go to** next time period $[(k+1)T_s, (k+2)T_s)$ (i.e., $k = k+1$), and go back to line 2.

the proposed distributed scheduling scheme selects the optimal adaptive ZOH-based event-triggered system of the CPS with the highest utility function value since it has the shortest BI and highest priority to access the shared communication network. In addition, the proposed algorithm can render the best performance schedules for each system of the overall CPS.

Proof. Assume the CPS lth system has the highest utility function value (i.e., $\Delta J_{l,k} = \max_{i \in \mathbf{G}_k} \Delta J_{i,k}$), then we have $\Delta J_{l,k} > \Delta J_{i,k}$ for any $i = \mathbf{G}_k, i \neq l$. Therefore,

$$e^{-\Delta J_{l,k}} < e^{-\Delta J_{i,k}} \qquad i = \mathbf{G}_k, i \neq l \tag{5.27}$$

Next, for any $i = \mathbf{G}_k, i \neq l$, the BI can be expressed as

$$\begin{aligned} BI_{l,k} &= \varsigma \times (e^{-\Delta J_{l,k}} + n_{l,k}) \\ &< \varsigma \times [e^{-\Delta J_{l,k}} + \min_{l,j \in [1,M]} (e^{-\Delta J_{l,k}} - e^{-\Delta J_{j,k}})] \\ &< \varsigma \times (e^{-\Delta J_{l,k}} + e^{-\Delta J_{i,k}} - e^{-\Delta J_{l,k}}) \\ &< \varsigma \times e^{-\Delta J_{i,k}} < \varsigma \times (e^{-\Delta J_{i,k}} + n_{i,k}) < BI_{i,k} \end{aligned} \tag{5.28}$$

Hence, $BI_{l,k} < BI_{i,k}$ for any $i = \mathbf{G}_k, i \neq l$. Based on the proposed distributed scheduling algorithm, the lth CPS system can be scheduled to use the shared communication network due to its shortest BI. Next, the cost-optimality of the proposed scheme will be proven by using the contradiction method. □

Assume that there exists another system j, and scheduling it can render better performance than scheduling the lth system even though the lth system has the shortest BI (i.e., cost function value $J_k^{sj} < J_k^{sl}$, but $BI_{l,k} < BI_{j,k}$). According to the definition of cost function (5.19), J_k^{sj} can be defined as

$$J_k^{sj} = \left(\sum_{i=1}^{M} J_{i,k}\right) - \Delta J_{j,k} \text{ for } j = 1, 2, \ldots, M \tag{5.29}$$

Since $BI_{l,k} < BI_{j,k}$ is given in the assumption above, we have $\Delta J_{j,k} < \Delta J_{l,k}$ by using (5.26) and (5.28). Meanwhile, the cost function of the jth CPS can be derived as

$$J_k^{sj} = \left(\sum_{i=1}^{M} J_{i,k}\right) - \Delta J_{j,k} > \left(\sum_{i=1}^{M} J_{i,k}\right) - \Delta J_{l,k} = J_k^{sl} \tag{5.30}$$

It is important to note that $J_k^{sj} > J_k^{sl}$ in (5.30) contradicts the assumption $J_k^{sj} < J_k^{sl}$. By contradiction, there is no other system which can obtain a better performance than scheduling the lth system which has the shortest BI. In other words, the proposed distributed scheduling scheme can render the best performance by scheduling the lth system with the shortest BI.

On the other hand, any system with a negative utility function value, $\Delta J_{i,k} < 0$, should not contend the shared communication network resource since it will degrade the performance. Next, the proof is given in detail. Assume the pth system with a negative utility function, $\Delta J_{p,k} < 0$ is scheduled, then the cost function with this decision can be expressed as

$$J_k^{sp} = \sum_{i=1}^{M} J_{i,k} - \Delta J_{p,k} > \sum_{i=1}^{M} J_{i,k} = J_k (\text{no pair is scheduled}) \tag{5.31}$$

Therefore, scheduling a CPS system with a negative utility function will degrade the performance.

REMARK 5.5 Fairness is an important factor to evaluate the performance of scheduling schemes. For the proposed distributed scheduling algorithm, an index given by $FI = \frac{\left(\sum_{i=1}^{M} \frac{R_i}{\sum_{j=0}^{\infty}(x_{i,j}^T Q_i x_{i,j} + u_{i,j}^T R_i u_{i,j}) + \beta_i R_i}\right)^2}{M * \sum_{i=1}^{M}\left(\frac{R_i}{\sum_{j=0}^{\infty}(x_{i,j}^T Q_i x_{i,j} + u_{i,j}^T R_i u_{i,j}) + \beta_i R_i}\right)^2}$ is defined to measure the fairness among different systems.

Until now, novel distributed scheduling has been proposed to improve the performance of both physical systems and network layers since the utility function includes information from both layers. Combined with the proposed optimal adaptive ZOH event-triggered control scheme in Section 5.2, the proposed cross-layer approach is completed.

5.5 Conclusions

In this work, through a novel cross-layer design for distributed CPS, it is demonstrated that the proposed scheme can optimize not only the performance of the control system, but also the shared communication network. The novel optimal adaptive ZOH event-triggered control scheme does not require the system dynamics of each system and uses event-based sampling instead of an inefficient periodic time-driven sampling in order to assist in the hardware implementation. The novel scheduling algorithm is distributed, simple, and requires less computation than centralized scheduling algorithms.

5.6 Future Work

In the future, additional network protocols such as routing and admission control will be considered. On the other hand, a general optimal design will be developed and implemented for distributed heterogeneous CPSs. The challenges include how to utilize cross-layer design for the development of different types of network protocols in the presence of an event-trigger control scheme; definition of the cost function that needs to be minimized; and safety and security aspects of the protocols together with the network control aspects within the CPS. While security and privacy of a computer network is known, however, the security of a cyber system in conjunction with a physical system is still in its infancy.

The security of a CPS modeled as a dynamic system is beginning to be discussed in the literature. The modeling of the CPS represented as a networked control system in this chapter will help overcome a few of the challenges. One such challenge is the development of cyber defense in the presence of a variety of attacks on the network.

5.7 Appendix: Example Problem

Consider a CPS with six systems which are located within a 300 m_2 area at random. To maintain the homogeneous property, all six systems are using similar control systems as [37]. The discrete-time representation is given as

$$x_{k+1} = \begin{bmatrix} 1.1138 & -0.0790 \\ 0.0592 & 0.8671 \end{bmatrix} x(t) + \begin{bmatrix} 0.2033 \\ 0.1924 \end{bmatrix} u(t) \tag{5.32}$$

The sampling interval is selected as $T_s = 0.15$ seconds, and the number of bits for the six quantized sensed data for the systems are defined as $N_{bit} = [10\ 8\ 6\ 7\ 8\ 4]$. The cost function weight matrix is selected as $Q_l = \begin{bmatrix} 1 & 0 \\ 0 & 1 \end{bmatrix}$, $R_l = 0.1$. The initial system state vectors are given by $x_{1,0} = [20 \quad 7]^T$, $x_{2,0} = [12 \quad 5]^T$, $x_{3,0} = [10 \quad 3]^T$, $x_{4,0} = [8 \quad 4]^T$, $x_{5,0} = [10 \quad 5]^T$, and $x_{6,0} = [0.1 \quad 6]^T$.

(a) Design the event-triggering condition and estimated optimal control.
(b) Design the distributed scheduling scheme.
(c) Plot the distributed CPS system state response.
(d) Plot the fairness of optimal distributed scheduling and distributed CPS cost function performance compared with embedded round robin (ERR) and greedy scheduling schemes.

The augment state is generated as $z_{l,k} = [x_{l,k}\ u_{l,k-1}\ u_{l,k-2}]^T \in \mathbb{R}^{4\times1}$ or $\chi_{l,k} = [(z_{l,k})^T\ u^T(z_{l,k})]^T \mathbb{R}^{6\times1}$. The regression function for the adaptive estimator is $\left\{\chi_{l,1}^2, \chi_{l,1}\chi_{l,2}, \chi_{l,1}\chi_{l,3}, \ldots, \chi_{l,2}^2, \ldots, \chi_{l,5}^2, \ldots, \chi_{l,6}^2\right\}$, as per (5.6). The tuning parameter for the adaptive estimator is selected as $\alpha_{l,h} = 10^{-6}$, while initial parameters for the adaptive estimator are set to zero at the beginning of the simulation. The parameter for the ZOH event-trigger condition is chosen as $\Gamma_l = 0.9$.

Without loss of generality, an average value of state regulation errors for the six subsystems is shown in Figure 5.5. The results indicate that the proposed optimal

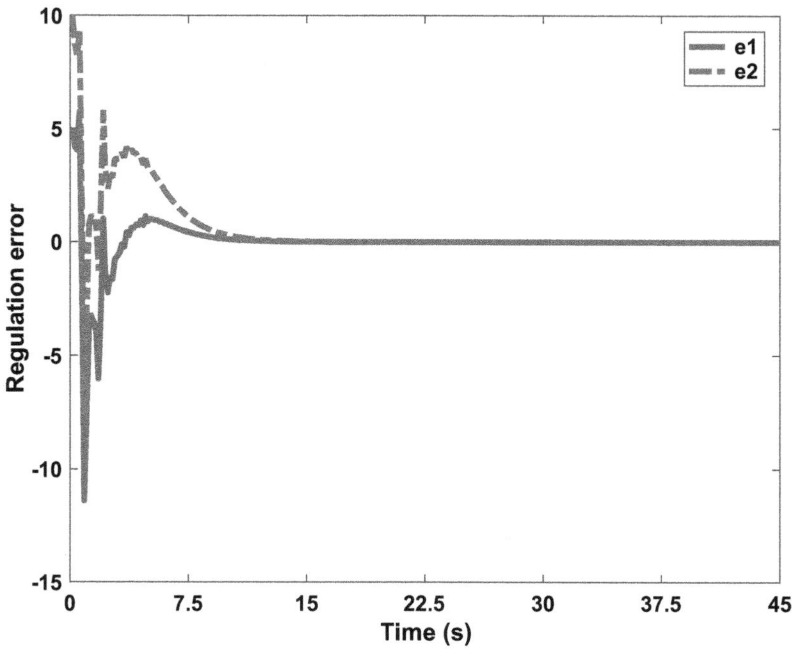

Figure 5.5 State regulation errors.

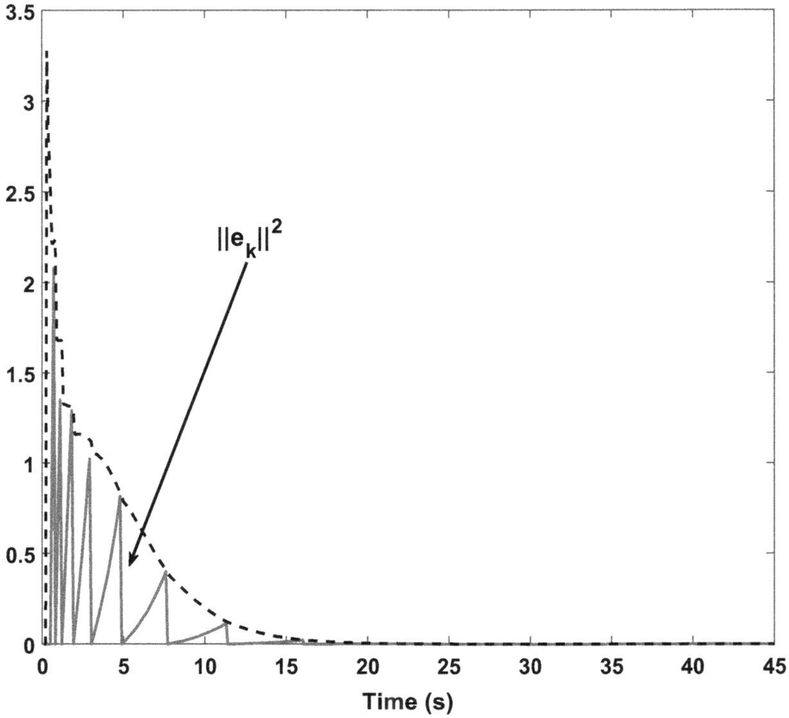

Figure 5.6 Event-trigger error norm $\|e_k\|^2$.

adaptive ZOH event-triggered control design can not only force the regulation errors to converge to zero asymptotically, but can also make the state regulation errors converge to zero quickly while ensuring all systems are stable. It is important to note that overshoots are observed at the beginning because the value function estimator tuning needs a brief adaptation phase.

Now, the performance of the optimal adaptive ZOH event-triggered control system is demonstrated. Without loss of generality, we pick the fifth optimal adaptive ZOH event-triggered subsystem to evaluate the performance. As shown in Figure 5.6, when an error event is triggered and scheduled, the estimation error will be reset to zero since the actual subsystem state will be received at the controller. On the other hand, if an error event is not triggered and scheduled, the estimation error increases due to the held system state in the ZOH event-triggered control system. Once the parameters of the value function estimator are assessed accurately, the estimation errors converge to zero. It is important to note that error events are triggered and scheduled more frequently at the beginning, since the value function estimator needs to be tuned often. After a short adaptation phase, the error events become less frequent, which would reduce network traffic.

The performance of the proposed cross-layer distributed scheduling is now evaluated. For comparison, the classical ERR [15] and greedy scheduling [15] schemes are considered. In Figure 5.7, the cost function of multiple subsystems of the distributed

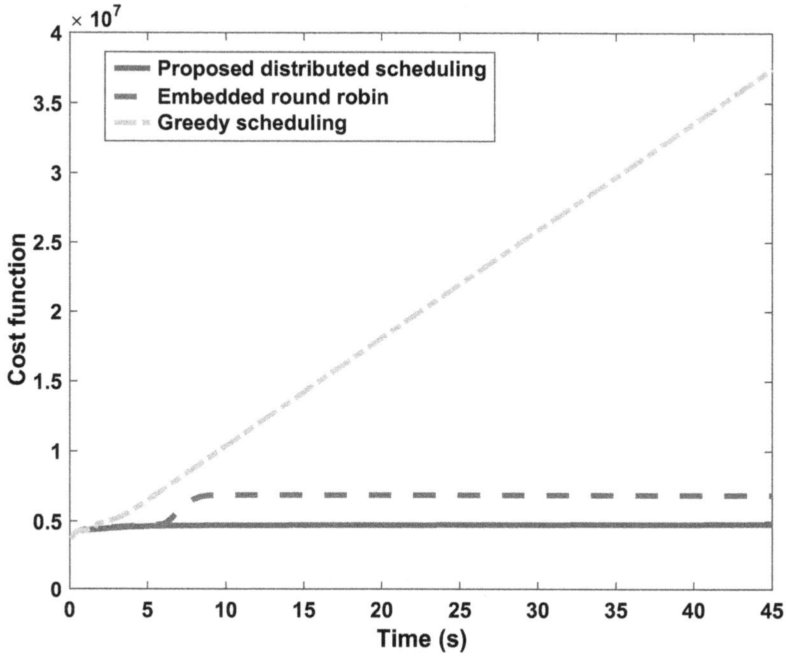

Figure 5.7 The cost function comparison.

CPS with three different scheduling schemes is compared. The proposed novel cross-layer distributed scheduling scheme maintains the lowest value while the cost function of the distributed CPS with ERR and greedy scheduling is much higher than that of the proposed scheduling scheme.

This indicates that the proposed distributed scheduling scheme can improve the performance of the distributed CPS much better than the widely used ERR and greedy scheduling. It is important to note that: (1) since ERR only guarantees that each CPS system can have the same probability to access the shared communication network resource, it could not assign communication network resources based on practical use of each system of the distributed CPS, which might waste precious network resources. Therefore, ERR cannot optimize performance of the multiple subsystems of the distributed CPS; and (2) since greedy scheduling only assigns network resources to the system with the best network layer performance (whereas physical system performance of each system in the distributed CPS is overlooked), so it also cannot optimize the performance of the distributed CPS.

The fairness of different scheduling schemes is now evaluated. As shown in Figure 5.8, fairness indices of the proposed cross-layer distributed scheduling and widely used ERR schemes are close and equal to 1, whereas that of greedy scheduling is much less than 1, indicating fair allocation of shared communication network resource for the proposed system while meeting overall performance requirements. The ERR method, though it is fair, has higher costs than the proposed one. According to the results illustrated in Figures 5.5 through 5.8, the proposed cross-layer design included

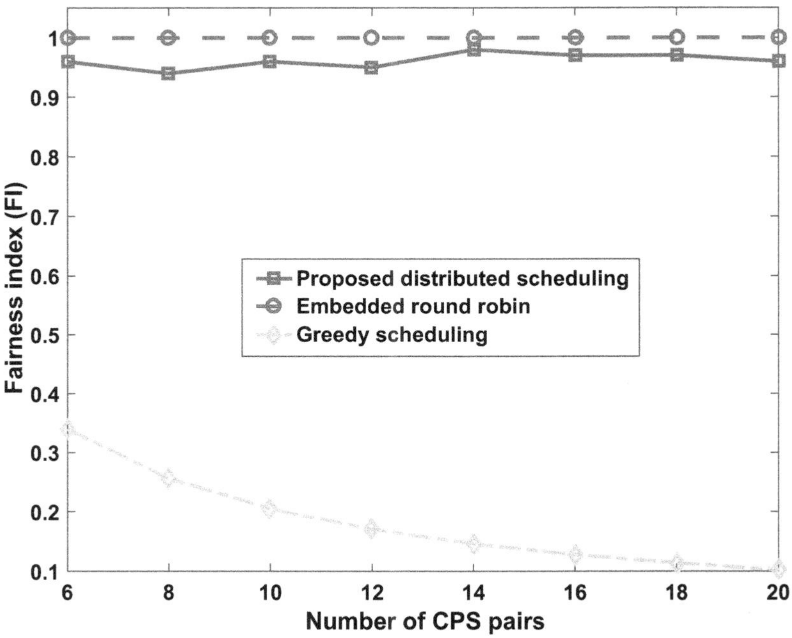

Figure 5.8 The fairness comparison.

optimal adaptive ZOH event-triggered control and a cross-layer distributed scheduling scheme to optimize the performance of both the communication network and the subsystems.

References

[1] G. Tian, Y. C. Tian, and C. Fidge, "Performance analysis of IEEE 802.11 DCF based WNCS networks," in *Proc. Local Computer Networks*, pp. 496–503, 2010.

[2] F. Xia, X. Kong, and Z. Xu, "Cyber-physical control over wireless sensor and actuator networks with packet loss," in *Wireless Networking Based Control*, S. K. Mazmumder, Ed., New York: Springer, 2012.

[3] F. Xia, A. Vinel, R. Gao, L. Wang, and T. Qiu, "Evaluating IEEE 802.15.4 for cyber-physical systems," *EURASIP J. Wireless Commun. Netw.*, vol. 2011, pp. 1–14, 2011.

[4] J. D. Day and H. Zimmermann, "The OSI reference model," *Proc. IEEE*, vol. 71, pp. 1334–1340, 1983.

[5] V. Srivastava and M. Motani, "Cross-layer design: A survey and the road ahead," *IEEE Commun. Mag.*, vol. 43, pp. 112–119, 2005.

[6] X. Guan, H. Wu, and S. Bi, "A game theory based obstacle avoidance routing protocol for wireless sensor networks," *Sensor*, vol. 11, pp. 9327–9343, 2011.

[7] N. Nie and C. Comaniciu, "Adaptive channel allocation spectrum etiquette for cognitive radio networks," *J. Mobile Netw. Appl.*, vol. 11, pp. 779–797, 2006.

[8] Q. Liu, S. Zhou, and G. B. Giannakis, "Queuing with adaptive modulation and coding over wireless links: Cross-layer analysis and design," *IEEE Trans. Wireless Commun.,* vol. 4, pp. 1142–1153, 2005.

[9] IEEE Standard 802.11a 1999 (R2003), "Wireless LAN medium access control (MAC) and physical layer (PHY) specification, high-speed physical layer in 5GHz band," 2003.

[10] J. G. Dai and B. Prabhakar, "The throughput of data switches with and without speedup," in *Proc. 31st IEEE Int. Conf. Computer Communications*, pp. 556–564, 2000.

[11] L. Jiang and J. Walrand, "A distributed CSMA algorithm for throughput and utility maximization in wireless networks," *IEEE/ACM Trans. Netw.*, vol. 18, pp. 960–972, 2010.

[12] N. Vaidy, A. Dugar, S. Gupta, and P. Bahl, "Distributed fair scheduling in wireless LAN," *IEEE Trans. Mobile Computing*, vol. 4, pp. 616–629, 2005.

[13] F. L. Lewis and V. L. Syrmos, *Optimal Control*, 2nd ed. New York: Wiley, 1995.

[14] H. Xu, S. Jagannathan, and F. L. Lewis, "Stochastic optimal control of unknown networked control systems in the presence of random delays and packet losses," *Automatica,* vol. 48, pp. 1017–1030, 2012.

[15] S. Jagannathan, *Wireless Ad-hoc and Sensor Network: Protocols, Performance, and Control.* Boca Raton, FL: CRC Press, 2007.

[16] M. Guinaldo, D. Lehmann, J. Sanchez, S. Dormido, and K. H. Johansson, "Distributed event-triggered control with network delays and packet losses," in *Proc. Control and Decision Conf.*, pp. 1–6, 2012.

[17] Y. Wei, J. Heidemann, and D. Estrin, "An energy-efficient MAC protocol for wireless sensor network," in *Proc. INFOCOM 2002*, pp. 1567–1576, 2002.

[18] Z. Zhang, M. Ma, and Y. Yang, "Energy-efficient multihop polling in clusters of two-layered heterogeneous sensor networks," *IEEE Trans. Comput.*, vol. 57, pp. 231–245, 2008.

[19] K. Lim and Y. H. Lee, "Optimal partitioning of heterogeneous traffic sources in mobile communication networks," *IEEE Trans. Comput.*, vol. 46, pp. 312–325, 1997

[20] D. Kumar, "Performance analysis of energy efficient clustering protocols for maximising lifetime of wireless sensor networks," *IET Wireless Sensor Syst.*, vol. 4, pp. 9–16, 2014.

[21] T. Salem, S. M. El-Kader, S. M. Ramadan, and M. Z. Abdel-Mageed, "Opportunistic spectrum access in cognitive radio ad hoc networks," *Int. J. Comput. Sci.,*, vol. 11, pp. 41–50, 2014.

[22] P. J. Werbos, "Approximate dynamic programming for real-time control and neural modeling," in *Handbook of Intelligent Control*, D. A. White and D. A. Sofge, Ed., New York: Van Nostrand Reinhold.

[23] F. Y. Wang, N. Jin, D. Liu, and Q. L. Wei, "Adaptive dynamic programming for finite horizon optimal control of discrete-time nonlinear system with ε-error bound," *IEEE Trans. Neur. Netw.*, vol. 22, pp. 24–36, 2011.

[24] A. Heydari and S. N. Balakrishan, "Finite horizon input-constrained nonlinear optimal control using single network adaptive critics," in *Proc. American Control Conf.*, pp. 3047–3052, 2011.

[25] T. Dierks and S. Jagannathan, "Online optimal control of affine nonlinear discrete-time systems with unknown internal dynamics by using time-based policy update," *IEEE Trans. Neur. Net. Learn. Syst.*, vol. 23, pp. 1118–1129, 2012.

[26] P. Tabuada, "Event-triggered real-time scheduling of stabilizing control tasks," *IEEE Trans. Autom. Contr.*, vol. 52, pp. 1680–1685, 2007.

[27] M. Donkers and W. Heemels, "Output-based event-triggered control with guaranteed L_∞-gain and improved and decentralised event-triggering," *IEEE Trans. Autom. Contr.*, vol. 57, pp. 1362–1376, 2012.
[28] E. Garcia and P. J. Antsaklis, "Model-based event-triggered control for systems with quantization and time-varying network delays," *IEEE Trans. Autom. Contr.*, vol. 58, pp. 422–434, 2013.
[29] M. Mazo, Jr. and P. Tabuada, "Decentralized event-triggered control over wireless sensor/actuator networks," *IEEE Trans. Autom. Control*, vol. 56, pp. 2456–2461, 2011.
[30] D. V. Dimarogonas and K. H. Johanson, "Event-triggered control for multi-agent systems," in *Proc. 48th IEEE Conf. Decision and Control*, pp. 713–719, 2009.
[31] A. Eqtami, D. V. Dimarogonas, and K. J. Kyriakopoulos, "Event-triggered control for discrete-time systems," in *Proc. American Control Conf.*, pp. 4719–4724, 2010.
[32] X. Wang and M. Lemmon, "Event-triggered in distributed networked control systems," *IEEE Trans. Autom. Control*, vol. 56, pp. 586–601, 2011.
[33] M. Green and J. B. Moore, "Persistency of excitation in linear systems," *Syst. Contr. Lett.*, vol. 7, pp. 351–360, 1989.
[34] S. Shakkottai, T. S. Rappaport, and P. C. Karlsson, "Cross-layer design for wireless networks," *IEEE Commun. Mag.*, vol. 41, pp. 74–80, 2003.
[35] P. Reena and L. Jacob, "Joint congestion and power control in UWB based wireless sensor network," in *Proc. 32nd IEEE Conf. Local Computer Network (LCN)*, pp. 911–918, 2007.
[36] D. Zheng, W. Ge, and J. Zhang, "Distributed opportunistic scheduling for ad hoc networks with random access: An optimal stopping approach," *IEEE Trans. Inf. Theor.*, vol. 55, pp. 205–222, 2009.
[37] E. Garcia and P. J. Antsaklis, "Model-based event-triggered control with time-varying network delays," in *Proc. 50th IEEE Conf. Decision and Control*, pp. 1650–1655, 2011.

6 Online Adaptive Network Design for Disturbance Rejection

Airlie Chapman, Eric Schoof, and Mehran Mesbahi

6.1 Introduction

The emergent area of consensus-type dynamics in networked multi-agent systems has become an important framework for problems such as formation control, distributed estimation, swarming, and multi-vehicle coordination [1–5]. Core to the appeal of this approach is its distributable and agent-anonymous nature. Consequently, consensus can efficiently be implemented on large-scale networks without requiring agent identifiers to be shared. The price paid for this flexibility and minimal memory usage is the exposure of the networked dynamics to unidentifiable disturbances. Examples of these include local agent perturbations and agent-to-agent link perturbations.

Agents in the network can experience local disturbances from inaccurately modeled dynamics and controls, as well as environmental perturbations. Consensus-type coordination describes a coupling across the agent dynamics and can inadvertently induce an error propagation pathway through the network links. Modeling disturbances as an additive noise input, network entropy measures can describe the ease of noise propagation through the network. Two such measures are loop entropy and the Kolmogorov–Sinai entropy, a generalization of Shannon entropy over graphs [6]. High entropy indicates poor noise dissipation and graph theoretically corresponds to a highly disordered network structure.

Disturbances on agent-to-agent links in the network can be caused by agent misidentification. Consensus algorithms based on inter-agent measurement observations rely on accurate identification of neighboring agents. In cluttered environments, agents can misidentify foreign objects as native agents (Figure 6.1). Knowledge of the number of neighboring agents indicates that a link disturbance has occurred, but the isolation of the misidentified neighbors may not be possible. Leader–follower dynamics can be used to model the addition of agent false positives into the network dynamics [3, 7]. The misidentified neighbors are encoded as foreign input signals, or corrupted leaders. A system-theoretic measure which encodes the impact of these foreign inputs on leader–follower dynamics is the open loop $\mathcal{H}_2$ norm. This measure describes the energy of the network states due to a unit impulse at the input.

There is a rich literature in control theory focusing on disturbance rejection in dynamic systems. Broadly, rejection techniques fall in two categories: passive control such as structural damping, and active control such as dynamic feedback. In a

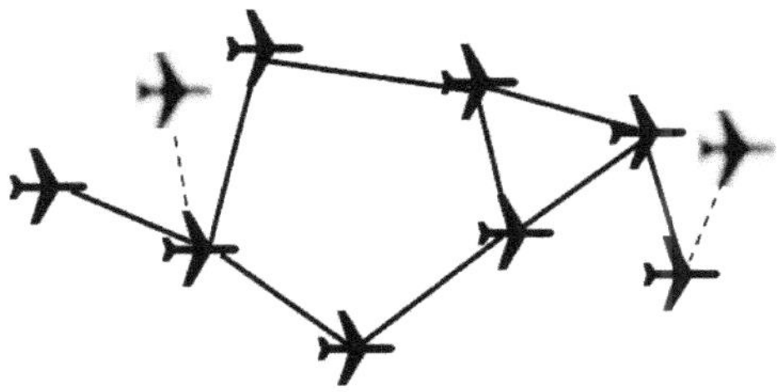

Figure 6.1 Visualization of the swarm of networked unmanned aerial vehicles (UAVs), with blurred images representing misidentified agents.

dynamic network setting, the network topology can be considered as a *passive* element. Strategic selection of edge weights in the network can vary the system's ability to reject disturbances, with the attraction that, for consensus-like algorithms, the agreement subspace is invariant to changes in positive weights. Previous work has examined the design of topological weights under different performance metrics. The second smallest and largest eigenvalue of the graph Laplacian was optimized by Ghosh and Boyd [8] and Wan et al. [9], respectively. Zelazo and Mesbahi [10] minimized the open loop $\mathcal{H}_2$ norm of the network exposed to link noise inputs. These topological selection problems can be cast as an optimization problem over a weighted graph. The discrete edge selection problem in which links in the network are either present or not becomes a more challenging network problem. Fading functions which approximate the on/off edge relationship and genetic algorithms have been employed by Kim and Mesbahi [11] and Wu and Wang [12], respectively, to maximize the second-smallest eigenvalue of the graph Laplacian. Optimizing for the open loop $\mathcal{H}_2$ norm over tree graphs, Chapman and Mesbahi [7] proposed a game-theoretic edge swap algorithm to trade edges between neighbors.

A less examined form of disturbance rejection for networked systems is *active* network topology control. Weights in the graph can be dynamically adapted in response to the disturbance. This is particularly challenging if topological weights must be implemented distributively. This chapter approaches the problem of distributed reweighting in two phases; local disturbance estimation and distributed online reweighting.

An external input signal can have a varying effect on an agent based on its proximity to the input location and the propagation pathways of the disturbance. Local disturbance estimation provides the necessary feedback to locally optimize the selection of edge weights. The gradient of the open loop $\mathcal{H}_2$ norm and entropy measures with respect to the edges describe the local contribution that varying an edge has on the measures. Calculating this gradient can be achieved by solving a system of linear equations. Efficient iterative methods to solve this problem are the conjugate gradient method formulated by Hestenes and Stiefel [13] and those based on the truncated Neumann series. Features of these methods are small storage requirements, fast theoretical convergence guarantees, and that only the range space of the underlying matrix needs to be sampled at each iteration. One significant property of the particular set of linear equations in question is that it encodes the network structure. Consequently, distributed versions of the conjugate gradient and truncated Neumann methods can be developed which are suitable for estimation of the gradient of the $\mathcal{H}_2$ norm.

To account for varying disturbances, an online reweighting method is coupled with the disturbance estimation. An online optimization approach provides performance guarantees in terms of the notion of *regret* under unpredictable time-varying cost functions. An online algorithm exhibiting sublinear regret indicates that an algorithm, provided only with cost function information in real time, performs as well as an offline algorithm provided with all cost information but constrained to a fixed-in-time decision. One of the simplest online algorithms, based on gradient descent or incremental gradient methods [14], has been proposed by Hazan et al. [15]. These are centralized algorithms with sublinear $O(\sqrt{T})$ and $O(\log(T))$ regret for convex functions and strongly convex functions, respectively. In this chapter, a distributable version of this online algorithm, suitable for the open loop $\mathcal{H}_2$ norm and entropy measures, is proposed to solve the distributed online network reweighting phase.

Yan et al. [16] presented an sublinear-regret online distributed optimization algorithm in which the cost function was decomposable into smaller cost functions. Stochastic subgradient algorithms and subgradient-based sequential convex optimization algorithms have also been adapted for online distributed implementation [17, 18]. Raginsky et al. [19] have proposed an algorithm with a similar network-of-agents approach and explored its robustness to network structure. Distributed dual averaging approaches for online learning have also been examined with respect to scalability, asynchronous communications, and certain classes of uncertainties [20, 21]. Shahrampour and Jadbabaie [22] examined relaxing the fixed-in-time benchmark appearing in the regret definition, termed dynamic regret, to allow for a time-varying benchmark.

Distinctive from the aforementioned works on online distributed optimization, non-decomposable global objectives have been explored by Lee et al. [23] using a dual-averaging approach, with each maintaining a private estimate of the optimal network state. Similarly, a distributed online gradient descent algorithm is proposed in this chapter, which does not assume that the cost function is decomposable. The gradient, however, is shown to be naturally decomposable when combined with distributed conjugate gradient techniques and truncated Neumann methods. This extension removes the requirement for a global private estimate to be computed by each agent.

Online adaptive network design draws on three distinctive research areas, represented in Figure 6.2. The necessary background material in the areas of online convex optimization and graph theory are presented in Section 6.2. Section 6.3 describes the agreement problem for multi-agent networks as well as the disturbance rejection problem for consensus-based, single and double integrator leader–follower dynamics. To quantify the disturbance rejection of a network, control-theoretic measures of the open loop $\mathcal{H}_2$ norm and entropy are explored in Section 6.4. These measures form the cost functions for the online optimization algorithms in Section 6.5. The local disturbance estimation and distributed online network reweighting methods are introduced in this section and applied to a multi-agent dynamic network. Concluding remarks are discussed in Section 6.6.

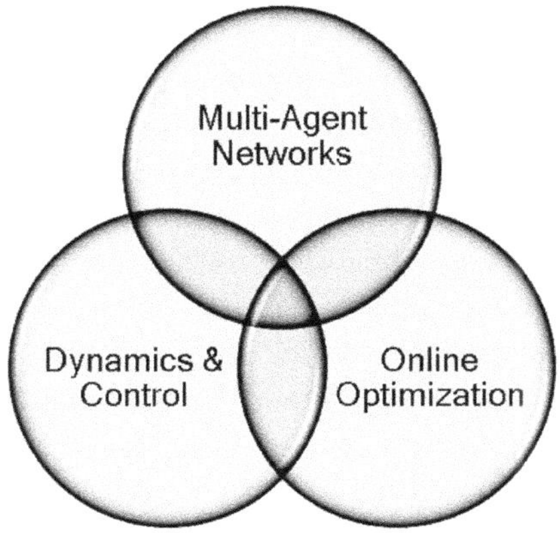

Figure 6.2 Research area intersection for online adaptive network design.

6.2 Background

6.2.1 Notation

In this section we will present concepts that are used throughout this chapter, including linear algebraic notation, descriptions of graphs, an overview of the consensus protocol, and a brief introduction to convex optimization and regret analysis.

We denote the 1-norm and the ∞-norm as $\|\cdot\|_1$ and $\|\cdot\|_\infty$, respectively. Similarly, $\|\cdot\|_2$, or simply $\|\cdot\|$, denotes the 2-norm. The cardinality of a set is denoted by $|\cdot|$.

To represent vectors, $v \in \mathbb{R}^p$ is a column vector and v_i or $[v]_i$ denotes its ith element; $\mathbf{1}$ is the column vector of all 1s, and e_i denotes the column vector which has 0s in every element except $[e_i]_i = 1$.

To represent matrices, $[M]_{ij}$ denotes the ijth element, the element in the ith row and jth column of the matrix M. The trace of a matrix is denoted $\mathbf{tr}(\cdot)$, and for matrices $M, N \in \mathbb{R}^{n \times n}$, the notation $N \preceq M$ indicates that $M - N$ is a positive semidefinite matrix, and I denotes the identity matrix. Given matrices $M, N \in \mathbb{R}^{p \times q}$, the Hadamard product, also called the entrywise product, is denoted by the symbol "$\circ$" as $[M \circ N]_{ij} = [M]_{ij} \cdot [N]_{ij}$.

6.2.2 Graphs

When analyzing communication networks, we often use an abstract representation of the network called a graph, in which communication points are called nodes, and links between communication points are referred to as edges. In a multi-agent network, the agents are thus represented as nodes and the inter-agent links (either communication or sensing) are represented as edges. We assume in this chapter that all inter-agent links are bidirectional, meaning that if agent i can communicate with agent j, then agent j can also communicate with agent i. Since all edges are in this way symmetric, we call the graph undirected.

More formally, we denote an undirected graph as $\mathcal{G} = (V, E, W)$, where V is the set of nodes with $|V| = n$ the number of nodes in the graph, E is the set of edges with $|E| = m$, and W is the set of edge weights. The edge set E contains $\{v_i, v_j\} \in E \subseteq [V]^2$ if there is an edge between nodes i and j. If there is an edge between nodes v_i and v_j, we also say that nodes v_i and v_j are adjacent with edge weight $w_{ij} \in W$. We define the neighborhood of node v_i, $N(v_i)$, as the set $\{v_j \in N \text{ s.t. } \{v_i, v_j\} \in E\}$. We denote the degree of node v_i as d_i, defined as the number of nodes adjacent to node v_i; note that $d_i = |N(v_i)|$. We denote the diameter D of a graph $\mathcal{G}$ as the cardinality of the longest path among all shortest paths between all pairs of nodes in V. We often wish to order edges, and do so through the index mapping function $\sigma(\cdot)$ such that $\ell = \sigma(i, j)$ when edge $\ell \in E$ corresponds to $\{v_i, v_j\} \in E$. If clear from the context, both forms e_ℓ and e_{ij} will be used interchangeably. We denote the weight matrix of a graph $\mathcal{G}$, $W(\mathcal{G}) \in \mathbb{R}^{m \times m}$, as a diagonal matrix with $[W(\mathcal{G})]_{\ell\ell}$ equal to the weight of edge ℓ, and zero otherwise. We denote the incidence matrix of a graph $\mathcal{G}$, $E(\mathcal{G}) \in \mathbb{R}^{n \times m}$, with columns a_{ij} for each edge $\{v_i, v_j\} \in E$ where $[a_{ij}]_i = 1$, $[a_{ij}]_j = -1$, and 0 otherwise. Finally, we denote the graph Laplacian of a graph $\mathcal{G}$ as $L(\mathcal{G}) = E(\mathcal{G})W(\mathcal{G})E(\mathcal{G})^T \in \mathbb{R}^{n \times n}$. Using the notation for the columns of the incidence matrix, the graph Laplacian can also be written $L(\mathcal{G}) = \sum_{\{v_i, v_j\} \in E} w_{ij} a_{ij} a_{ij}^T$.

6.2.3 Online Convex Optimization

Convex optimization seeks to find a point x which minimizes a convex function $f(x)$. In a so-called online scenario, the optimization is posed as a game where at each turn t of the game, the player is allowed to choose a point x_t, called an action, in some convex set $\mathcal{P}$ and only after the player has committed to their action, the opponent chooses a convex function f_t and the player incurs $f_t(x_t)$ as the penalty. The player's goal is to minimize the total penalty accumulated over time.

Since this game is inherently difficult due to the high degree of uncertainty from the online formulation, we need some way to judge the performance of the player's strategy. One common metric is called *regret*, defined as the difference between the player's accumulated penalty $\sum_{t=1}^{T} f_t(x_t)$ and the accumulated penalty $\sum_{t=1}^{T} f_t(x^*)$ when an oracle can only choose a single fixed point x^* but can also see all of the convex functions f_t a priori. We call x^* the best fixed action in hindsight. The regret $\mathcal{R}_T$ of an algorithm generating a sequence of actions $\{x_t\}$ is thus given as

$$\mathcal{R}_T = \sum_{t=1}^{T} [f_t(x_t) - f_t(x^*)]$$

A "good" online algorithm is one that exhibits low regret. More precisely, an algorithm that can guarantee sublinear regret, meaning $\mathcal{R}_T / T \to 0$, performs "on average" as well as the best fixed action in hindsight.

6.3 Dynamic Models

Networked multi-agent dynamics often adopt a two-layer dynamic model composed of an agent-level dynamics and a network-level dynamics. The agent-level dynamics, specific to the physical characteristics of the agent itself, stabilizes and tracks a reference signal provided by a network-level model. An example of such a model would be a plant and estimator/controller running on each UAV in a swarm able to track a commanded velocity reference signal using onboard sensors and actuators. The network-level dynamics coordinates cooperative mission tasks across the multi-agent network; it acts on the relative information between agents and provides a reference tracking signal for the agent-level dynamics. In the UAV swarming scenario with a rendezvous objective, the network-level dynamics would act on the relative position measurements between agents and provide a velocity reference signal for each individual agent to track. An assumption is that the agent-level dynamics operate much faster than the network-level dynamics, allowing each layer to be analyzed separately.

The upper portion of Figure 6.3 represents the agent-level dynamics with a velocity reference signal $v_{i,\mathrm{ref}}$. The output of the model is the position of the agent x_i. Relative sensors onboard each UAV are able to measure the relative position $x_j - x_i$ of those agents j in the neighborhood of i. These relative positions are fused through a gain matrix K_i within the network-level dynamics, providing a new reference signal $v_{i,\mathrm{ref}}$. The objective of this chapter is to find a method to effectively achieve such a network-level mission task via the judicious selection of the network-level gain matrix K_i.

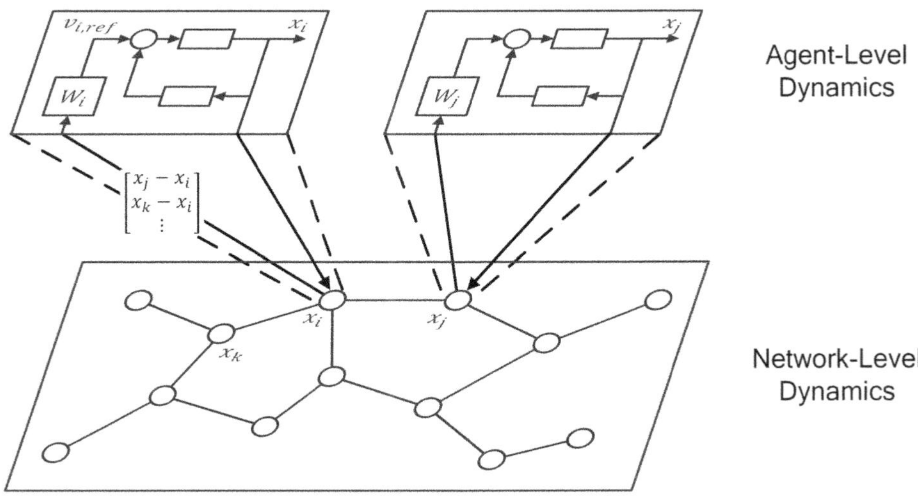

Figure 6.3 Block diagram of the two-layer dynamics of the networked multi-agent system. The agent-level dynamics running on all agents is indicated in the upper portion of the figure and constitutes the inner-loop dynamics, while the network-level dynamics, driven by relative state measurements between agents, forms the outer loop.

The following section describes the network-level single and double integrator consensus dynamics. These are augmented with noise terms that represent tracking errors within agent-level dynamics. Misidentified nodes are added to both of these models to form leader–follower versions of the network-level dynamics.

6.3.1 Noisy Consensus Dynamics

Multi-agent systems are often concerned with achieving a common goal or agreement on a set of states. The rendezvous problem whereby all agents must meet at a common location is an example of such an objective. Formation control is an extension of this whereby offset or virtual positions are introduced from this common location. Other desirable agreements are on velocity and bearing for applications like formation maintenance and pointing, respectively. While the centralized agreement problem is straightforward, when communication over the networked system is constrained to a communication graph, a distributed solution is needed. The consensus protocol is a well-studied distributed agreement protocol able to solve this problem. The single and double integrator versions are presented in the following subsections, accompanied by their noise-contaminated versions.

Noisy Single Integrator Consensus Dynamics

In this subsection, our objective is asymptotic agreement for a scalar state $x_i(t)$ for each node i at time t in the network. The network-level consensus control law defined over the graph $\mathcal{G} = (V, E, W)$ is

$$v_{i,\mathrm{ref}} = \sum_{\{v_i,v_j\}\in E} w_{ij}\left(x_j(t) - x_i(t)\right) \tag{6.1}$$

Specifically, the weights $w_{ij} \in W$, for $j \in N(i)$ denoted in Figure 6.3, compose the row vector K_i considered as a tunable gain feedback matrix. An attraction of the control law is that a node can formulate its own $v_{i,\mathrm{ref}}$ using only relative state measurements with respect to its neighbors.

To form the full network-level dynamics it is assumed, due to inaccuracies in modeling, controller implementation, and environmental perturbations at each node's agent-level dynamics, that $\dot{x}_i(t) - v_{i,\mathrm{ref}} = \omega_i(t)$ is a zero-mean, white-noise process for all agents. The dynamics can then be represented as a noisy continuous-time consensus [24],

$$\dot{x}_i(t) = \sum_{\{v_i,v_j\}\in E} w_{ij}\left(x_j(t) - x_i(t)\right) + \omega_i(t) \tag{6.2}$$

For a connected graph, the network dynamics, absent of noise $\omega_i(t)$, will converge to an agreement on the state, i.e., $x_1(t) = x_2(t) = \ldots = x_n(t) = \alpha$, for some α, for all initial conditions [2]. For an undirected connected graph $\mathcal{G}$, assumed here with $w_{ij} = w_{ji}$, the states will converge to the average of the initial positions of the agents, providing an unbiased rendezvous location.

The collective dynamics over the collective state $x(t) \in \mathbb{R}^n$, noise $\omega(t) \in \mathbb{R}^n$, and outputs $y(t) = x(t)$ can be compactly represented as

$$\dot{x}(t) = -L(\mathcal{G})x(t) + G\omega(t)$$
$$y(t) = Cx(t)$$

with $G = I$, $C = I$, and $L(\mathcal{G})$ being the graph Laplacian matrix of the underlying interaction topology described in Section 6.2.

Noisy Double Integrator Consensus Dynamics

A close relative of single integrator consensus is double integrator consensus, where the interactions drive the second derivative of the state x. The network-layer dynamics takes relative positions and velocity measurements and provides acceleration as a reference signal $a_{i,\text{ref}}$. The subsequent noisy double integrator consensus is defined as

$$\ddot{x}_i(t) = \sum_{\{v_i,v_j\}\in E} w_{ij}\left[\left(x_j(t) - x_i(t)\right) + \left(\dot{x}_j(t) - \dot{x}_i(t)\right)\right] + \omega_i(t) \tag{6.3}$$

where the weights w_{ij} and noise $\omega_i(t)$ are similarly defined as for the single integrator case (6.2). Here, the relative position and relative velocity of node i's neighbors are required to perform an update to the state of node i. The collective dynamics over the network state $x(t) \in \mathbb{R}^n$, noise $\omega(t) \in \mathbb{R}^n$, and outputs $y(t) = x(t)$ is represented as

$$\begin{bmatrix} \dot{x}(t) \\ \ddot{x}(t) \end{bmatrix} = \begin{bmatrix} 0 & I \\ -L(\mathcal{G}) & -L(\mathcal{G}) \end{bmatrix} \begin{bmatrix} x(t) \\ \dot{x}(t) \end{bmatrix} + G_d\,\omega(t)$$
$$y(t) = C_d \begin{bmatrix} x(t) \\ \dot{x}(t) \end{bmatrix}$$

with $G_d = \begin{bmatrix} 0 \\ I \end{bmatrix}$, $C_d = \begin{bmatrix} I & 0 \end{bmatrix}$, and $L(\mathcal{G})$ denoting the graph Laplacian of the underlying interaction topology described in Section 6.2. For a connected graph $\mathcal{G}$, the network dynamics, in the absence of noise $\omega(t)$, will converge to an agreement on the nodes' state and agreement on the nodes' state derivative. The dynamics reach agreement on position, but this position may be moving in space, i.e., $x_1(t) = x_2(t) = \ldots = x_n(t) = \beta t + \alpha$ and $\dot{x}_1(t) = \dot{x}_2(t) = \ldots = \dot{x}_n(t) = \beta$, for some α and β, and for all initial conditions [2].

6.3.2 Leader–Follower Consensus Dynamics

The ease of scalability is one of the attractive features of the consensus dynamics. A new individual agent dynamics augments the networked dynamics, appearing as a new node in the underlying graph, with those agents in communication range corresponding to new edges. This versatility comes at a price, as misidentification of agents introduces unwanted signals into the networked dynamics. The multi-agent system will subsequently "follow" this unwanted signal. The well-studied leader–follower consensus dynamics typically describes a favorable external signal from leader agents [3].

It is equally applicable for this unwanted signal from misidentified external agents to play the role of the leader.

The leader–follower consensus dynamics is formed by extending the underlying network graph $\mathcal{G}=(V,E,W)$ to include external signals into the networked dynamics. The external signals input location is encoded in the external agent triple $\mathcal{R}=(R,\mathcal{E}_R,W_R)$, where R is the r element foreign node set and $\mathcal{E}_R \subseteq V \times R$ is the set of false edges attached to the network and $W_R \in \mathbb{R}^r$ encodes the weights on the false edges.

Leader–Follower Single Integrator

The resulting leader–follower consensus system now assumes the form

$$\dot{x}_i(t) = \sum_{\{v_i,v_j\}\in E} w_{ij}\left(x_j(t) - x_i(t)\right) + \sum_{\{v_i,r_j\}\in \mathcal{E}_R} w^r_j\left(u_j(t) - x_i(t)\right) + \omega_i(t) \tag{6.4}$$

with the full dynamics

$$\dot{x}(t) = A(\mathcal{G},\mathcal{R})x(t) + B(\mathcal{R})u(t) + G\omega(t) \tag{6.5}$$

$$y(t) = Cx(t) \tag{6.6}$$

where $B(\mathcal{R}) \in \mathbb{R}^{n\times r}$ with $[B(\mathcal{R})]_{ij} = w^r_j$ when $\{v_i,r_j\} \in \mathcal{E}_R$ and $[B(\mathcal{R})]_{ij} = 0$ otherwise, and

$$A(\mathcal{G},\mathcal{R}) = -\left(L(\mathcal{G}) + \sum_{\{v_i,r_j\}\in\mathcal{E}_R} w^r_j e_i e_i^T\right) \in \mathbb{R}^{n\times n} \tag{6.7}$$

The matrix $A(\mathcal{G},\mathcal{R})$ in (6.7) is called the *Dirichlet matrix,* or grounded Laplacian [25, 26]. In the following discussion, only connected graphs are considered and hence $A(\mathcal{G},\mathcal{R})$ is always negative definite [27]. An example of the dynamics can be seen in Figure 6.4.

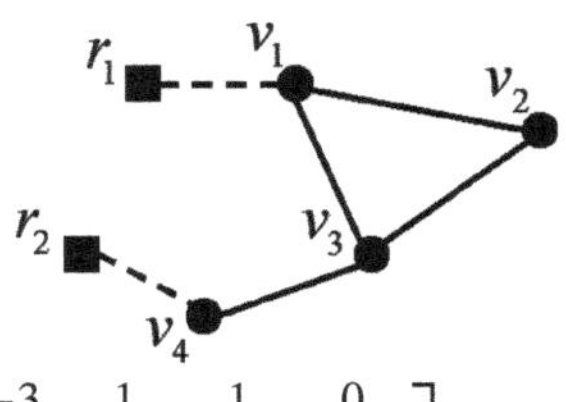

$$A(\mathcal{G},\mathcal{R}) = \begin{bmatrix} -3 & 1 & 1 & 0 \\ 1 & -2 & 1 & 0 \\ 1 & 1 & -3 & 1 \\ 0 & 0 & 1 & -2 \end{bmatrix}, \quad B(\mathcal{R}) = \begin{bmatrix} 1 & 0 \\ 0 & 0 \\ 0 & 0 \\ 0 & 1 \end{bmatrix}$$

Figure 6.4 Network graph with external (control) agents r_1 and r_2 attached to agents v_1 and v_4 respectively, leading to an altered Laplacian $A(\mathcal{G},\mathcal{R})$ and input matrix $B(\mathcal{R})$ of model (6.5). Here, all edges have weight 1.

Leader–Follower Double Integrator

A similar discussion to the leader–follower single integrator dynamics follows for the leader–follower double integrator dynamics with an unwanted control $u_j(t)$ included in the dynamics,

$$\begin{aligned}\ddot{x}_i(t) = &\sum_{\{v_i,v_j\}\in E} w_{ij}\left[\left(x_j(t)-x_i(t)\right)+\left(\dot{x}_j(t)-\dot{x}_i(t)\right)\right]\\ &+\sum_{\{v_i,r_j\}\in\mathcal{E}_R} w^r_j\left[\left(u_j(t)-x_i(t)\right)+\left(\dot{u}_j(t)-\dot{x}_i(t)\right)\right]+\omega_i(t)\end{aligned}$$

If we let $\bar{u}_j(t) = u_j(t) + \dot{u}_j(t)$, then the full dynamics is

$$\begin{aligned}\begin{bmatrix}\dot{x}(t)\\ \ddot{x}(t)\end{bmatrix} &= A_d(\mathcal{G},\mathcal{R})\begin{bmatrix}x(t)\\ \dot{x}(t)\end{bmatrix} + B_d(\mathcal{R})\bar{u}(t) + G_d\omega(t)\\ y(t) &= C_d\begin{bmatrix}x(t)\\ \dot{x}(t)\end{bmatrix}\end{aligned} \tag{6.8}$$

where the state and input matrices are

$$A_d(\mathcal{G},\mathcal{R}) = \begin{bmatrix}0 & I\\ A(\mathcal{G},\mathcal{R}) & A(\mathcal{G},\mathcal{R})\end{bmatrix} \text{ and } B_d(\mathcal{R}) = \begin{bmatrix}0\\ B(\mathcal{R})\end{bmatrix}$$

For a connected graph $\mathcal{G}$, $A_d(\mathcal{G},\mathcal{R})$ is always a Hurwitz matrix, i.e., all its eigenvalues have strictly negative real parts.

6.4 System Performance and Topology Measures

To characterize the performance of the system with respect to the undesired inputs and noise, we introduce two network measures; the open loop $\mathcal{H}_2$ norm and the network entropy. A family of topology-induced performance measures are also explored.

6.4.1 Open Loop $\mathcal{H}_2$ Norm

The transfer function $G(s) = C\,(sI - A)^{-1}\,B$ describes the control input to output mapping, in the Laplace domain, for the dynamics

$$\begin{aligned}\dot{x}(t) &= Ax(t) + Bu(t)\\ y(t) &= Cx(t)\end{aligned}$$

A measure of the amplification of the mapping is $\|G(s)\|_2$, referred to as the open loop $\mathcal{H}_2$ norm of the system. If the dynamics are initialized at the origin, $\|G(s)\|_2$ is the energy of the signal $y(t)$ when the input $u(t)$ is a unit impulse. If $u(t)$ is an undesirable signal, then decreasing $\|G(s)\|_2$ has the effect of dampening the impact of the disturbances on the system output. In other words, systems with a larger open loop $\mathcal{H}_2$ norm are more easily perturbed by an input $u(t)$. With this in mind, the focus of this work is to minimize $\|G(s)\|_2^2$ of the network.

The square of the open loop $\mathcal{H}_2$ norm can also be represented as

$$\|G(s)\|_2^2 = \mathbf{tr}\left(B^T P B\right) \tag{6.9}$$

where P is the observability Gramian, a positive definite matrix satisfying the Lyapunov equation

$$A^T P + PA = -C^T C \tag{6.10}$$

when A is Hurwitz.

Single Integrator Dynamics

For the single integrator dynamics (6.5), since the matrix A is invertible and symmetric and $C = I$, it is straightforward to check that $P = -\frac{1}{2}A(\mathcal{G},\mathcal{R})^{-1}$ satisfies (6.10) and is positive definite. Hence,

$$\|G(s)\|_2^2 = \mathbf{tr}\left(B^T P B\right) = -\frac{1}{2}\mathbf{tr}\left(B(\mathcal{R})^T A(\mathcal{G},\mathcal{R})^{-1} B(\mathcal{R})\right) \tag{6.11}$$

The $\mathcal{H}_2$ norm of the system has attractive graph theoretic properties linked through the concept of the effective resistance of the graph [7]. A consequence of this connection is that the weights on the edges can be considered as conductance in an equivalent electrical network realization of the graph [25]. Since the effective resistance always decreases with increasing conductance, $\|G(s)\|_2^2$ is non-decreasing with any increase in edge weights of $\mathcal{G}$. This useful property means that edge weights can be used as a tuning parameter to optimize $\|G(s)\|_2^2$. Another term for this measure is the variance resilience of the network [7]. With regard to edge weights in $\mathcal{R}$, from (6.7) and (6.11) the squared $\mathcal{H}_2$ norm can be written as

$$\|G(s)\|_2^2 = \frac{1}{2}\sum_{\{v_i,r_j\}\in\mathcal{E}_R} (w_j^r)^2 \left[\left(L(\mathcal{G}) + \sum_{\{v_i,r_j\}\in\mathcal{E}_R} w_j^r e_i e_i^T\right)^{-1}\right]_{ii}$$

Subsequently, decreases in edge weights $w_j^r \in W_R$ will decrease $\|G(s)\|_2^2$. We assume that w_j^r will be tuned to a designated minimum value. For convenience we shall select $w_j^r = 1$, and henceforth

$$A(\mathcal{G},\mathcal{R}) = -\left(L(\mathcal{G}) + \sum_{i\in R} e_i e_i^T\right)$$

Double Integrator Dynamics

The $\mathcal{H}_2$ norm associated with the double integrator dynamics (6.8) can be found similarly to the single integrator case. Applying the observability Gramian, derived in Proposition 6.1 of the Appendix, to the identity (6.9), one has

$$\begin{aligned}
\|G_d(s)\|_2^2 &= \mathbf{tr}\left(B_d(\mathcal{R})^T P(\mathcal{G},\mathcal{R}) B_d(\mathcal{R})\right) \\
&= \frac{1}{2}\mathbf{tr}\left(\begin{bmatrix} 0 & B(\mathcal{R})^T \end{bmatrix}\begin{bmatrix} I - A(\mathcal{G},\mathcal{R})^{-1} & -A(\mathcal{G},\mathcal{R})^{-1} \\ -A(\mathcal{G},\mathcal{R})^{-1} & A(\mathcal{G},\mathcal{R})^{-2} \end{bmatrix}\begin{bmatrix} 0 \\ B(\mathcal{R}) \end{bmatrix}\right) \\
&= \frac{1}{2}\mathbf{tr}\left(B(\mathcal{R})^T A(\mathcal{G},\mathcal{R})^{-2} B(\mathcal{R})\right)
\end{aligned}$$

This measure has been examined under the name of *coherence* [28, 29]. It shares many of the characteristics of single integrator dynamics $\mathcal{H}_2$ norm. One such relationship is that $\|G_d(s)\|_2^2$ is non-decreasing with increasing edge weights of $\mathcal{G}$ and non-decreasing edge weights in $\mathcal{R}$.

6.4.2 Network Entropy

Entropy is a measure of disorder or randomness of a system. Most often used in thermodynamics, this measure can be applied to system dynamics by considering the spectrum of the state matrix. Applying the Wang–Uhlenbeck entropy to a stable state matrix A defines the entropy as

$$H = -\log\det(-A) \tag{6.12}$$

The entropy relates to the system dynamics driven by the state matrix A. This connection is through the determinant of the controllability Gramian discussed in the following for the single and double integrator dynamics.

Single Integrator Dynamics

For the single integrator dynamics (6.5), (6.12) becomes

$$H = -\log\det(-A(\mathcal{G}, \mathcal{R})) \tag{6.13}$$

An alternative interpretation is to consider state dynamics (6.5) driven by white noise $w(t)$. The controllability Gramian $\bar{P}$ is the dual of the observability Gramian and describes the mapping from inputs to internal states. The controllability Gramian that describes the mapping from $\omega(t)$ to $x(t)$ is the positive definite $\bar{P}$ satisfying

$$A\bar{P} + \bar{P}A^T = -GG^T$$

For negative definite $A = A(\mathcal{G}, \mathcal{R})$ and $G = I$, $\bar{P} = -\frac{1}{2}A(\mathcal{G}, \mathcal{R})^{-1}$. The disruptiveness of the noise $w(t)$ can be represented as the volume of the parallelepiped whose edges are the column vectors of $\bar{P}$, or more succinctly $c(n)\sqrt[n]{\det \bar{P}}$ for a positive constant $c(n)$. Since

$$\det \bar{P} = \frac{1}{\det(-2A(\mathcal{G}, \mathcal{R}))} = \frac{1}{2^n \det(-A(\mathcal{G}, \mathcal{R}))} = -2^{-n}\exp(-H)$$

then minimizing $\det \bar{P}$ is equivalent to minimizing H.

Let the negative eigenvalues of $A(\mathcal{G}, \mathcal{R})$ be ordered as $0 > \lambda_1 \geq \lambda_2 \geq \ldots \geq \lambda_n$. The entropy can then be rewritten as $H(\mathcal{G}) = -\log\prod(-\lambda_i) = -\sum\log(-\lambda_i)$. As the eigenvalues of $A(\mathcal{G}, \mathcal{R})$ are non-increasing with increasing edge weights, the network entropy can be minimized by increasing the edge weights of $\mathcal{G}$.

Double Integrator Dynamics

For the double integrator dynamics (6.5), using the property that $\det \begin{bmatrix} A & B \\ C & D \end{bmatrix} = \det(AD - BC)$ when C and D commute, it follows that

$$\begin{aligned} H_d &= -\log\det \begin{bmatrix} \mathbf{0} & -I \\ -A(\mathcal{G},\mathcal{R}) & -A(\mathcal{G},\mathcal{R}) \end{bmatrix} \\ &= -\log\det(-A(\mathcal{G},\mathcal{R})) = H \end{aligned}$$

Similar to the single integrator dynamics, consider the controllability Gramian $\bar{P}_d$ associated with $\omega(t)$ for model (6.8). The positive definite $\bar{P}_d$ that satisfies the Lyapunov equation

$$A_d\bar{P}_d + \bar{P}_d A_d^T = -G_d G_d^T$$

is $\bar{P}_d = \frac{1}{2}\begin{bmatrix} I & 0 \\ 0 & -A(\mathcal{G},\mathcal{R})^{-1} \end{bmatrix}$. As the volume of the parallelepiped associated with $\bar{P}$ is proportional to

$$\det \bar{P}_d = \frac{1}{2^{2n}}\det(-A(\mathcal{G},\mathcal{R})^{-1}) = -2^{-2n}\exp(-H_d),$$

minimizing $\det \bar{P}_d$ is equivalent to minimizing H_d, and in turn H (6.13).

6.4.3 Topology Measures

Minimizing the proposed measures with respect to nonnegative edge weights in Section 6.4 to improve system performance requires increasing the edge weights. This is often unfavorable in consensus dynamics as the weights represent gains within the dynamics. In physical systems, overly high gains may be limited by bounds on the control authority, power requirements, bandwidth, or safety requirements.

Linear bounds on each weight w_{ij} can be introduced to enforce hard constraints. Another option is to introduce penalty measures to ensure that the weights are maintained within preferable regions. The following are a list of commonly used penalty measures which we wish to minimize [30]. Over all weights w_{ij} the measures are of the form

$$J_k(W) = \sum_{\{v_i,v_j\}\in E} \phi_k(w_{ij})$$

where $\phi_k(\cdot)$ is a convex function corresponding to a penalty of type k. As the weights must be nonnegative, each of the following penalty measures are defined for $\phi_k(\cdot) : \mathbb{R}_{\geq 0} \to \mathbb{R}_{\geq 0}$. Figure 6.5 depicts each of these penalty functions.

Linear

The linear penalty function $\phi_L(x) = x$ penalizes values uniformly proportional to their distance from the origin.

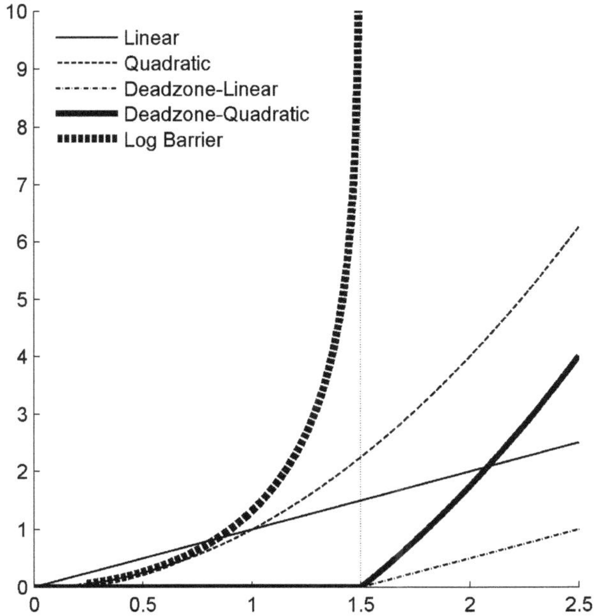

Figure 6.5 Plots of the five penalty functions. Where applicable, $a = 1.5$.

Quadratic

The quadratic function $\phi_Q(x) = x^2$ penalizes values far from the origin more heavily than values that are close, when compared with the linear weighting.

Deadzone-Linear and Deadzone-Quadratic

The deadzone-linear penalty function with deadzone width a is

$$\phi_{DL}(x) = \begin{cases} 0 & x \leq a \\ \pi(x) - \pi(a) & x > a \end{cases}$$

where $\pi(x) = \phi_L(x)$. The function places no penalty on values less than a and responds similar to $\phi_L(x)$ greater than a. The deadzone-quadratic function $\phi_{DQ}(x)$ follows similarly with $\pi(x) = \phi_Q(x)$.

Log Barrier

The log barrier penalty function

$$\phi_{LB}(x) = \begin{cases} -a^2 \log\left(1 - \left(\frac{x}{a}\right)^2\right) & x \leq a \\ \infty & x > a \end{cases}$$

places an infinite penalty for values farther from the origin than a and performs similarly to the quadratic penalty function otherwise.

6.5 Distributed Online Topology Design

Relative measurement state forms the basis of consensus-type and leader–follower dynamics with access to relative states dictated by constraints such as geographic proximity encoded as edges in the graph. On the other hand, the gains or weights placed on these measurements are not physically constrained, providing a method to dynamically adapt the coordination algorithm. Within the models in Section 6.3 the decision variables in question are the selection of a time-varying gain matrix $K(t)$, illustrated in Figure 6.3, or equivalently a time-varying weight $w_{ij}(t)$, for example in (6.1) for single integrator dynamics. The judicious choice of weights is motivated by improving the disturbance rejection properties of the networked dynamics. The measures to encode the performance in the network measures were discussed in Section 6.4.

To follow the distributed framework of the dynamics, it is desired to select the dynamics weights using only local knowledge of the network. This is particularly challenging as the location of external disturbances may not be in the local neighborhood of an agent. In fact, by the time information pertaining to a disturbance's location has propagated through the network, the disturbance may no longer be present. An online regret framework is well suited to formulate this dynamically changing problem. The dynamic change of the graph and the external input location appear as a time-varying graph $\mathcal{G}_t$ and foreign agent pair $\mathcal{R}_t$, respectively. This state model (6.5) is also adapted to have time-varying matrices as

$$\begin{aligned} \dot{x}(t) &= A_t x(t) + B_t u(t) + G\omega(t) \\ y(t) &= Cx(t) \end{aligned}$$

where A_t and B_t correspond to $A(\mathcal{G}_t, \mathcal{R}_t)$ and $B(\mathcal{R}_t)$, respectively. Similar time-varying dynamics can also be formed from the state model (6.8). The time-varying state matrix $A(\mathcal{G}_t, \mathcal{R}_t)$ represented in terms of the weights is

$$A(\mathcal{G}_t, \mathcal{R}_t) = -\sum_{\{v_i, v_j\} \in E} w_{ij} a_{ij} a_{ij}^T - \sum_{i \in R} e_i e_i^T$$

which is invertible for all positive weights so long as R is nonempty and $\mathcal{G} = (V, E, I)$ is connected.

The objective function $f_t(W)$ is selected to be relevant to the model the agents are running, i.e., single or double integrator, and the system performance and topology measures of interest based on the application, disturbance inputs, and noise characteristics.

The following is one such example with respect to the objective $f_t(W)$.

EXAMPLE 6.1 Consider the squared $\mathcal{H}_2$ norm of the single integrator leader–follower dynamics $\|G(s)\|_2^2 = -\frac{1}{2}\mathbf{tr}\left(B_t^T A_t^{-1} B_t\right)$. Any increase in the weights of the network will result in a decrease in the $\mathcal{H}_2$ norm. Large edge weights affect the responsiveness of the network and can cause large transient variations in dynamics. The following cost function's minimization promotes good disturbance rejection while penalizing large weights with a quadratic penalty function $J_Q(W)$,

$$f_t(W) = \|G(s)\|_2^2 + \frac{h}{2} J_Q(W) = -\frac{1}{2}\mathbf{tr}\left(B_t^T A_t^{-1} B_t\right) + \frac{h}{2}\mathbf{1}^T W^T W \mathbf{1},$$

where h is a positive constant.

The introduction of the quadratic weight penalty can result in near-zero weights in the system, practically rendering the graph disconnected. A hard constraint can avoid this behavior by requiring the weights to stay above some positive vector $q_{\sigma(ij)}$, i.e., $w_{ij} \geq q_{\sigma(ij)}$. Additional constraints relating to maximum edge deviations can also be considered. These constraints are specified via the vector $s_{\sigma(ij)} \in \mathbb{R}^m$, such that $w_{ij} \leq s_{\sigma(ij)}$. The optimization problem then becomes

$$\begin{aligned} &\min f_t(W) \\ &\text{s.t.} \quad q_{\sigma(ij)} \leq w_{ij} \leq s_{\sigma(ij)} \end{aligned} \tag{6.14}$$

The constraints form a linear constraint set denoted by $\mathcal{P}$.

The topology-induced measures discussed in Section 6.4.3 are well known Lipschitz continuous convex functions with easily computable gradients. The analysis of the derivative and Hessian of the measures in Section 6.4 with respect to the edge weights is relegated to the Appendix, proving their convexity via a positive semidefinite Hessian. We note that the sum of convex functions is convex, $f_t(W)$ is convex, and strictly convex when any of its terms are strictly convex.

EXAMPLE 6.2 Reexamining Example 6.1, $\|G(s)\|_2^2$ is a convex function and $J_Q(W)$ is strictly convex, therefore $f_t(W)$ is strictly convex with the derivative

$$\frac{\partial f_t(W)}{\partial w_{ij}} = -\frac{1}{2}\sum_{p \in R}\left(\left[A(\mathcal{G},\mathcal{R})^{-1}e_p\right]_i - \left[A(\mathcal{G},\mathcal{R})^{-1}e_p\right]_j\right)^2 + h w_{ij}$$

and the Hessian $\nabla^2 f_t(W) \succeq hI$.

We proceed to provide a coordinated version of the conjugate gradient method and truncated Neumann series to form $\nabla \|G(s)\|_2^2$, $\nabla \|G_d(s)\|_2^2$, ∇H, and ∇H_d distributively. The gradient is then applied to a distributed formulation of online gradient descent leading to a $\mathcal{O}(\sqrt{T})$ regret for problem (6.14) when $f_t(W)$ is convex, and a $\mathcal{O}(\log T)$ regret bound when $f_t(W)$ is strongly convex.

6.5.1 Distributed Gradient Updates

Updates in the gradient descent algorithm require the gradient of $f_t(W)$ with respect to edge weights w_{ij}. For distributed implementation, these gradients must be evaluated on the agent level. Referring to the gradient results in Propositions 6.2–6.4 in the

Appendix for $y^s := A(\mathcal{G},\mathcal{R})^{-1}e_s$ and $z^s := A(\mathcal{G},\mathcal{R})^{-2}e_s$ yields

$$\frac{\partial \|G(s)\|_2^2}{\partial w_{ij}} = -\frac{1}{2}\sum_{s\in R}\left(y_i^s - y_j^s\right)^2$$

$$\frac{\partial \|G_d(s)\|_2^2}{\partial w_{ij}} = -\sum_{s\in R}\left(z_i^s - z_j^s\right)$$

$$\left[\sum_{k\in N(i)} w_{ik}\left(z_i^s - z_k^s\right) + \sum_{k\in N(j)} w_{kj}\left(z_k^s - z_j^s\right) + \gamma_{ij}\left(z_i^s - z_j^s\right)\right]$$

where γ_{ij} is defined in Proposition 6.4, and

$$\frac{\partial H}{\partial w_{ij}} = \frac{\partial H_d}{\partial w_{ij}} = -\left(y_i^i - y_j^i\right) + \left(y_i^j - y_j^j\right)$$

Consequently, when using the measures $\|G(s)\|_2^2$ and $\|G_d(s)\|_2^2$, if every agent i has access to y_i^s and z_i^s, respectively, for all $s \in R$, the neighboring agents i and j can cooperatively calculate the gradient by sharing these terms with their neighbors. Similarly, for the entropy measures H and H_d, neighboring agents i and j can cooperatively compute the gradient with respect to w_{ij} if agent i has access to y_i^i and y_i^j, and agent j has access to y_j^i and y_j^j.

Exact Method: Coordinated Conjugate Gradient

To calculate the gradient, a distributed calculation of y_i^s and z_i^s is required for each agent i. We introduced a coordinated form of the conjugate gradient method to acquire these variables distributively for all $s \in R$. The conjugate gradient method, in its traditional centralized form, solves $A^{-1}b$ given $A \in \mathbb{R}^{n\times n}$ and $b \in \mathbb{R}^n$. For $A = A(\mathcal{G},\mathcal{R})$ and $b = e_s$, the method can be applied to find y^s. An attraction of the centralized algorithm is that many updates only require the vector b and the evaluation of Aq for a given $q \in \mathbb{R}^n$. For the case where A is encoded in the graph structure, this is ideal since each agent i only has knowledge of q_i and so $\left[Aq\right]_i$ can be calculated by agent i by communicating with its neighbors $N(i)$. The update's distributed form is

$$\left[A(\mathcal{G},\mathcal{R})q\right]_i = \sum_{j\in N(i)\bigcup\{i\}} a_{ij}q_j \tag{6.15}$$

Similarly, for the calculation of z^s, the evaluation of A^2q is required, which can be formed by querying its neighbors and neighbors' neighbors, as

$$\left[A(\mathcal{G},\mathcal{R})^2q\right]_i = \sum_{j\in N(i)\bigcup\{i\}} a_{ij}\left[A(\mathcal{G},\mathcal{R})q\right]_j = \sum_{j\in N(i)\bigcup\{i\}} a_{ij} \sum_{k\in N(j)\bigcup\{j\}} a_{jk}q_k \tag{6.16}$$

Algorithm 6.1 with time steps indexed by k describes the coordinated conjugate gradient algorithm with the output of the algorithm the calculation of y^s.[1] The esti-

[1] For the calculation of z^s, the summations in steps 4 and 9 can be replaced with (6.16).

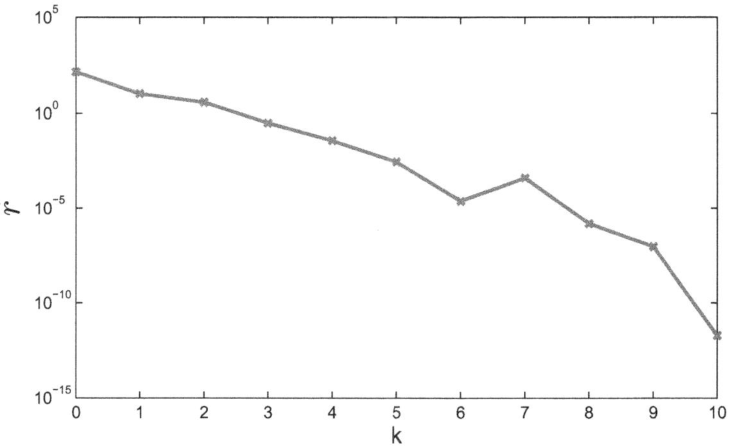

Figure 6.6 Average residue $\tilde{r}$ per time step k of the coordinated conjugate gradient method on a 20-node graph.

mate y, residues r, and conjugate p are the main components of the algorithm updates in lines 11, 12, and 15, respectively. The coordinated conjugate gradient method deviates from the centralized in the calculation of the two agreement variables $\tilde{p}$ and $\tilde{r}$ required per time step, shown in lines 9 and 13, respectively. Calculating these involves performing an accumulative sum on the values $p_i[k]r_i[k]$ and $r_i[k+1]$, respectively, over all the agents, and so can be calculated by accumulating values up a directed spanning tree rooted at agent p, then back down again.[2]Algorithm 6.1 terminates when the average residue $\tilde{r}$ falls below a small threshold value ε, which for our application is chosen as $\varepsilon = 1e - 6$. Convergence is typically fast, with theoretical guarantees to converge in fewer than n steps.

A factor that contributes to the rate of convergence of the algorithm is the spectrum of A, which in turn is closely tied to the eigenvalues of the graph Laplacian. Research on the conjugate gradient method has shown that clustering of eigenvalues improves the convergence rate [31]. Networks that exhibit this trait for the graph Laplacian are regular graphs and, in general, graphs with many symmetries [32]. In practice, the n step guarantee is not always met due to rounding errors. To mitigate this case, we found a restart criteria when $\tilde{r}$ increased significantly to be effective and easy to coordinate through the network. For further reading regarding the features of the conjugate gradient method, we direct the reader to the book by Nocedal and Wright [31].

A residue trajectory for the distributed conjugate gradient method is shown in Figure 6.6. The method was applied to a 20-node graph with four foreign agents. The algorithm converged in 10 time steps under a stopping variable of $\varepsilon = 1e - 10$. This is half the theoretical limit of $n = 20$ iterations. During the run the restart condition can be seen to be triggered at iteration seven as an increase in $\tilde{r}$ was detected.

[2] These steps can also be calculated approximately and distributively using a traditional information-based consensus model (6.2).

Algorithm 6.1: Distributed conjugate gradient method

Given $y_i[0]$ *for each node* $i \in V$
Initialize:
$k = 0$
$r_i[0] = \sum_{j \in N(i) \bigcup i} a_{ij} x_j[0] - b_i$
$p_i[0] = -r_i[0]$
$\widetilde{r}_i[0] \leftarrow \sum_{i=1}^{n} r_i[0]^2$
while $|\widetilde{r}_i[k]| > \varepsilon$ **do**
  $b_i[k] = \sum_{j \in N(i) \bigcup \{i\}} a_{ij} p_j[k]$
  $\widetilde{p}_i[k] \leftarrow \sum_{i=1}^{n} p_i[k] b_i[k]$
  $\alpha_i = \frac{\widetilde{r}_i[k]}{\widetilde{p}_i[k]}$
  $y_i[k+1] = y_i[k] + \alpha_i p_i[k]$
  $r_i[k+1] = r_i[k] + \alpha_i b_i[k]$
  $\widetilde{r}_i[k+1] \leftarrow \sum_{i=1}^{n} r_i[k+1]^2$
  $\beta_i = \frac{\widetilde{r}_i[k+1]}{\widetilde{r}_i[k]}$
  $p_i[k+1] = \beta_i p_i[k] - r_i[k+1]$
  $k = k + 1$
end

Inexact Method: Truncated Neumann Series

As mentioned, the conjugate gradient method requires a coordination step to accumulate states. To calculate this value exactly requires the accumulation of states across the network, which will take communication time of the order of the diameter of the graph. An alternative is to truncate the infinite series progression of the $A^{-1}b$ using a much smaller number of communication steps producing an inexact calculation.

For a non-singular matrix $\widetilde{A}$ with maximum singular value less than 1, i.e., $||\widetilde{A}|| < 1$, its inverse can be expressed by a Neumann series

$$\widetilde{A}^{-1} = \sum_{m=0}^{\infty} (I - \widetilde{A})^m.$$

As the matrix A does not necessarily have all its eigenvalues within the unit circle, it must be scaled before applying the series representation. The series can be applied to a scaled version of A instead, namely $\widetilde{A} = \frac{1}{\alpha} A$ for $\|A\| < \alpha$.

The corresponding series derivation can now be applied to the calculation of $A^{-1}b$ as

$$A^{-1}b = \frac{1}{\alpha} \widetilde{A}^{-1} b = \frac{1}{\alpha} \sum_{m=0}^{\infty} \left(I - \frac{1}{\alpha} A \right)^m b$$

By taking the first l terms of the series, a truncated approximation T_l of $A^{-1}b$ can be formed as

$$T_l = \frac{1}{\alpha}\sum_{m=0}^{l-1}\left(I - \frac{1}{\alpha}\right)^m b$$

Bounding the approximation error E_l of the truncated approximation T_l norm-wise using the smallest singular value σ_1 of A, we have

$$\begin{aligned}
E_l &\leq \left\| A^{-1}b - T_l \right\| \\
&\leq \left\| \frac{1}{\alpha}\sum_{m=l}^{\infty}\left(I - \frac{1}{\alpha}A\right)^m \right\| \|b\| \\
&\leq \frac{1}{\alpha}\sum_{m=l}^{\infty}\left(1 - \frac{1}{\alpha}\sigma_1\right)^m \|b\| \\
&\leq \frac{1}{\alpha}\frac{\left(1 - \frac{1}{\alpha}\sigma_1\right)^l}{1 - \left(1 - \frac{1}{\alpha}\sigma_1\right)} \|b\| \\
&\leq \frac{1}{\sigma_1}\left(1 - \frac{1}{\alpha}\sigma_1\right)^l \|b\|
\end{aligned}$$

For $A = A(\mathcal{G}, \mathcal{R})$ and $b = e_s$, the truncation method can be used to find an approximation of y^s. Similar to the conjugate gradient method, updates can be performed distributively using relation (6.15) with access only to the local information with the exception of a globally pre-assigned α. Unlike the conjugate gradient method, only those nodes within a path length distance of $l - 1$ from node s are involved in the calculation of y^s. This follows from the fact that $[T_l]_i = 0$ for node i, a path length distance of l or greater from node s. Further, the calculation can be performed with each node querying its neighbors at most l times instead of proportional to the diameter of the graph.

Algorithm 6.2: Distributed truncated Neumann calculation

For each node $i \in V(\mathcal{G})$
Idle:
while $\ell_i = 0$ **do**
  $\ell_i = \max_{j\in\mathcal{N}(i)} \ell_j$
end
Initialization:
$p_i[\ell_i] = 0, y_i = 0$
while $\ell_i > 0$ **do**
  $p_i[\ell_i - 1] = p_i[\ell_i] - \frac{1}{\alpha}\sum_{j\in\mathcal{N}(i)\bigcup\{i\}} a_{ij} p_j[\ell_i - 1]$
  $y_i[\ell_i - 1] = y_i[\ell_i] + \frac{p_i[\ell_i - 1]}{\alpha}$
  $\ell_i = \ell_i - 1$
end

The distributed truncated Neumann series is presented in Algorithm 6.2, designed to calculate y^s with time steps indicated by k.[3]

As $A(\mathcal{G},\mathcal{R})$ is symmetric, its smallest and largest magnitude eigenvalues are σ_1 and σ_n, respectively. From Proposition 6.5 in the Appendix, a suitable scaling for $A = A(\mathcal{G},\mathcal{R})$ is $\alpha = 2w_{\max}(\max_i d_i + 1)$ and the l-approximation error E_l bound is

$$E_l \leq \frac{n(1+|R|\,D)}{w_{\min}\,|R|}\left(1 - \frac{1}{2w_{\max}(\max_i d_i + 1)}\frac{w_{\min}\,|R|}{n(1+|R|\,D)}\right)^l$$

where D is the diameter of $\mathcal{G}$ and $w_{\min}$ and $w_{\max}$ are the minimum and maximum edge weights in the graph. The bound implies that the error is smaller for more steps k, fewer agents, more foreign agents, smaller graph diameter, larger minimum weight, smaller maximum weight, and smaller maximum degree. With respect to favorable undirected graph topologies, regular graphs exhibit smaller diameters and low maximum degrees [32] generating a smaller upper bound.

The error bound E_l can be subsequently used to bound the error in the gradient $\nabla(\|G(s)\|_2^2)$. Examining the 2-norm of the gradient

$$\begin{aligned}
\left\|\nabla(\|G(s)\|_2^2)\right\|_2 &\leq \sqrt{|E|}\left\|\nabla(\|G(s)\|_2^2)\right\|_1 \\
&\leq \sqrt{|E|}\sum_{\{v_i,v_j\}\in E}\frac{1}{2}\sum_{s\in R}\left(y_i^s - y_j^s\right)^2 \\
&\leq \frac{1}{2}\sqrt{|E|}\sum_{s\in R}\sum_{\{v_i,v_j\}\in E}\left(y_i^s - y_j^s\right)^2 \\
&\leq \frac{1}{2}\sqrt{|E|}\sum_{s\in R}\left\|E(\mathcal{G})^T y^s\right\|_2^2 \\
&\leq \frac{1}{2}\sqrt{|E|}\sum_{s\in R}\left\|E(\mathcal{G})^T\right\|_2^2\|y^s\|_2^2
\end{aligned}$$

Noting that $\|E(\mathcal{G})^T\|_2^2 = \|E(\mathcal{G})E(\mathcal{G})^T\|_2 = \|L(\mathcal{G}_u)\|_2 \leq 2\max_i d_i$, where $\mathcal{G}_u$ is the unweighted form of $\mathcal{G}$ and the final inequality follows from a proof similar to Proposition 6.5, it follows that

$$\left\|\nabla(\|G(s)\|_2^2)\right\|_2 \leq |R|\sqrt{|E|}\max_s\|y^s\|_2^2\max_i d_i$$

Let $\widetilde{\nabla}(\|G(s)\|_2^2)$ be the inexact gradient formed by using a truncated Neumann series T_l in place of y^s, then

$$\left\|\nabla(\|G(s)\|_2^2) - \widetilde{\nabla}(\|G(s)\|_2^2)\right\|_2 \leq |R|\sqrt{|E|}E_l^2\max_i d_i$$

Similar bounds can be formed for the inexact gradient calculations of $\nabla\|G_d(s)\|_2^2$, ∇H, and ∇H_d using the error bounds for calculating y^s and z^s.

[3] For the calculation of z^s, $A = A(\mathcal{G},\mathcal{R})^2$ and the summation in step 9 can be replaced with (6.16).

6.5.2 Online Algorithm

The time-varying nature of the foreign agent set $\mathcal{R}_t$ gives rise to a sequence of strictly convex functions $\{f_t(W)\}$. These functions are revealed at each time step as the new agent set is detected. The problem (6.14) is consequently well suited to being solved using an online convex algorithm. Here, the decision variable is the edge weight selection $W(t)$ at each time t, which will compose the gain matrices $K_i(t)$ for each agent i required for the network-level dynamic model.

In the following, we formulate a distributed version of the online gradient descent algorithm proposed by Hazan et al. [15]. The distributed version quoted in Algorithm 6.3 relies on the conjugate gradient algorithms and is applicable for the cost functions $f_t(W)$ introduced in the beginning of this section. At each time step $k+1$ the local gradient $g_{ij}[k]$ (line 5) of edge w_{ij} at time step k is revealed to agents i and j that perform a gradient descent step in line 6, using the local gradient calculation found in Section 6.5. An example of such a $g_{ij}\,[k]$ is provided in the following.

EXAMPLE 6.3 Pertaining to the cost function in Example 6.1, the gradient $g_{ij}[k]$ in line 5 of Algorithm 6.3 would be

$$g_{ij}[k] = -\frac{1}{2}\sum_{s\in R}\left(y_i^s[k] - y_j^s[k]\right)^2 + hw_{ij}[k]$$

To constrain the weights to the linear constraint set, a projection $\prod_{\mathcal{P}}(\cdot)$ corrects the step by projecting to the closest point on the constraint set $\mathcal{P}$ under the ∞-norm. For the linear constraints in problem (6.14) this can be accomplished distributively and cheaply.

Hazan et al. [15] proved that if the step size is chosen as $\eta[k] = \frac{1}{hk}$ then $\mathcal{R}_T$ is at most $O(\log(T))$ for strongly convex $f_t(\cdot)$ and $\mathcal{R}_T$ is at most $O\left(\sqrt{T}\right)$ for convex $f_t(\cdot)$. As the regret is sublinear in time, then the algorithm is referred to as having "low regret" and performs as well as a best fixed weight policy with a-priori knowledge of the sequence $\{f_t(W)\}$. Proposition 6.6 in the Appendix extends Hazan et al.'s proof for inexact gradients. It shows that given the weight selections $\{W_t\}$ from Algorithm 6.3 and any arbitrary fixed weight selection W^*, then

$$\lim_{T\to\infty}\frac{1}{T}\mathcal{R}_T = \lim_{T\to\infty}\frac{1}{T}\left(\sum_{t=1}^{T} f_t(W_t) - f_t(W^*)\right) = \mathcal{E}D_{\mathcal{P}}$$

where $\mathcal{E}$ is the 2-norm of the error in the gradient bound at each iteration and $D_{\mathcal{P}}$ is the diameter of the constraint set $\mathcal{P}$. In other words, Algorithm 6.3 without access to the function $\{f_t(\cdot)\}$ ahead of time performs within a margin of $\mathcal{E}D_{\mathcal{P}}$ as well as the best static feedback W chosen with the knowledge of all $\{f_t(\cdot)\}$ ahead of time.

EXAMPLE 6.4 Consider the random 50-node graph depicted in Figure 6.7 running the single integrator leader–follower dynamics. Four fixed foreign agents were detected by the agents marked with large circles. The agents in the network ran the distributed gradient descent algorithm with the cost function of Example 6.1. The

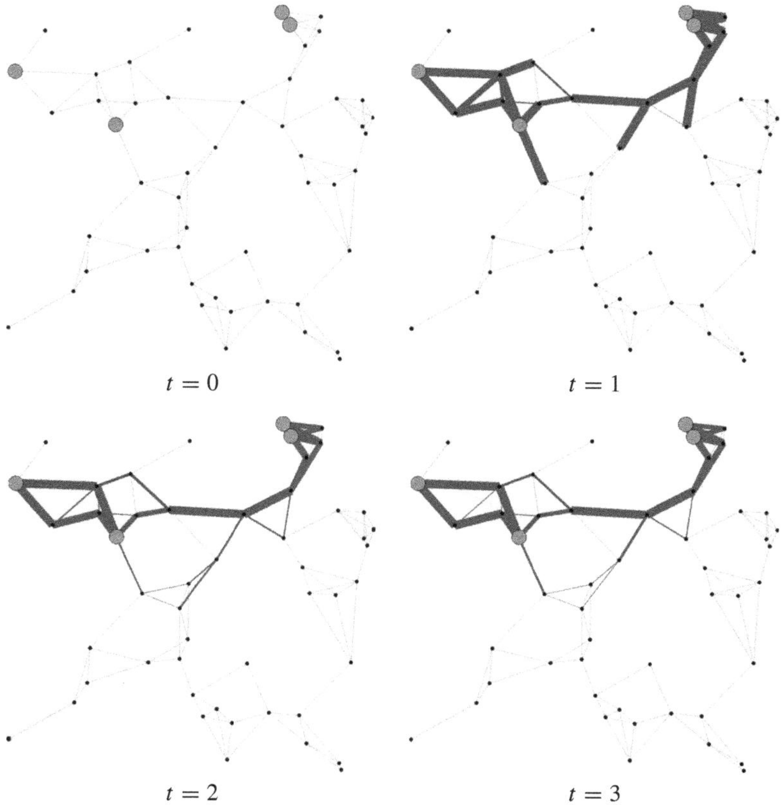

Figure 6.7 Distributed reweighting with a fixed leader set.

Algorithm 6.3: Distributed online gradient descent

Given convex set $\mathcal{P} \subset \mathbb{R}^n$ and $w_{ij}[1] \in \mathcal{P}$, for all $\{v_i, v_j\} \in E$
Initialize:
$k = 1$
for each $\{v_i, v_j\} \in E$ **do**
    g_{ij} is revealed using the distributed conjugate gradient
    $w_{ij}[k+1] = \prod_{\mathcal{P}} \left(w_{ij}[k] - \eta[k] g_{ij}[k]\right)$
    $k = k + 1$
    Here, $\prod_{\mathcal{P}}(z) = \arg\min_{x \in \mathcal{P}} \|x - z\|_2$.
end

---

edge weights were initialized uniformly as 0.25 and constrained such that $q_l = 0.05$ and $s_l = 1$ for all edges in E. The first few iterations are displayed in Figure 6.7. The notable characteristic is the increasing of edge weights on those edges close to the foreign agents. This aligns with the conductance analogy mentioned in Section 6.3, where additional conductance is added to edges in the neighborhood of the foreign

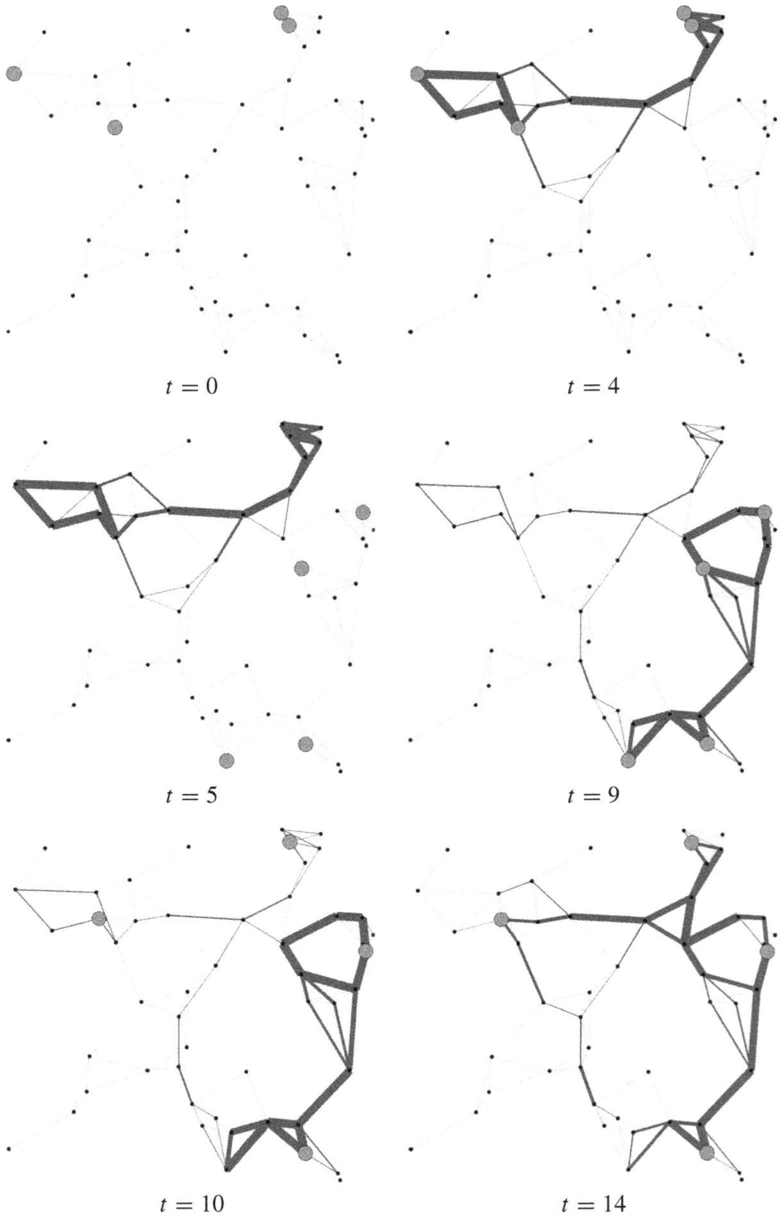

Figure 6.8 Distributed reweighting with a variable leader set.

agents. These edges are those most likely to dampen the effect of the foreign agents' signal.

Figure 6.8 shows the algorithm's evolution for varying foreign agent locations over time. The time evolution provides insight into the inner working of the online regret framework. Once a foreign agent location is revealed, the algorithm is penalized for the absence of heavily weighted edges in its vicinity, for example at time steps $t = 0$

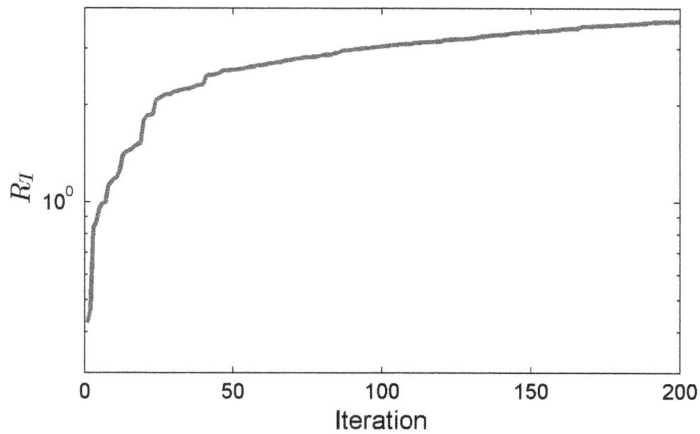

Figure 6.9 Regret trajectory over time of Algorithm 6.3.

and 5. The algorithm recovers by time steps $t = 4$ and 9, respectively. This process provides the algorithm with a "memory" of the foreign agent's location.

For comparison, the optimal static weight sequence for the graph was calculated with respect to the cost function in Example 6.1 and the sequence of foreign agents over the first 1000 iterations. This provides the value of $f_t(W^*)$ over time, which was used to calculate the regret of the algorithm and displayed in Figure 6.9. The regret curve aligns with the expectant sublinear performance found by Hazan et al. [15].

6.6 Conclusion

We have examined an online distributed network adaptation process suitable for multi-agent systems to dampen disturbances in noisy consensus and leader–follower dynamics. Edge weights were dynamically varied in response to changing input disturbance locations to reduce the system-theoretic measures of the open loop $\mathcal{H}_2$ norm and entropy. Posing the problem as a convex online optimization problem, distributed versions of the coordinated conjugate gradient method and the online gradient descent method were introduced to solve the problem. The result was an online reweighting algorithm that each agent in the network can apply to produce a small regret disturbance rejection solution. The chapter provides an example of the use of the network topology as a controller and the application of online optimization methods to tackle uncertain optimization problems in networked dynamics. Future research proposes to further these two strains of research to other types of topology control and uncertain networked dynamics.

Appendix

PROPOSITION 6.1 The observability Gramian of the double integrator dynamics (6.8) is

$$P = \frac{1}{2}\begin{bmatrix} I - A(\mathcal{G},\mathcal{R})^{-1} & -A(\mathcal{G},\mathcal{R})^{-1} \\ -A(\mathcal{G},\mathcal{R})^{-1} & A(\mathcal{G},\mathcal{R})^{-2} \end{bmatrix}$$

Proof. Schur's Lemma states that $A = \begin{bmatrix} A_{11} & A_{12} \\ A_{12}^T & A_{22} \end{bmatrix} \succ 0$ if and only if $A_{22} \succ 0$ and $A_{11} - A_{12}A_{22}^{-1}A_{12}^T \succ 0$ [33]. As $A(\mathcal{G},\mathcal{R})^{-1}$ is negative definite, $I - A(\mathcal{G},\mathcal{R})^{-1}$ is positive definite since it is the sum of two positive definite matrices. Further,

$$\begin{aligned}(I - A(\mathcal{G},\mathcal{R})^{-1}) - A(\mathcal{G},\mathcal{R})^{-1}A(\mathcal{G},\mathcal{R})^{2}A(\mathcal{G},\mathcal{R})^{-1} &= I - A(\mathcal{G},\mathcal{R})^{-1} - I \\ &= -A(\mathcal{G},\mathcal{R})^{-1} \succ 0\end{aligned}$$

Hence, P is positive definite. Validating that the Lyapunov equation (6.10) is satisfied as

$$\begin{aligned}A_d(\mathcal{G},\mathcal{R})^T P &= \frac{1}{2}\begin{bmatrix} 0 & A(\mathcal{G},\mathcal{R}) \\ I & A(\mathcal{G},\mathcal{R}) \end{bmatrix}\begin{bmatrix} I - A(\mathcal{G},\mathcal{R})^{-1} & -A(\mathcal{G},\mathcal{R})^{-1} \\ -A(\mathcal{G},\mathcal{R})^{-1} & A(\mathcal{G},\mathcal{R})^{-2} \end{bmatrix} \\ &= \frac{1}{2}\begin{bmatrix} -I & A(\mathcal{G},\mathcal{R})^{-1} \\ -A(\mathcal{G},\mathcal{R})^{-1} & 0 \end{bmatrix}\end{aligned}$$

and

$$\begin{aligned}A_d(\mathcal{G},\mathcal{R})^T P + PA_d(\mathcal{G},\mathcal{R}) &= \frac{1}{2}\begin{bmatrix} -I & A(\mathcal{G},\mathcal{R})^{-1} \\ -A(\mathcal{G},\mathcal{R})^{-1} & 0 \end{bmatrix} \\ &\quad + \frac{1}{2}\begin{bmatrix} -I & -A(\mathcal{G},\mathcal{R})^{-1} \\ A(\mathcal{G},\mathcal{R})^{-1} & 0 \end{bmatrix} \\ &= \begin{bmatrix} -I & 0 \\ 0 & 0 \end{bmatrix} = -C_d^T C_d\end{aligned}$$

□

PROPOSITION 6.2 The derivative with respect to w_{ij} and Hessian of $\|G(s)\|_2^2$ are

$$\frac{\partial \|G(s)\|_2^2}{\partial w_{ij}} = -\frac{1}{2}\sum_{p\in R}\left(\left[A(\mathcal{G},\mathcal{R})^{-1}e_p\right]_i - \left[A(\mathcal{G},\mathcal{R})^{-1}e_p\right]_j\right)^2$$

and

$$\nabla^2 \|G(s)\|_2^2 = -E^T A(\mathcal{G},\mathcal{R})^{-1}BB^T A(\mathcal{G},\mathcal{R})^{-T} E \circ E^T A(\mathcal{G},\mathcal{R})^{-1} E$$

Further, $\nabla^2 \|G(s)\|_2^2$ is positive semidefinite.

Proof. Let $U = A(\mathcal{G},\mathcal{R}) = -\sum_{\{v_i,v_j\}\in E} w_{ij}a_{ij}a_{ij}^T - \sum_{i\in R} e_i e_i^T$. The squared $\mathcal{H}_2$ norm is

$$\begin{aligned}\|G(s)\|_2^2 &= -\frac{1}{2}\mathbf{tr}\left(B^T A(\mathcal{G},\mathcal{R})^{-1}B\right) \\ &= -\frac{1}{2}\sum_{p\in R} e_p^T A(\mathcal{G},\mathcal{R})^{-1}e_p \\ &= -\frac{1}{2}\sum_{p\in R} g_p(\mathcal{G},\mathcal{R})\end{aligned}$$

where $g_p(\mathcal{G}, \mathcal{R}) = e_p^T U(w_{ij})^{-1} e_p$. Further,

$$\frac{\partial g_p(U(w_{ij}))}{\partial w_{ij}} = \mathbf{tr}\left[\left(\frac{\partial g_p(U)}{\partial U}\right)^T \frac{\partial U}{\partial w_{ij}}\right] \tag{6.17}$$

The relevant derivatives are

$$\frac{\partial g_p(U)}{\partial U} = \frac{\partial\left(e_p^T U^{-1} e_p\right)}{\partial U} = -U^{-T} e_p e_p^T U^{-T} \tag{6.18}$$

and

$$\frac{\partial U}{\partial w_{ij}} = \frac{\partial\left(-\sum_{\{i,j\}\in E} w_{ij} a_{ij} a_{ij}^T - \sum_{i\in R} e_i e_i^T\right)}{\partial w_{ij}} = -a_{ij} a_{ij}^T \tag{6.19}$$

Substituting (6.18) and (6.19) into (6.17), we have

$$\begin{aligned}\frac{\partial g_p\left(U\left(w_{ij}\right)\right)}{\partial w_{ij}} &= \mathbf{tr}\left[\left(-U^{-T} e_p e_p^T U^{-T}\right)^T a_{ij} a_{ij}^T\right] \\ &= \mathbf{tr}\left[U^{-1} e_p e_p^T U^{-1} a_{ij} a_{ij}^T\right] \\ &= \mathbf{tr}\left[\left(a_{ij}^T U^{-1} e_p\right)\left(e_p^T U^{-1} a_{ij}\right)\right] \\ &= \left(a_{ij}^T U^{-1} e_p\right)^2 \\ &= \left(a_{ij}^T A(\mathcal{G})^{-1} e_p\right)^2 \\ &= \left(\left[A(\mathcal{G},\mathcal{R})^{-1} e_p\right]_i - \left[A(\mathcal{G},\mathcal{R})^{-1} e_p\right]_j\right)^2\end{aligned}$$

Hence,

$$\frac{\partial \left\|G(s)\right\|_2^2}{\partial w_{ij}} = -\frac{1}{2}\sum_{p\in R}\left(\left[A(\mathcal{G},\mathcal{R})^{-1} e_p\right]_i - \left[A(\mathcal{G},\mathcal{R})^{-1} e_p\right]_j\right)^2$$

Examining the Hessian of g_p,

$$\frac{\partial g_p^2\left(U\left(w_{ij}\right)\right)}{\partial w_{ij}^2} = \frac{\partial}{\partial w_{ij}}\left(a_{ij}^T U^{-1} e_p\right)^2 = \mathbf{tr}\left[\left(\frac{\partial\left(\left(a_{ij}^T U^{-1} e_p\right)^2\right)}{\partial U}\right)^T \frac{\partial U}{\partial w_{ij}}\right] \tag{6.20}$$

The relevant derivatives are

$$\frac{\partial\left(\left(a_{ij}^T U^{-1} e_p\right)^2\right)}{\partial U} = 2 a_{ij}^T U^{-1} e_p \left(-U^{-T} a_{ij} e_p^T U^{-T}\right) \tag{6.21}$$

Substituting (6.19) and (6.21), (6.20) becomes

$$\begin{aligned}\frac{\partial g_p^2\left(U\left(w_{ij}\right)\right)}{\partial w_{ij}^2} &= \mathbf{tr}\left[\left(2a_{ij}^T U^{-1} e_p \left(U^{-T} a_{ij} e_p^T U^{-T}\right)\right)^T a_{ij} a_{ij}^T\right] \\ &= 2a_{ij}^T U^{-1} e_p \mathbf{tr}\left[U^{-1} e_p a_{ij}^T U^{-1} a_{ij} a_{ij}^T\right] \\ &= 2\left(a_{ij}^T U^{-1} e_p\right)\left(a_{ij}^T U^{-1} e_p\right)\left(a_{ij}^T U^{-1} a_{ij}\right) \\ &= 2\left(a_{ij}^T U^{-1} e_p e_p^T U^{-T} a_{ij}\right)\left(a_{ij}^T U^{-1} a_{ij}\right) \\ &= 2\left[E^T U^{-1} e_p e_p^T U^{-T} E\right]_{(\sigma(ij),\sigma(ij))}\left[E^T U^{-1} E\right]_{(\sigma(ij),\sigma(ij))}\end{aligned}$$

as well as

$$\begin{aligned}\frac{\partial g_p^2\left(U\left(w_{ij}\right)\right)}{\partial w_{ij} w_{kl}} &= \mathbf{tr}\left[\left(2a_{ij}^T U^{-1} e_p \left(U^{-T} a_{ij} e_p^T U^{-T}\right)\right)^T a_{kl} a_{kl}^T\right] \\ &= 2a_{ij}^T U^{-1} e_p \mathbf{tr}\left[U^{-1} e_p a_{ij}^T U^{-1} a_{kl} a_{kl}^T\right] \\ &= 2\left(a_{ij}^T U^{-1} e_p\right)\left(a_{kl}^T U^{-1} e_p\right)\left(a_{ij}^T U^{-1} a_{kl}\right) \\ &= 2\left(a_{ij}^T U^{-1} e_p e_p^T U^{-1} a_{kl}\right)\left(a_{ij}^T U^{-1} a_{kl}\right) \\ &= 2\left[E^T U^{-1} e_p e_p^T U^{-T} E\right]_{(\sigma(ij),\sigma(kl))}\left[E^T U^{-1} E\right]_{(\sigma(ij),\sigma(kl))}\end{aligned}$$

Consequently,

$$\nabla^2 g_p = 2E^T U^{-1} e_p e_p^T U^{-T} E \circ E^T U^{-1} E$$

and hence

$$\begin{aligned}&\nabla^2 \|G(s)\|_2^2 \\ &\quad = -E^T U^{-1} B B^T U^{-T} E \circ E^T U^{-1} E \\ &\quad = -E^T A(\mathcal{G},\mathcal{R})^{-1} B B^T A(\mathcal{G},\mathcal{R})^{-T} E \circ E^T A(\mathcal{G},\mathcal{R})^{-1} E\end{aligned}$$

Now, $E^T A(\mathcal{G},\mathcal{R})^{-1} B B^T A(\mathcal{G},\mathcal{R})^{-T} E = \left(B^T A(\mathcal{G},\mathcal{R})^{-T} E\right)^T \left(B^T A(\mathcal{G},\mathcal{R})^{-T} E\right) \succcurlyeq 0$, and as $A(\mathcal{G},\mathcal{R})$ is negative definite then $-E^T A(\mathcal{G},\mathcal{R})^{-1} E \succcurlyeq 0$. Finally, the Hadamard product boasts the property that the Hadamard product of two positive semidefinite matrices is positive semidefinite [33] and so the result follows. □

PROPOSITION 6.3 The derivative with respect to w_{ij} of $\|G_d(s)\|_2^2$ is

$$\begin{aligned}\frac{\partial \|G_d(s)\|_2^2}{\partial w_{ij}} = &-\sum_{p\in R}\left(\left[A(\mathcal{G},\mathcal{R})^{-2} e_p\right]_i - \left[A(\mathcal{G},\mathcal{R})^{-2} e_p\right]_j\right) \\ &\times\left[\sum_{q\in N(i)} w_{iq}\left(\left[A(\mathcal{G},\mathcal{R})^{-2} e_p\right]_i - \left[A(\mathcal{G},\mathcal{R})^{-2} e_p\right]_q\right)\right.\end{aligned}$$

$$+\sum_{q\in N(i)} w_{pj}\left(\left[A(\mathcal{G},\mathcal{R})^{-2}e_p\right]_p - \left[A(\mathcal{G},\mathcal{R})^{-2}e_p\right]_j\right)$$
$$+\gamma_{ij}\left(\left[A(\mathcal{G},\mathcal{R})^{-2}e_p\right]_i - \left[A(\mathcal{G},\mathcal{R})^{-2}e_p\right]_j\right)\Bigg]$$

where $\gamma_{ij} = 2$ if $i, j \in R$, $\gamma_{ij} = 1$ if $i \in R$ or $j \in R$ and 0, otherwise. Further, $\nabla^2 \|G(s)\|_2^2$ is positive semidefinite.

Proof. Following the proof of Proposition 6.2, let $U = A(\mathcal{G},\mathcal{R})^2 = \left(-\sum_{\{v_i,v_j\}\in E} w_{ij}a_{ij}a_{ij}^T - \sum_{i\in R} e_i e_i^T\right)^2$. The squared $\mathcal{H}_2$ norm is

$$\|G_d(s)\|_2^2 = -\frac{1}{2}\sum_{p\in R} g_p(\mathcal{G},\mathcal{R}) \tag{6.22}$$

where $g_p(\mathcal{G},\mathcal{R}) = e_p^T U(w_{ij})^{-1} e_p$. Further,

$$\frac{\partial g_p(U)}{\partial U} = -U^{-T} e_p e_p^T U^{-T}$$

and

$$\frac{\partial U}{\partial w_{ij}} = \frac{\partial\left(-\sum_{(p,q)\in E} w_{pq}a_{pq}a_{pq}^T - \sum_{i\in R} e_i e_i^T\right)^2}{\partial w_{ij}}$$
$$= \frac{\partial\left(w_{ij}a_{ij}a_{ij}^T\left(\sum_{(p,q)\in E} w_{pq}a_{pq}a_{pq}^T + \sum_{i\in R} e_i e_i^T\right) + \left(\sum_{(p,q)\in E} w_{pq}a_{pq}a_{pq}^T + \sum_{i\in R} e_i e_i^T\right) w_{ij}a_{ij}a_{ij}^T\right)}{\partial w_{ij}}$$
$$= a_{ij}a_{ij}^T\left(\sum_{q\in N(i)} w_{iq}a_{iq}a_{iq}^T + \sum_{p\in N(j)} w_{pj}a_{pj}a_{pj}^T + \sum_{i\in R} e_i e_i^T\right)$$
$$+\left(\sum_{q\in N(i)} w_{iq}a_{iq}a_{iq}^T + \sum_{p\in N(j)} w_{pj}a_{pj}a_{pj}^T + \sum_{i\in R} e_i e_i^T\right) a_{ij}a_{ij}^T$$
$$= a_{ij}\left(\sum_{q\in N(i)} w_{iq}a_{iq}^T + \sum_{p\in N(j)} w_{pj}a_{pj}^T\right) + \left(\sum_{q\in N(i)} w_{iq}a_{iq} + \sum_{p\in N(j)} w_{pj}a_{pj}\right) a_{ij}^T + \gamma_{ij}a_{ij}a_{ij}^T$$

Hence,

$$\frac{\partial g_p\left(U\left(w_{ij}\right)\right)}{\partial w_{ij}} = \mathbf{tr}\left[\left(\frac{\partial g_p(U)}{\partial U}\right)^T \frac{\partial U}{\partial w_{ij}}\right]$$
$$= \mathbf{tr}\Bigg[\left(-U^{-T}e_p e_p^T U^{-T}\right)^T (a_{ij}\left(\sum_{q\in N(i)} w_{iq}a_{iq}^T + \sum_{p\in N(j)} w_{pj}a_{pj}^T\right)$$
$$+\left(\sum_{q\in N(i)} w_{iq}a_{iq} + \sum_{p\in N(j)} w_{pj}a_{pj}\right) a_{ij}^T + \gamma_{ij}a_{ij}a_{ij}^T)\Bigg]$$
$$= 2\mathbf{tr}\Bigg[\sum_{q\in N(i)} w_{iq}\left(a_{iq}^T U^{-1} e_p\right)\left(e_p^T U^{-1} a_{ij}\right)$$

$$+\sum_{q\in N(i)} w_{pj}\left(a_{pj}^T U^{-1}e_p\right)\left(e_p^T U^{-1}a_{ij}\right)+\gamma_{ij}\left(e_p^T U^{-1}a_{ij}\right)^2\Bigg]$$

$$=2\Bigg[a_{ij}^T U^{-1}e_p^T\Bigg(\sum_{q\in N(i)} w_{iq}\left(a_{iq}^T U^{-1}e_p\right)$$

$$+\sum_{q\in N(i)} w_{pj}\left(a_{pj}^T U^{-1}e_p\right)+\gamma_{ij}\left(a_{ij}^T U^{-1}e_p\right)\Bigg)\Bigg]$$

and applying (6.22) and the definition of U, the result with respect to the derivative of $\|G_d(s)\|_2^2$ follows. For brevity, the Hessian derivation is omitted but follows similarly to the proof of Proposition 6.2. □

PROPOSITION 6.4 The derivative with respect to w_{ij} of H and H_d is

$$\frac{\partial H}{\partial w_{ij}}=\frac{\partial H_d}{\partial w_{ij}}=-\left(\left[A(\mathcal{G},\mathcal{R})^{-1}e_i\right]_i-\left[A(\mathcal{G},\mathcal{R})^{-1}e_i\right]_j\right)$$
$$+\left(\left[A(\mathcal{G},\mathcal{R})^{-1}e_j\right]_i-\left[A(\mathcal{G},\mathcal{R})^{-1}e_j\right]_j\right)$$

Further, $\nabla^2 H$ and $\nabla^2 H_d$ is positive definite.

Proof. Following the proof of Proposition 6.2, let $U=A(\mathcal{G},\mathcal{R})=\sum_{\{v_i,v_j\}\in E} w_{ij}a_{ij}a_{ij}^T-\sum_{i\in R} e_ie_i^T$. The entropy is $H=g(\mathcal{G},\mathcal{R})$, where $g(\mathcal{G},\mathcal{R})=-\log\det(-U)$. Further,

$$\frac{\partial g_p(U)}{\partial U}=U^{-T}$$
$$\frac{\partial U}{\partial w_{ij}}=-a_{ij}a_{ij}^T$$

and

$$\begin{aligned}\frac{\partial g\left(U\left(w_{ij}\right)\right)}{\partial w_{ij}}&=\mathbf{tr}\left[\left(U^{-T}\right)^T-a_{ij}a_{ij}^T\right]\\&=-\mathbf{tr}\left[a_{ij}^TU^{-1}a_{ij}\right]\\&=-a_{ij}^TU^{-1}a_{ij}\\&=-a_{ij}^TA(\mathcal{G},\mathcal{R})^{-1}\left(e_i-e_j\right)\\&=-\left(\left[A(\mathcal{G},\mathcal{R})^{-1}e_i\right]_i-\left[A(\mathcal{G},\mathcal{R})^{-1}e_i\right]_j\right)\\&\quad-\left(\left[A(\mathcal{G},\mathcal{R})^{-1}e_j\right]_i-\left[A(\mathcal{G},\mathcal{R})^{-1}e_j\right]_j\right)\end{aligned}$$

For brevity, the Hessian derivation is omitted but follows similarly to the proof of Proposition 6.2. □

PROPOSITION 6.5 The smallest magnitude eigenvalue λ_1 and largest magnitude eigenvalue λ_n of $-A(\mathcal{G},\mathcal{R})=-(L(\mathcal{G})+\sum_{\{v_i,r_j\}\in\mathcal{E}_R} w_j^r e_ie_i^T)\in\mathbb{R}^{n\times n}$, where $\mathcal{G}$ is a connected graph with smallest positive weights $w_{\min}$ and $w_{\max}$, respectively, are bounded as

$$\lambda_1 \geq \frac{w_{\min}\,|R|}{n(1+|R|\,D)} \quad \text{and} \quad \lambda_n \leq w_{\max} \max_i(2d_i+1)$$

Proof. Using the variational characterization of the eigenvalues of $-A(\mathcal{G},\mathcal{R})$ in terms of the Rayleigh quotient of $-A(\mathcal{G},\mathcal{R})$. Let f be an arbitrary function that assigns to each vertex v of $\mathcal{G}$ a real value $f(v)$ and consider f the vector of these functions. Then

$$\frac{f^T(-A(\mathcal{G},\mathcal{R}))f}{f^Tf} = \frac{f^TL(\mathcal{G})f + f^T\sum_{\{v_k,r_j\}\in\mathcal{E}_R} w^r_j e_k e_k^T f}{f^Tf}$$
$$= \frac{\sum_{\{v_i,v_j\}\in E} w_{ij}(f(v_i)-f(v_j))^2 + \sum_{\{v_k,r_j\}\in\mathcal{E}_R} w^r_j f(v_k)^2}{\sum_{i=1}^n f(v_i)^2}$$

We note that

$$\lambda_1 = \inf_{f\neq 0} \frac{\sum_{\{v_i,v_j\}\in E} w_{ij}(f(v_i)-f(v_j))^2 + \sum_{\{v_k,r_j\}\in\mathcal{E}_R} w^r_j f(v_k)^2}{\sum_{i=1}^n f(v_i)^2}$$

Let v_1 and v_n be the nodes corresponding to the smallest and largest function values, respectively, i.e., $f(v_1)\leq f(v_i)\leq f(v_n)$ and $w_{\min}\leq w_{ij}$ for all i, then

$$\lambda_1 \geq \inf_{f\neq 0} \frac{w_{\min}\sum_{\{v_i,v_j\}\in E}(f(v_i)-f(v_j))^2 + |R|\,w_{\min} f(v_1)^2}{nf(v_n)^2}$$

As $\mathcal{G}$ is connected, there is a path $\mathcal{P} = (V_p, E_P)$ connecting v_1 and v_n; as such, removing all but the edges in $\mathcal{P}$ we have that

$$\lambda_1 \geq \frac{w_{\min}}{n} \inf_{f\neq 0} \frac{\sum_{(v_i,v_j)\in E_p}(f(v_i)-f(v_j))^2 + |R|\,f(v_1)^2}{f(v_n)^2}$$

Noting that $\sum_{i=1}^m (a_i - a_{i+1})^2 \geq \sum_{i=1}^m (\frac{a_n-a_1}{m})^2 = \frac{1}{m}(a_n-a_1)^2$, where $a_i \in [a_1, a_n]$ for all i, then

$$\lambda_1 \geq \frac{w_{\min}}{n} \inf_{f\neq 0} \frac{\frac{1}{|E_p|}(f(v_n)-f(v_1))^2 + |R|\,f(v_1)^2}{f(v_n)^2}$$

As $\mathcal{P}$ is at most the diameter of the graph, then $\left|E_p\right| \leq D$; minimizing the subsequent quadratic, then

$$\lambda_1 \geq \frac{w_{\min}}{n} \inf_{f(v_1),f(v_n)} \frac{1}{D}\left(1-\frac{f(v_1)}{f(v_n)}\right)^2 + |R|\left(\frac{f(v_1)}{f(v_n)}\right)^2$$
$$\geq \frac{w_{\min}}{n}\left(\frac{|R|}{|R|\,D+1}\right)$$

The maximum eigenvalue can be similarly represented using the Rayleigh quotient

$$\lambda_n = \sup_{f\neq 0} \frac{\sum_{\{v_i,v_j\}\in E} w_{ij}(f(v_i)-f(v_j))^2 + \sum_{\{v_k,r_j\}\in\mathcal{E}_R} w^r_j f(v_k)^2}{\sum_{i=1}^n f(v_i)^2}$$

Applying the property that $(a_i - a_j)^2 \leq 2(a_i^2 + a_j^2)$ and $w_{ij} \leq w_{\max}$, then

$$\begin{aligned}\lambda_n &\leq w_{\max} \sup_{f \neq 0} \frac{\sum_{\{v_i,v_j\}\in E} 2(f(v_i)+f(v_j))^2 + \sum_{k\in R} f(v_k)^2}{\sum_{i=1}^n f(v_i)^2} \\ &= w_{\max} \sup_{f\neq 0} \frac{2\sum_{i=1}^n d_i f(v_i)^2 + \sum_{k\in R} f(v_k)^2}{\sum_{i=1}^n f(v_i)^2} \\ &\leq w_{\max} \sup_{f\neq 0} \frac{\max_i(2d_i+1)\sum_{i=1}^n f(v_i)^2}{\sum_{i=1}^n f(v_i)^2} \\ &\leq w_{\max} \max_i (2d_i + 1)\end{aligned}$$

□

PROPOSITION 6.6 Let $\nabla_t(x)$ and $\nabla_t^2(x) \succeq HI$ be the gradient and Hessian of strongly convex $f_t(x)$, $\mathcal{P}$ a convex set with diameter $D_{\mathcal{P}}$, i.e., $\|x-y\| \leq D_{\mathcal{P}}$ for all $x, y \in \mathcal{P}$ and $\|\nabla_t(x)\| \leq G$ for $x \in \mathcal{P}$. Initializing $x_1 \in \mathcal{P}$ and applying the updates,

$$x_{t+1} = \prod_{\mathcal{P}}(x_t - \eta_t \widetilde{\nabla}_t)$$

where $\widetilde{\nabla}_t = \nabla_t(x_t) - \varepsilon_t$, $\|\varepsilon_t\| \leq \mathcal{E}$, and $\eta_t = \frac{1}{Ht}$, then

$$\mathcal{R}_T \leq \mathcal{E}D_{\mathcal{P}}T + \frac{1}{2H}(G^2+\mathcal{E}^2)\log T + \frac{1}{2}\left(G^2 + \mathcal{E}^2 + \frac{1}{H}D_{\mathcal{P}}^2\right)$$

Proof. Using the Taylor series approximation and strong convexity,

$$\begin{aligned}f(x^*) &\geq f(x_t) + \nabla_t(x_t)^T(x^*-x_t) + \frac{H}{2}\|x^*-x_t\|^2 \\ f(x_t) - f(x^*) &\leq \nabla_t(x_t)^T(x_t - x^*) - \frac{H}{2}\|x^*-x_t\|^2 \\ &\leq (\widetilde{\nabla}_t + \varepsilon_t)^T(x_t-x^*) - \frac{H}{2}\|x^*-x_t\|^2 \\ &\leq \widetilde{\nabla}_t^T(x_t-x^*) - \frac{H}{2}\|x^*-x_t\|^2 + \|\varepsilon_t\|\,\|x^*-x_t\|\end{aligned}$$

From update $x_{t+1} = \prod(x_t - \eta_{t+1}\widetilde{\nabla}_t)$, then

$$\begin{aligned}\|x_{t+1}-x^*\|^2 &= \left\|\prod_{\mathcal{P}}(x_t - \eta_t\widetilde{\nabla}_t) - x^*\right\|^2 \\ &\leq \|x_t - \eta_t\widetilde{\nabla}_t - x^*\|^2 \\ &\leq \|x_t - x^*\|^2 + \eta_t^2\|\widetilde{\nabla}_t\|^2 - 2\eta_t\widetilde{\nabla}_t^T(x_t-x^*) \\ 2\eta_t\widetilde{\nabla}_t^T(x_t-x^*) &\leq \|x_t-x^*\|^2 - \|x_{t+1}-x^*\|^2 + \eta_{t+1}^2\|\widetilde{\nabla}_t\|^2 \\ 2\widetilde{\nabla}_t^T(x_t-x^*) &\leq \frac{1}{\eta_t}\left(\|x_t-x^*\|^2 - \|x_{t+1}-x^*\|^2\right) + \eta_t\|\widetilde{\nabla}_t\|^2 \\ &\leq \frac{1}{\eta_t}\left(\|x_t-x^*\|^2 - \|x_{t+1}-x^*\|^2\right) + \eta_t(G^2 + \|\varepsilon_t\|^2)\end{aligned}$$

with the final inequality following from $||\widetilde{\nabla}_t||^2 \leq \|\nabla(x_t)\|^2 + \|\varepsilon_t\|^2 = G^2 + \|\varepsilon_t\|^2$. The regret calculation becomes

$$\begin{aligned}
2\mathcal{R}_T &= 2\left(\sum_{t=1}^{T} f_t(x_t) - \sum_{t=1}^{T} f_t(x^*)\right) \\
&\leq \sum_{t=1}^{T} 2\widetilde{\nabla}_t^T(x_t - x^*) - H\left\|x^* - x_t\right\|^2 + 2\left\|\varepsilon_t\right\|\left\|x^* - x_t\right\| \\
&\leq 2\sum_{t=1}^{T} \widetilde{\nabla}_t^T(x^* - x_t) - H\left\|x^* - x_t\right\|^2 + 2\left\|\varepsilon_t\right\|\left\|x^* - x_t\right\| \\
&\leq \sum_{t=1}^{T} \frac{1}{\eta_t}\left(\left\|x_t - x^*\right\|^2 - \left\|x_{t+1} - x^*\right\|^2\right) + \eta_t(G^2 + \|\varepsilon_t\|^2) \\
&\quad - H\left\|x^* - x_t\right\|^2 + 2\left\|\varepsilon_t\right\|\left\|x^* - x_t\right\| \\
&\leq \sum_{t=2}^{T} \left\|x_t - x^*\right\|^2\left(\frac{1}{\eta_t} - \frac{1}{\eta_{t-1}} - H\right) + \sum_{t=1}^{T}\left(G^2 + \|\varepsilon_t\|^2\right)\eta_t \\
&\quad + 2\sum_{t=1}^{T} \left\|\varepsilon_t\right\|\left\|x^* - x_t\right\| + \frac{1}{\eta_1}\left\|x_1 - x^*\right\|^2
\end{aligned}$$

Setting $\eta_t = \frac{1}{Ht}$ and applying bounds $\sum_{t=2}^{T} \eta_t \leq \frac{1}{H}(1 + \log T)$, $\|\varepsilon_t\| \leq \mathcal{E}$, and $\|x^* - x_t\| \leq D_{\mathcal{P}}$, then

$$\begin{aligned}
2\mathcal{R}_T &\leq \sum_{t=2}^{T} \left\|x_t - x^*\right\|^2 (Ht - H(t-1) - H) \\
&\quad + (G^2 + \mathcal{E}^2)(1 + \log T) + 2\mathcal{E}D_{\mathcal{P}}T + \frac{1}{H}D_{\mathcal{P}}^2
\end{aligned}$$

and thus

$$\mathcal{R}_T \leq \mathcal{E}D_{\mathcal{P}}T + \frac{1}{2H}(G^2 + \mathcal{E}^2)\log T + \frac{1}{2}\left(G^2 + \mathcal{E}^2 + \frac{1}{H}D_{\mathcal{P}}^2\right) \qquad \square$$

Acknowledgments

This research of the authors was supported by AFOSR grant FA9550-16-1-0022 and ONR grant N00014-12-1-1002.

References

[1] H. G. Tanner, G. J. Pappas, and V. Kumar, "Leader-to-formation stability," *IEEE Trans. Robotics Autom.*, vol. 20, no. 3, pp. 443–455, 2004.

[2] R. Olfati-Saber, J. A. Fax, and R. M. Murray, "Consensus and cooperation in networked multi-agent systems," *Proc. IEEE*, vol. 95, no. 1, pp. 215–233, 2007.

[3] M. Mesbahi and M. Egerstedt, *Graph Theoretic Methods in Multiagent Networks*. Princeton, NJ: Princeton University Press, 2010.

[4] A. Jadbabaie, J. Lin, and A. S. Morse, "Coordination of groups of mobile autonomous agents using nearest neighbor rules," *IEEE Trans. Autom. Control*, vol. 48, no. 6, pp. 988–1001, 2003.

[5] Y. Hatano and M. Mesbahi, "Agreement over random networks," *IEEE Trans. Autom. Control*, vol. 50, no. 11, pp. 1867–1872, 2005.

[6] M. Hudoba de Badyn, A. Chapman, and M. Mesbahi, "Network entropy: A system-theoretic perspective," in *54th IEEE Conf. Decision and Control*, 2015, pp. 5512–5517.

[7] A. Chapman and M. Mesbahi, "Semi-autonomous consensus: Network measures and adaptive trees," *IEEE Trans. Autom. Control*, vol. 58, no. 1, pp. 19–31, 2013.

[8] A. Ghosh and S. Boyd, "Growing well-connected graphs," in *Proc. 45th IEEE Conf. Decision and Control*, 2006, pp. 6605–6611.

[9] Y. Wan, S. Roy, and A. Saberi, "Network design problems for controlling virus spread," in *Proc. 46th IEEE Conf. Decision and Control*, 2007, pp. 3925–3932.

[10] D. Zelazo and M. Mesbahi, "Edge agreement: Graph-theoretic performance bounds and passivity analysis," *IEEE Trans. Autom. Control*, vol. 56, no. 3, pp. 544–555, 2011.

[11] Y. Kim and M. Mesbahi, "On maximizing the second smallest eigenvalue of a state-dependent graph Laplacian," *IEEE Trans. Autom. Control*, vol. 51, no. 1, pp. 116–120, 2006.

[12] Z. Wu and R. Wang, "The consensus in multi-agent system with speed-optimized network," *Int. J. Modern. Phys. B*, vol. 23, no. 10, pp. 2339–2348, 2009.

[13] M. R. Hestenes and E. Stiefel, "Methods of conjugate gradients for solving linear systems," *J. Res. Natl. Inst. Stand.*, vol. 49, pp. 409–436, 1952.

[14] D. Bertsekas, "Incremental gradient, subgradient, and proximal methods for convex optimization: A survey," *Optimization for Machine Learning*, S. Sra, S. Nowozin, and S. Wright, Eds., Cambridge, MA, MIT Press, 2011.

[15] E. Hazan, A. Agarwal, and S. Kale, "Logarithmic regret algorithms for online convex optimization," *Mach. Learn.*, vol. 69, no. 2–3, pp. 169–192, 2007.

[16] F. Yan, S. Sundaram, S. Vishwanathan, and Y. Qi, "Distributed autonomous online learning: Regrets and intrinsic privacy-preserving properties," in *IEEE Trans. Knowl. Data Eng.*, vol. 25, no. 11, 2013, pp. 2483–2493.

[17] S. Sundhar Ram, A. Nedić, and V. V. Veeravalli, "Distributed stochastic subgradient projection algorithms for convex optimization," *J. Optim. Theor. Appl.*, vol. 147, no. 3, pp. 516–545, 2010.

[18] D. Mateos-Núñez and J. Cortés, "Distributed online convex optimization over jointly connected digraphs," *IEEE Trans. Netw. Sci. Eng.*, vol. 1, no. 1, pp. 23–37, 2014.

[19] M. Raginsky, N. Kiarashi, and R. Willett, "Decentralized online convex programming with local information," in *Proc. American Control Conf.*, 2011, pp. 5363–5369.

[20] K. I. Tsianos, S. Lawlor, and M. G. Rabbat, "Consensus-based distributed optimization: Practical issues and applications in large-scale machine learning," in *50th Ann. Allerton Conf. Communication, Control, and Computing*, 2012, pp. 1543–1550.

[21] S. Hosseini, A. Chapman, and M. Mesbahi, "Online distributed convex optimization on dynamic networks," *IEEE Trans. Autom. Control*, vol. 61, no. 11, pp. 3545–3550, 2016.

[22] S. Shahrampour and A. Jadbabaie, "Distributed online optimization in dynamic environments using mirror descent," *IEEE Trans. Autom. Control*, vol. 63, no. 3, pp. 714–725, 2017.

[23] S. Lee, A. Nedić, and M. Raginsky, "Coordinate dual averaging for decentralized online optimization with nonseparable global objectives," *IEEE Trans. Control Netw. Syst.*, vol. 5, no. 1, pp. 34–44, 2016.
[24] F. Lin, M. Fardad, and M. R. Jovanović, "Algorithms for leader selection in stochastically forced consensus networks," *IEEE Trans. Autom. Control*, vol. 59, no. 7, pp. 1789–1802, 2014.
[25] P. Barooah and J. P. Hespanha, "Graph effective resistance and distributed control: Spectral properties and applications," in *Proc. 45th IEEE Conf. Decision and Control*, 2006, pp. 3479–3485.
[26] S. Salsa, *Partial Differential Equations in Action: From Modelling to Theory*. New York: Springer, 2008.
[27] A. Chapman and M. Mesbahi, "System theoretic aspects of influenced consensus: Single input case," *IEEE Trans. Autom. Control*, vol. 57, no. 6, pp. 1505–1511, 2012.
[28] S. Patterson and B. Bamieh, "Network coherence in fractal graphs," in *IEEE Conf. Decision and Control and European Control Conf.*, 2011, pp. 6445–6450.
[29] B. Bamieh, M. R. Jovanović, P. Mitra, and S. Patterson, "Coherence in large-scale networks: Dimension-dependent limitations of local feedback," *IEEE Trans. Autom. Control*, vol. 57, no. 9, pp. 2235–2249, 2012.
[30] S. Boyd and L. Vandenberghe, *Convex Optimization*. New York: Cambridge University Press, 2004.
[31] J. Nocedal and S. J. Wright, *Numerical Optimization*. New York: Springer, 2006.
[32] B. Bollobás, *Modern Graph Theory*. New York: Springer, 1998.
[33] R. A. Horn and C. R. Johnson, *Matrix Analysis*. New York: Cambridge University Press, 1990.

7 Meanings and Applications of Structure in Networks of Dynamic Systems

Vasu Chetty and Sean Warnick

7.1 Introduction: Dynamics and Structure

Dynamics and structure are two of the most important properties of a system. A system's dynamics describe its *behavior*, that is, how it constrains allowed combinations of manifest variables and defines what is possible and impossible [1]. So, for example, a deterministic signal-processing system restricts the allowed output trajectory that corresponds to a given input trajectory; this allowed combination is possible, while other combinations of different output trajectories with this input trajectory are impossible. A considerable body of literature over the last century has addressed the representation, analysis, and design of a system's dynamics, and a rich theory is now well established for doing so, especially for systems in feedback [2–4].

This focus on feedback highlights the power of systems theory to describe problems involving the interconnection of systems, and thus to describe and address questions of *structure*. Indeed, results demonstrating how a feedback structure can systematically compensate for model uncertainty are among the most important system-design concepts available. Understanding structure can also help describe information constraints in cyber-physical systems (CPSs), that is, constraints characterizing what information is available to which parts of a system at various times [5, 6]. Such constraints are not only useful for characterizing distributed systems, but they also play an important role in describing uncertainty about a system, including the uncertainty arising from possibilities of cyber-attacks and other security problems [7, 8]. Furthermore, understanding the specific structure of a particular system can play an important role in reverse engineering how the system realizes its behavior – that is, how it actually achieves the observed behavior.

Systems theory offers a powerful language for interconnecting systems into composite systems, and therefore it naturally describes system structure as the interconnection of component subsystems [9]. Nevertheless, its varied descriptions of systems lead to varied notions of system structure, some of which are more meaningful in certain domains than others. For example, two representations of a controlled, causal, linear time-invariant system include its impulse-response matrix (or, equivalently, its transfer function matrix) and a state space realization. These different representations of the same system lead to different perspectives on the structure of the system. On the one hand, the sparsity pattern of the impulse-response matrix indicates the existence of *paths* from each input to each output within the system, while the sparsity pattern of the state matrices reveal details about how the system *captures*, *stores*, and *retrieves* information. The critical lessons are that *structure is a property of a mathematical model of a system, not of the system itself*, and *any given system will have multiple structures available for consideration and analysis*.

This chapter reviews four notions of system structure, three of which are contextual and classic (i.e., the complete computational structure linked to a state space model, the sparsity pattern of a transfer function, and the interconnection of subsystems) and one which is relatively new (i.e., the *signal structure* of a system's *dynamical structure function*). Although each of these structural concepts apply to the nonlinear and stochastic setting, this work will focus on linear time-invariant systems to distill the key concepts and make their relationships clear. We then discuss three applications of the newest structural form (the signal structure of a system's dynamical structure function): network reconstruction, vulnerability analysis, and a more recent result in distributed control that guarantees the synthesis of a stabilizing controller with a specified structure, or proves that no such controller exists.

7.1.1 What Is System Structure?

Many physical systems have a natural notion of structure; loosely speaking, it's the way things are put together. Certainly buildings, bridges, and other edifices are excellent examples of the idea; who doesn't immediately recognize the characteristic structure of the Eiffel Tower, the Parthenon, or the Golden Gate Bridge? These structures maintain their distinct appearances because of the ways their respective components have been assembled; the idea of structure, in these cases, is intimately connected to the interconnection pattern of components.

Certainly this notion of structure, the interconnection pattern of components, is equally meaningful for a number of other engineered systems. Four wheels or sweptback wings are familiar structural choices for many land or air vehicles. Clocks, watches, locks, musical instruments, incandescent light bulbs, LEDs, televisions, radios, satellites, circuit boards, server farms, and even the Internet all have meaningful notions of structure characterized by the interconnection of components or subsystems. Each of these systems results in a characteristic physical appearance and a corresponding interconnection pattern, at various scales, that we may come to think of as its "structure."

But what about chemical processes, ocean currents, or even the weather? It doesn't seem to make sense to talk about the way the "weather" is put together. Does the *fluidic* aspect of these kinds of systems remove them from meaningful interpretations of structure? Certainly the resulting amorphous quality of these systems does make it difficult to conceptualize compartmentalized subsystems and think about their interconnection. Nevertheless, another notion of structure, based on how manifest variables affect each other, can be very meaningful for such systems.

We call this newer notion of structure a system's *signal structure*, since it describes how manifest signals affect each other. It is defined for all systems, even those with obvious component subsystems, and sometimes it coincides with the interconnection pattern of subsystems. Nevertheless, frequently – even when a system exhibits a meaningful pattern of interconnected subsystems, and especially when no such interconnection pattern exists – the signal structure offers a unique perspective on a system's internal interactions and lends insight about the system not available otherwise. Consider, for example, the market demand for the offering of goods available at a particular store. Prices of the goods each day may be viewed as inputs to the system, and corresponding daily sales can be seen as outputs. Although it is difficult to imagine compartmentalized subsystems interconnected to produce the resulting market demand, it is, on the other hand, quite natural to consider how interrelatedness of products shapes the resulting sales dynamic, leading to particular dependencies among observed sales. Signal structure captures such dependencies and can reveal elegant structure in complex systems. Information diffuses and flows, much like a fluid, so CPSs benefit from both subsystem and signal structural views. In this chapter we review standard notions of system structure, including the interconnection of subsystems, and we present a detailed treatment of signal structure and its corresponding mathematical representation, the dynamical structure function.

7.1.2 Why Does System Structure Matter?

There are many situations in which a particular structure of a system directly impacts its dynamic behavior, and thus the need, for example, for a specific shape in a ship's hull, or a design of a complex freeway interchange, is well understood. In these cases we simply choose a structure that yields a system with the desired behavior.

Nevertheless, what about situations where very different structural choices yield exactly the same behavior, such as is frequently the case with software, electronics, and a variety of other systems? In these cases, is one structural choice preferred over others? What criteria should one use to evaluate different structural options when the system dynamics are otherwise equivalent? Consider the following:

- **Implementation cost**. The fact that a given transfer function has many state realizations, some of which may be much more sparse than others, illustrates the important idea that the same manifest behavior of a system can often be realized from implementations with significantly different numbers of internal components. In situations in which the number of components is proportional to the cost of the implementation, as is the case for many physical systems, implementation cost then becomes an important reason for understanding the structural choices available to realize a specific dynamic design.
- **Understandability**. The internal structure of a system can be important to help one understand (or hinder an outsider from understanding) how the system works. Hierarchy and modularity of subsystems are examples of methods for organizing designs so that complex systems can be more easily understood. This understandability can have a major impact on other aspects of system management, such as making the system easier to:
  - visualize
  - promote situational awareness

- verify
- diagnose for component failure
- facilitate targeted access to system components
- maintain.

On the other hand, making structural choices that reduce the understandability of a system can help to secure the system from various types of infiltration, including espionage and sabotage.

- **Learning from data**. Since different mathematical models can describe the same system, and some of these models correspond to more detailed notions of structure than others, models with coarser structural descriptions are easier to learn from data than are others. Choosing a mathematical representation of a system consistent with the information available to identify it from data is critical to accurately infer its network structure; input–output data can identify the sparsity structure of a system's transfer function, but more information about the system must be known, a priori, to identify more detailed models. Sometimes, identifying structurally richer models of a system is called *reverse engineering*, *network inference*, or *network reconstruction*. More about network reconstruction is discussed in Section 7.3.1.
- **Attack modeling**. Different mathematical models of a system exhibit different notions of system structure, and these various structures expose different parts of the system as "links" or "nodes" in an appropriately defined structural graph. Under the assumption that the likelihood of attacks involving multiple links in the system is inversely proportional to the number of participating links (e.g., that single-link attacks or failures are more likely than coordinated, multi-link strikes), a particular structural representation of the system also characterizes a class of anticipated perturbations with associated risks. Analyzing the robustness of system properties such as stability, controllability, and observability with respect to perturbations in this class leads to a method for systematically characterizing system vulnerability. More on this topic of vulnerability analysis is discussed in Section 7.3.2.
- **Constraint modeling**. As the size and complexity of engineered systems grow, the need to make judicious choices about how to move information from one part of the system to another becomes increasingly important, since communication costs or delays may have a significant impact on system performance. These choices suggest the need for structural analyses of the system, and various notions of structure can effectively model different types of information constraints of the system. More on this topic, with respect to the design of distributed stabilizing controllers, is discussed in Section 7.3.3.

Although there are many purposes for developing a rich theory of system structure, this handful of reasons describes much of the motivation behind the work presented in this chapter. The next section reviews classical notions of system structure as a context for introducing dynamical structure functions and the signal structure. It is followed by a discussion of three motivating applications of signal structure: network reconstruction, vulnerability analysis, and the design of stabilizing, distributed controllers.

7.2 Mathematical Representations of Systems and Their Structures

This section describes four different mathematical representations of systems and discusses the definition and meaning of the corresponding structure for each: the generalized state space model with its complete computational structure; the transfer function and the input–output sparsity structure, structured linear fractional transformations, and the subsystem structure; and the dynamical structure function with its signal structure. Each of these system representations completely characterize the dynamic behavior of a system, and thus they are equivalent from a behavioral perspective. Nevertheless, they retain varying degrees of structural information, and thus these system representations can be ordered based on their structural informativity.

In this work, a "structure" is a directed graph. We will see that different system representations specify different structural graphs, and each structural graph carries with it a unique interpretation, or meaning. We will restrict our attention to finite-dimensional, causal, deterministic linear time-invariant (LTI) systems defined over continuous time, but the concepts extend naturally to the nonlinear and stochastic settings with different types of independent variables.

7.2.1 State Space Models and the Complete Computational Structure

State space models are the most structurally informative system representation considered here. The standard state space model for LTI systems is given by:

$$\begin{aligned} \dot{x} &= Ax + Bu \\ y &= Cx + Du \end{aligned} \tag{7.1}$$

where $x(t) \in \mathbb{R}^n$ represents a vector of n system state variables, each defined over $t \in \mathbb{R}$; $\dot{x}(t) \in \mathbb{R}^n$ represents the time derivative of these state variables; $u(t) \in \mathbb{R}^m$ are controlled inputs into the system; and $y(t) \in \mathbb{R}^p$ are measured outputs. Recall that states have a particular meaning, being the information necessary to characterize the future evolution of the system. That is to say, given the values of x at some time (which will be labeled $t = 0$ without loss of generality), only future values of the input, $u(t)$ for $t \geq 0$, are needed to completely specify the evolution of the system for all times $t \geq 0$. This representation is sufficiently detailed to completely characterize both the transfer function and the dynamical structure function of a system, with their corresponding structures.

Nevertheless, this standard state space model does not differentiate between systems with different subsystem structures. For example, consider two systems in feedback. One can easily compute the closed-loop dynamics of such an interconnection and represent them with a single standard state space model. Nevertheless, if presented with this closed-loop model, one cannot determine what the two subsystems are that generate it. This failure to distinguish different subsystem structures comes from the standard state space model's lack of representation power to distinguish between the composition of functions (see Example 7.1).

To distinguish different subsystem structures, we need to differentiate between behaviorally equivalent computations such as (1) $f(x) = x$, (2) $f(x) = 2(0.5x)$, and (3) $f(x) = 0.3x + 0.7x$. We accomplish this by introducing *auxiliary variables*, w, that represent intermediate stages of computation. In this way we can differentiate (1) $f(x) = x$ from (2) $f(x) = 2w$ and $w = 0.5x$ or (3) $f(x) = w_1 + w_2$ and $w_1 = 0.3x$ and $w_2 = 0.7x$, since each of these different ways of computing the same functional relationship involve zero, one, or two auxiliary variables, respectively. The auxiliary variables that are specified, say, in a system's "blueprint" or manifest directly to observers help us distinguish the system's actual computational structure from others we could imagine.

Introducing auxiliary variables into the standard state space model characterizes a differential-algebraic system of equations capable of uniquely specifying all three of the other system representations discussed here. We call this modified system of equations the *generalized state space model* of a system, and represent it as

$$\begin{aligned} \dot{x} &= Ax + \hat{A}w + Bu \\ w &= \bar{A}x + \widetilde{A}w + \bar{B}u \\ y &= Cx + \bar{C}w + Du \end{aligned} \tag{7.2}$$

where $w \in \mathbb{R}^l$, $\hat{A} \in \mathbb{R}^{n\times l}$, $\bar{A} \in \mathbb{R}^{l\times n}$, $\widetilde{A} \in \mathbb{R}^{l\times l}$, $\bar{B} \in \mathbb{R}^{l\times m}$, and $\bar{C} \in \mathbb{R}^{p\times l}$. The number of auxiliary variables, l, is called the *intricacy* of the generalized state space model. Choosing $\widetilde{A}$ so that $I - \widetilde{A}$ is invertible yields a differentiability index of zero. This ensures that the auxiliary variables can always be algebraically eliminated from the system, producing a dynamically equivalent standard state space model (7.1). We call this equivalent standard state space model the *zero-intricacy* realization or representation of a given generalized state space model (7.2).

EXAMPLE 7.1 Consider the feedback interconnection of two systems, given by

$$\begin{aligned} \dot{x}_1 &= A_1x_1 + B_1r_1 \qquad & \dot{x}_2 &= A_2x_2 + B_2r_2 \\ y_1 &= C_1x_1 & y_2 &= C_2x_2 \end{aligned}$$

with $r_1 = u_1 + y_2$ and $r_2 = u_2 + y_1$, where u_1 and u_2 are exogenous inputs to the closed-loop system, and y_1 and y_2 are measured outputs from the closed-loop system. Defining $w_1 = y_1$ and $w_2 = y_2$, we obtain the following generalized state space model of the feedback interconnection:

$$\begin{aligned} \begin{bmatrix} \dot{x}_1 \\ \dot{x}_2 \end{bmatrix} &= \begin{bmatrix} A_1 & 0 \\ 0 & A_2 \end{bmatrix}\begin{bmatrix} x_1 \\ x_2 \end{bmatrix} + \begin{bmatrix} 0 & B_1 \\ B_2 & 0 \end{bmatrix}\begin{bmatrix} w_1 \\ w_2 \end{bmatrix} + \begin{bmatrix} B_1 & 0 \\ 0 & B_2 \end{bmatrix}\begin{bmatrix} u_1 \\ u_2 \end{bmatrix} \\ \begin{bmatrix} w_1 \\ w_2 \end{bmatrix} &= \begin{bmatrix} C_1 & 0 \\ 0 & C_2 \end{bmatrix}\begin{bmatrix} x_1 \\ x_2 \end{bmatrix} + \begin{bmatrix} 0 & 0 \\ 0 & 0 \end{bmatrix}\begin{bmatrix} w_1 \\ w_2 \end{bmatrix} + \begin{bmatrix} 0 & 0 \\ 0 & 0 \end{bmatrix}\begin{bmatrix} u_1 \\ u_2 \end{bmatrix} \\ \begin{bmatrix} y_1 \\ y_2 \end{bmatrix} &= \begin{bmatrix} 0 & 0 \\ 0 & 0 \end{bmatrix}\begin{bmatrix} x_1 \\ x_2 \end{bmatrix} + \begin{bmatrix} 1 & 0 \\ 0 & 1 \end{bmatrix}\begin{bmatrix} w_1 \\ w_2 \end{bmatrix} + \begin{bmatrix} 0 & 0 \\ 0 & 0 \end{bmatrix}\begin{bmatrix} u_1 \\ u_2 \end{bmatrix} \end{aligned} \tag{7.3}$$

Note that $I - \widetilde{A}$ is invertible, thus enabling us to easily eliminate w from the equations. Doing so yields the zero-intricacy representation of the feedback interconnection:

$$\begin{aligned} \begin{bmatrix} \dot{x}_1 \\ \dot{x}_2 \end{bmatrix} &= \begin{bmatrix} A_1 & B_1C_2 \\ B_2C_1 & A_2 \end{bmatrix} \begin{bmatrix} x_1 \\ x_2 \end{bmatrix} + \begin{bmatrix} B_1 & 0 \\ 0 & B_2 \end{bmatrix} \begin{bmatrix} u_1 \\ u_2 \end{bmatrix} \\ \begin{bmatrix} y_1 \\ y_2 \end{bmatrix} &= \begin{bmatrix} C_1 & 0 \\ 0 & C_2 \end{bmatrix} \begin{bmatrix} x_1 \\ x_2 \end{bmatrix} \end{aligned} \tag{7.4}$$

Although these representations are dynamically equivalent, meaning that (7.3) and (7.4) generate identical state and output trajectories if they are given the same initial condition x_o and input trajectory $u(t)$, (7.3) encodes information to uniquely specify the original subsystems and their feedback interconnection structure, while (7.4) does not.

Example 7.1 illustrates a generalized state space model and the corresponding zero-intricacy realization of a system composed of the interconnection of multiple subsystems. In fact, whenever $I - \widetilde{A}$ is invertible, every generalized state space model has a unique, well-defined zero-intricacy realization. Likewise, every zero-intricacy state space model is dynamically equivalent to a rich variety of generalized state space models of any positive intricacy; these generalized state space models differ only in how their computations are performed, or in their underlying computational structure. We call this structure of the most refined generalized state space description of a system, even zero-intricacy ones, the *complete computational structure*, and all other notions of system structure discussed in this work can be derived directly from it.

DEFINITION 7.1 (*Complete computational structure*) Given a generalized state space model, as in (7.2), its *complete computational structure* is a weighted directed graph $\mathscr{C}$ with vertex set $V(\mathscr{C})$ and edge set $E(\mathscr{C})$ given by:

- $V(\mathscr{C}) = \{u_1, \ldots, u_m, x_1, \ldots, x_n, w_1, \ldots, w_l, y_1, \ldots, y_p\}$, and
- $E(\mathscr{C})$ is specified by the nonzero entries of the adjacency matrix $\mathscr{A}(\mathscr{C})$, where

$$\mathscr{A}(\mathscr{C}) = \begin{bmatrix} 0 & 0 & 0 & 0 \\ B & A & \hat{A} & 0 \\ \bar{B} & \bar{A} & \widetilde{A} & 0 \\ D & C & \bar{C} & 0 \end{bmatrix}^T \tag{7.5}$$

That is to say, a potential edge from $v_i \in V(\mathscr{C})$ to $v_j \in V(\mathscr{C})$ has weight $\mathscr{A}(\mathscr{C})_{ij}$, but we only recognize the existence of edges with nonzero weight.

The generalized state space model (7.2) encodes information about how the system performs the computations necessary to realize its dynamic behavior. It is like an *information blueprint* of how specific components are interconnected to access information from input signals; how this information is represented (in a specific coordinate system) and combined with other data retrieved from memory; how these new calculations are stored; and how all of these data combine to produce measurable output signals. The meaning, then, of the complete computational structure characterized by (7.5) is the *information architecture* of a very specific computation system: how information is represented, transformed, and flows through the system. Note that there is a distinction between "physical structure" and state space models; in some cases, the particular basis specified by a state space model is more detailed than the physical structure may suggest. For example, consider an inertial mass. This mass behaves like a second-order system according to Newton's Second Law of Motion, but it is not clear whether states of the system are necessarily position and velocity, or whether they are some linear combinations of position and velocity. Exactly how some systems represent and store information may be unclear, but if it were known, state space models are capable of representing this refined level of structural knowledge. These models (the generalized state space model and its associated complete computational structure) then become the most refined knowledge of our system, ground truth from which all other representations can be compared.

Note that because intricacy variables can always be eliminated from a generalized state description without changing its dynamics, the most refined generalized state space model, with intricacy $l > 0$, immediately defines a particular sequence of state space models indexed by their intricacies, $l-1, \ldots, 0$. Each of these coarser models has a structure associated with it that we call a *computational structure*, but we reserve the descriptor *complete computational structure* for the most refined structural specification of the system; once the complete computational structure is specified, even if it has zero intricacy, all other hypothetical refinements are considered fictitious while any agglomerative structure derived from it is a valid notion of structure for the system.

EXAMPLE 7.2 Making Example 7.1 concrete, consider the following two systems:

$$\begin{bmatrix}\dot{x}_1\\ \dot{x}_2\end{bmatrix} = \begin{bmatrix}-1 & 2\\ 0 & -2\end{bmatrix}\begin{bmatrix}x_1\\ x_2\end{bmatrix} + \begin{bmatrix}2 & -1\\ -1 & 1\end{bmatrix}\begin{bmatrix}r_1\\ r_2\end{bmatrix} \qquad \begin{bmatrix}\dot{x}_3\\ \dot{x}_4\\ \dot{x}_5\end{bmatrix} = \begin{bmatrix}-5 & -4 & 2\\ 3 & 2 & -1\\ 0 & 0 & -3\end{bmatrix}\begin{bmatrix}x_3\\ x_4\\ x_5\end{bmatrix} + \begin{bmatrix}0 & -1\\ 0 & 1\\ 1 & 0\end{bmatrix}\begin{bmatrix}r_3\\ r_4\end{bmatrix}$$

$$\begin{bmatrix}y_1\\ y_2\end{bmatrix} = \begin{bmatrix}1 & 2\\ 1 & 1\end{bmatrix}\begin{bmatrix}x_1\\ x_2\end{bmatrix}, \qquad \begin{bmatrix}y_3\\ y_4\end{bmatrix} = \begin{bmatrix}1 & 2 & 0\\ 1 & 1 & 0\end{bmatrix}\begin{bmatrix}x_3\\ x_4\\ x_5\end{bmatrix}$$

interconnected in feedback, so that

$$\begin{bmatrix}r_1\\ r_2\end{bmatrix} = \begin{bmatrix}y_3\\ y_4\end{bmatrix} + \begin{bmatrix}u_1\\ u_2\end{bmatrix}, \qquad \begin{bmatrix}r_3\\ r_4\end{bmatrix} = \begin{bmatrix}y_1\\ y_2\end{bmatrix} + \begin{bmatrix}u_3\\ u_4\end{bmatrix}$$

leading to the following generalized state space model:

$$\begin{aligned}
\begin{bmatrix}\dot{x}_1\\ \dot{x}_2\\ \dot{x}_3\\ \dot{x}_4\\ \dot{x}_5\end{bmatrix} &= \begin{bmatrix}-1 & 2 & 0 & 0 & 0\\ 0 & -2 & 0 & 0 & 0\\ 0 & 0 & -5 & -4 & 2\\ 0 & 0 & 3 & 2 & -1\\ 0 & 0 & 0 & 0 & -3\end{bmatrix}\begin{bmatrix}x_1\\ x_2\\ x_3\\ x_4\\ x_5\end{bmatrix} + \begin{bmatrix}0 & 0 & 2 & -1\\ 0 & 0 & -1 & 1\\ 0 & -1 & 0 & 0\\ 0 & 1 & 0 & 0\\ 1 & 0 & 0 & 0\end{bmatrix}\begin{bmatrix}w_1\\ w_2\\ w_3\\ w_4\end{bmatrix} + \begin{bmatrix}2 & -1 & 0 & 0\\ -1 & 1 & 0 & 0\\ 0 & 0 & 0 & -1\\ 0 & 0 & 0 & 1\\ 0 & 0 & 1 & 0\end{bmatrix}\begin{bmatrix}u_1\\ u_2\\ u_3\\ u_4\end{bmatrix}\\
\begin{bmatrix}w_1\\ w_2\\ w_3\\ w_4\end{bmatrix} &= \begin{bmatrix}1 & 2 & 0 & 0 & 0\\ 1 & 1 & 0 & 0 & 0\\ 0 & 0 & 1 & 2 & 0\\ 0 & 0 & 1 & 1 & 0\end{bmatrix}\begin{bmatrix}x_1\\ x_2\\ x_3\\ x_4\\ x_5\end{bmatrix} + \begin{bmatrix}0 & 0 & 0 & 0\\ 0 & 0 & 0 & 0\\ 0 & 0 & 0 & 0\\ 0 & 0 & 0 & 0\end{bmatrix}\begin{bmatrix}w_1\\ w_2\\ w_3\\ w_4\end{bmatrix} + \begin{bmatrix}0 & 0 & 0 & 0\\ 0 & 0 & 0 & 0\\ 0 & 0 & 0 & 0\\ 0 & 0 & 0 & 0\end{bmatrix}\begin{bmatrix}u_1\\ u_2\\ u_3\\ u_4\end{bmatrix}\\
\begin{bmatrix}y_1\\ y_2\\ y_3\\ y_4\end{bmatrix} &= \begin{bmatrix}0 & 0 & 0 & 0 & 0\\ 0 & 0 & 0 & 0 & 0\\ 0 & 0 & 0 & 0 & 0\\ 0 & 0 & 0 & 0 & 0\end{bmatrix}\begin{bmatrix}x_1\\ x_2\\ x_3\\ x_4\\ x_5\end{bmatrix} + \begin{bmatrix}1 & 0 & 0 & 0\\ 0 & 1 & 0 & 0\\ 0 & 0 & 1 & 0\\ 0 & 0 & 0 & 1\end{bmatrix}\begin{bmatrix}w_1\\ w_2\\ w_3\\ w_4\end{bmatrix} + \begin{bmatrix}0 & 0 & 0 & 0\\ 0 & 0 & 0 & 0\\ 0 & 0 & 0 & 0\\ 0 & 0 & 0 & 0\end{bmatrix}\begin{bmatrix}u_1\\ u_2\\ u_3\\ u_4\end{bmatrix}.
\end{aligned} \tag{7.6}$$

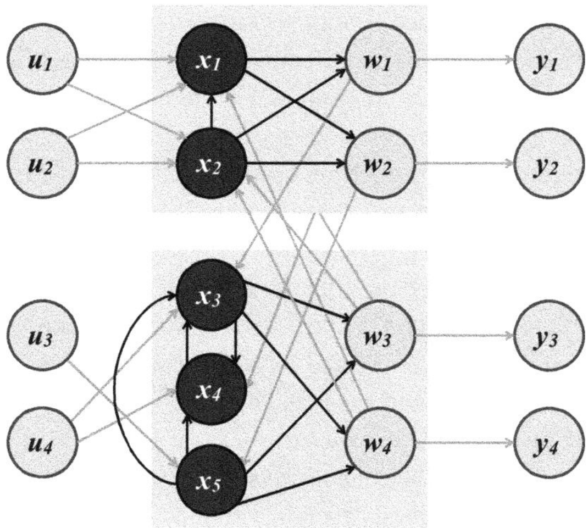

Figure 7.1 Complete computational structure of the generalized state space model from (7.6). Light nodes are manifest variables, while dark nodes indicate hidden variables. Notice that the original feedback structure of subsystems, reflected by gray boxes, is preserved, since the only interaction between subsystems is through manifest variables.

The complete computational structure of this system, given in (7.6), is shown in Figure 7.1. The zero-intricacy realization of this generalized state space model, (7.6), is then given by:

$$\begin{bmatrix} \dot{x}_1 \\ \dot{x}_2 \\ \dot{x}_3 \\ \dot{x}_4 \\ \dot{x}_5 \end{bmatrix} = \begin{bmatrix} -1 & 2 & 1 & 3 & 0 \\ 0 & -2 & 0 & -1 & 0 \\ -1 & -1 & -5 & -4 & 2 \\ 1 & 1 & 3 & 2 & -1 \\ 1 & 2 & 0 & 0 & -3 \end{bmatrix} \begin{bmatrix} x_1 \\ x_2 \\ x_3 \\ x_4 \\ x_5 \end{bmatrix} + \begin{bmatrix} 2 & -1 & 0 & 0 \\ -1 & 1 & 0 & 0 \\ 0 & 0 & 0 & -1 \\ 0 & 0 & 0 & 1 \\ 0 & 0 & 1 & 0 \end{bmatrix} \begin{bmatrix} u_1 \\ u_2 \\ u_3 \\ u_4 \end{bmatrix}$$

$$\begin{bmatrix} y_1 \\ y_2 \\ y_3 \\ y_4 \end{bmatrix} = \begin{bmatrix} 1 & 2 & 0 & 0 & 0 \\ 1 & 1 & 0 & 0 & 0 \\ 0 & 0 & 1 & 2 & 0 \\ 0 & 0 & 1 & 1 & 0 \end{bmatrix} \begin{bmatrix} x_1 \\ x_2 \\ x_3 \\ x_4 \\ x_5 \\ x_6 \end{bmatrix} + \begin{bmatrix} 0 & 0 & 0 & 0 \\ 0 & 0 & 0 & 0 \\ 0 & 0 & 0 & 0 \\ 0 & 0 & 0 & 0 \end{bmatrix} \begin{bmatrix} u_1 \\ u_2 \\ u_3 \\ u_4 \end{bmatrix} \tag{7.7}$$

The computational structure of the zero-intricacy realization, given in (7.7), is shown in Figure 7.2. Notice the differences with the complete computational structure shown in Figure 7.1. For example, the complete computational structure has nodes for auxiliary variables, w, while the computational structure of the zero-intricacy realization does not. Also, original subsystem structure is preserved in the complete computational structure, highlighted by the background gray boxes, while it is lost in the computational structure of the zero-intricacy realization, resulting in no distinguishable subsystems.

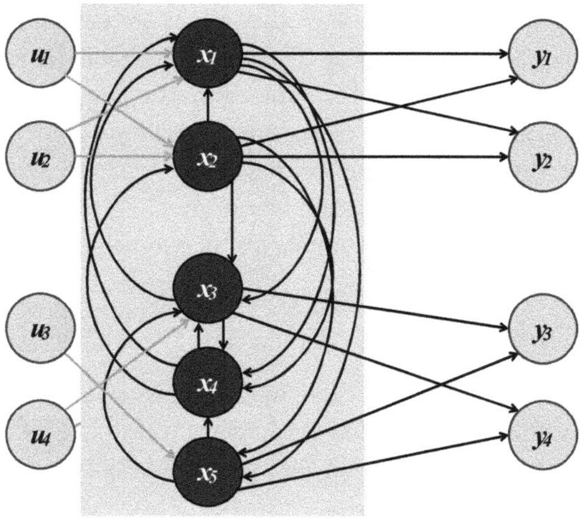

Figure 7.2 Computational structure of the zero-intricacy realization (7.7) of the generalized state space model in (7.6). Like Figure 7.1, light nodes indicate manifest variables while dark nodes are hidden variables. Notice that the original subsystem structure is lost, and only a single subsystem remains visible from manifest variables.

7.2.2 Functional System Descriptions and the Manifest Structure

While state space models are the most structurally informative system representations, functional system descriptions, such as convolution models or transfer functions, are at the other end of the spectrum. These "black box" representations of a system are capable of describing the same dynamic behavior as their state space counterparts, yet they do not model the detailed interactions among system components the way state space representations do.[1]

This inability to convey detailed structural information is not necessarily a weakness, however. For example, functional representations need fewer parameters to characterize a given dynamic behavior, making them easier to learn from data (called *system identification* [10, 11]) than their state space counterparts. Moreover, their parsimonious description of a system's dynamics creates an important distinction between a system's behavior and how it realizes that behavior, enabling a concerted focus on the design of a system's dynamics without worrying about implementation.

Just as high-level programming languages abstract many of the details of the computer they run on, functional system descriptions are high-level abstractions of state space models. In particular, the specific processes a state realization uses to decide which information is stored in which parts of the state vector correspond to memory management activities that are completely invisible to a functional description of a system. This distinction is further exemplified by noting that state space models are imperative descriptions of a system, encoding computations in terms of the time

[1] Although transfer functions and convolution models of LTI systems assume zero initial conditions, the impact of a nonzero initial condition is easily modeled with the addition of an appropriately designed external disturbance.

evolution of the system state, while functional descriptions are inherently declarative, specifying what the system does without prescribing how it should do it.

The result of this high-level/low-level relationship between functional system descriptions and state space models is a one-to-many relationship between the two model classes. That is, every state space model has a zero-intricacy realization, as in (7.1), that identifies a unique functional system description, whether it be the impulse response matrix of a convolution model or a transfer function matrix, given by:

$$\begin{aligned} y(t) &= h(t) * u(t) & Y(s) &= H(s)U(s) \\ h(t) &= Ce^{At}B + D\delta(t), & H(s) &= C(SI - A)^{-1}B + D \end{aligned} \tag{7.8}$$

where $*$ denotes convolution, $\delta(t)$ is the Dirac delta function, $h(t)$ is the system's $p \times m$ impulse response matrix, $Y(s)$ and $U(s)$ are the Laplace transforms of $y(t)$ and $u(t)$, and $H(s)$ is the system's $p \times m$ transfer function matrix – which is also the Laplace transform of $h(t)$.

Note, however, that there are many state space models that specify the same impulse response or transfer function; each of these state space models specifies a different implementation (or realization) of the same dynamic behavior. Among all these state realizations of a given functional description of a system, some have fewer states than others. In fact, systems with functional descriptions that can be described by finite-dimensional LTI state space models[2] have a unique integer, n, associated with them, called the *Smith–McMillan degree*. This degree is the minimal number of states necessary for any state space realization of the system. Nevertheless, even restricting attention to state space models with order equal to the Smith–McMillan degree does not yield a unique state realization; given a minimal realization (A, B, C, D) of a transfer function $H(s)$, any $n \times n$ transformation, T, yields another minimal realization $(\hat{A}, \hat{B}, \hat{C}, \hat{D})$ given by:

$$\hat{A} = TAT^{-1} \quad \hat{B} = TB \quad \hat{C} = CT^{-1} \quad \hat{D} = D \tag{7.9}$$

such that $C(sI - A)^{-1}B + D = H(s) = \hat{C}(sI - \hat{A})^{-1}\hat{B} + \hat{D}$. Thus, even among minimal realizations, there are infinitely many implementations of a given dynamic behavior specified by a functional description such as $H(s)$, and these implementations differ only in their structural properties.

The functional description of a system, however, retains only the structural properties that are common among all of its state realizations, which is precisely the mathematical structure of the functional description itself. This structure describes the internal closed-loop relationships among manifest variables, and therefore is called the *manifest structure*.

[2] Although all LTI state space models have transfer functions, not all transfer functions have state space realizations. This is because the imperative nature of state space models demands that they are *causal*, meaning that future values of manifest variables only depend on past and present values of manifest variables. Transfer functions that are *proper* rational functions of the Laplace variable correspond to causal finite dimensional LTI systems; we do not concern ourselves with other kinds in this work.

DEFINITION 7.2 (*Manifest structure*) Given a generalized state space model, as in (7.2), identified by a functional system description, as in (7.8), its *manifest structure* is a weighted directed graph $\mathscr{M}$ with vertex set $V(\mathscr{M})$ and edge set $E(\mathscr{M})$ given by:

- $V(\mathscr{M}) = \{u_1, \ldots, u_m, y_1, \ldots, y_p\}$, each representing a manifest signal of the system, and
- $E(\mathscr{M})$ has an edge from u_i to y_j, labeled by either H_{ji} or h_{ji}, provided they are nonzero.

Note that when a system's manifest variables partition naturally into inputs and outputs, then its manifest structure is a bipartite graph, with directed edges from inputs to outputs.

An alternative definition of the manifest structure characterizes $\mathscr{M}$ directly from $\mathscr{C}$ using only graphical properties (which is useful when extending these results to the nonlinear setting). In that case, we say $\mathscr{M}$ has an edge from u_i to y_j if the net impact of all paths in $\mathscr{C}$ from u_i to y_j is nonzero, or, equivalently, if every equivalent realization of the system, specified by a transformation T as in (7.9), with complete computational structure $\mathscr{C}_T$, has a path from u_i to y_j.

EXAMPLE 7.3 Consider the zero-intricacy state space model in (7.7) from Example 7.2. The corresponding transfer function is given by:

$$H(s) = C(sI - A)^{-1}B + D =$$

$$\begin{bmatrix} 0 & \frac{s^2+5s+6}{s^3+6s^2+11s+5} & \frac{1}{s^3+6s^2+11s+5} & 0 \\ \frac{s+1}{s^2+3s+1} & \frac{s^3+6s^2+11s+8}{s^5+9s^4+30s^3+44s^2+26s+5} & \frac{3s+3}{s^5+9s^4+30s^3+44s^2+26s+5} & \frac{1}{s^2+3s+1} \\ \frac{1}{s^2+3s+1} & \frac{s^2+7s+10}{s^5+9s^4+30s^3+44s^2+26s+5} & \frac{2s^2+6s+5}{s^5+9s^4+30s^3+44s^2+26s+5} & \frac{s+2}{s^2+3s+1} \\ 0 & \frac{1}{s^3+6s^2+11s+5} & \frac{s+1}{s^3+6s^2+11s+5} & 0 \end{bmatrix} \quad (7.10)$$

The manifest structure corresponding to this transfer function (7.10) that represents the internal closed-loop pathways from inputs to outputs of the system in (7.2) is given in Figure 7.3.

Note that in some cases, although a pathway exists from an input to an output in the system's complete computational structure, it is possible that the corresponding transfer function from the input to the output is zero. For example, notice that although paths exist from every input to every output in the computational structure of the zero-intricacy realization generating H (Figure 7.2), H_{11}, H_{41}, H_{14}, and H_{44} are nevertheless all zero. Thus, we see that the existence of paths from u_i to y_j is not sufficient for H_{ij} to be nonzero; the closed-loop, net effect of *all* paths from u_i to y_j must be nonzero for H_{ij} to be nonzero; exact cancellations, which can be common in software and other engineered systems, can generate zeros in the functional description.

7.2.3 Structured Linear Fractional Transformations and the Subsystem Structure

Having identified the complete computational structure as the most informative structural representation, and the manifest structure as the least, we now explore the most

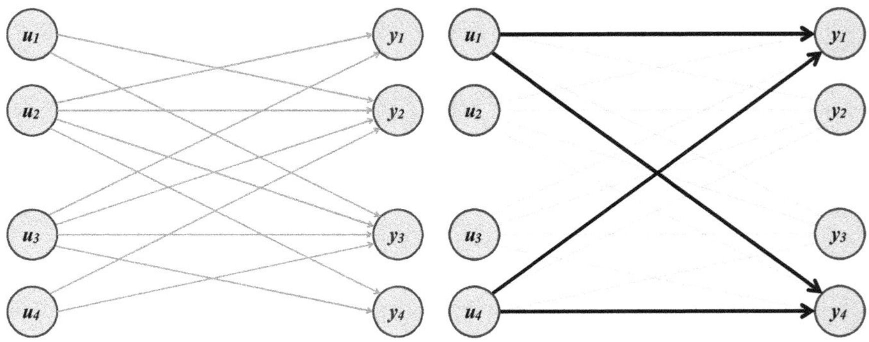

(a) Manifest structure of the system with transfer function (7.10)

(b) Missing edges in the manifest structure, corresponding to zero elements in H.

Figure 7.3 Manifest structure of the same system from Figures 7.1 and 7.2. Notice the lack of edges from u_1 to y_1 and y_4, and from u_4 to y_1 and y_4, corresponding to associated zeros in $H(s)$. These missing links are highlighted in Figure 7.3(b). Note that these links are missing in the manifest structure even though paths exist in Figure 7.2 from every input to every output.

common intermediate structural representation: the interconnection of subsystems. Subsystem structure is less informative than the complete computational structure because it does not reveal the internal structure of subsystems. On the other hand, subsystem structure can be more informative than manifest structure because it reveals the interconnection pattern among subsystems.

To isolate and represent the interconnection pattern of subsystems for a given system, begin by considering a set of q subsystems, $S = \{S_1 \quad S_2 \quad \dots \quad S_q\}$, interconnected into a composite system, H. It is conceivable that each of these subsystems are themselves divisible into constituent subsystems, or that not all of the q subsystems are discernible from H's manifest variables, so we specify the level of modeling abstraction by:

1. Modeling each of the q constituent subsystems with a suitable functional description, such as a proper or strictly proper transfer function $S_i(s)$, $i = 1, 2, \dots, q$, or a *single-subsystem* state space realization, characterized as a generalized state space model with subsystem structure consisting of a single subsystem, so that no further division of the subsystems is possible.
2. Ensuring that each of the subsystem's outputs, w_i, is a measured output of the composite system H, so $y = [w_1^T \quad w_2^T \quad \cdots \quad w_q^T]^T$, where y is the output of H.

Note that each subsystem is *distinct*, meaning that state variables internal to one subsystem are different from those of the other subsystems, yielding no mechanism for interaction except through their respective manifest variables. Let u be a vector of external inputs; v_i and w_i be the vectors of inputs and outputs for system S_i; and v and w be the stacked inputs and outputs from all systems, $v = [v_1^T \quad v_2^T \quad \cdots \quad v_q^T]^T$ and $w = [w_1^T \quad w_2^T \quad \cdots \quad w_q^T]^T$, so that $w = y$. Interconnecting these systems then means defining binary matrices L and K such that:

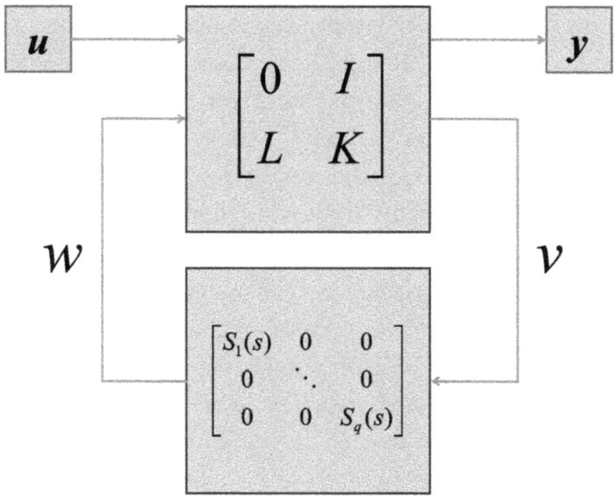

Figure 7.4 A structured linear fractional transformation revealing the interconnection structure among subsystems in binary matrices L and K.

$$\begin{bmatrix} L & K \end{bmatrix} \begin{bmatrix} u \\ w \end{bmatrix} = v \tag{7.11}$$

Our convention is that the process of interconnection only allows the selection of particular signals and possibly adding them together, thus restricting the interconnection matrices, L and K, to have elements with values of either 0 or 1; all other computations are part of the systems in S. Further, we assume that the resulting interconnection is *well-posed*, meaning that all signals within H are uniquely specified for any value of external inputs and underlying state variables [12]. This assumption ensures that the proposed interconnection is physically sensible and not merely a mathematical artifact.

The composite system, H, is then clearly defined by the structured linear fractional transformation (LFT), as in Figure 7.4, given by:

$$\begin{aligned} N \begin{bmatrix} u \\ w \end{bmatrix} &= \begin{bmatrix} y \\ v \end{bmatrix} \\ w &= Sv \end{aligned} \tag{7.12}$$

where

$$N = \begin{bmatrix} 0 & I \\ L & K \end{bmatrix}, \qquad S = \begin{bmatrix} S_1 & 0 & \cdots & 0 \\ 0 & S_2 & & 0 \\ \vdots & & \ddots & \vdots \\ 0 & & \cdots & S_q \end{bmatrix} \tag{7.13}$$

and S_i can be represented by either a suitable functional description, such as a proper or strictly proper transfer function matrix or the associated impulse response matrix of a convolution model, or by any single-subsystem generalized state realization. The symbol S is overloaded, representing both the set of subsystems and the decoupled operator of subsystem models in (7.13), but the appropriate meaning should always be clear from context. Equations (7.12) and (7.13) characterize H as a structured LFT in terms of S. Combining these equations yields, for example, $Y(s) = [S(s)(I - KS(s))^{-1}L]U(s)$, implying that $H(s) = S(s)(I - KS(s))^{-1}L$, where $Y(s)$ and $U(s)$ are the Laplace transforms of $y(t)$ and $u(t)$, respectively; similarly, a well-specified expression can be obtained for h directly in the time domain. The functional description of the composite system, H, in either the time or frequency domain, is completely specified by the structured LFT description in (7.12) and (7.13).

Although the structured LFT completely specifies the functional description of the composite system, H, the structured LFT does not have enough structural information to specify H's complete computational structure or its associated generalized state space description. To do so, it would need information about the "true" structure of each constituent subsystem. This point may be clear when S is specified by a functional description for each subsystem, such as its transfer function, but it becomes more subtle when S is specified by a generalized state space model for each subsystem. In this case, it is important to understand that the state space model for each subsystem in S can be *any* single-subsystem realization of the associated transfer function, $S_i(s)$, since the structured LFT does not use any information about the internal structure of its subsystems. To realize the "true" generalized state description of H and its associated complete computational structure, one must have accurate descriptions of the complete computational structures for each constituent subsystem to complement the "interconnection" information in the structured LFT.

The structured LFT reveals the interconnection structure among subsystems, encoded in the binary *interconnection matrix*, N, in general, and in L and K in particular. Note that the interconnection structure in N is unaffected by whether the subsystems in S are represented by state space models or transfer functions. The internal computational structure of subsystems, revealed by state models of subsystems but not transfer function representations of subsystems, is not used when representing the subsystem structure of a system – only the interconnection structure *among* subsystems, not *within* subsystems, is relevant for this representation.

Aggregating L and K appropriately to account for the potentially multi-input multi-output nature of the constituent subsystems yields adjacency matrices from which the composite system's subsystem structure can be built. To accomplish this, let e_{v_i} denote the vector of 1s with length equal to the length of vector v_i. We then define the aggregation matrices

$$A_v = \begin{bmatrix} e_{v_1}^T & 0 & \cdots & 0 \\ 0 & e_{v_2}^T & & 0 \\ \vdots & & \ddots & \vdots \\ 0 & 0 & \cdots & e_{v_q}^T \end{bmatrix}, \qquad A_w = \begin{bmatrix} e_{w_1}^T & 0 & \cdots & 0 \\ 0 & e_{w_2}^T & & 0 \\ \vdots & & \ddots & \vdots \\ 0 & 0 & \cdots & e_{w_q}^T \end{bmatrix} \tag{7.14}$$

and use them to create the adjacency matrices:

$$\mathscr{A}(L) = \text{sgn}(A_v L)^T, \qquad \mathscr{A}(K) = \text{sgn}(A_v K A_w^T)^T \tag{7.15}$$

where sgn(·) denotes the sign function, yielding a value of 1 for positive entries, 0 for zero, and −1 for negative entries (which can never occur in this case). With these definitions, we are now prepared to characterize a system's *subsystem structure*:

DEFINITION 7.3 (*Subsystem structure*) Given a generalized state space model, as in (7.2), identified by a structured LFT, (N, S), as in (7.12) and (7.13) and with associated aggregation matrices as in (7.14) and adjacency matrices as in (7.15), its *subsystem structure* is a weighted directed graph $\mathscr{S}$ with vertex set $V(\mathscr{S})$ and edge set $E(\mathscr{S})$ given by:

- $V(\mathscr{S}) = \{u_1, \ldots, u_m, S_1, \ldots, S_q, y_1, \ldots, y_p\}$, representing input signals, subsystems, and output signals, respectively.
- $E(\mathscr{S})$ has an edge from
  - u_i to S_j if $\mathscr{A}(L)_{ij} = 1$, labeled u_i;
  - S_i to S_j if $\mathscr{A}(K)_{ij} = 1$, labeled w_i;
  - S_i to y_j if $(A_w)_{ij} = 1$, labeled y_j.

Note that the subsystem structure is qualitatively different from either the complete computational structure or the manifest structure in a few ways. First, while all the nodes of either the complete computational structure or the manifest structure represent *signals*, the nodes of the subsystem structure represent *systems*, namely the subsystems and exosystems associated with the generation of each input or measurement of each output signal. As a result, we often denote the nodes in the subsystem structure with a different shape, e.g., rectangles instead of circles, to highlight this distinction (see Figure 7.5(e)). Also, the edges in both the complete computational structure and the manifest structure are labeled to represent *systems*, while the edges in the subsystem structure are labeled with the names of *signals*. These distinctions make it clear that the subsystem structure carries the interpretation of a *block diagram*, while the other structures are *signal flow graphs*.

The definition of subsystem structure given above characterizes the graph in terms N and S. Nevertheless, the subsystem structure can be obtained directly from the complete computational structure, which not only lends a graphical interpretation to the concept of a subsystem, but naturally facilitates the extension of the definitions to the nonlinear and stochastic setting. We achieve this by first extending the definition of a *manifest node* or *manifest signal* of $\mathscr{C}$ to include any node representing a signal identically equal to a manifest signal, u_i or y_j. We then consider the subgraph of $\mathscr{C}$ obtained by (1) removing all input nodes and any outgoing edges leaving them; (2) removing all output nodes and any incoming edges entering them; and (3) removing all outgoing edges leaving any remaining manifest nodes. This subgraph $\mathscr{H}$ is the *hidden structure* of $\mathscr{C}$, and it immediately reveals its subsystems and their interconnection, as follows:

THEOREM 7.1 Consider a system H characterized by a structured LFT, (N, S). Construct a complete computational structure for H, as in (7.2), by realizing each subsystem in S with a single-subsystem state space model, and let $\mathscr{C}$ be the resulting complete computational structure. Then every connected component of $\mathscr{H}$, the hidden structure of $\mathscr{C}$, corresponds to a distinct subsystem in S.

Proof. Since each subsystem is realized by a single-subsystem state space realization, variables internal to each subsystem correspond to nodes of $\mathscr{C}$ that are connected to each other. Moreover, since all outputs of S are manifest, and S is diagonal, these connected components can only be interconnected by manifest signals. By removing all outgoing edges from internal manifest nodes in $\mathscr{C}$, as well as removing all input and output nodes and their associated edges, $\mathscr{H}$ isolates each subsystem so the remaining connected components of $\mathscr{H}$ correspond to the subsystems in S. □

The next example illustrates this procedure of obtaining a system's subsystem structure directly from its complete computational structure.

EXAMPLE 7.4 Consider the generalized state space model of two subsystems in feedback from Example 7.2, given by (7.6). Figures 7.1 and 7.5(a) illustrate the system's complete computational structure, $\mathscr{C}$, and we can generate its subsystem structure by identifying the connected components in the hidden structure of $\mathscr{C}$, as demonstrated in Figure 7.5.

The process of constructing a system's subsystem structure from its complete computational structure involves (1) identifying all manifest nodes in $\mathscr{C}$; (2) removing all input and output nodes and their adjacent edges; and (3) removing all outgoing edges from any remaining manifest nodes. These three steps construct the hidden structure $\mathscr{H}$, and each connected component in $\mathscr{H}$ corresponds to a subsystem. Compress these connected components into single subsystem nodes and replace the input and output nodes as exosystems (instead of signals). Replace all removed edges following the convention that if a node in $\mathscr{C}$ is no longer in $\mathscr{S}$, connect the edge to the corresponding subsystem node. This may lead to multiple edges between nodes in $\mathscr{S}$ (e.g., between subsystems), so we compress these edges into a single edge and change the label to be a vector label, reflecting the multiple signals on that edge.

Now, compare the resulting subsystem structure with the results we obtain if we work directly from the equations defining the original subsystems in Example 7.2 leading up to (7.6). If we find the transfer function of each subsystem individually, build the associated subsystem matrix S, and then interconnect appropriately, we recover the following structured LFT:

$$N = \begin{bmatrix} 0 & 0 & 0 & 0 & 1 & 0 & 0 & 0 \\ 0 & 0 & 0 & 0 & 0 & 1 & 0 & 0 \\ 0 & 0 & 0 & 0 & 0 & 0 & 1 & 0 \\ 0 & 0 & 0 & 0 & 0 & 0 & 0 & 1 \\ 1 & 0 & 0 & 0 & 0 & 0 & 0 & 0 \\ 0 & 1 & 0 & 0 & 0 & 0 & 0 & 0 \\ 0 & 0 & 0 & 0 & 0 & 0 & 1 & 0 \\ 0 & 0 & 0 & 0 & 0 & 0 & 0 & 1 \\ 0 & 0 & 1 & 0 & 0 & 0 & 0 & 0 \\ 0 & 0 & 0 & 1 & 0 & 0 & 0 & 0 \\ 0 & 0 & 0 & 0 & 1 & 0 & 0 & 0 \\ 0 & 0 & 0 & 0 & 0 & 1 & 0 & 0 \end{bmatrix} \tag{7.16}$$

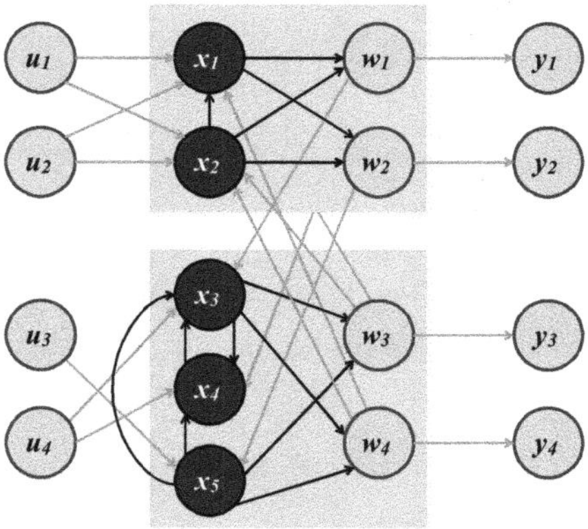

(a) Step 1: Identify the manifest variables (lighter nodes).

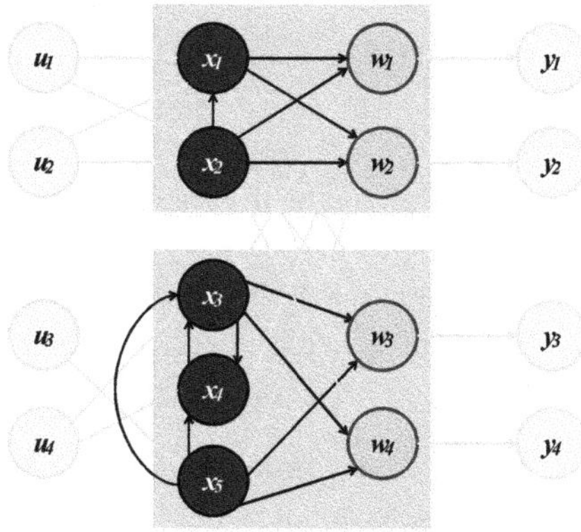

(b) Step 2: Remove all input and output nodes, along with any edges adjacent to these nodes.

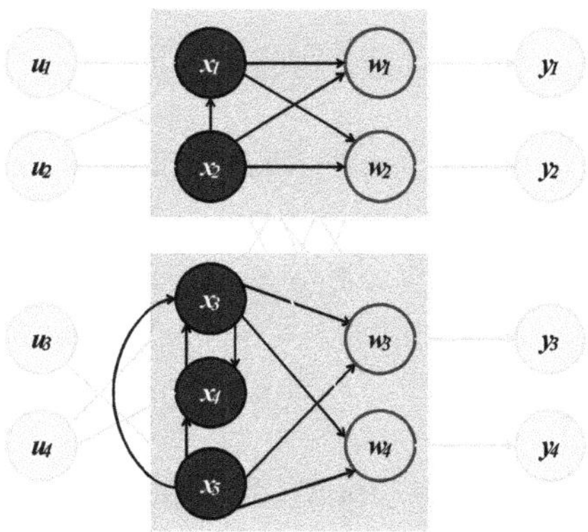

(c) Step 3: Remove any outgoing edges from any remaining manifest variables.

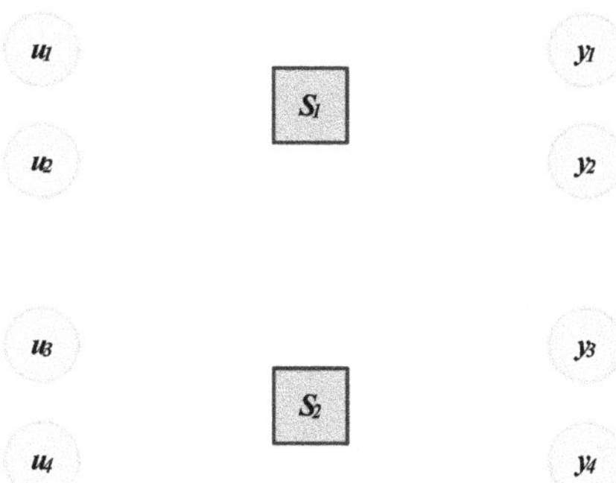

(d) Step 4: The remaining connected components correspond to subsystems.

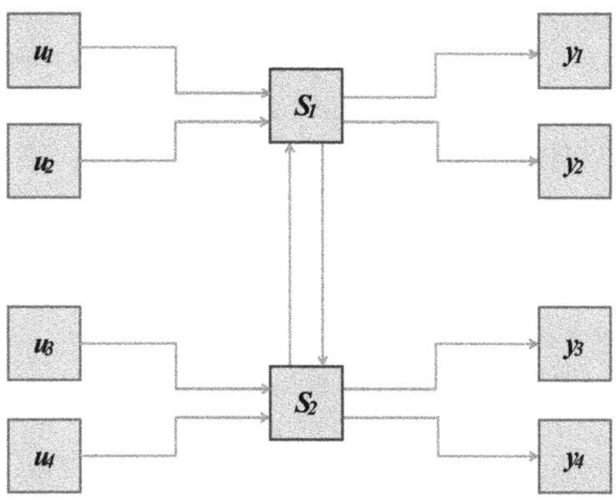

(e) Step 5: Reintroduce the input and output variables as exosystem nodes, and replace all removed edges, compressing any duplicate edges into a single edge.

Figure 7.5 Subsystem structure, built from a system's complete computational structure.

$$S = \frac{1}{s^2+3s+2}\left[\begin{array}{cccc|cccc} 0 & s+2 & 0 & s+2 & 0 & 0 & 0 & 0 \\ s+1 & 1 & s+1 & 1 & 0 & 0 & 0 & 0 \\ \hline 0 & 0 & 0 & 0 & \frac{2}{s+3} & s+2 & \frac{2}{s+3} & s+2 \\ 0 & 0 & 0 & 0 & \frac{s+1}{s+3} & 0 & \frac{s+1}{s+3} & 0 \end{array}\right]$$

Compare the results of the structured LFT with the signal structure in Figure 7.5(e). Notice that building the subsystem structure according to Definition 7.3 leads to the same result; both processes construct the same graph. Nevertheless, building subsystem structure directly from $\mathscr{C}$ sheds insight into the meaning of subsystems, as the connected components of the hidden structure of $\mathscr{C}$.

These procedures uniquely specify (N, S) and $\mathscr{S}$ from a generalized state space model and its complete computational structure, $\mathscr{C}$. This implies that the system models and their associated structural representations considered so far produce a totally ordered set with respect to the relation "uniquely specified by." These are, in order of increasing structural informativity:

1. functional system descriptions and the manifest structure; which are uniquely specified by
2. structured LFTs and the subsystem structure; which are uniquely specified by
3. generalized state space models and the complete computational structure.

The next section considers an alternative approach for representing systems, focusing on the interaction among manifest signals as opposed to the interconnection among subsystems.

7.2.4 Dynamical Structure Functions and the Signal Structure

One of the difficulties in learning a system's subsystem structure from data is that it necessarily perfectly partitions the system states into subsystem groups, so one must be able to identify the correct subsystem for each state variable – even those that are "hidden," or not directly manifest. This section considers a system representation that precisely characterizes the interaction between manifest signals without drawing any conclusions about "hidden" variables. This ability to remain agnostic about the structural role of hidden variables not only makes this representation easier to learn from data, but also makes it extremely useful for describing systems with a "fluidic" component that makes the very idea of subsystems difficult to conceptualize, such as chemical reaction processes or market behavior.

This representation, called the *dynamical structure function* (DSF), like the structured LFT and the subsystem structure, is part of a totally ordered set with respect to the relation "uniquely specified by." This is, in order of increasing structural informativity:

1. functional system descriptions and the manifest structure; which are uniquely specified by
2. dynamical structure functions and the signal structure; which are uniquely specified by
3. generalized state space models and the complete computational structure.

Note that representations of the signal structure similar to the dynamical structure function (DSF) are found across the literature, especially in the area of network reconstruction, [13–15]. Here, we define a system's DSF by considering the zero-intricacy realization of each subsystem in a generalized state space model, as in (7.1). The overall DSF of a system is determined by finding the DSF of each subsystem and then interconnecting them using block diagram algebra.

Without loss of generality, the DSF of a subsystem characterized by the zero-intricacy realization (A, B, C, D) is determined if we let $p_1 \leq p$ be the rank of C and assume without loss of generality that the outputs $y = [y_1' \quad y_2']'$, $y_1 \in \mathbb{R}^p$, and $y_2 \in \mathbb{R}^{(l-p)}$, are ordered so the first p rows of C are linearly independent, i.e.,

$$C = \begin{bmatrix} C_1 \\ C_2 \end{bmatrix} \tag{7.17}$$

with $C_1 \in \mathbb{R}^{p\times n}$ being full row rank. The DSF of the system with respect to y_1 is then given by a pair of $(l \times p)$ and $(l \times m)$ real rational matrix functions $(\hat{Q}(s), \hat{P}(s))$ defined over the Laplace variable $s \in \mathbb{C}$.

In order to determine a view of the system from the perspective of the manifest variables, first create the $(n \times n)$ state transformation:

$$T = \begin{bmatrix} C_1' & E_1 \end{bmatrix}' \tag{7.18}$$

where $E_1 \in \mathbb{R}^{n\times(n-p)}$ is any basis of the null space of C_1, with

$$T^{-1} = \begin{bmatrix} R_1 & E_1 \end{bmatrix} \tag{7.19}$$

where $R_1 = C_1'(C_1C_1')^{-1}$.

Change the basis such that $z = Tx$, yielding $\hat{A} = TAT^{-1}$, $\hat{B} = TB$, $\hat{C} = CT^{-1}$, and $\hat{D} = D$, and partition commensurate with the block partitioning of T and T^{-1} to give

$$\begin{aligned} \begin{bmatrix} \dot{z}_1 \\ \dot{z}_2 \end{bmatrix} &= \begin{bmatrix} \hat{A}_{11} & \hat{A}_{12} \\ \hat{A}_{21} & \hat{A}_{22} \end{bmatrix} \begin{bmatrix} z_1 \\ z_2 \end{bmatrix} + \begin{bmatrix} \hat{B}_1 \\ \hat{B}_2 \end{bmatrix} u \\ \begin{bmatrix} y_1 \\ y_2 \end{bmatrix} &= \begin{bmatrix} I & 0 \\ \hat{C}_{21} & 0 \end{bmatrix} \begin{bmatrix} z_1 \\ z_2 \end{bmatrix} + \begin{bmatrix} \hat{D}_1 \\ \hat{D}_2 \end{bmatrix} u \end{aligned} \tag{7.20}$$

Note that while it is easily seen that $C_1R_1 = I$ and $C_1E_1 = 0$, the fact that $C_2E_1 = 0$ may demand some reflection. The reason this is true is because every row of C_2 is in the row space of C_1. If it were not so, then either the rank of C would be greater than p or C_1 would not be composed of p linearly independent rows. Being in the row space of C_1, each row in C_2 is thus also orthogonal to every vector in E_1, which spans the orthogonal complement of the row space of C_1.

Assume zero initial conditions, take Laplace transforms, and solve for Z_2, yielding

$$sZ_1 = \left[\hat{A}_{11} + \hat{A}_{12}(sI - \hat{A}_{22})^{-1}\hat{A}_{21}\right] Z_1 + \left[\hat{B}_1 + \hat{A}_{12}(sI - \hat{A}_{22})^{-1}\hat{B}_2\right] U \tag{7.21}$$

$$\begin{bmatrix} Y_1 \\ Y_2 \end{bmatrix} = \begin{bmatrix} I & 0 \\ \hat{C}_{21} & 0 \end{bmatrix} \begin{bmatrix} Z_1 \\ Z_2 \end{bmatrix} + \begin{bmatrix} \hat{D}_1 \\ \hat{D}_2 \end{bmatrix} U \tag{7.22}$$

where Z, U, and Y denote the Laplace transforms of z, u, and y respectively.

For notational simplicity, define:

$$W(s) = \hat{A}_{11} + \hat{A}_{12}(sI - \hat{A}_{22})^{-1}\hat{A}_{21} \tag{7.23}$$

$$V(s) = \hat{B}_1 + \hat{A}_{12}(sI - \hat{A}_{22})^{-1}\hat{B}_2 \tag{7.24}$$

and let $D_W(s) = \text{diag}(W(s))$ be a diagonal matrix function composed of the diagonal entries of $W(s)$.

Define $Q(s) = (sI - D_W)^{-1}(W - D_W)$ and $P(s) = (sI - D_W)^{-1}V$, yielding

$$Z_1 = Q(s)Z_1 + P(s)U \tag{7.25}$$

$$\begin{bmatrix} Y_1 \\ Y_2 \end{bmatrix} = \begin{bmatrix} I & 0 \\ \hat{C}_{21} & 0 \end{bmatrix}\begin{bmatrix} Z_1 \\ Z_2 \end{bmatrix} + \begin{bmatrix} \hat{D}_1 \\ \hat{D}_2 \end{bmatrix} U \tag{7.26}$$

Noting from (7.25) that $Z_1 = Y_1 - D_1U$, the DSF of a zero-intricacy state space realization of the form (7.1) with respect to y_1 is then given by:

$$\hat{Q}(s) = \begin{bmatrix} Q(s) \\ C_{21} \end{bmatrix}, \; \hat{P}(s) = \begin{bmatrix} P(s) + (I - Q(s))D_1 \\ D_2 - C_{21}D_1 \end{bmatrix} \tag{7.27}$$

which satisfies

$$\begin{bmatrix} Y_1 \\ Y_2 \end{bmatrix} = \hat{Q}(s)Y_1 + \hat{P}(s)U \tag{7.28}$$

The graphical representation of the DSF is known as the *signal structure* of a system and is denoted $\mathcal{W}$, with a vertex set $V(\mathcal{W})$ and edge set $E(\mathcal{W})$ [16]. The elements of a system's signal structure are defined to be:

- $V(\mathcal{W}) = \{u_1, \ldots, u_m, y_{11}, \ldots, y_{1p_1}, y_{21}, \ldots, y_{2p_2}\}$, each representing a manifest variable of the system with $p_2 = p - p_1$; and
- $E(\mathcal{W})$ contains an edge from $v_i \in V(\mathcal{W})$ to $v_j \in V(\mathcal{W})$ if the associated entry of $\overline{Q}$ and $\overline{P}$ is nonzero.

Unlike the subsystem structure, the signal structure uses circular nodes to denote signals rather than systems, while the edges between these signals represent systems since it is a condensation graph of the signal flow representation of the complete computational structure.

A key aspect of the derivation of the DSF is the transformation of the form (7.18) used to put C in the desired form. This transformation has several important properties.

The first property of the transformed system is that it is invariant to a change of basis on the hidden states, much in the way a transfer function is invariant to state transformations. This essentially means that transforming the internal structure of hidden states does not change the input–output dynamics of the hidden states. This concept is formalized in Lemma 7.1.

LEMMA 7.1 (*Invariance to a class of block diagonal transformations*) Given a system (A,B,C,D) of the form (7.20) with DSF $(\hat{Q},\hat{P})$, then $(\hat{Q},\hat{P})$ is invariant to block diagonal state transformations; that is, the set of systems characterized by block diagonal state transformations,

$$S=\left\{(MAM^{-1},MB,CM^{-1},D)\mid M=\begin{bmatrix} I_{p\times p} & 0\\ 0 & M_{22}\end{bmatrix}\right\}$$

with M_{22} any invertible matrix of appropriate size, all share the same DSF, $(\hat{Q},\hat{P})$.

Proof. Transforming the given system, $\bar{z}=Mz$, yields

$$\begin{aligned}\begin{bmatrix}\dot{\bar{z}}_1\\ \dot{\bar{z}}_2\end{bmatrix}&=\begin{bmatrix}A_{11} & A_{12}M_{22}^{-1}\\ M_{22}A_{21} & M_{22}A_{22}M_{22}^{-1}\end{bmatrix}\begin{bmatrix}\bar{z}_1\\ \bar{z}_2\end{bmatrix}+\begin{bmatrix}B_1\\ M_{22}B_2\end{bmatrix}u\\ \begin{bmatrix}y_1\\ y_2\end{bmatrix}&=\begin{bmatrix}I & 0\\ C_{21} & 0\end{bmatrix}\begin{bmatrix}\bar{z}_1\\ \bar{z}_2\end{bmatrix}+\begin{bmatrix}D_1\\ D_2\end{bmatrix}u\end{aligned}\tag{7.29}$$

which leads to

$$\begin{aligned}W(s)&=A_{11}+A_{12}M_{22}^{-1}(sI-M_{22}A_{22}M_{22}^{-1})^{-1}M_{22}A_{21}\\ &=A_{11}+A_{12}(sI-A_{22})^{-1}A_{21}\end{aligned}\tag{7.30}$$

$$\begin{aligned}V(s)&=B_1+A_{12}M_{22}^{-1}(sI-M_{22}A_{22}M_{22}^{-1})^{-1}M_{22}B_2\\ &=B_1+A_{12}(sI-A_{22})^{-1}B_2\end{aligned}\tag{7.31}$$

Since $W(s)$ and $V(s)$ are invariant to M_{22}, $(\hat{Q},\hat{P})$ also remain unchanged with respect to M_{22}. □

Note that Lemma 7.1 shows that the DSF is invariant to transformations on the manner in which hidden states interact with other hidden states, not on how hidden states interact with measured states.

The second property of the state transformation (7.18) is that the choice of basis of the null space does not change the DSF. This is important because it means an optimal basis for the determination of the DSF is not required and, more importantly, that the DSF only depends on the states chosen to represent the view of the system. Note that this means the DSF is sensitive to the choice of C_1 in (7.17). This concept is formalized in Theorem 7.2.

THEOREM 7.2 (*Invariance to basis of the null space*) Given a system (A,B,C,D) as in (7.1), consider two distinct bases of the null space of C, $E\neq\overline{E}$, with corresponding state transformations:

$$T=\begin{bmatrix}C_1\\ E'\end{bmatrix},\quad \overline{T}=\begin{bmatrix}C_1\\ \overline{E}'\end{bmatrix}$$

as in (7.18), and each leading to its corresponding DSF, $(\hat{Q},\hat{P})$ and $(\overline{Q},\overline{P})$ as in (7.27). Then $(\hat{Q},\hat{P})=(\overline{Q},\overline{P})$.

Proof. Let $z = Tx$ and $\bar{z} = \bar{T}x$. Then $\bar{z} = \bar{T}T^{-1}z$:

$$\bar{T}T^{-1} = \begin{bmatrix} C_1 \\ \bar{E}' \end{bmatrix} \begin{bmatrix} R_1 & E \end{bmatrix} = \begin{bmatrix} I & 0 \\ 0 & \bar{E}'E \end{bmatrix} \tag{7.32}$$

where $R_1 = C_1'(C_1C_1')^{-1}$. The block diagonal structure of $\bar{T}T^{-1}$ then ensures, by Lemma 7.1, that the DSF produced for $\bar{z}$ is the same as that for z, i.e., $(\hat{Q}, \hat{P}) = (\bar{Q}, \bar{P})$. □

The last property explored here is the invariance of the DSF to state permutations; this means that permuting the states of the DSF reveals the same DSF with a renumbering of the measured states. This concept is formalized in Theorem 7.3.

THEOREM 7.3 (*Invariance to state permutations*) Consider a system as in (7.1) with state matrices (A, B, C, D) and DSF $(\hat{Q}, \hat{P})$. Then $(\hat{Q}, \hat{P})$ is invariant to state permutations; that is, the set of systems characterized by state permutations,

$$S = \{(MAM^{-1}, MB, CM^{-1}, D) \mid M \text{ is a permutation matrix}\}$$

all share the same DSF, $(\hat{Q}, \hat{P})$, up to a permutation.

Proof. The result follows from the fact that the state transformation selected in Step 1 of constructing the DSF transforms each system in the set S to the same system for Step 2; the resulting DSF is thus the same, up to a permutation. To see this, consider the transformation T constructed for the unpermuted system (A, B, C, D):

$$T = \begin{bmatrix} C_1 \\ E_1' \end{bmatrix}, \quad T^{-1} = \begin{bmatrix} R_1 & E_1 \end{bmatrix}$$

and compare with the transformation T_M constructed for any permuted system, $(MAM^{-1}, MB, CM^{-1}, D)$:

$$T_M = \begin{bmatrix} C_1 \\ E_1' \end{bmatrix} M^{-1} = TM^{-1}$$
$$T_M^{-1} = M \begin{bmatrix} R_1 & E_1 \end{bmatrix} = MT^{-1}$$

Applying each set of transformations to their respective systems yields the same transformed system, thus producing the same DSF:

$$(T_M MAM^{-1}T_M^{-1}, T_M MB, CM^{-1}T_M^{-1}, D) = (TAT^{-1}, TB, CT^{-1}, D).$$ □

EXAMPLE 7.5 Given the generalized state space model in (7.6) with complete computational structure shown in Figure 7.1, the DSF is given in (7.33), the procedure for determining the corresponding signal structure from a system's generalized state space model is then outlined in Figure 7.6.

$$Q = \begin{bmatrix} 0 & 0 & 0 & \frac{1}{s+1} \\ \frac{1}{s+2} & 0 & \frac{1}{s+2} & 0 \\ 0 & \frac{2}{s+1} & 0 & \frac{1}{s+1} \\ 0 & 0 & \frac{1}{(s+2)(s+3)} & 0 \end{bmatrix}, \; P = \begin{bmatrix} 0 & \frac{1}{s+1} & 0 & 0 \\ \frac{1}{s+2} & 0 & 0 & 0 \\ 0 & 0 & 0 & \frac{1}{s+1} \\ 0 & 0 & \frac{1}{(s+2)(s+3)} & 0 \end{bmatrix} \tag{7.33}$$

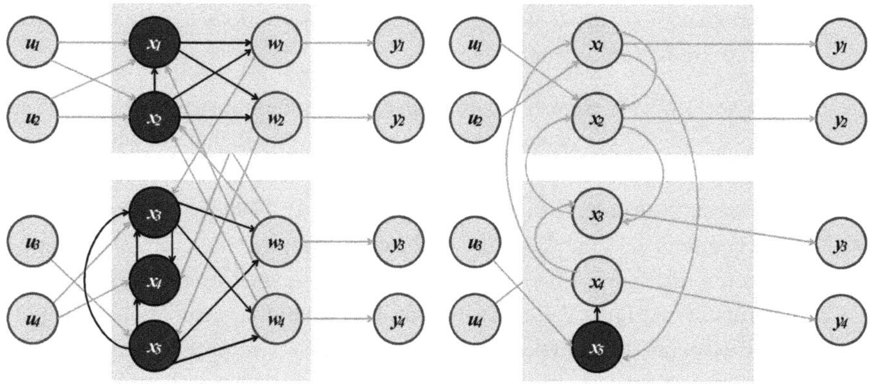

(a) Step 1: Begin with a generalized complete computational structure.

(b) Step 2: Transform each subsystem so that $C_1 = [I \quad 0]$

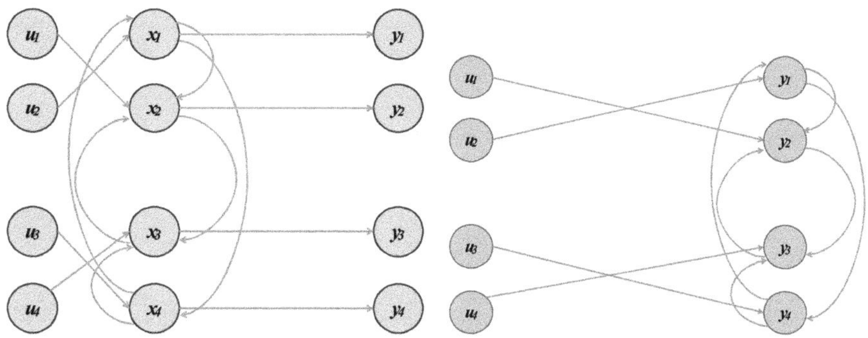

(c) Step 3: Remove hidden nodes from the subsystems, maintaining pathways from manifest variable i to manifest variable j through the removed hidden nodes.

(d) Step 4: Rename manifest state variables, x, to their corresponding output variables, y, while removing the edges from manifest states to outputs along with the corresponding node.

Figure 7.6 Signal structure, built from a system's complete computational structure.

7.2.5 Relationship to Other System Representations

The transfer function associated with a given DSF is given by

$$H(s) = (I - Q)^{-1} P \tag{7.34}$$

which is found easily from (7.28) when Y_2 is of length zero. We note that $I - Q$ is invertible since Q is a square, hollow transfer function matrix, so $I - Q$ will always have full rank. Necessary and sufficient conditions for determining a DSF given a system's transfer function were developed in [17]. This process is known as *network reconstruction* and is discussed in detail in Section 7.3.1.

Comparing the signal structure to subsystem structure, Example 7.6 shows that it is possible for a system's signal structure to be consistent with two or more subsystem structure representations. Example 7.7 shows it is also possible for a system's subsystem structure to be consistent with two or more DSFs. The implication of this result is that these two partial structure system representations denote two different notions of structure within a system.

EXAMPLE 7.6 Given the complete computational structure shown in Figure 7.7(a), the associated subsystem structure was found to be that shown in Figure 7.7(b).

Given a second complete computational structure in Figure 7.7(c), the associated subsystem structure is shown in Figure 7.7(d). Note that this complete computational structure is the same structure as the computational structure of the zero-intricacy realization of Figure 7.7(a). The two are distinguished by the fact that the zero-intricacy structure in Figure 7.2 is a computational structure, meaning that it is not complete and the associated complete computational structure required auxiliary variables to model various compositions of functions. The complete computational structure given in Figure 7.7(c), however, has the same structure, but no auxiliary variables were utilized for composition of functions, so the structure is considered complete.

The corresponding signal structure for both complete computational structures is then given in Figure 7.7(e); thus we have shown that it is possible for a single signal structure to be consistent with multiple subsystem structures.

EXAMPLE 7.7 Given the complete computational structure in Figure 7.8(a), which is the complete computational structure from Figure 7.1 with an edge removed (highlighted by a bounding box), the associated subsystem structure (shown in Figure 7.8(b)) does not change.

Transforming the system to get $C = \begin{bmatrix} I & 0 \end{bmatrix}$ in each subsystem yields the structure given in Figure 7.8(c), which is similar to the transformed structure given in Figure 7.6(b), with an extra edge, highlighted by a bounding box. The associated signal structure is then given in Figure 7.8(d), also containing an extra edge, meaning the subsystem structure given in Figure 7.8(b) is consistent with multiple signal structures.

One of the properties of the signal structure of a system that distinguishes it from the subsystem structure is known as shared hidden states.

DEFINITION 7.4 A shared hidden state is a state within a system that is not manifest, i.e., part of the hidden structure, that has either multiple pathways from it that lead toward a manifest structure or multiple pathways that come from a manifest structure or both.

When a system contains a shared hidden state, the associated signal structure is agnostic to that state, meaning that it allows for hidden states to be shared across system edges. In contrast, the subsystem structure does not allow for hidden states to be shared across systems.

Therefore, when shared hidden states exist in a system, the signal structure contains more structural information than the subsystem structure, as shown in Figure 7.9. Moreover, since the signal structure is agnostic to shared hidden states, the process of determining a unique DSF from a system's transfer function has reasonable conditions (see Section 7.3.1), unlike the subsystem structure of the system.

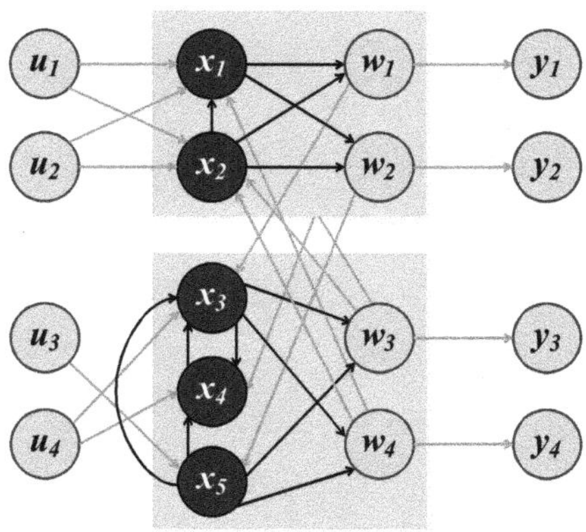

(a) The complete computational structure of a system.

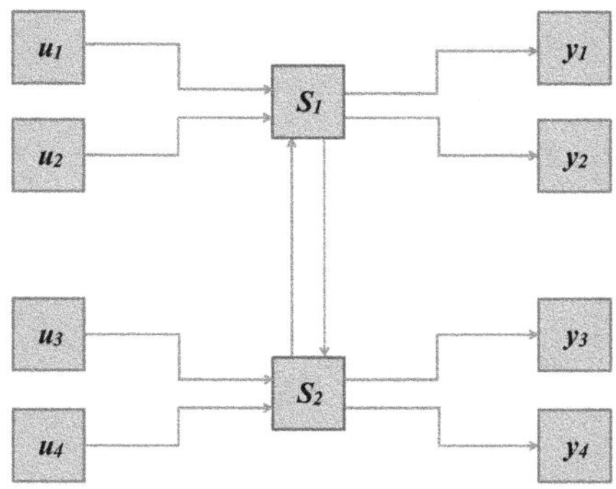

(b) This yields a subsystem structure of two systems in feedback.

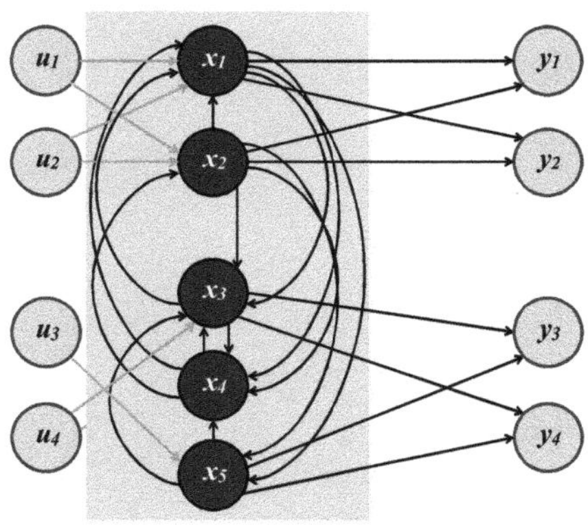

(c) A complete computational structure without any intricacy variables.

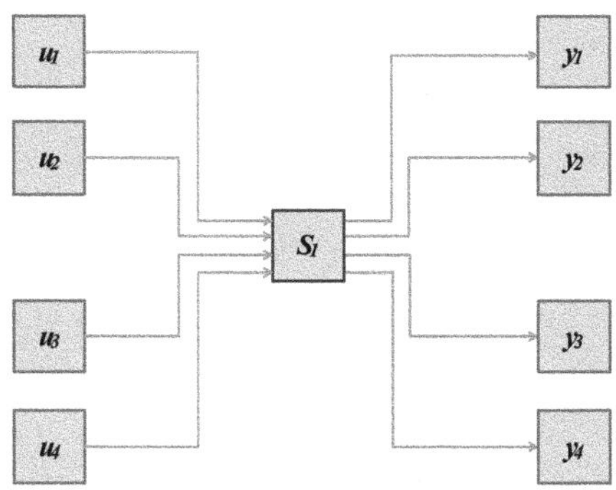

(d) This yields a new subsystem structure with a single subsystem.

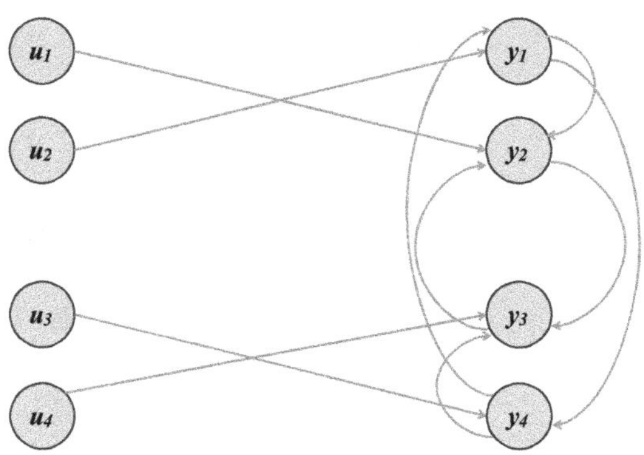

(e) The signal structure is the same for both complete computational structures.

Figure 7.7 Signal structure, Figure 7.7(e), consistent with two subsystem structures, Figures 7.7(b) and 7.7(d).

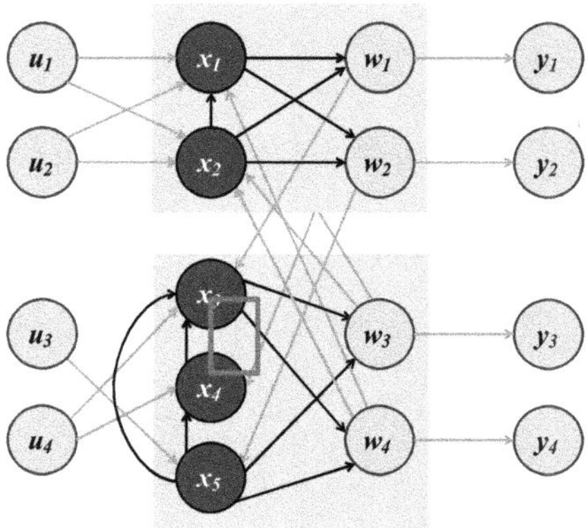

(a) Remove an edge from the complete computational structure.

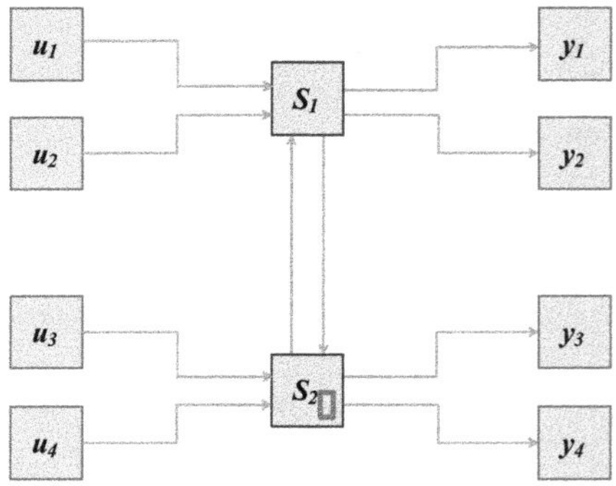

(b) The subsystem structure remains the same, with two subsystems in feedback.

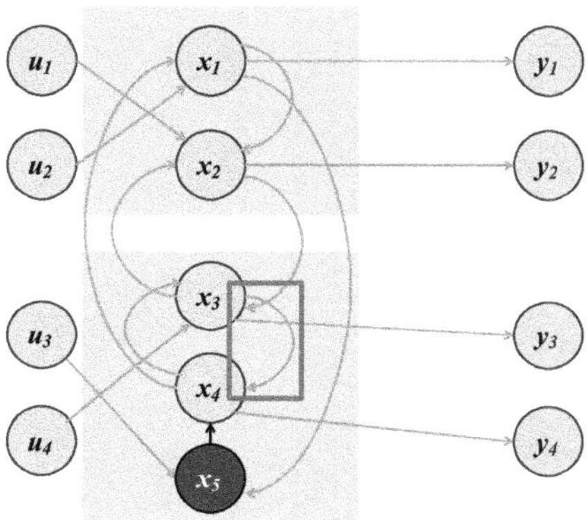

(c) The transformed internal structure with an extra edge.

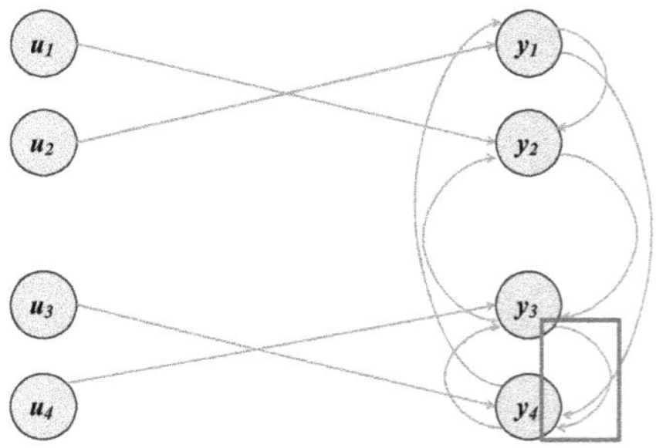

(d) The signal structure with an extra edge that does not exist in Figure 7.7(e).

Figure 7.8 Subsystem structure, Figure 7.8(b), consistent with two signal structures, Figures 7.7(e) and 7.8(d).

(a) Complete computational structure with shared hidden node.

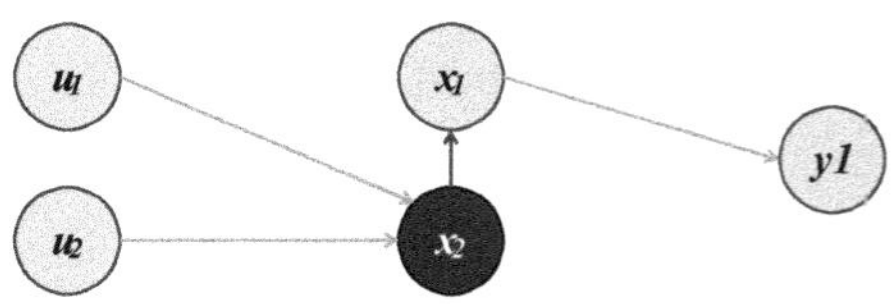

(b) Signal structure containing paths that represent two separate systems.

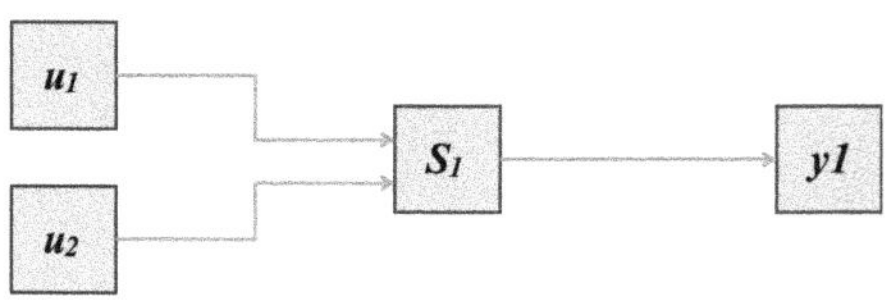

(c) Subsystem structure containing only a single subsystem.

Figure 7.9 Shared hidden state.

The DSF of a system is uniquely defined, given a zero-intricacy state space model, as derived previously in this section. Determining a unique state space model given a system's DSF is an ill-posed problem, though a procedure for determining a minimal state space realization given a DSF (Q, P) was given in [18].

7.3 Applications of Dynamical Structure Functions

The DSF is a versatile system representation and in this section we develop several features of the DSF, including network reconstruction, vulnerability analysis, and distributed controller design. Note that the applications are built from the DSF of zero-intricacy state space realizations of a system for simplicity of exposition.

7.3.1 Network Reconstruction

Network reconstruction is the process of determining the structure of an unknown system [17, 19–22]. The network reconstruction process detailed here determines a unique DSF, given a system's transfer function H. Given H and noting the relationship in (7.34), we can reorder the equation to get

$$\begin{bmatrix} I & H^T \end{bmatrix} \begin{bmatrix} P^T \\ Q^T \end{bmatrix} = H^T \tag{7.35}$$

where A^T represents the transpose of A. Noting that

$$AX = B \iff \text{blckdiag(A, \ldots, A)}\overleftarrow{x} = \overleftarrow{b}$$

where $\text{blckdiag}(A, \ldots, A) = \begin{bmatrix} A & 0 & 0 \\ 0 & \ddots & 0 \\ 0 & 0 & A \end{bmatrix}$ and $\overleftarrow{a}$ is a vector of the stacked columns of the matrix A and defining $X = \begin{bmatrix} P^T & Q^T \end{bmatrix}$, (7.35) can be rewritten as

$$\begin{bmatrix} I & \text{blckdiag}(H^T, \ldots, H^T) \end{bmatrix} \overleftarrow{x} = \overleftarrow{h} \tag{7.36}$$

Since the diagonal elements of Q are identically zero and the dimensions of P, Q, and H are $p \times m$, $p \times p$, and $p \times m$ respectively, where p is the number of outputs from the system and m is the number of inputs, then exactly p elements of $\overleftarrow{x}$ are always zero. Abusing notation, $\overleftarrow{x}$ can be defined with these zero elements removed, reducing Equation (7.36) to

$$\begin{bmatrix} I & \text{blckdiag}(H^T_{-1}, H^T_{-2}, \ldots, H^T_{-p}) \end{bmatrix} \overleftarrow{x} = \overleftarrow{h} \tag{7.37}$$

where H^T_{-i} is the matrix H^T with the ith column removed.

Equation (7.37) reveals the mapping from elements of the DSF, contained in $\overleftarrow{x}$, to its associated transfer function, represented by $\overleftarrow{h}$. The mapping is a linear transformation represented by the matrix operator $L = \begin{bmatrix} I & \text{blckdiag}(H^T_{-1}, H^T_{-2}, \ldots, H^T_{-p}) \end{bmatrix}$.

This matrix has dimensions $(pm) \times (pm + p^2 - p)$, and thus is not injective. This means the problem of network reconstruction from input–output dynamics is ill-posed and other information about the system is required a priori in order to determine a unique DSF.

Certain elements of the vector $\overleftarrow{x}$ need to be known a priori in order to reduce the transformation to an injective map. To accomplish this, consider the $(pm + p^2 - p) \times k$ transformation T such that

$$\overleftarrow{x} = Tz \tag{7.38}$$

where z is an arbitrary vector of size k.

Letting $M = LT$, which makes M a $pm \times k$ matrix, M will be injective if and only if rank$(M) = k$, i.e., M has full column rank. Observing that M is the mapping from unidentified model parameters to the system's transfer function, we see that if M is injective, one can clearly solve for z given H and then construct the DSF from $\overleftarrow{x}$. This means that a T that ensures the rank of M is equal to k is precisely the a-priori system information that is necessary and sufficient for reconstruction of a unique DSF, given a system's transfer function.

EXAMPLE 7.8 Given the following transfer function of a system

$$H = \begin{bmatrix} \frac{s+3}{s^2+3s+2} & -\frac{(s+3)}{s^3+6s^2+10s+5} \\ \frac{1}{s+1} & -\frac{(s^2+5s+6)}{s^3+6s^2+10s+5} \end{bmatrix}$$

we attempt to find the DSF (Q, P) of the system

$$Q = \begin{bmatrix} 0 & Q_{12} \\ Q_{21} & 0 \end{bmatrix} \text{ and } P = \begin{bmatrix} P_{11} & P_{12} \\ P_{21} & P_{22} \end{bmatrix}$$

yielding the vector of unknowns $\vec{x} = \begin{bmatrix} P_{11} & P_{12} & P_{21} & P_{22} & Q_{12} & Q_{21} \end{bmatrix}'$. This gives us $L\vec{x} = \vec{b}$ as

$$\begin{bmatrix} 1 & 0 & 0 & 0 & \frac{1}{s+1} & 0 \\ 0 & 1 & 0 & 0 & -\frac{(s^2+5s+6)}{s^3+6s^2+10s+5} & 0 \\ 0 & 0 & 1 & 0 & 0 & \frac{s+3}{s^2+3s+2} \\ 0 & 0 & 0 & 1 & 0 & -\frac{(s+3)}{s^3+6s^2+10s+5} \end{bmatrix} \begin{bmatrix} P_{11} \\ P_{12} \\ P_{21} \\ P_{22} \\ Q_{12} \\ Q_{21} \end{bmatrix} = \begin{bmatrix} \frac{s+3}{s^2+3s+2} \\ -\frac{(s+3)}{s^3+6s^2+10s+5} \\ \frac{1}{s+1} \\ -\frac{(s^2+5s+6)}{s^3+6s^2+10s+5} \end{bmatrix}$$

Without additional a-priori structural information, the system cannot be reconstructed. Suppose, however, that a-priori information details that P has the form

$$P = \begin{bmatrix} P_{11} & 0 \\ P_{21} & -P_{11} \end{bmatrix}$$

Using this information the vector $\vec{x}$ can then be decomposed into the form $T\vec{z}$ where

$$T = \begin{bmatrix} 1 & 0 & 0 & 0 \\ 0 & 0 & 0 & 0 \\ 0 & 1 & 0 & 0 \\ -1 & 0 & 0 & 0 \\ 0 & 0 & 1 & 0 \\ 0 & 0 & 0 & 1 \end{bmatrix} \text{ and } \vec{z} = \begin{bmatrix} P_{11} \\ P_{21} \\ Q_{12} \\ Q_{21} \end{bmatrix}$$

Replacing $\vec{x}$ with $T\vec{z}$ above yields $M\vec{z} = \vec{b}$, where $M = LT$, as

$$\begin{bmatrix} 1 & 0 & \frac{1}{s+1} & 0 \\ 0 & 0 & -\frac{(s^2+5s+6)}{s^3+6s^2+10s+5} & 0 \\ 0 & 1 & 0 & \frac{s+3}{s^2+3s+2} \\ -1 & 0 & 0 & -\frac{(s+3)}{s^3+6s^2+10s+5} \end{bmatrix} \begin{bmatrix} P_{11} \\ P_{22} \\ Q_{12} \\ Q_{21} \end{bmatrix} = \begin{bmatrix} \frac{s+3}{s^2+3s+2} \\ -\frac{(s+3)}{s^3+6s^2+10s+5} \\ \frac{1}{s+1} \\ -\frac{(s^2+5s+6)}{s^3+6s^2+10s+5} \end{bmatrix}$$

In this case, M is full rank, which means that the system is reconstructible. By solving for $\vec{x} = M^{-1}\vec{b}$ we get the DSF

$$Q = \begin{bmatrix} 0 & \frac{1}{s+2} \\ \frac{1}{s+3} & 0 \end{bmatrix} \text{ and } P = \begin{bmatrix} \frac{1}{s+1} & 0 \\ \frac{1}{s+2} & -\frac{1}{s+1} \end{bmatrix} \tag{7.39}$$

The signal structure corresponding to the DSF in (7.39) is given in Figure 7.10.

Note that robust reconstruction methods for reconstructing the DSF of a system in the face of noise and nonlinearities were first developed in [23]. A more efficient polynomial time algorithm for network reconstruction from noisy data was developed in [24]. Further improvements to the robust reconstruction process, including a more efficient algorithm that also allowed for the reconstruction of more systems, were detailed in [25].

This network reconstruction process is useful because it allows us to determine a notion of structure from a representation with little structural information under very reasonable conditions. Once the structure of the system is determined, more analysis can be performed on the system, such as the vulnerability of the system to various classes of attacks.

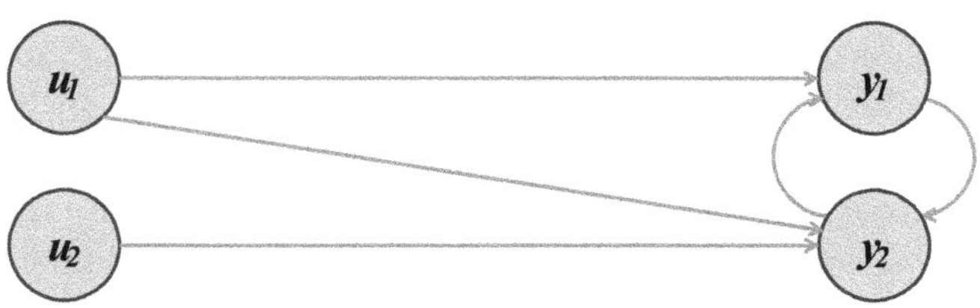

Figure 7.10 Reconstructed signal structure.

7.3.2 Vulnerability Analysis

Vulnerability analysis on a system attempts to determine the effects of external attacks or accidental component failures on the overall dynamics of the system. Conducting a vulnerability analysis of a system is important in the design stages of system construction, because it allows for the system to be built to be robust to both internal and external perturbations [26–30].

In order to conduct a vulnerability analysis on a system's DSF, the derivation of the DSF must be extended to include external disturbances. This can be done by first redefining the state space model (7.1) to include an external disturbance term

$$\begin{bmatrix}\dot{y}\\ \dot{z}\end{bmatrix}=\begin{bmatrix}A_{11} & A_{12}\\ A_{21} & A_{22}\end{bmatrix}\begin{bmatrix}y\\ z\end{bmatrix}+\begin{bmatrix}B_1\\ B_2\end{bmatrix}u+\begin{bmatrix}F_1\\ F_2\end{bmatrix}\psi$$
$$y=\begin{bmatrix}I & 0\end{bmatrix}\begin{bmatrix}y\\ z\end{bmatrix} \qquad (7.40)$$

Following a similar process to that found in [31], we find the corresponding DSF to be

$$Y = QY + PU + \Delta\Psi \qquad (7.41)$$

where

$$Q = (sI - D)^{-1}(W - D)$$
$$P = (sI - D)^{-1}V$$
$$\Delta = (sI - D)^{-1}N$$

with $W = A_{11}+A_{12}(sI-A_{22})^{-1}A_{21}$, $D = \text{diag}(W_{11},\dots,W_{pp})$, $V = B_1+A_{12}(sI-A_{22})^{-1}B_2$, and $N = F_1 + A_{12}(sI - A_{22})^{-1}F_2$. Equation (7.41) is then a generalized attack model in the DSF domain.

Focusing on a class of destabilizing attacks, assume that the system being analyzed is stable and only consider attack models that use the existing communication structure to conduct an attack. This is not an unreasonable assumption, since creating new links within a system may be a difficult or expensive task for an attacker.

Vulnerability to a Single Link Attack

Starting with (7.41) and solving for Y in terms of U and Ψ yields

$$Y = (I - Q)^{-1}PU + (I - Q)^{-1}\Delta\Psi \qquad (7.42)$$

where the input–output relationship is given by $H = (I - Q)^{-1}P$ and the transfer function describing how Ψ affects the exposed states, Y, is $(I - Q)^{-1}\Delta$. Given that the system is stable, no bounded input can destabilize the system, so the case when $\Psi = Y$ is analyzed since it means that an attacker is using some combination of additive perturbations on existing communication links to destabilize the system.

In [32], it is stated that a stable additive perturbation Δ on a link Q_{ij} or P_{ij} is able to destabilize the system if and only if the transfer function, M_{ij}, seen by Δ is nonzero.

This means that the link Q_{ij} or P_{ij} is in feedback with some series of links in Q or P. Note that M_{ij} is the transfer function from ΔY_j to Y_j. The transfer function seen by a perturbation ΔY is then given in (7.42) as $(I - Q)^{-1}$. In particular, if we want to determine the vulnerability to a single link attack on a link Q_{ij}, we know this can be modeled as $\Delta_{ij} = (sI - D_{ii})^{-1}N_{ij}$, with the rest of the entries in Δ equal to zero. Then, the transfer function seen by the perturbation on the link Q_{ij} is found from

$$\begin{bmatrix} Y_1 \\ \vdots \\ Y_{j-1} \\ Y_j \\ Y_{j+1} \\ \vdots \\ Y_p \end{bmatrix} = K \begin{bmatrix} 0 \\ \vdots \\ 0 \\ \Delta_{ij} Y_j \\ 0 \\ \vdots \\ 0 \end{bmatrix} \tag{7.43}$$

where $K = (I - Q)^{-1}$. From (7.43), we can see that $Y_j = K_{ji}\Delta_{ij}Y_j$ since $\Delta_{ij}Y_j$ is in the ith row of the vector given in (7.43). Therefore, the vulnerability of a single link can be defined as

$$v_{ij} = ||K_{ji}||_\infty$$

which means that the vulnerability of the entire system is

$$V = \max_{Q_{ij} \neq 0 \in Q} ||K_{ji}||_\infty \tag{7.44}$$

which is simply the maximum possible vulnerability across all links.

Vulnerability to a Multiple Link Distributed Attack

Consider now an attack in which multiple attackers are simultaneously performing unique single link attacks in the system and are not sharing information, known as a distributed attack. This is modeled by the concatenation of several single link attacks on the system and by application of the small gain theorem, the vulnerability, $v_{ij,\ldots,kl}$ to this type of an attack is the structured singular value, $\mu_{ij,\ldots,kl}$, of the matrix

$$R_{ij,\ldots,kl} = \begin{bmatrix} K_{ji} & 0 & \cdots & 0 \\ 0 & \ddots & \ddots & \vdots \\ \vdots & \ddots & \ddots & 0 \\ 0 & \cdots & 0 & K_{lk} \end{bmatrix} \tag{7.45}$$

That is,

$$v_{ij,\ldots,kl} = \mu(R_{ij,\ldots,kl}, \Pi)$$

The overall vulnerability of the system to a distributed attack is

$$V = \max_{R_{links} \in \mathscr{R}} \mu_{links}$$

where $\mathscr{R}$ is the set of matrices of the form (7.45) over the set of all possible combinations of links, $\mathscr{L}$, and μ_{links} is the structured singular value of R_{links}.

Vulnerability to a Multiple Link Coordinated Attack

A multiple link coordinated attack is another generalization of a single link attack and is similar to a distributed attack, except that it models either communication between multiple attackers or a single attacker targeting multiple links. The transfer function seen by a perturbation on multiple links when allowing for communication in the attack is then given by

$$T_{ij,\ldots,kl} = \begin{bmatrix} K_{ij} & \cdots & K_{il} \\ \vdots & \ddots & \vdots \\ K_{kj} & \cdots & K_{kl} \end{bmatrix} \tag{7.46}$$

In this case, the vulnerability to a multiple link coordinated attack is

$$v_{ij,\ldots,kl} = ||T_{ij,\ldots,kl}||_{\infty}$$

and the overall vulnerability of the system to a coordinated attack is

$$V = \max_{links \in \mathscr{L}} ||T_{links}||_{\infty}$$

Reducing Vulnerability in Open-Loop Systems

Since the vulnerability of any given link in a system is the transfer function seen by a perturbation on that link, the vulnerability of a system to bounded perturbations is nonzero if and only if feedback exists within the system. Therefore, a completely secure architecture is one in which no links in Q exist.

Note that since $G = (I - Q)^{-1}P$, when $Q = 0$, then $P = G$. Since links in P are never in feedback for open-loop systems in which attackers cannot create links, they are never vulnerable (see Example 7.9). Thus, the overall vulnerability of a system with $Q = 0$ is $V = 0$, meaning there does not exist a finite additive perturbation on a link in the system that can destabilize the system under the assumption that the attacker can only use the existing communication network of the system [32].

EXAMPLE 7.9 Consider the following DSF

$$Q = \begin{bmatrix} 0 & \frac{1}{s+1} & 0 \\ 0 & 0 & \frac{1}{s+2} \\ \frac{1}{s+3} & 0 & 0 \end{bmatrix}, \quad P = \begin{bmatrix} \frac{1}{s+1} & 0 & 0 \\ 0 & \frac{1}{s+1} & 0 \\ 0 & 0 & \frac{1}{s+1} \end{bmatrix} \tag{7.47}$$

The corresponding transfer function for (7.47) is

$$G = \frac{1}{s^3 + 6s^2 + 11s + 5} \begin{bmatrix} s^2 + 5s + 6 & \frac{s^2+5s+6}{s+1} & \frac{s+3}{s+1} \\ 1 & s^2 + 5s + 6 & s + 3 \\ s + 2 & \frac{s+2}{s+1} & s^2 + 5s + 6 \end{bmatrix} \tag{7.48}$$

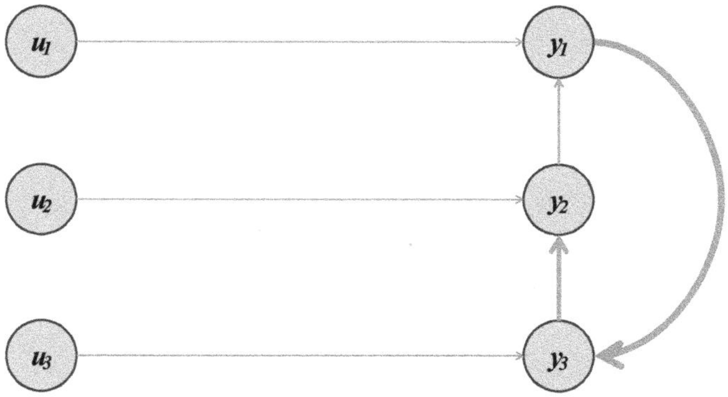

Figure 7.11 Vulnerable structure: The size of each link corresponds to its vulnerability. Remember that links in P are not vulnerable.

By the small gain theorem, the smallest perturbation that could destabilize the system is $\frac{1}{||G||_\infty} = 0.4152$, which makes the vulnerability of the input–output system $V = ||G||_\infty = 2.4085$.

The signal structure of the system in (7.47), pictured in Figure 7.11, is a ring structure with a feedback loop in Q. As mentioned previously, this system structure is vulnerable to destabilizing attacks that target specific links. Given $H = (I - Q)^{-1}$, we can determine the vulnerabilities of each link Q:

$$\begin{aligned} v_{12} &= 0.2 \\ v_{23} &= 0.4 \\ v_{31} &= 0.6 \end{aligned}$$

The overall vulnerability of the system to single link attacks is $V = v_{31} = 0.6 < ||G||_\infty$, which means the smallest perturbation on a single link that can destabilize the system is $\frac{1}{V} = 1.67$. This perturbation is smaller than the perturbation required to destabilize the input–output representation since it restricts attacks to only perturb one link within the system, rather than perturbations that affect the entire system.

Also discussed previously is that one possible secure structure for the system in (7.47) is one in which we remove all the links in Q, which removes all feedback from Q. This system would have a DSF of the form

$$\bar{Q} = \begin{bmatrix} 0 & 0 & 0 \\ 0 & 0 & 0 \\ 0 & 0 & 0 \end{bmatrix}, \quad \bar{P} = \frac{1}{s^3 + 6s^2 + 11s + 5} \begin{bmatrix} s^2 + 5s + 6 & \frac{s^2+5s+6}{s+1} & \frac{s+3}{s+1} \\ 1 & s^2 + 5s + 6 & s + 3 \\ s + 2 & \frac{s+2}{s+1} & s^2 + 5s + 6 \end{bmatrix} \tag{7.49}$$

The signal structure of the system in (7.49), pictured in Figure 7.12, has no feedback in Q while still maintaining all the pathways from inputs to measured states/outputs that existed in the original system from (7.47), making it a secure structure without compromising the input–output dynamics of the system. Note that by *secure* we mean that the vulnerability of the system to both single link and multiple link attacks is 0, although the system may still be vulnerable to other types of attacks.

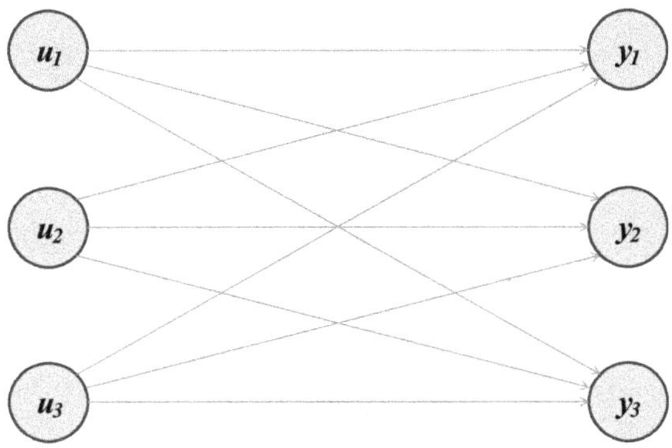

Figure 7.12 Secure structure: $P = G$ and $Q = 0$, meaning that no links in the system are in feedback, so no links in the system are vulnerable to single link attacks that use the existing communication network.

Unfortunately, in some systems removing all internal links is infeasible and, in some cases, feedback in a system is necessary. Consider an unstable plant in feedback with a stabilizing controller; the feedback is necessary to keep the system stable, but this produces vulnerable links within the closed-loop system. The process of minimizing vulnerability in the face of feedback is still an open problem, although preliminary work in this area was conducted in [33].

7.3.3 Structured Controller Design

Another interesting problem involving stabilizing an unstable system is the problem of structured controller design, which refers to the design of a stabilizing controller in which the structure of the controller is restricted by certain conditions [34–38]. For example, a controller could be restricted to using existing communication links within a system or the controller must be designed to reduce the vulnerability of links to an attack.

Denoting (Q^{bin}, P^{bin}) to be the Boolean structure of allowable links in a controller and H the transfer function of the unstable plant, the following procedure was developed in [39] for generating a stabilizing controller with the desired structure

Procedure $\mathbb{P}$

1. Choose an undesigned link p_{ij} such that $p_{ij}^{bin} = 1$.
2. Design p_{ij} to stabilize h_{ji} such that there is no pole zero cancellation in PG. That is, the controller link is designed such that it stabilizes the transfer function it sees, and there is no pole-zero cancellation.
3. After adding p_{ij}, if the closed-loop system (H, P) is still unstable, repeat for all p_{xy}, where $p_{xy}^{bin} = 1$, so that the added link attempts to stabilize the plant as well as all the previously added controller links.

4. If the closed loop system S, formed by adding P in feedback with H, is still unstable, add links in Q^{bin} such that there is no pole-zero cancellation between Q and S. Again, each added link attempts to stabilize the plant H along with the previously added links of P and Q.

Furthermore, it was shown in [39] that if this procedure does not create a stabilizing controller, then no such controller given the restrictions (Q^{bin}, P^{bin}) exists in the system. In particular, we note that this procedure works if the unstable modes of the plant in the system are both observable and controllable by the controller with the required structure. So, for example, a controller with a diagonal structure, i.e.,

$$P^{bin} = \begin{bmatrix} 1 & 0 & \cdots & 0 \\ 0 & 1 & \ddots & \vdots \\ \vdots & \ddots & \ddots & 0 \\ 0 & \cdots & 0 & 1 \end{bmatrix} \text{ and } Q^{bin} = 0$$

which represents a completely distributed controller, can only stabilize a system if the system's unstable modes that are controllable from input i are also observable from output i.

Another interesting controller structure is the cycle structure

$$P^{bin} = \begin{bmatrix} 1 & 0 & \cdots & 0 \\ 0 & 1 & \ddots & \vdots \\ \vdots & \ddots & \ddots & 0 \\ 0 & \cdots & 0 & 1 \end{bmatrix} \text{ and } Q^{bin} = \begin{bmatrix} 0 & 1 & 0 & \cdots & 0 \\ \vdots & \ddots & \ddots & \ddots & \vdots \\ \vdots & & \ddots & \ddots & 0 \\ 0 & \cdots & \cdots & 0 & 1 \\ 1 & 0 & \cdots & \cdots & 0 \end{bmatrix}$$

which can stabilize any unstable plant as long as the plant is detectable and stabilizable, since the cyclic structure of the controller allows every unstable mode to be observable from any output and controllable from any input. A similar result for the subsystem structure was later determined and published in [40].

EXAMPLE 7.10 Given the following unstable system

$$\dot{x} = \begin{bmatrix} 1 & 0 & 0 \\ 1 & 2 & 3 \\ 1 & 0 & 3 \end{bmatrix} x + \begin{bmatrix} 1 & 0 \\ 0 & 1 \\ 0 & 0 \end{bmatrix} u$$

$$y = \begin{bmatrix} 1 & 0 & 0 \\ 0 & 1 & 0 \end{bmatrix} x$$

with associated DSF

$$Q = \begin{bmatrix} 0 & 0 \\ \frac{s}{s^2-5s+6} & 0 \end{bmatrix}, \quad P = \begin{bmatrix} \frac{1}{s-1} & 0 \\ 0 & \frac{1}{s-2} \end{bmatrix}$$

Figure 7.13 Unstable plant structure.

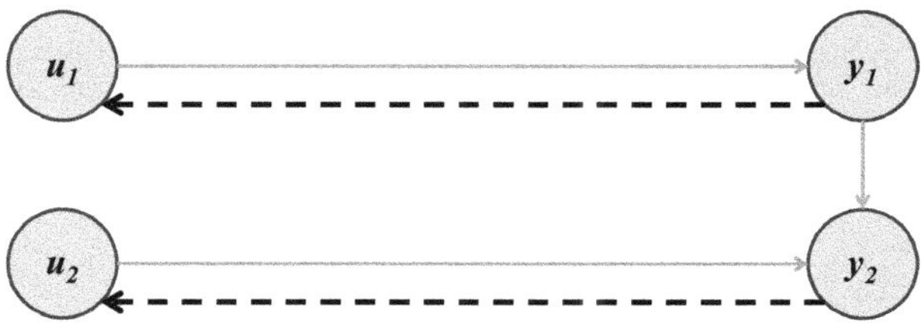

Figure 7.14 Unstable plant with diagonal controller structure shown by dashed edges.

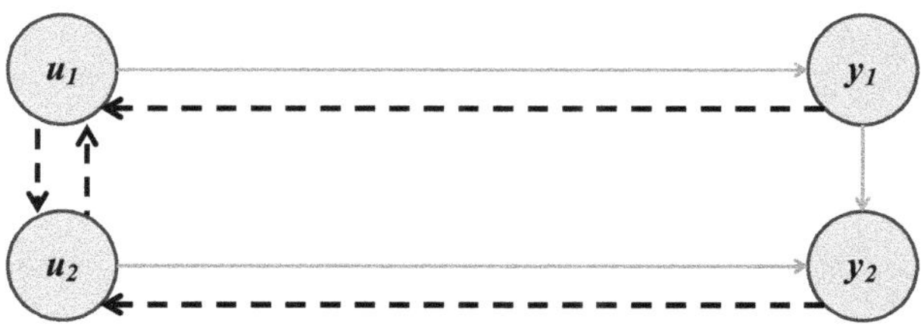

Figure 7.15 Unstable plant with stabilizing cyclic controller structure shown by dashed edges.

with a signal structure as shown in Figure 7.13, restrict the controller structure to be a diagonal controller, shown in Figure 7.14.

Noting that the system has modes $\{1, 2, 3\}$, the Popov–Belevitch–Hautus test for controllability and observability shows that mode 3 is controllable only from input u_1, but observable only from output y_2, which means no stabilizing diagonal controller exists. However, since the system is both stabilizable and detectable, the system can be stabilized by a cyclic controller, shown in Figure 7.15.

7.4 Conclusion

This chapter explored four different mathematical system representations and their associated structures. Three of these representations are standard for LTI systems: transfer functions, state space models, and the interconnection of subsystems. The fourth representation, the dynamical structure function, and its associated structure, the signal structure, are relatively new.

The dynamical structure function and its signal structure were then used to discuss three important problems: network reconstruction, vulnerability analysis, and the design of distributed stabilizing controllers. These applications highlight the practicality of a theory of structures for networks of dynamic systems.

References

[1] J. C. Willems. "Paradigms and puzzles in the theory of dynamical systems." *IEEE Trans. Autom. Control*, vol. 36, no. 3, pp. 259–294, Mar. 1991.

[2] J. Hespanha. *Linear Systems Theory*. Princeton, NJ: Princeton University Press, 2009.

[3] K. Astrom and R. Murray. *Feedback Systems: An Introduction for Scientists and Engineers*. Princeton, NJ: Princeton University Press, 2008.

[4] G. E. Dullerud and F. G. Paganini. *A Course in Robust Control Theory: A Convex Approach*. New York: Springer-Verlag, 1999.

[5] L. Chen, S. Roy, and A. Saberi. "On the information flow required for tracking control in networks of mobile sensing agents." *IEEE Trans. Mob. Comput.*, vol. 4, pp. 519–531, 2010.

[6] W. Ren, R. W. Beard, and E. M. Atkins. "Information consensus in multivehicle cooperative control." *Control Syst.*, vol. 27, no. 2, pp. 71–82, 2007.

[7] S. Roy, M. Xue, and S. K. Das. "Security and discoverability of spread dynamics in cyber-physical networks." *IEEE Trans. Parallel Distrib. Syst.*, vol. 23, no. 9, pp. 1694–1707, 2012.

[8] F. Pasqualetti, F. Dorfler, and F. Bullo. "Attack detection and identification in cyber-physical systems." *IEEE Trans. Autom. Control*, vol. 58, no. 11, pp. 2715–2729, 2013.

[9] R. W. Beard, J. Lawton, F. Y. Hadaegh, et al. "A coordination architecture for spacecraft formation control." *IEEE Trans. Control Syst. Technol.*, vol. 9, no. 6, 777–790, 2001.

[10] T.S. Soderstrom and P. Stoica. *System Identification*. Cambridge: Cambridge University Press, 1989.

[11] R. Johansson. *System Modeling and Identification*. Upper Saddle River, NJ: Prentice Hall, 1993.

[12] K. Zhou, J. Doyle, and K. Glover. *Robust and Optimal Control*. Upper Saddle River, NJ: Prentice Hall, 1995.

[13] D. Materassi and M. V. Salapaka. "On the problem of reconstructing an unknown topology via locality properties of the Wiener filter." *IEEE Trans. Autom. Control*, vol. 57, no. 7, pp. 1765–1777, 2012.

[14] A. Dankers, P. M. J. Van den Hof, P. S. C. Heuberger, and X. Bombois. "Dynamic network structure identification with prediction error methods: Basic examples," in *IFAC Symp. System Identification*, 2012, pp. 876–881.

[15] C. Quinn, N. Kiyavash, and T. P. Coleman. "Directed information graphs." arXiv preprint arXiv:1204.2003, 2012.

[16] E. Yeung, J. Gonçalves, H. Sandberg, and S. Warnick. "Representing structure in linear interconnected dynamical systems," presented at the *Conf. Decision and Control*, Atlanta, GA, Dec. 15–17, 2010.

[17] J. Adebayo, T. Southwick, V. Chetty, et al. "Dynamical structure function identifiability conditions enabling signal structure reconstruction," presented at the *Conf. Decision and Control*, Wailea Beach, HI, Dec. 10–13, 2012.

[18] Y. Yuan, G.-B. Stan, S. Warnick, and J. M. Gonçalves. "Minimal dynamical structure realisations with application to network reconstruction from data," presented at the *Conf. Decision and Control*, Shanghai, China, Dec. 16–18, 2009.

[19] D. Materassi and G. Innocenti. "Topological identification in networks of dynamical systems." *IEEE Trans. Autom. Control*, vol. 55, no. 8, pp. 1860–1871, 2010.

[20] A. S. Bazanella, M. Gevers, and L. Miškovic. "Closed-loop identification of MIMO systems: A new look at identifiability and experiment design." *Eur. J. Control*, vol. 16, no. 3, pp. 228–239, 2010.

[21] A. Chiuso and G. Pillonetto. "A Bayesian approach to sparse dynamic network identification." *Automatica*, vol. 48, no. 8, pp. 1553–1565, 2012.

[22] B. M. Sanandaji, T. L. Vincent, and M. B. Wakin. "Exact topology identification of large-scale interconnected dynamical systems from compressive observations," in *American Control Conf.*, 2011, pp. 649–656.

[23] Y. Yuan, G.-B. Stan, S. Warnick, and J. Gonçalves. "Robust dynamical network structure reconstruction." *Automatica*, special issue on Systems Biology, 2011.

[24] D. Hayden, Y. Yuan, and J. Gonçalves. "Robust network reconstruction in polynomial time," presented at the *Conf. Decision and Control*, Wailea Beach, HI, Dec. 10–13, 2012.

[25] V. Chetty, D. Hayden, S. Warnick, and J. Gonçalves. "Robust signal-structure reconstruction." presented at the *Conf. Decision and Control*, Florence, Italy, May 15, 2013.

[26] M. Long, C.-H. J. Wu, and J. Y. Hung. "Denial of service attacks on network-based control systems: Impact and mitigation." *IEEE Trans. Ind. Inf.*, vol. 1, no. 2, pp. 85–96, 2005.

[27] S. Amin, A. A. Cárdenas, and S. S. Sastry. "Safe and secure networked control systems under denial-of-service attacks." In *Hybrid Systems: Computation and Control*, R. Alur, Ed. New York: Springer, 2009, pp. 31–45.

[28] V. Pichai, M.E. Sezer, and D.D. Šiljak. "Vulnerability of dynamic systems." *Int J Control*, vol. 34, no. 6, pp. 1049–1060, 1981.

[29] Y. Liu, P. Ning, and M. K. Reiter. "False data injection attacks against state estimation in electric power grids." *ACM Trans. Inf. Syst. Secur.*, vol. 14, no. 1, p. 13, 2011.

[30] Y. Mo, E. Garone, A. Casavola, and B. Sinopoli. "False data injection attacks against state estimation in wireless sensor networks," in *Conf. Decision and Control*, 2010, pp. 5967–5972.

[31] J. Gonçalves and S. Warnick. "Necessary and sufficient conditions for dynamical structure reconstruction of LTI networks." *IEEE Trans. Autom. Control*, vol. 53, no. 7, pp. 1670–1674, 2008.

[32] A. Rai, D. Ward, S. Roy, and S. Warnick. "Vulnerable links and secure architectures in the stabilization of networks of controlled dynamical systems," in *American Control Conf.*, 2012, pp. 1248–1253.

[33] N. Woodbury. "Vulnerability analysis of closed-loop systems." Undergraduate honor's thesis, Brigham Young University, 2013.

[34] K.-K. Oh, M.-C. Park, and H.-S. Ahn. "A survey of multi-agent formation control." *Automatica*, vol. 53, pp, 424–440, 2014.

[35] A.S.M. Vamsi and N. Elia. "Design of distributed controllers realizable over arbitrary directed networks," in *Conf. Decision and Control*, 2010, pp. 4795–4800.

[36] S. Knorn, Z. Chen, and R. Middleton. Overview: collective control of multiagent systems. *IEEE Trans. Control Netw. Syst.*, vol. 3, no. 4, pp. 334–347, Dec. 2016.

[37] R. M. Murray. "Recent research in cooperative control of multivehicle systems." *J Dyn. Syst. Meas. Contr.*, vol. 129, no. 5, pp. 571–583, 2007.

[38] G. Fazelnia, R. Madani, and J. Lavaei. "Convex relaxation for optimal distributed control problem," presented at the *Conf. Decision and Control*, Los Angeles, CA, December 15–17, 2014.

[39] A. Rai and S. Warnick. "A technique for designing stabilizing distributed controllers with arbitrary signal structure constraints," presented at the *European Control Conf.*, Zurich, Switzerland, 2013.

[40] A. Alavian and M. Rotkowitz. "Stabilizing decentralized systems with arbitrary information structure," presented at the *Conf. Decision and Control*, Los Angeles, CA, December 15–17, 2014.

Part II Exercises

1. Chapter 4 is concerned with remote estimation. What CPS application motivates the authors' work in this area? Can you think of two other CPS applications where the presented remote estimation techniques would be valuable?
2. Remote estimation in CPSs may form part of a scheme to enact feedback controls over communication channels. For the applications that you listed in Problem 1, please envision what feedback controls may need to be implemented, and how the developed remote estimation schemes may be leveraged for this.
3. Chapter 4 discusses solutions for several different remote estimation problems. Compare and contrast these problems. Also, complete the following exercises, which require a detailed understanding of the methods used to solve the various remote estimation problems:
   a. Solve the Imer–Basar problem with $\sigma^2 = 1$ and $N = 4$ and $M = 2$. Construct the table containing the solutions of the two-dimensional DP recursion in 4.10.
   b. Obtain an upper bound for the performance of the Xu–Hespanha problem by computing and evaluating the performance of the suboptimal policy proposed by Cogill et al., shown at the end of 4.3 with

$$A = \begin{bmatrix} 1 & 0 \\ 0 & 1 \end{bmatrix},\ Q = \begin{bmatrix} 2 & 1 \\ 1 & 2 \end{bmatrix}, \Sigma = \begin{bmatrix} 1 & 0.1 \\ 0.1 & 1 \end{bmatrix} \text{ and } \rho = 1 \tag{1}$$

   c. Compute the optimal thresholds τ_1^* and τ_2^* for the Lipsa–Martins problem with $N = 2$, $d = 1$, $a = 0.5$, $\sigma^2 = 1$, and $\rho = 1$.
   d. Obtain the Snell envelope for the Rabi–Moustakides–Baras problem with $a = 0$, $b = 1$, $T = 1$.
   e. Obtain an optimal communication policy for estimation over the single-user collision channel of Figure 4.6 with $\beta = 0.5$, $\rho = 1$, and $\sigma^2 = 1$ by solving

$$\text{minimize} \int_{x \in [a,b]} (x - \hat{x}_0)^2 f_X(x)dx + \int_{x \notin [a,b]} \Big[\beta(x - \hat{x}_1)^2 + \rho\Big] f_X(x)dx \tag{2}$$

   with variables $a, b, \hat{x}_0, \hat{x}_1 \in \mathbb{R}$.

4. The presented remote estimation methods are focused on canonical models (e.g., scalar diffusion-type models). What challenges might arise in generalizing the methods to more sophisticated models that often arise in real-world CPSs?
5. Consider a CPS with six systems randomly located within a 300×300 m^2 area. To maintain the homogeneous property, all six systems are using similar control system as [37] in Chapter 4. The discrete-time representation is given as

$$x_{k+1} = \begin{bmatrix} 1.1138 & -0.0790 \\ 0.0592 & 0.8671 \end{bmatrix} x(t) + \begin{bmatrix} 0.2033 \\ 0.1924 \end{bmatrix} u(t) \tag{3}$$

Moreover, the sampling interval is selected as $T_s = 0.15$ seconds, the number of bits for the six quantized sensed data for the systems are defined as $N_{bit} = [10\ 8\ 6\ 7\ 8\ 4]$. Further, the cost function weight matrix is selected as $Q_l = \begin{bmatrix} 1 & 0 \\ 0 & 1 \end{bmatrix}$, $R_l = 0.1$. The initial system state vectors are given by $x_{1,0} = [20 \quad 7]^T$, $x_{2,0} = [12 \quad 5]^T$, $x_{3,0} = [10 \quad 3]^T$, $x_{4,0} = [8 \quad 4]^T$, $x_{5,0} = [10 \quad 5]^T$, and $x_{6,0} = [0.1 \quad 6]^T$.

   a. Design the event-triggering condition and estimated optimal control.
   b. Design distributed scheduling scheme.
   c. Plot the distributed CPS system state response.
   d. Compared with embedded round robin (ERR) and greedy scheduling schemes, plot the fairness of optimal distributed scheduling and distributed CPS cost function performance.

   Note: Please see the Hint of Solution section at the end of the Part II Exercises for the solution to this exercise.
6. Chapters 5 and 6 are both concerned with event-triggered controls. Why are such controls relevant to CPS applications? What are particular applications where the methods may be brought to bear?
7. What are the main innovations related to event-triggered controls introduced in Chapters 5 and 6?
8. Explain, conceptually, how trigger-based controls can be designed for resilience to failures and attacks on a CPS's communications infrastructure. What are some applications where such resilience is critical?
9. Event-triggered control techniques for CPSs are often concerned with networked or distributed dynamical processes. What are some of the challenges associated with event-triggered controls in these settings? What strategies can be used to address these challenges?
10. Repeat the design exercise developed in Chapter 7 for a different CPS model of your own choosing. How do the results change for a different dynamical model?
11. In Chapter 7, two-tiered multi-agent network models, which encapsulate agent-level and network-level dynamical interactions, are considered. Describe CPS applications where these models are apt.
12. Chapter 7 proposes a distributed and adaptive method for designing the topology of a network, to optimize a performance measure (disturbance-rejection

measure). What are the possible benefits of such distributed and adaptive methods in CPS applications? Are there any possible drawbacks?

13. The results described in Chapter 7 contribute to a research effort on designing multi-agent systems to shape a synchronization or consensus dynamics. Undertake a review of this effort on designing consensus, and discuss its relevance to CPSs.
14. Contrast the notions of subsystem structure and signal structure introduced in Chapter 7. In what settings are the two notions of structure appropriate?
15. Computer scientists and engineers both consider state machines and signal flows, yet the representations that they use for state and signal evolution are rather different. Explain how a common language for system states and signals that spans the computing and engineering communities may be used in CPS applications.
16. Choose a control or estimation problem for an example CPS (e.g., temperature control in a smart refrigerator). For the example, please explain what the manifest variables of the system are, and how a description of the interactions among the manifest variables can support controller or estimator design.
17. Several of the chapters in Part II are concerned with the responses of networked CPSs to disturbances. Discuss how the structure of a networked CPS impacts its ability to withstand disturbances or disruptions.

Hint of Solution

The augment state is generated as $z_{l,k} = [x_{l,k}\ u_{l,k-1}\ u_{l,k-2}]^T \in \mathbb{R}^{4\times 1}$ or $\chi_{l,k} = [(z_{l,k})^T\ u^T(z_{l,k})]^T \mathbb{R}^{6\times 1}$. Moreover, the regression function for the adaptive estimator is $\left\{\chi_{l,1}^2, \chi_{l,1}\chi_{l,2}, \chi_{l,1}\chi_{l,3}, \ldots, \chi_{l,2}^2, \ldots, \chi_{l,5}^2, \ldots, \chi_{l,6}^2\right\}$ as per (6.6). Further, the tuning parameter for the adaptive estimator is selected as $\alpha_{l,h} = 10^{-6}$, while initial parameters for the adaptive estimator are set to zeros at the beginning of the simulation. The parameter for the ZOH event-trigger condition is chosen as $\Gamma_l = 0.9$.

First, the performance of the proposed optimal adaptive ZOH event-triggered control is shown. Without loss of generality, an average value of state regulation errors for the six subsystems is shown in Figure 1. The results indicate that proposed optimal

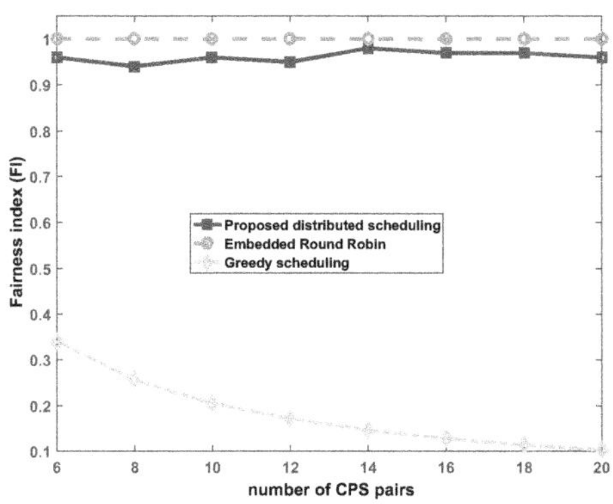

Figure 1 State regulation errors.

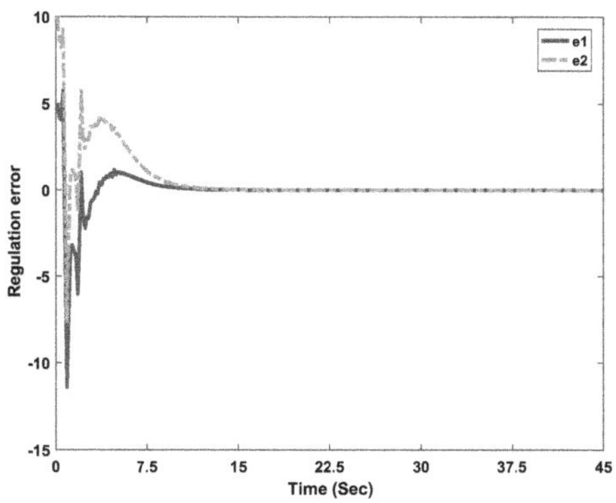

Figure 2 Event-trigger error norm $\|e_k\|^2$.

adaptive ZOH event-triggered control design can not only force the regulation errors to converge to zero asymptotically, but also make the state regulation errors converge to zero quickly while ensuring all systems are stable. It is important to note that overshoots are observed at the beginning because the value function estimator tuning needs a brief adaptation phase.

Then, the performance of the optimal adaptive ZOH event-triggered control system is demonstrated. Without loss of generality, we pick the fifth optimal adaptive ZOH event-triggered subsystem to evaluate the performance. As shown in Figure 2, when an error event is triggered and scheduled, the estimation error will be reset to zero since the actual subsystem state will be received at the controller. On the other hand, if an error event is not triggered and scheduled, the estimation error increases due to the held system state in the ZOH event-triggered control system. Once the parameters of the value function estimator are assessed accurately, the estimation errors converge to zero. It is important to note that error events are triggered and scheduled more frequently at the beginning, since the value function estimator needs to be tuned often. After a short adaptation phase, the error events become less frequent, which would reduce network traffic.

Next, the performance of the proposed cross-layer distributed scheduling has been evaluated. For comparison, classical embedded round robin (ERR) [15] and greedy scheduling [15] schemes are considered. In Figure 3, the cost function of multiple subsystems of a distributed CPS with three different scheduling schemes is compared. The proposed novel cross-layer distributed scheduling scheme maintains a lowest value while the cost function of the distributed CPS with ERR and greedy scheduling is much more than the proposed scheduling scheme.

This indicates that the proposed distributed scheduling scheme can improve the performance of the distributed CPS much better than the widely used ERR and greedy scheduling. It is important to note that: (1) Since ERR only guarantees that each CPS system can have the same probability to access the shared communication network resource, it could not assign communication network resources based on practical usage of each system of the distributed CPS, which might waste precious network resources. Therefore, ERR cannot optimize the multiple subsystems of the distributed

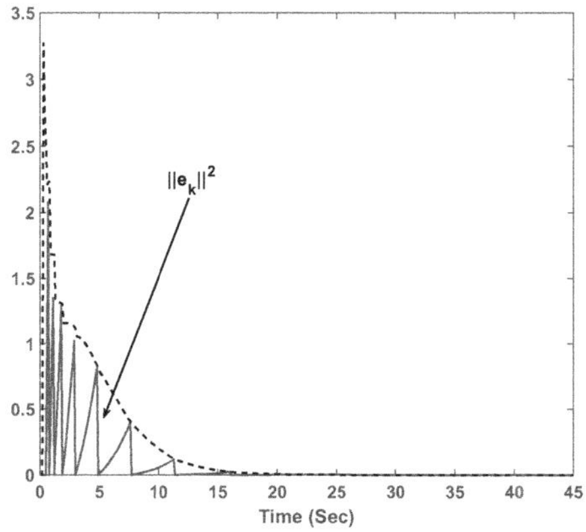

Figure 3 The cost function comparison.

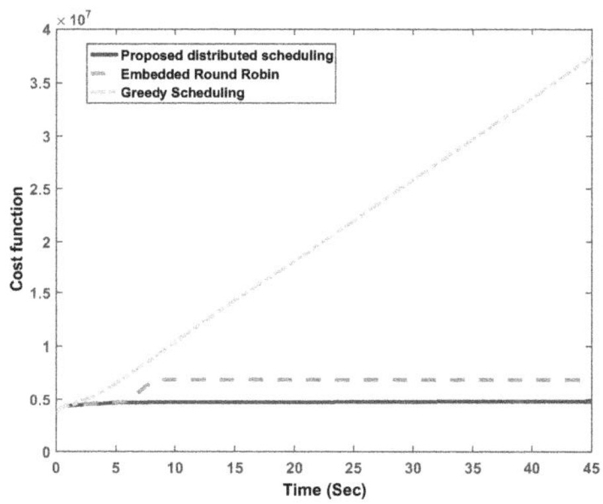

Figure 4 The fairness comparison.

CPS's performance. And (2) since greedy scheduling only assigns network resources to the system with the best network layer performance, whereas the physical system performance of each system in the distributed CPS is overlooked, it also cannot optimize the performance of the distributed CPS.

The fairness of different scheduling schemes has been evaluated. As shown in Figure 4, fairness indices of the proposed cross-layer distributed scheduling and widely used ERR schemes are close and equal to 1, whereas that of greedy scheduling is much less than 1 thus indicating fair allocation of shared communication network resource for the proposed one while meeting overall performance. The ERR method, though fair, has higher cost than the proposed one. According to results illustrated in Figures 1 through Figure 4, the proposed cross-layer co-design included optimal adaptive ZOH event-triggered control and a cross-layer distributed scheduling scheme to optimize the performance of both communication network and the subsystems.

Part III

Privacy, Security, and Protection

Designers and operators of physical-world systems naturally account for diverse unusual conditions and disruption events, which alter the physical responses of the system. Computer scientists have devised a wide array of techniques for protecting information and maintaining fidelity of cyber systems. Despite these methodologies, however, protecting and securing cyber-physical systems (CPSs) remains extremely challenging. Crucially, the tight interface of cyber and physical components necessitates considering both information violation and physical disruption in protecting and securing the system. Additionally, in CPSs, primarily information-flow events in the cyber world may implicate complicated physical-world responses, and conversely physical-world disruptions may leak into the cyber system. Additionally, the networking or coupling of devices caused by cyber-enablement further complicates threat assessment and mitigation. The complex and propagative impact of threats to CPSs has been a significant concern recently: Consider, for instance, the implications of the Stuxnet worm, computer failures and attacks in the air-transportation system, credit-card cyber-theft events, etc. Based on these examples, it is clear that threat assessment and mitigation needs are pervasive in diverse CPSs ranging from cyber-enabled devices like cars, to large-scale infrastructures and the global Internet of Things. Thus, achieving security, privacy, and threat mitigation in CPSs is a keystone challenge.

Part III of the book introduces principles and methods for secure and safe operation of CPSs. The individual chapters introduce diverse problems and issues related to safe, secure, and private operation of CPSs, and develop techniques that are targeted to address these problems. In doing so, the chapters treat simultaneously the security of information and the protection of physical-world processes. They also study how cyber and physical components can be used in tandem for protection of holistic dynamics. Each chapter also discusses application(s) of the presented methods to specific infrastructure-network security and protection problems.

The contributions of the chapters are as follows. Chapter 8, by Lesieutre et al., develops a means for securing the cyber-models of physical-world infrastructures, specifically of the power transmission network, via a computational masking procedure. Chapter 9 introduces a framework and methods for evaluation of CPS security, focusing particularly on situational awareness in the face of diverse threats; the chapter particularly addresses security and situational awareness for the network's communication infrastructure. Chapter 10, by McMillin et al., introduces a formal approach for analyzing security of the cyber-physical power-distribution infrastructure, which simultaneously captures physical and cyber elements in the network. Finally, Chapter 11, by Preciado et al., addresses resource allocation for protection of cyber-physical networks from spread processes, using a biology-inspired framework.

8 Preserving Confidentiality of Critical Energy Infrastructure Information

Bernard Lesieutre, Alex Borden, and Parameswaran Ramanathan

8.1 Introduction

The electric power grid is one of the nation's critical infrastructures and there is considerable concern about its vulnerability to malicious physical and cyber attacks. It constitutes a closely coupled cyber-physical system (CPS) in which the grid that literately spans the continent is controlled at relatively few discrete locations using dedicated computing facilities that rely on accurate sensing and communication networks. The core computational functions of control centers are to estimate current grid conditions, make adjustments for reliability and for economic optimal use of controllable resources, and to anticipate future disturbance scenarios that may risk blackouts. Event security analyses are run continuously, and optimal economic dispatch programs result in system adjustments on a time scale of every five minutes for some markets. These computational tasks all involve a sophisticated analysis of a large nonlinear model of the electrical network represented by the power flow equations.

There are many aspects of this cyber-physical network that can be examined in terms of vulnerabilities and network security. Much work can and has been done on examining the vulnerabilities to the physical infrastructure, and many studies have been directed toward understanding cascading outages resulting in blackouts (e.g., [1, 2]). The potential threat of a malicious attack leading to a widespread blackout raises the cyber-physical security of the power grid to one of national importance. There is concern that an unsuccessful attack took place on April 16, 2013 at a major substation near San Jose, California after someone destroyed communication links and then critically damaged equipment, resulting in the loss of the facility [3]. While the attack challenged the grid in the immediate region, a blackout was averted. In that instance the attack was direct and physical, yet a similar or worse result could be achieved through a cyber-attack on systems that control elements of the physical grid. This cyber security concern has been noted by the President of the United States [4], the Secretary of Defense [5], and the Director of the Department of Homeland Security in the United States [6].

In the private sector, the power grid in many areas supports competitive markets for electricity. In addition to the common interest in a secure, reliable physical network, market participants desire economic operation that shields their proprietary data. Combining physical and market needs, an ideal system would clear and operate without revealing any network or market data to market participants and network operators beyond their own knowledge of their own systems and data.

The focus of this chapter is the topic of securing the information, data, and models of the power network that are used in the computing environment that controls the grid. Understandably, the detailed data describing grid topology and components are considered sensitive, and in the United States can only be shared through a Critical Energy Infrastructure Information (CEII) and similar nondisclosure agreements. This data control is intended to help mitigate the risk that a detailed analysis of grid models could reveal the weakest points and vulnerabilities. In particular, major outages are observed to be initiated by many smaller events, and with detailed grid models, multiple simultaneous physical and cyber attacks could be considered and simulated, which is considerably more difficult without the model. With regard to current market and grid operations, physical grid models are shared among grid operators as these are absolutely necessary for coordinated, reliable operation across the coupled network. The network models are not shared publicly and detailed economic data are not shared publicly or among participants, except in a masked form and after a long delay. This tightly controlled access to power grid information limits advances in power system research and power system operations. Researchers who work with the data must mask their results in publications, and other researchers cannot openly and independently verify their results. Similarly, the care and control of critical information requires security safeguards when using shared computing platforms such as cloud computing.

These security and privacy concerns beg a question that defines a current theme of research: how can the power system models be effectively jointly shared and jointly analyzed without revealing critical information about the underlying electrical grid or corresponding market data? In this chapter we provide a tutorial on basic power system optimization models used in markets and operations, and on techniques for masking optimization models drawing from research in cloud computing. We then extend the cloud computing techniques to apply to power system problems.

Current research considers a linear power system representation that is commonly used in optimization problems associated with electricity markets. The models are solved to match generator supply and expected demand while observing limits on generator capacity and transmission line flows. The results direct facility-specific generation schedules and location-specific electricity prices which are related to dual variables in the optimization. A primary objective is to seek a transformation to mask the optimization problem so that it may be shared and solved publicly, and whose results can be transformed back to yield dispatch direction and prices. Ideally, with multiple participants, they would be able to retrieve their private results from the public solution, but not those of others. For the purpose of sharing data and models among researchers, we may also choose to restrict the form of the masked, transformed model to take the form of another power system optimization model. The ultimate purpose of this structure-preserving approach is to enable the sharing of like-models among researchers to allow the verification of methods and results. This is achieved by using system models that are provably equivalent to true, critical infrastructure system models through a transformation only known to the transformer. Ideally this will lead to a set of publicly available standard models for use in research.

8.2 Power System Models

The purpose of the electric power grid is to enable the transport of electric energy from suppliers to end-users, and most models contain some representation of the physical network topology and its electric characteristics. The models used for making decisions are usually cast as optimization problems in which an economic objective is considered under a variety of physical and engineering constraints. There are a large number of questions posed in this context. Most common in the operating time scale, over time frames of five minutes to one day, the cost of producing electricity from available generating suppliers is minimized, accounting for network power flow and component operating constraints. The decision variables are the generator power outputs, and the optimization provides a schedule for these resources. Over a longer time scale, up to one week, an optimization problem will determine which generators will be selected to be available to supply energy, again minimizing costs. There are additional facility-specific constraints concerning maximum/minimum operating limits, ramping limits, minimum run/down times, and also security constraints to ensure adequate supply if one or more facilities is lost during operation. The primary result of this type of optimization is the "unit commitment" schedule for available generating facilities. Over a longer planning horizon, from months to years, optimization problems are cast to determine the best locations for candidate transmission upgrades, again attempting to minimize the cost of the upgrade while meeting expected future operational needs.

The complexity of these optimization problems varies from nonlinear, nonconvex, mixed-integer programs (security-constrained, unit-commitment, ACOPF [AC optimal power flow]) to a simpler linear program (market clearing DCOPF [DC optimal power flow]). In this chapter we introduce some basic modeling concepts and initial research in the masking of power system models. Here, we used the market-motivated linear program formulation. For the interested reader, we also briefly discuss the basics of the nonlinear model.

The flow of electric power along branches of the network is driven by voltage differences at the terminals. The active (average) power and reactive power between two connected buses (nodes) are related by the voltage magnitudes (V), voltage phase angles (δ), and the electrical characteristics of the network element. For example, the active power (P) and reactive power (Q) flow along a line from node i to node j; as measured at node i this is given by the following equation, assuming line electrical conductance (g) and susceptance (b):

$$P_{12} = V_1^2 g - V_1 V_2 g \cos(\delta_1 - \delta_2) + V_1 V_2 b \sin(\delta_1 - \delta_2)$$
$$Q_{12} = -V_1^2 b + V_1 V_2 b \cos(\delta_1 - \delta_2) + V_1 V_2 g \sin(\delta_1 - \delta_2)$$

The power "injected" at a particular bus (node) is the simple sum of the branch flows emanating from that node: $P_i = \sum_j P_{ij}$ and $Q_i = \sum_j Q_{ij}$.

These nonlinear equations, combining associated limits and a cost function, constitute the so-called ACOPF. The linear program that approximates this is derived using the following assumptions:

1. The conductance is negligible. i.e., $g \ll b$.
2. The voltages are all at nominal levels, i.e., $V = 1.0$ per unit.
3. Phase angle differences are small, i.e., $\sin(\delta_k - \delta_m) \approx \delta_k - \delta_m$.
4. The focus is on active power, ignoring reactive power equations.

With these assumptions, the active power flow along a branch is approximated to be $P_{ij} = b(\delta_i - \delta_j)$. Mathematically, the representation of all the branch and node equations can be written in vector form using a bus-branch incidence matrix representing topological connections, $\mathbf{A}_{\text{inc}}$. The number of rows and columns of $\mathbf{A}_{\text{inc}}$ equal the number of branches and nodes in the network, respectively. Each row contains two nonzero entries, a $+1$ element at the branch "from" bus and a -1 element at the branch "to" bus. The vector of branch flows is given by $\mathbf{P}_{\text{br}} = \text{diag}\,(()\,\mathbf{b}_{\text{br}})\mathbf{A}_{\text{inc}}\delta$, and the vector of power injections is given by

$$\begin{aligned}\mathbf{P}_{\text{inj}} &= \mathbf{A}_{\text{inc}}\mathbf{P}_{\text{br}}\\ &= \mathbf{A}_{\text{inc}}{}^{T}\,\text{diag}\,(()\,\mathbf{b}_{\text{br}})\mathbf{A}_{\text{inc}}\delta\\ &\equiv \mathbf{B}\delta\end{aligned}$$

With an objective function and constraints, the DCOPF is written as

Minimize $\mathbf{c}^T\mathbf{P}_g$
Subject to:

$$\begin{aligned}-\mathbf{P}_g + \mathbf{B}\delta &= -\mathbf{P}_L\\ -\mathbf{P}_{flow,max} &\le \text{diag}\,(()\,\mathbf{b}_{\text{br}})\mathbf{A}_{inc}\delta \le \mathbf{P}_{flow,max}\\ \mathbf{P}_{g,min} &\le \mathbf{P}_g \le \mathbf{P}_{g,max}\\ \delta_{ref} &= 0\end{aligned} \tag{8.1}$$

where the vector includes generator power taken as decision variables in the optimization, vector $\mathbf{P}_L$ are load demands taken as fixed quantities, and $\mathbf{P}_{g,min}$, $\mathbf{P}_{g,max}$, and $\mathbf{P}_{flow,max}$ are capacity limits on generator output and transmission line flows. Since all the power flows are defined in terms of angle differences, an absolute angle reference is needed, and one element of the vector of angles is set to zero. Also note that the generator powers are typically specified as positive numbers for injection into the network, while the load power (elements of $\mathbf{P}_L$) are also usually specified as positive quantities but represent power leaving the network. The first equality constraint in the model represents conservation of energy (power) at each bus. The difference between generation and load at a node must equal the power flows emanating from that node. The first inequality constraint represents engineering limits placed on the amount of power allowed to flow along a network branch. They are not physical limits in the same sense as conservation of energy, but are operational limits intended to protect equipment from overload and to ensure grid stability in cases of disturbances. Similarly, the second inequality represents operational limits on the output of generators.

We have specified the model here with a linear objective function. In the literature it is common to see this, as well as piecewise linear, quadratic, and sometimes higher-order polynomials. In electricity markets, a piecewise linear cost function is employed. We begin with the linear model and include reference to extensions using quadratic and piecewise linear models when appropriate.

This DCOPF can be solved for generator dispatches, $\mathbf{P}_g$, and the Lagrange multipliers (dual variables in the linear programming context) associated with bus power injection constraints (the equality constraints) define the electricity prices by location. While these solutions are important in terms of use, we return to our original issue: The quantities defining the network, $\mathbf{A}_{\text{inc}}$, $\mathbf{B}$, $\mathbf{b}_{\text{br}}$, demand $\mathbf{P}_L$, generator and line capacity limits, and cost function weights $\mathbf{c}^T$ are not publicly shared, for reasons of security and privacy. We examine methods to transform and mask this model to enable sharing it without revealing sensitive details.

8.3 Confidentiality-Preserving Transformations

There is an active and evolving research area focusing on techniques for masking linear optimization programs for cloud computing [7, 8]. We are inspired by the original work of Dreier and Kerschbaum [7], who developed a technique for masking a generic linear program through a sequence of transformations. To make their approach applicable to DCOPF, some extensions were necessary. First, the DCOPF description is inherently sparse and for computational reasons this sparsity needs to be retained in the masked model. In the original work, the transformations do not maintain a sparse system model. In [8], Shojaei et al. build on [7] to preserve the sparsity of the transformed model. Second, in an OPF, we require the values of the dual variables as these are used as prices in settling the market. In [9], we extended the work of [7] and [8] to include the transformation of the dual variables as well, and applied the technique to power system examples. These methods generally apply to all linear programs. In our particular power system, we may wish to obscure coarse features of the system model in addition to the details. These may include, for example, the number and nature of facilities in the model: the number of generators, buses, and lines.

Further extensions are needed to mask proprietary economic information among multiple participants. In some of our most recent work, outlined in this chapter, we extend the methods of [7–9] to enable independent masking of subnetworks, which are then combined and solved as a large coupled optimization problem. The individual results for each subsystem are then only available to that subsystem through the transformation only known to itself.

A thorough analysis of these methods must consider attacks to reveal the sensitive models and information. Recent literature has posed some powerful attacks that can break certain masked optimization models [10–13]. Specifically, in [13], Mangasarian considers the problem of solving a linear program jointly formulated by multiple collaborating parties. Each party has a set of constraints on the optimization variables which it does not want to share with any other party. Mangasarian proposes a masked optimization approach that preserves the privacy requirement of each party in solving the linear program. However, recent papers [10–12] propose several different attacks on the solution approach proposed in [13]. They also propose solutions to alleviate these attacks. These attacks are not directly feasible in our models because of one key difference; our models are either single-party-based linear programs with privacy constraints against a cloud computer, or multiparty vertically partitioned linear programs. In contrast, the attacks in [10–12] are for the multiparty horizontally partitioned model of [13]. In multiparty vertically partitioned linear programs, all parties have the same set of constraints, but each party has a privacy constraint on a subset of the optimization variables. While these models differ in important details to our own, modifications of these attacks need be considered in designing our transformations.

In the next subsection we review a general transformation approach to masking a linear program, cognizant of sparsity, and tracking dual variables.

8.3.1 Masking Transformations for the Linear Program

The sequence of transformations followed here derive from [7], with additional consideration of sparsity [8], and additional recovery of dual variables [9]. We then provide an DCOPF example from our prior work [9] and a larger example. The primal model notation in this section is from [7] for a standard linear program to which we add the dual form. Variables x are primal variables, and u and v are the dual variables.

Primal program:

$$\begin{aligned} \min_{x} \ & \mathbf{c}^T\mathbf{x} \\ \text{s.t.} \ \ & \mathbf{M}_1 x = \mathbf{b}_1 \\ & \mathbf{M}_2 x \leq \mathbf{b}_2 \\ & \mathbf{x} \geq 0 \end{aligned} \tag{8.2}$$

Dual program:

$$\begin{aligned} \max_{\mathbf{u},\mathbf{v}} \ & \mathbf{b}_1^T\mathbf{u} + \mathbf{b}_2^T\mathbf{v} \\ \text{s.t.} \ \ & \mathbf{M}_1^T\mathbf{u} + \mathbf{M}_2^T\mathbf{v} \leq \mathbf{c} \\ & \mathbf{v} \leq 0 \end{aligned} \tag{8.3}$$

Recall that a monomial matrix is one that contains exactly one nonzero entry in every row and column. For example, a permutation is a monomial matrix with nonzero elements equal to 1. Let $\mathbf{Q}$ be a random positive monomial matrix and let $\mathbf{r}$ be a random positive vector. Define $\mathbf{z} = \mathbf{Q}^{-1}\mathbf{x} + \mathbf{r}$. Then, (8.2) can be rewritten as

$$\begin{aligned} \min_{x} \ & \mathbf{c}^T\mathbf{Q}(\mathbf{Q}^{-1}\mathbf{x} + \mathbf{r}) \\ \text{s.t.} \ \ & \mathbf{M}_1\mathbf{Q}(\mathbf{Q}^{-1}\mathbf{x} + \mathbf{r}) = \mathbf{b}_1 + \mathbf{M}_1\mathbf{Q}\mathbf{r} \\ & \mathbf{M}_2\mathbf{Q}(\mathbf{Q}^{-1}\mathbf{x} + \mathbf{r}) \leq \mathbf{b}_2 + \mathbf{M}_2\mathbf{Q}\mathbf{r} \\ & \mathbf{Q}^{-1}\mathbf{x} + \mathbf{r} \geq \mathbf{r} \end{aligned} \tag{8.4}$$

which in turn can be further rewritten as

Primal program:

$$
\begin{aligned}
&\min_{\mathbf{z}} \mathbf{c}^T \mathbf{Qz} \\
\text{s.t.} \quad & \mathbf{M}_1\mathbf{Qz} = \mathbf{b}_1 + \mathbf{M}_1\mathbf{Qr} \\
& \mathbf{M}_2\mathbf{Qz} \le \mathbf{b}_2 + \mathbf{M}_2\mathbf{Qr} \\
& \mathbf{Sz} \ge \mathbf{Sr}
\end{aligned}
\tag{8.5}
$$

where $\mathbf{S}$ is a positive monomial matrix. The corresponding dual optimization problem is

Dual program:

$$
\begin{aligned}
&\max_{\mathbf{u},\mathbf{v}} (\mathbf{b}_1 + \mathbf{M}_1\mathbf{Qr})^T\mathbf{u} + (\mathbf{b}_2 + \mathbf{M}_2\mathbf{Qr})^T \\
\text{s.t.} \quad & (\mathbf{M}_1\mathbf{Q})^T\mathbf{u} + (\mathbf{M}_2\mathbf{Q})^T\mathbf{v} \le (\mathbf{c}^T\mathbf{Q})^T \\
& \mathbf{v} \le 0
\end{aligned}
\tag{8.6}
$$

Note that, in the above reformulation, the value of the cost function is shifted by a constant $\mathbf{c}^T\mathbf{Qr}$ and will not affect the optimization problem. Further, since the monomial matrix $\mathbf{Q}$ has only positive entries, the inequalities retain their orientation.

Next the inequality constraints are converted to equality constraints using the usual approach of introducing slack variables $\mathbf{z}_{sl}$. Define $\mathbf{z}'$ by $\mathbf{z}'^T = [\mathbf{z}^T \mathbf{z}_{sl}^T]$. To maintain the correct dimension, the cost weight vector is padded with zero entries associated with the slack variables: $\mathbf{c}'^T = [\mathbf{c}^T\mathbf{Q}\ 0 \ldots 0]$. For the dual variable vectors, we define $\mathbf{u}'^T = [\mathbf{u}^T \mathbf{v}^T]$. We define

$$
\mathbf{M}' = \begin{pmatrix} \mathbf{M}_1\mathbf{Q} & 0 \\ \mathbf{M}_2\mathbf{Q} & \\ -\mathbf{S} & \mathbf{A} \end{pmatrix}, \quad \mathbf{b}' = \begin{pmatrix} \mathbf{b}_1 + \mathbf{M}_1\mathbf{Qr} \\ \mathbf{b}_2 + \mathbf{M}_2\mathbf{Qr} \\ -\mathbf{Sr} \end{pmatrix}
$$

where $\mathbf{A}$ is another random positive monomial matrix. Then the primal and dual programs can be written as

Primal program:

$$
\begin{aligned}
&\min_{\mathbf{z}'} \mathbf{c}'^T\mathbf{z}' \\
\text{s.t.} \quad & \mathbf{M}'\mathbf{z}' = \mathbf{b}' \\
& \mathbf{z}' \ge 0
\end{aligned}
\tag{8.7}
$$

Dual program:

$$
\begin{aligned}
&\max_{\mathbf{u}'} \mathbf{b}'^T\mathbf{u}' \\
\text{s.t.} \quad & \mathbf{M}'^T\mathbf{u}' \le \mathbf{c}'
\end{aligned}
\tag{8.8}
$$

While the transformations appear complicated, we note that all the transforming matrices are positive monomial matrices, which in the steps above serve to scale and permute columns in the original linear program. At this stage, the transformations are vulnerable to attack – if not the actual values, at least the form, which in the power system context could yield sensitive information about topology. The next step applies a nonsingular matrix $\mathbf{P}$ (any nonsingular matrix) and a random positive monomial matrix $\mathbf{T}$ so that $\mathbf{M}'' = \mathbf{P}\mathbf{M}'T$, $\mathbf{b}'' = \mathbf{P}\mathbf{b}'$, and $\mathbf{c}''\mathbf{T} = \mathbf{c}'^T\mathbf{T}$. The left multiplication of the $\mathbf{P}$ matrix makes linear combinations of the constraint equations, i.e., row combinations. The $\mathbf{T}$ matrix scales and permutes the columns of matrix $\mathbf{M}'$ and vector $\mathbf{c}'^T$. The new primal decision variable vector is $\mathbf{z}'' = \mathbf{T}^{-1}\mathbf{z}'$, and the new dual decision variable vector is $\mathbf{u}'' = (\mathbf{P}^T)^{-1}\mathbf{u}'$. The final masked linear program is given by

Primal program:

$$\begin{aligned} &\min_{\mathbf{z}''} \mathbf{c}''^T\mathbf{z}'' \\ \text{s.t.} \quad &\mathbf{M}''\mathbf{z}'' = \mathbf{b}'' \\ &\mathbf{z}'' \geq 0 \end{aligned} \tag{8.9}$$

Dual program:

$$\begin{aligned} &\max_{\mathbf{u}''} \mathbf{b}''^T\mathbf{u}'' \\ \text{s.t.} \quad &\mathbf{M}''^T\mathbf{u}'' \leq \mathbf{c}'' \end{aligned} \tag{8.10}$$

This masked problem can be shared and solved via cloud computing. The original variables are readily obtained from the masked problem by inverting the transformations:

Primal solution:

$$\mathbf{T}\mathbf{z}''^* = \mathbf{z}'^* = \left[\mathbf{z}^{*T}\mathbf{z}_{sl}^{*\,T}\right]^T \quad \text{and} \quad \mathbf{x}^* = \mathbf{Q}(\mathbf{z}^* - \mathbf{r}) \tag{8.11}$$

Dual solution:

$$\mathbf{P}^T\mathbf{u}''^* = \mathbf{u}'^* = \left[\mathbf{u}^{*T}\mathbf{v}^{*T}\right]^T \tag{8.12}$$

We comment here and later that particular care should be taken in choosing a $\mathbf{P}$ matrix in the transformations. It is useful to maintain sparsity when the original problem is sparse, and it is necessary to introduce enough row operations with appropriate scalings to avoid susceptibility to attacks.

DCOPF Masking

We now proceed with the masking of the DCOPF. For discussion purposes, we keep track of the number of buses, generators, and branch elements in the model, n_b, n_g, and n_{br}, respectively. This allows us to discuss the dimensions of vectors and matrices, and in the end we will also discuss ways to mask these numbers as they provide some coarse information about the system.

The first step in masking the DCOPF in the manner described in the previous subsection is to convert the model to standard form. In (8.1) $\mathbf{P}_g$ and δ are unconstrained, whereas the standard form constrains the variables to be nonnegative. To place the system of equations in standard form, we represent $\mathbf{P}_g$ and δ as the difference between two nonnegative variables: $\mathbf{P}_g = \mathbf{P}_g^+ - \mathbf{P}_g^-$ and $\delta = \delta^+ - \delta^-$, where $\mathbf{P}_g^+$, $\mathbf{P}_g^-$, δ^+, $\delta^- \geq 0$.

In the standard form, the decision variable vector is $\mathbf{x}^T = \left[\mathbf{P}_g^{+T}\ \mathbf{P}_g^{-T}\ \delta^{+T}\ \delta^{-T}\right]$. Note that there are $2n_g + 2n_b$ elements in vector x. The linear cost weighting vector is $\mathbf{c}^T = [\mathbf{d}^T\ \ -\mathbf{d}^T\ \ 0\ldots 0]$. There is a slight abuse of notation here as $\mathbf{d}$ is the linear generator cost weights $\mathbf{c}$ in (8.1). This repetition of matrix forms continues with

$$\mathbf{M}_1 = \begin{bmatrix} -\mathbf{E}_g & \mathbf{E}_g & \mathbf{B} & -\mathbf{B} \\ 0 & 0 & \mathbf{e}_{ref} & -\mathbf{e}_{ref} \end{bmatrix}$$

which has $n_b + 1$ rows. The $n_b \times n_g$ matrix $\mathbf{E}_g$ is akin to an indicator matrix with a $+1$ entry in each row corresponding to a bus with a generator present. The $n_b \times n_b$ matrix $\mathbf{B}$ is the DCOPF bus susceptance matrix and by construction is a sparse weighted Laplacian matrix. The row vector $\mathbf{e}_{ref}$ is used to indicate the reference angle and it has a single $+1$ in the column corresponding to the reference bus. In (8.2) the load powers are given by $\mathbf{b}_1 = [-\mathbf{P}_L^T\ 0]^T$ in which the first n_b rows help enforce power balance at each bus. The last row of $\mathbf{M}_1$ enforces the reference bus angle, $\delta_{ref} = \delta^+_{ref} - \delta^-_{ref} = 0$. For the inequality constraints, define

$$\mathbf{M}_2 = \begin{bmatrix} \mathbf{I}_g & -\mathbf{I}_g & 0 & 0 \\ -\mathbf{I}_g & \mathbf{I}_g & 0 & 0 \\ 0 & 0 & \text{diag}\,(()\,\mathbf{b}_{br})\mathbf{A}_{\text{inc}} & -\text{diag}\,(()\,\mathbf{b}_{br})\mathbf{A}_{\text{inc}} \\ 0 & 0 & -\text{diag}\,(()\,\mathbf{b}_{br})\mathbf{A}_{\text{inc}} & \text{diag}\,(()\,\mathbf{b}_{br})\mathbf{A}_{\text{inc}} \end{bmatrix}$$

This matrix has $2n_g + 2n_br$ rows. Matrix $\mathbf{I}_g$ is an $n_g \times n_g$ identity matrix and these rows correspond to generator capacity limits. Vector $\mathbf{b}_{br}$ contains the branch susceptances associated with the electrical characteristics of the branch, matrix $\mathbf{A}_{\text{inc}}$ is the bus-to-branch incidence matrix, and diag $(\mathbf{b}_{br})\mathbf{A}_{\text{inc}}$ is an $n_{br} \times n_b$ matrix associated with branch flows. These rows represent branch flow capacity limits specified by $\mathbf{b}_2 = [\mathbf{P}^T_{g,max}\ \ -\mathbf{P}^T_{g,min}\ \ \mathbf{P}^T_{flow,max}\ \ \mathbf{P}^T_{flow,max}]^T$.

These matrices apply to the dual program in the obvious way. The masking steps outlined above are easily followed with the appropriate transformation matrices. Matrices $\mathbf{Q}$, $\mathbf{A}$, $\mathbf{S}$, and $\mathbf{T}$ and vector $\mathbf{r}$ can be generated randomly, and do not require particular attention, although some consideration might be given on the magnitude of the positive scaling elements in these matrices in order to mask values in the original model. The nonsingular $\mathbf{P}$ matrix does take more care in design. We would like the choice to preserve sparsity, and to mask the system model to make it invulnerable to attack.

In our work, we have investigated two approaches to generating suitable $\mathbf{P}$ matrices for transformation. In the first approach we developed a routine that both enforces a user-specified level of sparsity (i.e., limit on the percentage of nonzero elements in the transformed $\mathbf{M}$ matrix), and distributes the nonzero elements to appear random, with similar numbers of entries in the rows and columns. This method is effective, but it has a serious drawback: The computation required to perform the masking is high. If one were able to use the same masking matrices for a number of analyses on the same original system under different operating conditions, then this may not be a concern. However, if only a single optimization problem is run, then one must consider the level of computation required for the purpose of the problem. If it takes more computational

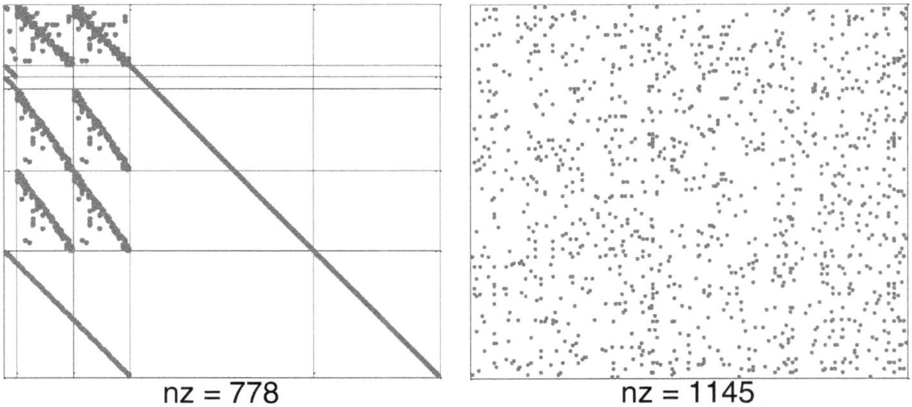

(a) Before (b) After

Figure 8.1 Comparison of structure of elements in DCOPF formulation for a IEEE 30-bus test system before and after the masking procedure.

power to mask the system model than to solve it, then the value of off-site cloud computing is diminished. As a second approach, we consider randomly generating sparse **P** matrices, which is computationally attractive. It does appear to also mask the original system model while retaining a sparse model. Further investigation into attacks is needed to compare the two methods.

We have applied the masking techniques to the several test systems that are provided in the MATPOWER distribution [14]. Since the problem is entirely set up as a transformation of variables, and there is no approximation in the transformations, the results from analysis of the original system, and the results obtained by inverting the transformation from the solution of the masked system, must be equal, and our tests confirm this. More interesting is a view of the pattern of entries in the original and transformed models. The entries of **M** and **M** matrices are compared using the spy function in MATLAB.

In Figure 8.1 we compare the structure of the original and masked formulations for the DCOPF for the IEEE 30-bus system. In the original model, shown on the left, with 778 nonzero elements, the structure is immediately obvious. One familiar with power systems can easily back out all the information about the system from this matrix. The masked system, shown on the right, is still sparse but has 1145 nonzero entries with no obvious structure on the locations of the nonzero entries. However, as discussed in [10, 11], attacks based on the variance of the nonzero values in each column may reveal columns corresponding to the slack variables, which in turn can be used to recover part of the structure of the original model. However, as discussed above, the transformation matrices can be chosen to enforce a level of sparsity, spread of nonzero entries, and the variance of the nonzero values in each column. Similar results on the sparsity and the spread of the nonzero entries for the larger 3375-bus system are shown in Figure 8.2. Since the structure of the original model is destroyed in the masking process, the time required to solve the problem may increase. However, our experience shows that if sparsity is adequately preserved, then the increase in the time required

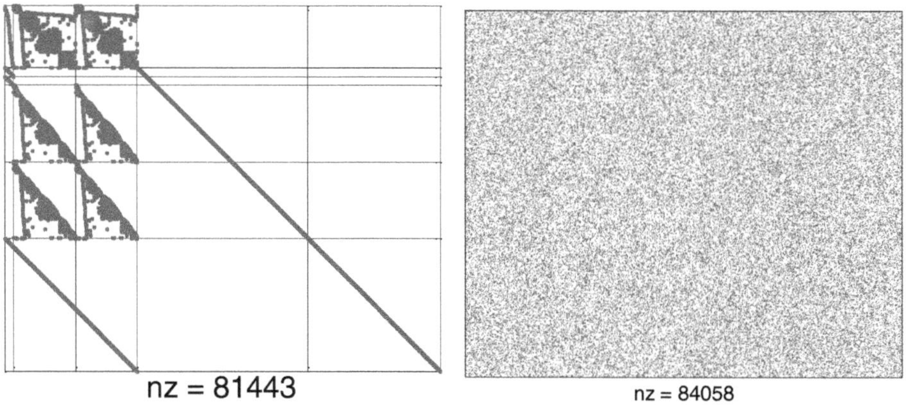

(a) Before (b) After

Figure 8.2 Comparison of structure of elements in DCOPF formulation for a 3375-bus system, before and after the masking procedure.

to solve the model is marginal. For example, for the 3375-bus system, the original model solved in 11.6 seconds on a desktop with common reconfiguration, whereas the transformed model required 16.6 seconds to solve. The increase in computational time may be even less of an issue if more powerful cloud computers are used to solve the masked model.

We note that in the masking process so far, no effort has been made to mask the coarse information about the number of different types of facilities. Some of this information may simply be inferred from the dimensions of the matrices. Matrix $\mathbf{M}$ as constructed above has $m = n_b + 2n_g + 2n_{br} + 1$ rows and $n = 2n_b + 4n_g + 2n_{br}$ columns. The difference between twice the number of rows and the number of columns, $2m - n = 2n_{br} + 2$, reveals the number of branches in the model. These dimensions can be altered if we assume that the generation is only positive and omit $\mathbf{P}_g^-$ variables from the model (a reasonable assumption), and if some branches do not have capacity limits assigned to them those rows can be omitted as well. Other ways of hiding high-level information involve additional changes to the original model, such as introducing fictitious generator/load pairs at certain buses, so that they exactly cancel locally and don't impact the solution of the entire model. Fictitious buses can be introduced by splitting transmission lines into parallel lines. By these means, the original system can be adjusted to obscure the number of each type of facility before the masking steps. If these additional masking steps are included, then attacks that succeed in identifying the slack variables will not be able to successfully recover sensitive information such as the number of different types of facilities.

8.4 Power System-Preserving Transformations

In this subsection, we consider an additional constraint on the transformations such that the transformed model retains the structure of a power system model. There are a few reasons why this task is important. First, there has been considerable research and software development that exploits power system structures in solving OPF problems, so there could be computational advantages. Second, there is a need for accepted synthetic models for power system analysis, and a model that is related to an actual system will have more credibility, even masked, than some model a researcher invents on his or her own. Third, if it is possible to transform between power system models, then there exists a family of power system models that are indistinguishable from an attacker's perspective. This would prove a very high level of data security, because without knowledge of the transformation actually used, an attacker cannot distinguish them among an entire and potentially large family of possible models.

Generically, it turns out that there is no reason to expect that there exists a large family of power system models that are related through the transformation sequence presented in the previous subsection. To see this, consider seeking the P and T matrices needed to transform a specific $\mathbf{M}'$ matrix to a specific $\mathbf{M}''$ matrix, both with power system structures, possibly with the same number of generators, buses, and lines. The target matrix has $m \times n$ entries that need to be matched exactly, where $m = n_b + 2n_g + 2n_{br} + 1$ and $n = 2n_b + 4n_g + 2n_{br}$. The combined number of entries whose values can be adjusted in the $\mathbf{P}$ and $\mathbf{T}$ matrices is $m \times m + n$, which is smaller than $m \times n$. While this argument appears to preclude generic transformation from one specific system model to another specific system model, it is not a proof, as the special characteristics of power system models might allow such transformations. Nevertheless we find it useful to consider other transformations with higher degrees of freedom.

The degrees of freedom are limited above because matrix $\mathbf{T}$ is constrained to the form of a positive monomial matrix. It is chosen this way to ensure all the variables in the standard form remain positive through the transformation. We note here that this positive monomial restriction on the form of $\mathbf{T}$ is sufficient for this purpose, but is not necessary. In this subsection we consider whether there exist transformations between power system models when we allow $\mathbf{T}$ to take on any values, and we don't force the model in standard form, i.e., the model variables may take on any values, and the slack variables are still required to be nonnegative.

With these model adjustments, the DCOPF can be written in the form

$$\begin{aligned} &\min_{\mathbf{x}} \mathbf{c}^T \mathbf{x} \\ &\text{s.t.} \quad \mathbf{M}\mathbf{x} = \mathbf{b} \\ &\qquad \mathbf{I}_{sl}\mathbf{x} \geq 0 \end{aligned} \tag{8.13}$$

where $x = [\mathbf{P}_g^T \ \delta^T \ \mathbf{x}_{sl}^T]^T$. The subset of variable vector $\mathbf{x}$ associated with $\mathbf{P}_g$ and δ are free variables, and the subset of vector $\mathbf{x}$ associated with the slack variables are held nonnegative through the constraint $\mathbf{I}_{sl}\mathbf{x} \geq 0$. The form of a typical matrix $\mathbf{M}$ and vector $\mathbf{b}$ in this formulation is shown in Figure 8.3.

We pause here to note the dimensions of $\mathbf{M}$ in this formulation: it has $m = n_b + 2n_g + 2n_{br} + 1$ rows and $n = n_b + 3n_g + 2n_{br}$ columns. Here, the difference between number of rows and columns clearly identifies the number of generators in the model.

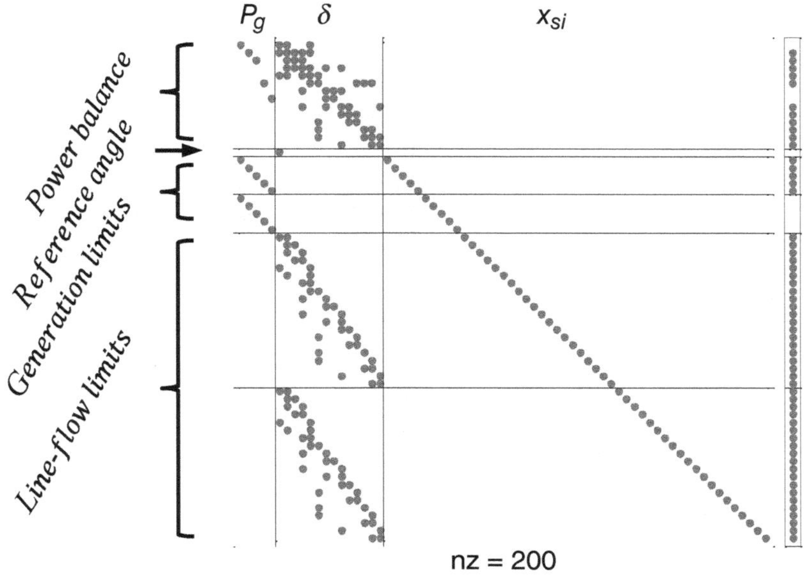

Figure 8.3 **M** matrix and **b** vector for IEEE 14-bus system.

Some of the methods suggested above for obscuring the number of facilities may be applied if this coarse-level information is to be hidden in the model. Also, we note that with the angle reference included, the non-square **M** matrix is usually full-rank in DCOPF applications.

Next we investigate the question of whether there exist transformation matrices that will convert one power system DCOPF to a different power system DCOPF. While doing so, we do not restrict any structure on the transformation matrices, i.e., **T** need not be positive monomial. Also, we do not require the feasible spaces to map exactly, rather we impose the constraint that the optimal solutions are correctly transformed. To study this question, we begin with two different DCOPF formulations and attempt to find matrices to transform between them. The two models are described by (8.13) and (8.14):

$$\begin{aligned} \min_{\mathbf{x}} \ & \mathbf{c}'^T \mathbf{x}' \\ \text{s.t.} \quad & \mathbf{M}'\mathbf{x}' = \mathbf{b}' \\ & \mathbf{I}_{sl}\mathbf{x}' \geq 0 \end{aligned} \tag{8.14}$$

In this exercise with two predefined DCOPFs, the matrices **M** and **M**′ and vectors **c**, **c**′, **b**, and **b**′ are known, and the optimal solutions $\mathbf{x}^*$ and $\mathbf{x}'^*$ are computed and are used to show the existence of the transformation matrices. Specifically, we seek transformation matrices **P** and **T** and vectors **w** and **r** such that

$$\mathbf{M}' = \mathbf{PMT} \tag{8.15}$$

$$\mathbf{c}'^T = \mathbf{c}^T\mathbf{T} + \mathbf{w}^T\mathbf{M}' \tag{8.16}$$

$$\mathbf{b}' = \mathbf{P}\mathbf{b} + \mathbf{M}'\mathbf{r} \tag{8.17}$$

$$\mathbf{x}^* = \mathbf{T}\left(\mathbf{x}'^* - \mathbf{r}\right) \tag{8.18}$$

If such transformation matrices and vectors exist that satisfy (8.15)–(8.18), then there exists a transformation between the two DCOPF problems that preserves the optimal solution (8.18). Equation (8.16) is used to map the cost functions in the models. We note that (8.15) and (8.17) correspond to the problem constraints, and as discussed above, if $\mathbf{T}$ is restricted to be positive monomial, there do not appear to be enough degrees of freedom to satisfy these equations. With $\mathbf{T}$ free, however, there appear to be more than enough degrees of freedom, and a transformation, if any exist, may not be unique.

This non-uniqueness is clear in (8.15). If there exist any $\mathbf{P}_0$ and $\mathbf{T}_0$ that satisfy it, then there exists a space of $\mathbf{T} = \mathbf{T}_0 + \mathbf{NQ}$ matrices that satisfy it, where the $n \times (n-m)$ full-rank matrix $\mathbf{N}$ satisfies $\mathbf{MN} = 0_{m\times(n-m)}$. Furthermore with $\mathbf{P}_0$ nonsingular, $\mathbf{T}_0 = \mathbf{M}^\dagger \mathbf{P}_0^{-1}\mathbf{M}'$ (where $\dagger$ denotes the Moore–Penrose pseudoinverse) provides a $\mathbf{P}_0$-dependent solution, again emphasizing the non-uniqueness of possible solutions. Note that $\mathbf{T}_0$ is not necessarily sufficient for the transformation. It is necessary for $\mathbf{T}$ to be nonsingular, therefore matrix $\mathbf{Q}$ must be full-rank. Matrix $\mathbf{Q}$ cannot be chosen arbitrarily as it will influence the cost function in (8.15).

Next, relationship (8.16) is expanded by substituting $\mathbf{T} = \mathbf{T}_0 + \mathbf{NQ}$ as shown in (8.19), where matrix $\mathbf{Q}$ and vector $\mathbf{w}$ are unknown:

$$\mathbf{c}'^T = \mathbf{c}^T\mathbf{T}_0 + \mathbf{c}^T\mathbf{NQ} + \mathbf{w}^T\mathbf{M}' \tag{8.19}$$

The solution to this relation is also not unique. One method for determining $\mathbf{Q}$ and $\mathbf{w}$ is described in [15]. Here, we present a much simpler relation:

$$\mathbf{Q} = \mathbf{Q}_0 + \mathbf{Q}_1 \tag{8.20}$$

where, assuming $\mathbf{c}^T\mathbf{NN}^T\mathbf{c} \neq 0$,

$$\mathbf{Q}_0 = \frac{(\mathbf{c}^T\mathrm{N})^T(\mathbf{c}'^T - \mathbf{c}^T\mathbf{T}_0 - \mathbf{w}^T\mathbf{M}')}{(\mathbf{c}^T\mathbf{NN}^T\mathbf{c})} \tag{8.21}$$

for any $\mathbf{w}$ and any $\mathbf{Q}_1$ satisfying $(\mathbf{c}^T\mathrm{N})\mathbf{Q}_1 = 0$. Because we need $\mathbf{Q}$ to be full-rank, and $\mathbf{Q}_0$ is rank 1, $\mathbf{Q}_1$ should be chosen to complete the rank.

The remaining two relations (8.17) and (8.18) are satisfied by $\mathbf{r} = \mathbf{r}_0 + \mathbf{N}'\mathbf{q}$ where full-rank $\mathbf{N}'$ is such that $\mathbf{M}'\mathbf{N}' = 0$,

$$\mathbf{r}_0 = \mathbf{M}'^\dagger(\mathbf{b}' - \mathbf{Pb}) \tag{8.22}$$

and

$$\mathbf{q} = \mathbf{N}'^\dagger(\mathbf{x}'^* - \mathbf{T}^{-1}\mathbf{x}^* - \mathbf{r}_0). \tag{8.23}$$

In summary, the two models are transformed using the relations described above while preserving the optimal solutions. The transformations are not unique.

8.4.1 Example

In this section, we revisit an example from [15, 16] that is based on the IEEE 14-bus system [17]. The number of buses and the lines remain the same, but only three generators are used instead of the five in the original data. This allows us to easily plot and compare the feasible spaces in terms of the scheduled power from just two generators. The third generator's output is directly tied to the other two through conservation of energy. The topologies for the two systems are seen in Figure 8.4. In Figure 8.4a, the one-line diagram corresponds to the original IEEE 14-bus system data. In Figure 8.4b, another 14-bus system is presented with different topology, different electrical characteristics, and different generator placement.

The generator cost coefficients used in the system cost function, and the upper limits on generation, are shown in Tables 8.1 and 8.2. The electrical loads for the system shown in Figure 8.4b are provided in Table 8.3, and the network electrical susceptances are shown in Table 8.4. Those of the original system shown in Figure 8.4a,

Table 8.1 Generator data for the first OPF problem

Generator bus	Cost coefficient ($ MWh	Maximum power (MW)
1	20	330
2	30	140
8	25	50

Table 8.2 Generator data for the second OPF problem

Generator bus	Cost coefficient ($ MWh	Maximum power (MW)
1	25	150
8	10	40
13	20	150

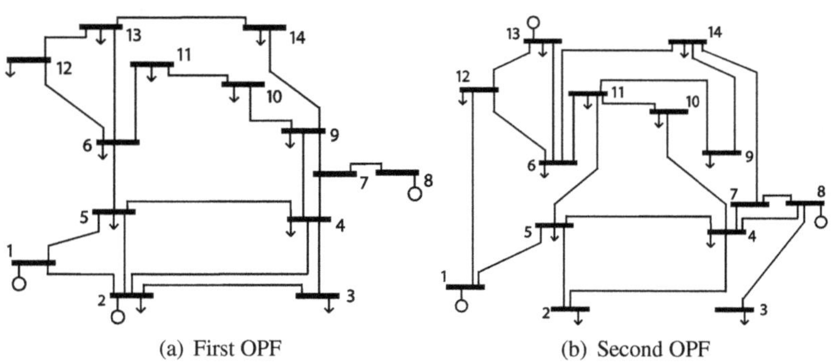

Figure 8.4 One-line diagram for the first and second OPF problems.

Table 8.3 Load data for the second OPF problem

Bus	1	2	3	4	5	6	7
Load (MW)	0	20.2	87.8	49.9	7.9	10.9	0
Bus	8	9	10	11	12	13	14
Load (MW)	0	28.7	9.5	3.5	5.6	14.3	15.4

Table 8.4 Line susceptances for the second OPF problem

From bus	1	1	8	2	2	8	4	4	4	5
To bus	12	5	3	4	5	4	5	7	10	11
Susceptance (p.u.)	17.04	3.85	5.76	5.79	5.49	5.63	22.27	5.27	1.85	3.54
From bus	6	6	6	7	7	9	9	10	12	6
To bus	11	12	13	8	14	11	14	11	13	14
Susceptance (p.u.)	4.29	3.96	8.32	5.82	9.85	13.0	4.32	6.01	5.02	3.44

are contained in the standard data set [17]. As industry practice, the electrical line parameters are specified in per-unit form, whereas the load powers are specified in megawatts (MW). In performing the calculations, the powers are converted to per-unit quantities using a 100 MVA base. For example, the 20.2 MW load is converted to 0.202 per-unit in the equations. The line-flow limits for the first OPF (Figure 8.4a) are all set to 100 MW. The line-flow limits for the second OPF (Figure 8.4b) are all set to 90 MW.

It is useful to discuss the geometric properties of the two optimization problems. In Figure 8.5a, the feasible space for the first system is projected on the space of generation for generator 1 and generator 2 in per-unit (located at buses 1 and 2, respectively). The output of the third generator at bus 8 is necessarily the difference between the load demand and the sum of the output of the other two generators, and is not shown. In this case, the feasible space is bounded by four constraints. The shaded parallel lines represent the cost function, with lighter lines corresponding to lower-cost solutions. The optimal solution is indicated with an asterisk. Similarly, the feasible space and solution for the second system is shown in Figure 8.5b. This system is bounded by three constraints instead of four. We then compute a set of transformation matrices to map the solution from one system to the solution of the other. As described above, the transformations were specifically chosen to force the mapping of the optimal solution. Using the transformation matrices, we map the projection of the feasible space of the first system onto the coordinates of the second system, and this is shown in Figure 8.6. The larger polytope is the feasible space corresponding to the first system. The optimal points overlap, as designed.

These results prove that there exist transformations that relate different power system DCOPF problems that preserve a transformable optimal solution. There are key

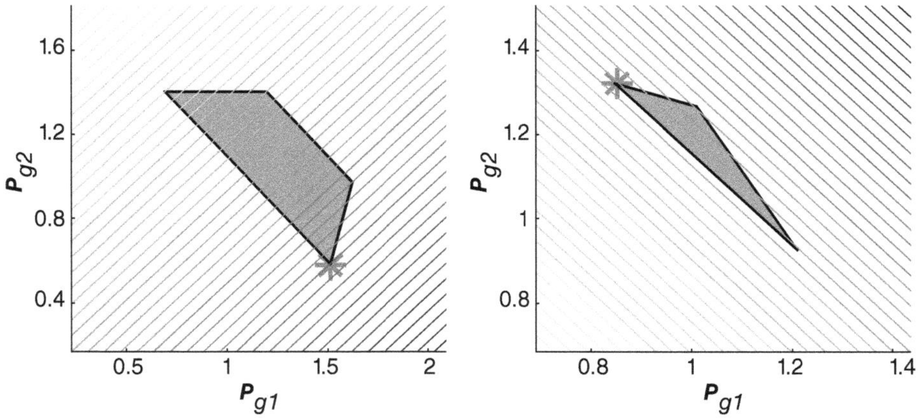

(a) First OPF (b) Second OPF

Figure 8.5 Feasible space of power generation for the first and second OPF problems.

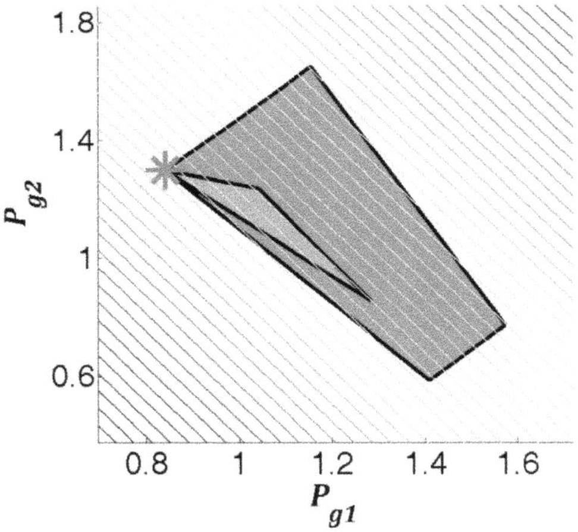

Figure 8.6 Feasible space of the first and second OPF problems in the coordinates of the second OPF problem.

differences to this approach and that in which we required a positive monomial form for certain matrices. When the monomial matrices were used, the entire feasible spaces map to each other. In the example presented here, which preserves a power system structure, the feasible spaces are not preserved, while the solution is. This is not an isolated case. In Figure 8.7, we add four more 14-bus systems that can be transformed from the original, and hence from any one to any other. In terms of security, it is impossible to determine which model is the original and which are transformed versions of the original.

The challenge, and a topic of ongoing research, is to determine appropriate transformations that preserve power system structure and the optimal solution without knowing the solutions a priori, as was done here to show existence.

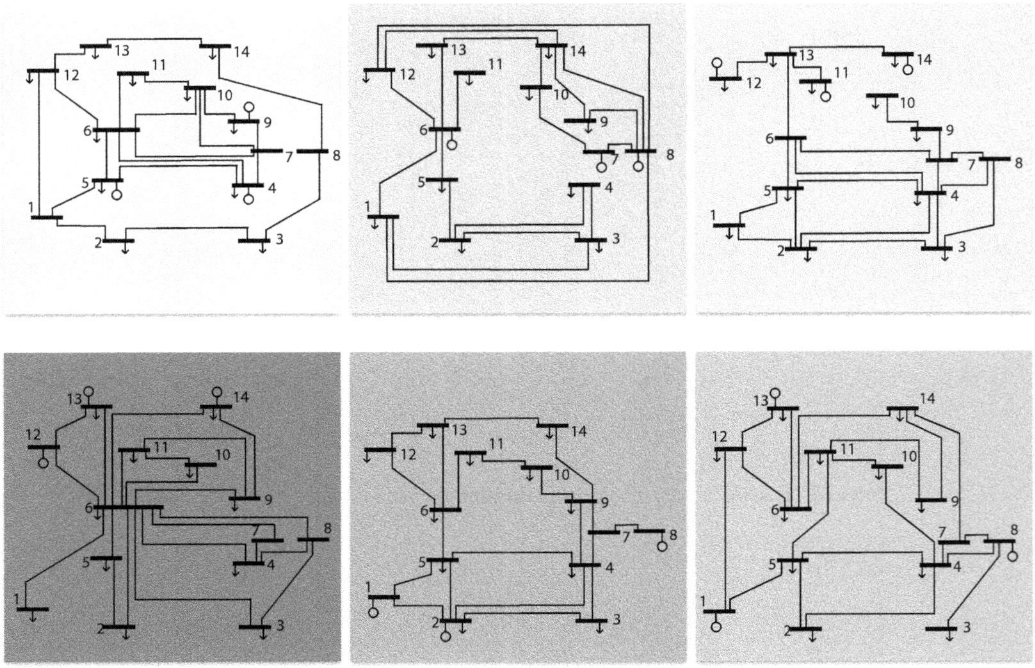

Figure 8.7 A selection of 14-bus systems that are all transformable from one to another. Given one system model, it is impossible to know whether it is the original, or a transformed model.

8.5 Conclusions

Research into the masking of optimization problems through the mathematical transformations of models is a relatively new and active field. There is an obvious need for continued research in methods, and importantly into potential attacks on these methods. With regard to power system applications, we have presented a goal of a structure preserving transformation; however, this may not be needed in all applications. It could be useful for the development of acceptable synthetic models, but for the purposes of computation it may not be necessary. Nevertheless, if it is possible to devise a transformation between models, then this would provide a theoretical result concerning the strength of a transformed model against attack.

The security attacks in the literature that strive to recover transformation matrices rely on the vertical partitioning structure of the linear programming problem and hence are not directly applicable to the horizontally partitioned structure inherent in the DCOPF. However, further research is needed to determine whether similar attacks can be developed for horizontally partitioned linear programming problems. To the best of our knowledge, no such attacks are known at this time.

In this chapter we have not addressed the multiparty extension mentioned in the Section 8.1. In addition to security, there is a desire for privacy in the market for electricity. We point out that with the very mild assumption that different power system operating regions know which of their branches connect to other regions, and vice versa, then the multiparty masking problem is readily solved through straightforward extensions of the models presented here.

The process of preserving power system structure through transformation requires more research. We have shown the existence of transformations, but only by solving both an original and arbitrary target models a priori. This may have value in the development of synthetic models for sharing and research, but little else. A method is needed to develop transformation between power system models without having solved them a priori.

Given the importance of the masking of models, for security and privacy, this continued research has tremendous value.

References

[1] U.S.-Canada Power System Outage Task Force, *Final Report on the August 14, 2003 Blackout in the United States and Canada: Causes and Recommendations*, 2004.

[2] G. Andersson, P. Donalek, R. Farmer, et al., "Causes of the 2003 major grid blackouts in North America and Europe, and recommended means to improve system dynamic performance," *IEEE Trans. Power Syst.*, vol. 20, no. 4, pp. 1922–1928, 2005.

[3] R. Smith, "Assault on California power station raises alarm on potential for terrorism," *Wall Street Journal*, February 5, 2014.

[4] B. Obama, "2013 state of the union address," February 12, 2013. [Online]. Available: www.whitehouse.gov/the-press-office/2013/02/12/remarks-president-state-union-address.

[5] E. Bumiller and T. Shankar, "Panetta warns of dire threat of cyberattack on U.S." *New York Times*, October 11, 2012.

[6] T. Romm, "Janet Napolitano warns of cyberattack on utilities," November 1, 2012. [Online]. Available: www.politico.com/news/stories/1012/83124.html.

[7] J. Dreier and F. Kerschbaum, "Practical privacy-preserving multiparty linear programming based on problem transformation," in *Int. Conf. Privacy, Security, Risk and Trust*, 2011, pp. 916–924.

[8] H. Shojaei, A. Davoodi, and P. Ramanathan, "Confidentiality preserving integer programming for global routing," in *DAC Design Automation Conf.*, 2012, pp. 709–716.

[9] A. R. Borden, D. Molzahn, B. Lesieutre, and P. Ramanathan, "Confidentiality-preserving optimal power flow for cloud computing," in *Proc. Ann. Allerton Conf. Communication, Control, and Computing*, 2012, pp. 1300–1307.

[10] Y. Hong and J. Vaidya, "An interference-proof approach to privacy-preserving horizontally partitioned linear programs," *Optim. Lett.*, vol. 8, no. 1, pp. 267–277, Oct. 2012.

[11] Y. Hong, J. Vaidya, and H. Lu, "Secure and efficient distributed linear programming," *J. Comput. Security*, vol. 20, pp. 583–634, 2012.

[12] P. Laud and A. Pankova, "New attacks against transformation-based privacy-preserving," in *International Workshop on Security and Trust Management*, R. Accorsi and S. Ranise, Eds. New York: Springer, 2013.

[13] O. L. Mangasarian, "Privacy-preserving horizontally partitioned linear programs," *Optim. Lett.*, vol. 6, no. 3, pp. 431–436, 2012.

[14] R. D. Zimmerman, C. E. Murillo-Sanchez, and R. Thomas, "MATPOWER: Steady-state operations, planning and analysis tools for power systems research and education," *IEEE Trans. Power Syst.*, vol. 26, no. 1, pp. 12–19, 2011.

[15] A. R. Borden, D. Molzahn, P. Ramanathan, and B. Lesieutre, "Power system structure and confidentiality preserving transformation of optimal power flow problem," in *Proc. Ann. Allerton Conf. Communication, Control, and Computing*, 2013, pp. 1021–1028.

[16] A. Borden, "Power system reliability tools for modal analysis and model security," Ph.D. dissertation, University of Wisconsin-Madison, 2014.

[17] "Power system test case archive." [Online]. Available: www.ee.washington.edu/research/pstca/.

9 A Framework for Cyber-Physical System Security Situation Awareness

Wei Yu, Guobin Xu, Khanh Pham, Erik Blasch, Genshe Chen, Dan Shen, and Paul Moulema

9.1 Introduction

A cyber-physical system (CPS) is a system that features tight integration between the system's computational and physical elements [1]. Application areas include a number of next-generation critical infrastructures, including: (1) smart transportation, smart electrical power grids, smart medical systems, and smart buildings; and (2) next-generation air-traffic management, advanced manufacturing systems, and complex information sensor systems. The geographically distributed sensors, actuators, and controllers are tightly integrated through cyber-communication networks, which enable the effective control of the physical system. To be specific, sensor nodes obtain measurements from the physical system and transmit the measured data to the controller through cyber-communication networks. The controller then derives the states of physical systems and sends control commands to the actuators, which further control the operation of physical systems. It is worth noting that this chapter focuses on CPSs that integrate communication and networking capabilities with other functionalities. Our contributed framework can be used to explore the space of attacks and help in developing countermeasures.

Through the integration of computational intelligence, communication networks, and sensor control, CPSs can transform our world, with systems having numerous benefits such as responding quickly (e.g., autonomous collision avoidance), providing large-scale and distributed coordination (e.g., automated traffic control), being more precise (e.g., robotic surgery, nano-tolerance manufacturing), working in dangerous or inaccessible environments (e.g., autonomous systems for search and rescue, firefighting, exploration), becoming highly efficient (e.g., zero net energy buildings), augmenting human capabilities (e.g., assistive technologies), and enhancing societal well-being (e.g., ubiquitous healthcare monitoring and delivery) [2].

Nonetheless, CPSs face more threats than traditional real-time embedded systems because of the integration of cyber-communication networks [3]. In traditional real-time embedded systems, the computational cores, sensors, actuators, and the physical system are colocated. Although researchers and system designers put more emphasis on the computational elements, much less emphasis has been put on the interaction between the computational and physical elements. For example, CPSs in a wired/wireless communication network are used to connect the computational elements, sensors, and actuators, which may be geographically distributed. In addition to the faults and errors that may occur at computational elements, sensors, and actuators, the communication network that is responsible for the information exchange between these CPS elements may be attacked and/or fail, leading to the malfunction of the entire CPS. In addition, CPSs for supporting next-generation critical infrastructures, including the electrical power grid, transportation systems, and others, will be targets for cyber-adversaries. Therefore, addressing CPS network security is an urgent and critical issue.

In this chapter, we first conduct a literature review of CPSs and CPS security and then present a generic layered model for CPSs, which consists of four layers: *physical*, *sensor*, *network*, and *service* (PSNS). Based on the proposed layered architecture for CPSs, we develop a taxonomy of cyber-threats, including: attack venues on different CPSs (e.g. smart grid intelligent transportation system), attack objectives (e.g. confidentiality, integrity, availability), and attack targets on different layers of CPSs (e.g. PSNS layers). With the exploitation of the threat taxonomy, we then develop defensive strategies by categorizing defense approaches into three orthogonal defense dimensions: *objectives*, *scope*, and *mechanisms*. Based on the developed defensive taxonomy, we discuss methods of trust to secure CPSs and a defense system for conducting CPS security situational awareness. With explored attack and defense strategies, we consider development of an integrated simulation and emulation framework, and tools to evaluate its effectiveness.

The remainder of the chapter is organized as follows: In Section 9.2, we provide background and a literature review of existing CPSs and security research efforts. In Section 9.3, we present the generic layered CPS architecture. In Section 9.4, we present the taxonomy of cyber-threats. In Section 9.5, we present the defensive taxonomy and the root of trust model to secure CPSs. We present our prototypical detection system for CPS security situational awareness in Section 9.6. In Section 9.7, we present the need for an integrated simulation and emulation environment to evaluate the impact of attacks and effectiveness of defense mechanisms. Finally, we give conclusions in Section 9.8.

9.2 Related Work

A CPS is an engineered system that is built upon the synergy of computation, cyber-communication, and physical components [4]. When designing an effective CPS, one must efficiently and securely integrate in the physical system the computational and communication capabilities, along with monitoring and control functions. Unlike traditional real-time embedded systems, a CPS is integrated, monitored, and controlled through an intelligent computational core with the support of highly efficient and secured communication networks [5]. Since the initiation of CPSs from the US National Science Foundation (NSF) in 2006, researchers from different fields have discussed the related concepts, technologies, applications, and challenges during CPS meetings [6–8] and various international conferences on CPS subjects [9, 10]. A number of CPSs, including the smart grid [11], process control systems [12], transportation designs [13], and healthcare organizations [14], are expected to be developed using advanced computing and communication technologies [15].

Cyber-security is a critical issue for CPSs and a number of research efforts have investigated the security of CPSs [16–22]. For example, the challenges in the security of CPSs have been partially addressed in [23, 24]. Shafi [25] conducted a review of existing research efforts on CPS security. Zhu et al. [26] described a layered architecture toward a secure CPS, which helps to identify research problems and challenges in each layer and to build models for designing security measures for the control systems of critical infrastructures. The CPS security problem comprises CPS detection of element faults and errors, assaults on cyber-communication networks (e.g. operating systems, applications, data), and attacks against the communication between cyber and physical components. The available techniques only consider or address a part of the CPS security problem. For example, in [27], the communication between cyber and physical components is considered as ideal (i.e., without attacks nor faults). Ground tracking control systems in the satellite-based space system is an example of a typical CPS [28–31]. In order to take advantage of CPS advances to support key elements of future missions, the Ames Center Chief Technologist (CCT) Office has established the "Cyber-Physical Systems Modeling and Analysis (CPSMA) Initiative" to apply CPS technologies to specific requirements for long-duration human spaceflight [32, 33]. In addition, the Idaho National Laboratory published "Control Systems Cyber Security: Defense in Depth Strategies" [34], where a number of cyber-security issues, including cyber-attacks and wireless attacks, were presented.

To secure CPSs, game theory, resilient control, attack detection, forensic techniques, and others have been studied [27, 35–42]. For example, Li et al. [35] presented a game-theoretic method for sensor attack avoidance. Pasqualetti et al. [36] proposed a unified framework to analyze the resilience of CPSs against threats. Cárdenas et al. [37] presented a game-theoretical model to defend against bad data and protect the privacy of users in the smart grid. Chris et al. [38] presented a number of game theoretic formulations of attack and defense aspects in CPSs with different cost and utility functions for the attacker and defender. Zhu et al. [39] presented a game-theoretic-based resilient control framework to deal with attacks on cyber components and faults of actuators with perfect sensors and communication. Jin et al. [27] applied resilient control techniques in CPSs to deal with component failures. Mitchell et al. [40] analyzed the effects of intrusion detection and response on CPSs comprising sensors, actuators, control units, and physical objects for controlling and protecting physical infrastructures.

9.3 A Modeling Framework

Generally speaking, a CPS is a large, distributed, and complex system, featuring tight integration between cyber computation, communication, and physical components. To enable effective integration, a number of geographically dispersed components need to be effectively managed and controlled. Nonetheless, the intrinsic

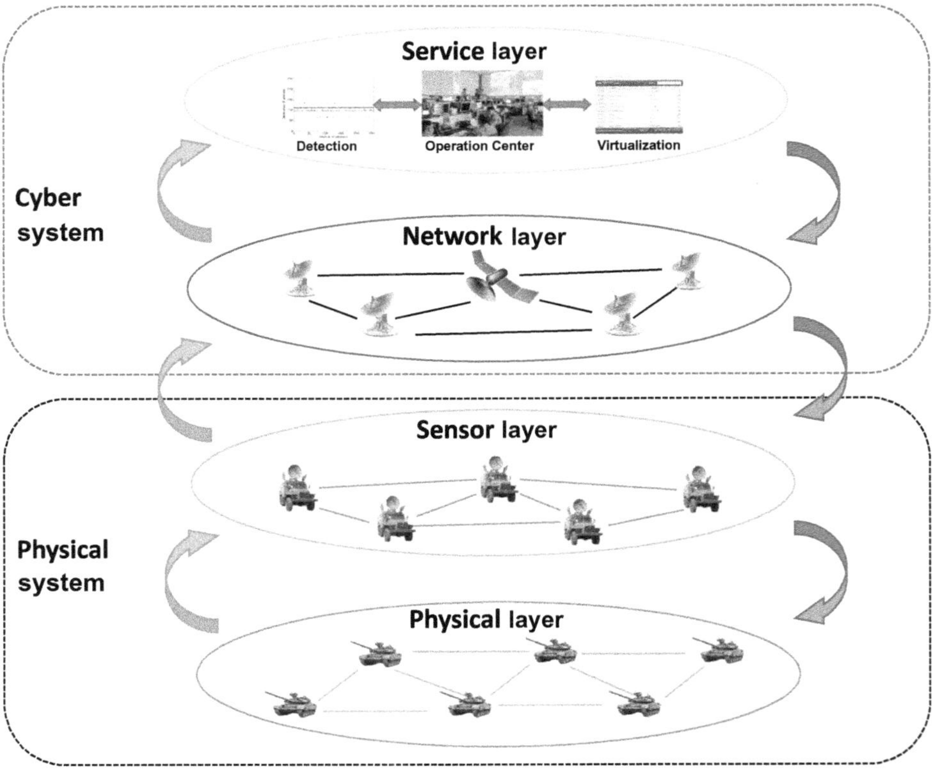

Figure 9.1 A layered architecture for CPSs.

complexity, heterogeneity, and sensitivity to the system performance make the modeling and design of a CPS challenging. As an example, the space CPS is a typical instance of a CPS, which tends to effectively provide global coverage for supporting heterogeneous mobile devices (e.g., people, vehicles, and manned/unmanned aircraft) and delivering services (e.g., voice, video, and data transmission) to end-users.

To address integrating cyber and physical components, we propose a generic layered framework for CPSs as shown in Figure 9.1. Our proposed layered framework consists of four layers: physical layer, sensor layer, network layer, and service layer. In the *physical layer*, there are hardware devices composed of electronics, actuators, and memory states. These elements are combined for computation, processing, and various functions. They can be viewed as agents of sensors.

In the *sensor layer*, sensors are connected through cyber-communication networks to measure the characteristics associated with the physical components. Because sensor nodes may be deployed in unattended or even harsh environments, a lack of tamper-resistant hardware increases the possibility of nodes being compromised by cyber-adversaries. Due to the limited resources, both lossless and lossy aggregation techniques can be used in sensor networks to reduce resource usage (e.g., energy consumption and bandwidth) for transmitting information through the network while preserving the desired accuracy for CPS operations [43]. To this end, two lossless data aggregation techniques can be considered: compression-based and event-based aggregation [43]. The compression-based aggregation can effectively remove redundant information and reduce the data size, whereas event-based aggregation can compress the data based on a predefined event.

In the *network layer*, the nodes can establish self-organizing networks. For a large CPS, the nodes can self-organize as a cluster in a lower level. Each cluster has a cluster-header master, which can aggregate activities in the cluster and some cluster-header nodes can organize as a cluster in a higher level. In each cluster, sensors can directly communicate with each other in the same cluster. The cluster-header node with the highest level can directly communicate with the operation center. Using the space CPS as an example, the ground station can be a cluster-header in its coverage area to aggregate the traffic from mobile hosts. Some ground stations can organize into a high-level cluster to transmit data to a low Earth orbit (LEO) satellite. The measurement information will be transmitted through multiple forwarding or aggregating nodes through a series of bandwidth-constrained wireless links and long-haul bandwidth-constrained satellite links.

In the *service layer*, the operation center is responsible for coordinating nodes distributed over the network and carrying out network situational awareness. The operation center will receive and analyze the measurement information from the nodes in a CPS. The operation center would also provide tools to enable a cyber-analyst to monitor, visualize, and query the status of the secured operations within a CPS.

9.4 Cyber-Threat Taxonomy

Based on our proposed layered architecture for a CPS, we now study the threat taxonomy of CPSs. It is worth noting that CPS security comprises CPS element faults and errors, threats on cyber-communication networks (e.g., operating systems and applications on end hosts, network devices, and information transmitted across multiple components), and attacks against the interactions between the cyber-communication network and the physical system.

As cyber-security is importance in a CPS, the challenges for CPS cyber-security must be fully understood and addressed. There are several challenges. First, the CPS may operate in hostile environments. Sensors and actors lacking tamper-resistant hardware may increase the possibility of being compromised and further disrupt CPS operations. The adversary may compromise sensor nodes, including both sensing nodes and forwarding nodes. Once a node is compromised, the restricted information stored in the node becomes visible to the adversary. The adversary can inject false measurement reports to the operation center through compromised nodes. This leads to the operation center deriving misleading system states [44] or sending control commands with misleading information to actuators, posing a danger to CPS operations. Second, a CPS must interwork with existing physical infrastructures, which comprise a large number of legacy devices that possibly lack appropriate security mechanisms. Third,

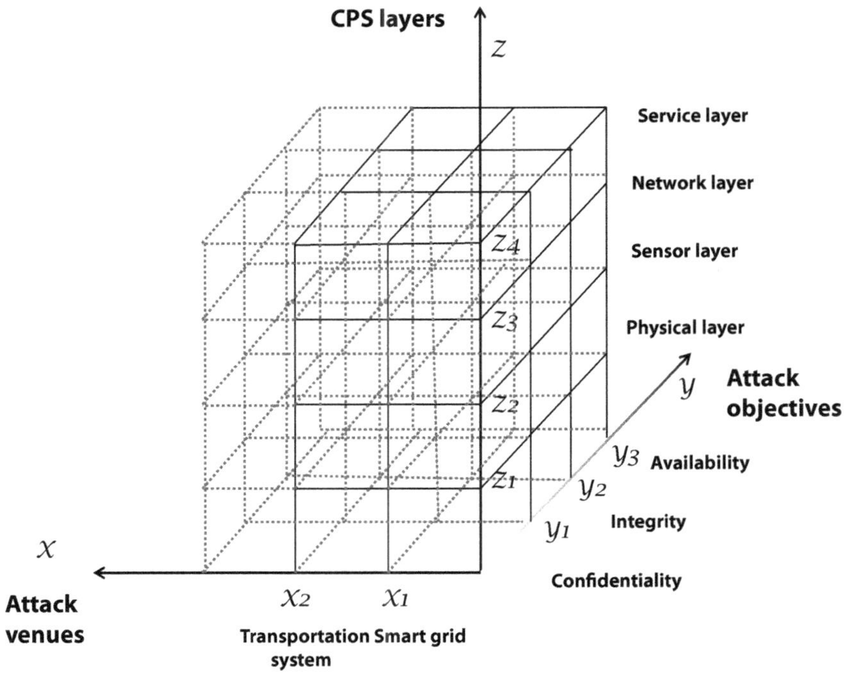

Figure 9.2 3D attack space.

a CPS generally needs high-performance guarantees with real-time requirements to enable the stability of system operations. Hence, the identification and mitigation of malicious threats in both cyber and physical components must be studied in order to develop an efficient and secure CPS [45].

To address these issues to enable a secure CPS, we propose a cyber-threat taxonomy for a CPS which is illustrated in Figure 9.2. As we can see from the figure the x-axis represents attack venues on different CPSs, such as a smart grid or an intelligent transportation system. The y-axis represents attack objectives, where Y_1 refers to the attacks that aim at compromising confidentiality, Y_2 refers to the attacks that compromise integrity, and Y_3 refers to the attacks that compromise availability. The z-axis represents attack objectives on different layers of a CPS (Z_1: physical layer, Z_2: sensor layer, Z_3: network layer, and Z_4: service layer). In the following subsections, we will detail these dimensions.

9.4.1 Attack Venues

In the X dimension, attack venues on different CPSs need to be investigated, such as for CPS for a smart grid or an intelligent transportation system.

(1) Smart grid: Smart grids [11, 46–48] are expected to be developed using advanced cyber computer and communication technologies to achieve a more efficient, reliable, secure, and resilient system operation and provide better service

to customers than traditional power grids [15]. On the transmission and distribution side, supervisory control and data acquisition (SCADA) systems collect real-time information that provides wide-area situational awareness of the power grid status. On the user side, more precise real-time estimates of anticipated usage through an advanced metering system (AMI) will enable demand response controls, which will increase efficiency. Nonetheless, to improve the operational efficiency and reliability of power grids through the use of cyber computing and communication technologies, the risks of cyberspace breach of power grid systems need to be seriously studied before massively deploying the smart grid technologies. There have been growing concerns in the smart grid about protection against malicious cyber-threats [44, 49, 50]. The operation and control of a smart grid depends on a complex cyberspace of computers, software, and communication technologies. Because the measurement component supported by smart equipment (e.g., smart meters and sensors) plays an important role, it can be a target for attacks. Because those measuring devices may be connected through open network interfaces and lack tamper-resistance hardware, there is an increased possibility they will be compromised by an adversary [51]. The adversary may modify data and compromise measuring components by injecting malicious code into the memories of these components [44, 52, 53]. For example, Liu et al. [44] proposed false-data injection attacks, which can bypass existing bad data detection schemes and arbitrarily manipulate the state estimation of grid systems, causing controllers to make incorrect decisions on power grid operations. For the smart grid security, we also explored the vulnerabilities of distributed energy management and studied several false-data injection attacks against distributed energy transmission [54], dynamic state estimation using Kalman filtering [55], and cascading failures and an energy storage-based mitigation scheme [56].

(2) Intelligent transportation system: The intelligent transportation system is one type of CPS that can provide a more efficient/low-latency, effective, reliable, and safe driving experience that minimizes congestion and improves traffic flow management [57, 58]. To fulfill the data collection requirements, a set of resource-constrained sensor nodes can form a sensor network and a communication network. For example, the data collected include the average speed of the vehicles, the weather, the numbers of vehicles passed in a time interval, and others. The sensors can be classified as two types: passive and active. The passive sensors are usually embedded into the roadside and used to collect road-related traffic data (e.g., loop detector, radar, video recorder, RFID transponder, and others). Active sensors are monitoring agents installed on the vehicles, which dynamically record vehicle-related information, including GPS, speed, collisions, and more. In a transportation system, the high mobility of vehicles poses the challenge of trustworthiness between sensors. In addition, some sensors are usually deployed in unattended or even harsh environments (e.g., road sensors) and the lack of tamper-resistant hardware increases the possibility of sensors being compromised by adversaries. Once sensors are compromised, the adversary can manipulate information, posing serious threats to transportation system operations. For example, the false data injected by compromised sensors can affect the decisions made by the intelligent transportation system. An adversary could inject false data to cause

traffic congestion or collisions. They could also launch an eavesdropping attack to secretly track the location of a vehicle, which would pose a threat to the driver's privacy. An adversary may learn a driver's political and religious affiliations based on their locations, such as where the driver regularly visits [59, 60]. In our preliminary study, we systematically reviewed cyber-threats against transportation systems [61] and performed a simulation study to evaluate the performance of different routing algorithms in transportation systems [62].

9.4.2 Attack Objectives

In the *Y* dimension, the cyber-attacks on a CPS mainly focus on the objectives of confidentiality, integrity, and availability.

(1) Confidentiality: Confidentiality is referred to as the ability of the system to protect information from disclosure to unauthorized individuals while granting access to authorized individuals. When data are collected from the physical layer, users' privacy-related information or the state information associated with physical systems may be included. The adversary can launch eavesdropping attacks between layers in a CPS to obtain sensitive information [63]. In addition, packet-sniffing techniques can be leveraged to capture data packets transferred through the network [64].

(2) Integrity: Integrity is referred to as the ability of the system to guarantee that information is accurate, reliable, and tamper-proof. As such, a CPS needs to maintain the consistency, accuracy, and trustworthiness of its data and to make sure that data can be changed only by authorized individuals. After the integrity of data is compromised, the information is no longer reliable nor accurate. Hence, integrity will affect the accuracy of data processing and decision-making processes. As an example, the adversary can launch false-data injection attacks through compromised nodes, directly posing a serious threat to layers in a CPS or affecting the interactions between layers in a CPS. In particular, the false data injected by compromised nodes can lead to false events and affect the decision made at the controller, which could pose dangerous threats to CPS operations. As a large amount of false data injected by compromised nodes is transmitted to the controller through multiple-hop routes, the communication overload of energy-limited nodes is increased, ultimately shortening the sensors' lifetimes. For example, Mo et al. [65] studied the effects of false-data injection attacks on a control system with injected fake sensor measurements. In [66], Chabukswar et al. analyzed the effect of integrity attacks on SCADA systems, which are composed of computers and remote station control devices used for data acquisition.

(3) Availability: Availability is referred to as the ability of the system to ensure that information, services, and CPS assets can be accessible. There are two types of availability failures: (1) data and system resources can be unavailable due to loss or physical damage; and (2) access to information and resources can be denied or delayed by a denial-of-service attack. For example, a denial-of-service (DoS) attack aims to make a host or a network resource unavailable to normal users by sending multiple service requests from compromised systems toward the attached target. One example of a DoS attack is the distributed DoS (DDoS) attack, which can be launched

from a number of compromised computers in different geographical locations. In the physical layer, one particular type of DoS attack is the jamming attack [67], which can disrupt efficient communication between different CPS layers. Jamming attacks can especially be used to disrupt data transmission in the physical layer. For example, they can be used to degrade radio communications, disrupting the control of mobile devices [67]. With the generated high noise power or overwhelming frequencies in the attack traffic, CPS communications can be disrupted and services will be ultimately denied. Control signals used to regulate the operation of physical devices can be jammed, leading to the malfunction of those physical devices [68].

9.4.3 Attacks in CPS Layers

After integrating with attack dimensions X and Y, we now study attacks on different CPS layers in dimension Z, including attacks against physical components, sensors, communication networks, and service layers, along with their interactions. Table 9.1 summarizes possible attacks on each layer. For example, attacks on the physical layer can cause physical process distrust and physical components deployed in an open environment can be physically damaged, resulting in failed data collection. Attacks on sensors can manipulate measurement data by maliciously injecting misleading information. In addition, wireless sensor networks are vulnerable to spoofing attacks [69]. In the network layer, network protocols are commonly used to coordinate the entities in a CPS. In a CPS, SCADA systems have been widely used for control and communication needs for critical infrastructures. SCADA systems may operate in hostile environments where vulnerabilities of sensor and communication networks increase the chance to be compromised by adversaries. One example is *Stuxnet*, which is a publicly known malware used to exploit vulnerabilities in and attack SCADA systems [70–74]. In addition, there are still numerous vulnerabilities, such as segmentation faults and stack, heap, or buffer overflows [74], in common SCADA protocols (e.g., MODBUS and DNP3). When the service layer is attacked, the data analysis result will be misleading resulting in incorrect decisions and disrupted CPS operations. In addition, the adversary can disrupt communications between layers by flooding or desynchronization attack methods [75, 76].

Table 9.1 Threats in CPS Layers

Layer	Element	Threats
Service	Operation center	DoS, spoofing
Network	Servers, routers, switches, firewall	Misleading system states, DoS, protocol attacks, man-in-the-middle attacks, eavesdropping, spoofing
Sensor	Hardware (e.g., sensors)	False-measurement reports, jamming, eavesdropping, spoofing
Physical	Devices (e.g., actuators)	Failed system states, jamming, physical failure

In terms of attack impact, performance metrics used to measure the impact of attacks on both the cyber and physical components need to be studied to understand the consequences. For example, in the smart grid, performance metrics (e.g., energy supply loss, outages, energy prices) and system stability (e.g., outage time, electrovalence) can be used to measure attack impact. In transportation systems, we can evaluate physical performance with metrics (e.g., vehicle travel time, vehicle travel distance, the numbers of collisions) and evaluate network performance with metrics (e.g., throughput, end-to-end delay, packet delivery rate). The metrics can be helpful for the design of effective and commensurate defensive schemes to mitigate attacks.

9.5 Defensive Taxonomy

Based on our explored taxonomy of cyber-threats, we now investigate the defensive taxonomy, as illustrated in Figure 9.3. As shown in the figure, we propose categorizing corresponding defensive strategies into the following three orthogonal dimensions: (1) protective objectives, (2) defensive scope, and (3) counter mechanisms. In the following subsections we introduce these three dimensions.

9.5.1 Protective Objectives

From the specific angle of CPS cyber-security, defense mechanisms involve the following objectives: confidentiality, availability, integrity.

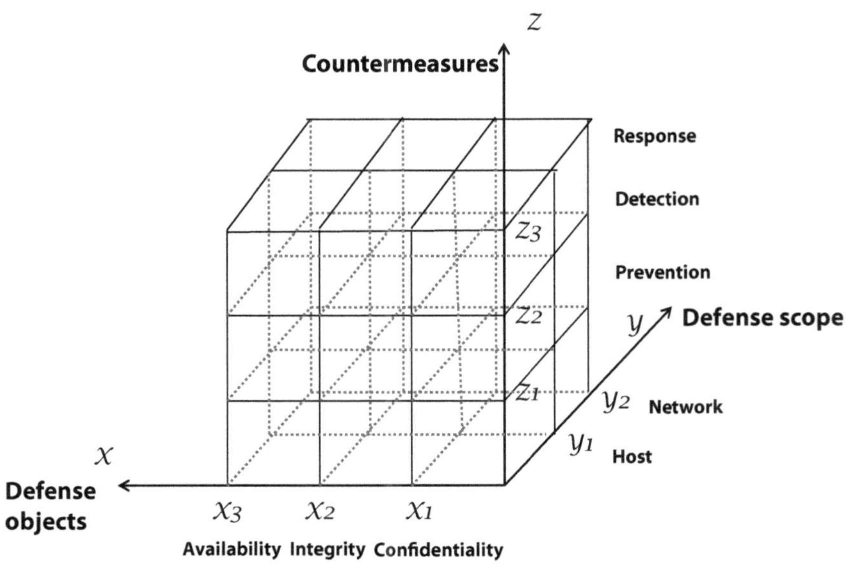

Figure 9.3 3D defense space.

(1) Confidentiality: In the cyber domain, the most common countermeasure against attacks that violate information confidentiality is the encryption of exchanged data and commands within the CPS by using an efficient symmetric or asymmetric cryptography mechanism. It is worth noting that cryptography in a CPS must be highly efficient because many CPS applications demand real-time communication. In the physical domain, mechanisms such as secure authentication and strong protection of unlimited certificates can preserve confidentiality by securing devices in a CPS.

(2) Integrity: A protective mechanism for preventing integrity violations consists of stopping data tampering or data alteration. For example, a polynomial-based en-route filtering scheme [77] can quickly filter falsely injected data and achieve high resilience against a number of compromised devices in a CPS. A trusted platform module (TPM) embedded in a CPS device ensures strong machine identity and integrity by verifying and reporting the integrity of platforms, including basic input/output (I/O) system (BIOS), operating systems application software, and memory content of devices, thereby ensuring that no unauthorized changes have occurred [78]. Tamper-resistant hardware attached to a device or a computer system in a CPS will significantly increase the cost of compromising a system because an adversary would need to use sophisticated hardware techniques [79].

(3) Availability: DoS attacks are the main violation of security involving the availability of information and physical resources. To this end, over-provisioning (e.g., redundancy) is an effective mechanism to mitigate availability violations. With additional cost, over-provisioning mechanisms can make the CPS more resilient to attacks by reducing the magnitude of attacks. For example, having a backup system or an off-site location ready to restore services will significantly reduce the downtime in the event of a severe attack.

9.5.2 Defense Deployment Opportunities

The deployment opportunities of defensive strategies can generally be categorized as host-based defense and network-based defense, which are further explained as follows.

(1) Host-based defense: The hosts in a CPS can be deployed with threat monitoring and detection agent software (denoted as a host-based intrusion detection system [IDS]). The agent collects the information in real time from system logs, security logs, application logs, and other logs for security situation awareness analysis [43]. The detection agent may have a small memory requirement and a small CPU footprint by default, not affecting system usage. For example, OSSEC (an open-source, host-based network IDS) [80] and other open-source host-based IDS tools (e.g., Advanced Intrusion Detection Environment [AIDE] [81]) can be deployed as detection agents in a CPS.

(2) Network-based defense: In a network-based defense system, both passive and active network sensors need to be deployed [82]. The passive network sensors, which are deployed with intrusion detection modules and distributed over the network, are used to monitor network traffic and identify suspicious attack behaviors. As a

complementary component, active network sensors can also be deployed, such as honeypots, to directly and actively interact with attacks and collect more useful information from these attacks. Generally speaking, a honeypot is considered a trap in cyberspace and the illegal user is supposed to access it. Any activities taking place on it are inherently suspicious and likely a probe, attack, or compromise. In this way, honeypots have few false alarms in detecting intrusions, whereas other intrusion detection mechanisms often suffer from high numbers of false alarms because they are trying to isolate intrusions within a high volume of routine events from host and network traffic.

9.5.3 Counter Mechanisms

Defensive schemes to protect, detect, and respond to attacks in CPSs should consider both the cyber and physical domains to ensure information security, infrastructure security, and application security.

(1) Prevention: In a CPS, some of the critical sensors need to be made more resilient to prevent attacks so that the attack effort can be significantly increased. One example of physical prevention is to design a high-resilience physical and cyber communication topology to increase the difficulty of launching successful attacks [83]. To ensure resilient time synchronization in communication networks, over-provisioning and associated mechanisms (e.g., redundancy and fail-over, as well as load balancing) need to be considered in the system architecture design.

(2) Detection: To accurately detect attacks, effective and efficient detection schemes should be investigated. For example, in order to attack sensors, the adversary may launch malicious traffic over the network and inject the malicious code into internal memory locations or modify parts of firmware code. For this purpose, the correlation and the integration of both software behavior and network traffic can significantly improve the accuracy of IDS detection. In monitored systems, the runtime behavior of a CPS and its applications can use machine-learning schemes (e.g., support vector machines, Naive Bayes, neural networks) to learn the behavior of attacks and detect unknown attacks effectively.

(3) Response: Once an attack is detected, the compromised nodes should be identified and isolated from a CPS. To achieve this goal, one scheme is to adopt efficient watermarking-based traceback mechanisms [41, 42, 84], which embed secret signals (bits of 1 or 0) into the datastream. If the datastream is manipulated by any node during the transmission path, the receiver can correlate the received datastream with the secret signal and detect whether the datastream has been manipulated. By repeating the process over the transmission path, the origin of the data manipulation can be traced.

9.6 Elements of Trust in CPS Security Design

Based on the explored defensive taxonomy in Section 9.4, as shown in Figure 9.4 the elements of trust include mechanisms to enable secured, trustworthy, and dependable

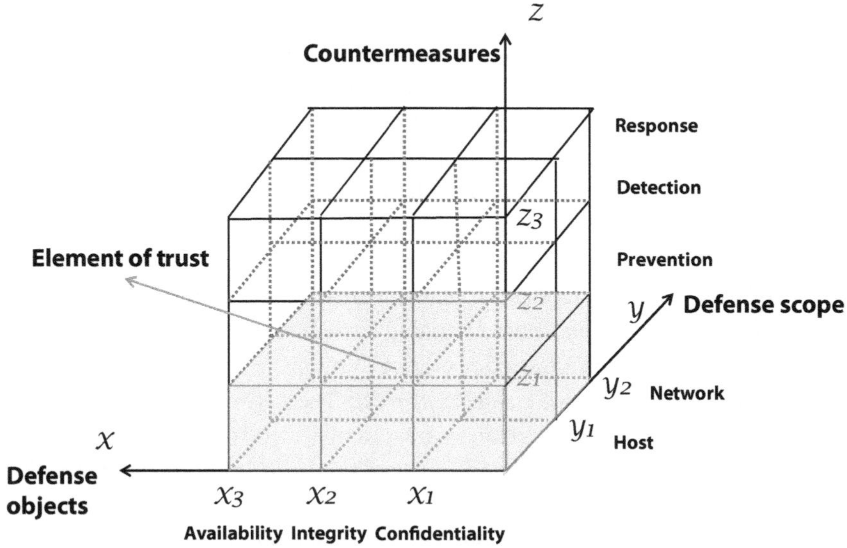

Figure 9.4 Elements of trust in defense space.

CPSs [85]. Elements of trust are the key concept in trustworthy computing, defined as hardware and software components consisting of a set of unconditional trusted functions, leading to tamper-resistant actions [86]. The trusted elements provide services, including the verification of the system, the integrity of the software and data, confidentiality, and data integrity attestation between trusted devices in a system or a network [79]. As a prevention strategy, we present elements of trust in a CPS through two dimensions: their scope and their objectives.

9.6.1 Elements of Trust Scope

The scope for end-to-end elements of trust can be categorized as host-based and network-based elements of trust.

(1) Host-based elements of trust: Elements of trust can be implemented in a single host of a CPS to verify the integrity of applications on different layers of abstraction. Elements of trust installed in a CPS lower layer (e.g., hardware or firmware) verify whether or not the application in the upper layer has been tampered with. If the integrity of the application is validated, the integrity of the next application in the layer above can be checked. This verification mechanism will result in a trusted chain and tampering with the application will be detected. Elements of trust for measurement are located in the BIOS (firmware) chip and perform measurements of all components involved in the boot process [87]. Practically, at boot, the elements of trust attest to the integrity of the BIOS of the physical system where the BIOS measures and validates the integrity of the bootloader prior to handing over control. The bootloader measures and gives a trustworthy description of the operating system loader of the physical

system and then the operating system loader measures and hands control over to the operating system. Finally, the operating system of a physical system will authenticate and validate applications. As the initial elements of trust of physical systems, the BIOS is the main target of cyber-attacks, such as BIOS rootkits [88], Mebromi malware [89], and the Chernobyl virus [90]. In these cyber-attacks, the operation mechanism attempts to reflash the BIOS configuration or modify the advanced configuration and power interface (ACPI) tables set up by the BIOS in order to take control of the device or disrupt device behavior. For example, the INTEL TXT [88] is a host-based root of trust. As such, it attests to the authenticity of a platform and its operating system, ensuring that an authentic operating system starts in a trusted environment and thus can be considered a trusted operating system.

(2) Network-based elements of trust: The network-based elements of trust enable the establishment of a chain of trust to cyber and physical devices distributed across the network. For example, Schiffman et al. [91] proposed a network-based root of trust for installation. When a physical system needs to install applications or conduct configurations, the request shall be sent to the operation center. If the operation center trusts the installer, it can trust systems derived from such an installation. When the installation is complete, each physical system will report its state to the operation center for attestation.

9.6.2 Elements of Trust Objectives

Elements of trust can be implemented for diverse objectives and services, including integrity, availability, and confidentiality.

(1) Confidentiality: Elements of trust enables the secure authentication and strong protection of unlimited certificates. In a CPS, physical devices can be equipped with a trusted platform module (TPM) which is a hardware chip wired to the computer platform that provides basic cryptographic functionality [87, 91] and embedded crypto capability that supports user, application, and machine authentication [78]. This built-in authentication can be leveraged in network settings to ensure that only authorized users and authorized devices can connect to the network. A TPM can also be used as a cost-effective secure vault for certificates, keys, and passwords. Elements of trust should protect critical data and systems against a variety of attacks. For example, Self-Encrypting Drive (SED) is defined as an open standard solution for integrated encryption and access control within the protected hardware of the drive, which can offer the advantage of full disk encryption, providing data protection when the machines or drives are lost, stolen, or reformatted [78].

(2) Integrity: Elements of trust establishes strong machine identity and integrity. The TPM embedded in a CPS measures and reports on the integrity of a platform, including its BIOS, disk master boot record, boot sector, operating system, and application software, to ensure that no unauthorized changes have occurred [78]. Hence, the TPM-enabled components of a CPS are capable of checking the integrity of devices during the boot process through embedded protection and detection mechanisms running at pre-boot, and then during the secure boot process [92].

Integrity checks can also be performed in a network setting. For example, Schiffman et al. [93] proposed architecture for building runtime integrity proofs for general-purpose systems in distributed computing architectures based on an approximation of the Clark–Wilson integrity model.

(3) Availability: In addition to confidentiality and integrity, the availability of information and data is a critical requirement for operating safe and dependable CPSs. Concerning elements of trust solutions, software or hardware updates can affect the availability and accessibility of a TPM with sealed data. In addition, the complete migration of certified keys from one TPM to another platform can affect the availability of services provided by the TPM because migration authorities (MAs) must explicitly certify every key that the platform owner may want to migrate [94]. To this end, we shall investigate possibilities and standards that guarantee availability of services and data using the above circumstances.

In addition, elements of trust can be designed on the hardware or software of a CPS, which depends on the targeted level of security and other constraints, such as costs, upgradability, and others. The hardware-based root of trust is a tamper-resistant piece of hardware attached to a computer system. The advantage of the hardware-based root of trust is that it is physically secure and the cost of compromising it is very high as the adversary needs sophisticated hardware techniques [79]. A software-based root of trust is a tuple of algorithms and constraints ($H; INT; EVC$), where H is an implementation of a set of computationally hard problems, INT is a subalgorithm to extract integrity information, and EVC is a set of external verifiable constraints for measurements that do not rely on the existence of tamper-resistant hardware [79].

9.7 A Detection System for CPS Security Situational Awareness

We introduce a generic detection system in defense space for carrying out CPS security situational awareness, as shown in Figure 9.5. To monitor and control both passive and active sensors, a detection center is deployed to receive the detection logs and alerts from both the passive and active sensors, and stores the detection information in the database. Based on the collected information, anomaly-based detection techniques using digital signal processing and image processing-based schemes can be implemented to detect malicious activities [95]. To help the network administrator carry out cyber security situational awareness, a visualization component is designed and implemented to display the geolocation and other information relevant to attacks, displayed on Google Map and Google Earth.

As a preliminary result, we implemented a prototypical system using both Windows Servers and Linux Servers to emulate attacks on the cyber-communications network in the CPS [95]. To show the security vulnerability exploits and to validate the effectiveness of the detection schemes, we chose an unpatched Windows Server and used BackTrack Linux, which is a Linux distribution with various hacking tools to emulate real-world cyber-attacks. Our testbed consisted of four computers (i.e., User, VMware host, XEN host dom0, and a detection center) and one router. On the XEN

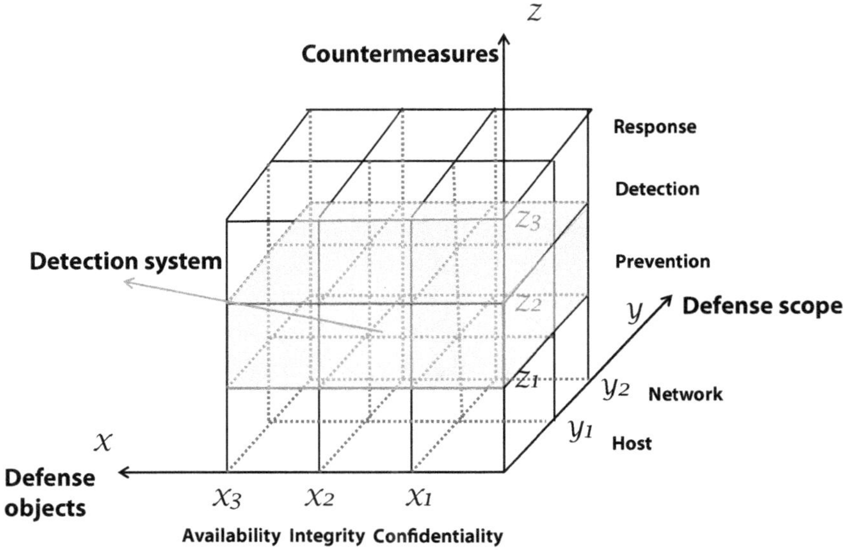

Figure 9.5 A detection system in defense space.

host machine dom0, we installed six virtual machines to emulate the end-hosts in the CPS. A HoneyD machine was configured as an active sensor. In addition, a Snort-based IDS (to act as a passive sensor) was installed on dom0 so that it could monitor the traffic associated with the emulated CPS communication network.

In the proposed detection system, anomaly detection mechanisms (e.g., image-based schemes) can be leveraged against cyber-attacks on the CPS through detecting anomalies in network traffic. There are three types of attacks considered in the detection system: worm/malware propagation, denoted as many-to-many attacks; DDoS, denoted as many-to-one attacks; and port-scanning attacks, denoted as one-to-many attacks. For the visualization component, several visualization features to display network traffic in the CPS shall be considered, including the time series of network traffic volume, the correlation between source internet protocol (IP) and destination IP addresses that can be used to identify various attack patterns, and the display of the geographical attributes of adversaries and victims.

9.8 Integrated Simulation and Emulation Environment

It is worth noting that our developed attack and defense taxonomies can establish a foundation to further explore possible attacks and defensive mechanisms. Orthogonal to mathematical and logical methods, we can derive attack scenarios and defensive schemes based on our proposed taxonomies. To understand the consequence of attacks and study the effectiveness of countermeasures, we should develop an integrated simulation (cosimulation) and emulation environment. Using the smart grid as an example, an integrated cosimulation platform should be developed to simulate and emulate both physical (e.g., smart meters, inverters, sensors, relays, data acquisition devices, PMUs [power management units]) and cyber components (e.g., SCADA and AMI). High-performance servers can be deployed to execute a large number of virtual machines to build a large-scale emulation testbed. Simulink [96], the virtualized software-based PMU [97], and network emulator (ns-3) [98] can be integrated into the testbed. In each virtual machine, the specific software (e.g., Powerworld [99], RTDS [100], Trilliant [101]) can be deployed to emulate energy-related information. Based on this integrated platform, a set of use-case scenarios derived from our developed taxonomies can be evaluated. In this way, the attack impact on both the power grid and cyber-networks and the effectiveness of countermeasures to mitigate attack impacts can be evaluated.

9.9 Conclusion

In this chapter we have addressed cyber-security issues in CPSs. We first conducted a literature review of existing research efforts on CPS security and then presented a generic layered framework for CPSs. We then developed a taxonomy of cyber-attacks (e.g., attack venues on different CPSs, objectives, and targets on different CPS layers). With the exploitation of the threat taxonomy, we also developed a defensive taxonomy by categorizing defense objectives, defense scope, and countermeasures in terms of prevention, detection, and response. Based on the developed defensive taxonomy, we presented a root of trust model to secure CPSs and a prototypical detection system for conducting security situational awareness in CPSs. In addition, we discussed the need for an integrated simulation and emulation environment for CPSs, which can be used to evaluate the impact of cyber-attacks on CPSs and the effectiveness of defensive schemes.

References

[1] Wikipedia, "Cyber-physical system." [Online] Available: http://en.wikipedia.org/wiki/Cyber-physical_system.

[2] J. Sztipanovitz and J. Stankovic, "Cyber-physical systems: A national priority for federal investment in infrastructure and competitiveness," 2008. [Online]. Available: http://cra.org/ccc/docs/init/Cyber-Physical_Systems.pdf.

[3] X. Yang, J. Lin, W. Yu, et al., "On situational aware en-route filtering against injected false data in cyber physical systems," in *Situational Awareness in Computer Network Defense: Principles, Methods and Applications*, C. Onwubiko, Ed. Hershey, PA: IGI-Global, 2011.

[4] National Science Foundation (NSF), "Cyber-physical systems (CPS), " 2013. [Online]. Available: www.nsf.gov/pubs/2013/nsf13502/nsf13502.htm#toc.

[5] T. Znati, "Security for emerging cyber-physical systems research challenges and directions," in *Proc. First International Workshop on Data Security and Privacy*, 2010.

[6] "CPS Weeks 2010," 2010. [Online]. Available: www.cpsweek.org.

[7] "CPS Weeks 2012," 2011. [Online]. Available: http://cpsweek2011.cs.illinois.edu.

[8] "CPS Weeks 2012," 2012. [Online]. Available: http://www.cpsweek.org.

[9] M. D. Ilić, L. Xie, and U. A. Khan, "Modeling future cyber-physical energy systems," in *Proc. Power and Energy Society General Meeting: Conversion and Delivery of Electrical Energy in the 21st Century*, 2008, pp. 1–9.

[10] L. Sha, S. Gopalakrishnan, X. Liu, and Q. Wang, "Cyber-physical systems: A new frontier," in *Proc. IEEE Int. Conf. Sensor Networks, Ubiquitous, and Trustworthy Computing*, 2008, pp. 1–9.

[11] "SMART GRID." [Online]. Available: www.oe.energy.gov/smartgrid.htm.

[12] A. A. Cárdenas, S. Amin, G. Schwartz, R. Dong, and S. Sastry, "A game theory model for electricity theft detection and privacy-aware control in AMI systems," in *Proc. 2012 Annual Allerton Conference on Communication, Control, and Computing*, 2012, pp. 1830–1837.

[13] A. Gokhale and M. P. McDonald, "A cyber physical systems perspective on the realtime and reliable dissemination of information in intelligent transportation systems," *Int. J. Netw. Protocol Algorithm*, vol. 2, no. 3, pp. 116–136, 2010.

[14] L. T. X. Phan, I. Lee, and O. Sokolsky, "Compositional analysis of real-time embedded systems," in *Proc. IEEE Int. Conf. Compilers, Architectures and Synthesis of Embedded Systems (CASES)*, 2011, pp. 237–238.

[15] J. Wang, D. Li, Y. Tu, P. Zhang, and F. Li, "A survey of cyber physical systems", in *Proc. IEEE Int. Conf. Cyber Technology in Automation, Control, and Intelligent Systems*, 2011, pp. 1–6.

[16] M. Xue, S. Roy, Y. Wan, and S. K. Das, "Security and vulnerability of cyber-physical infrastructure network: a control-theoretic approach," in *Handbook on Securing Cyber-Physical Critical Infrastructure*, San Francisco, CA: Morgan Kaufmann, 2012, p. 5.

[17] S. Roy, M. Xue, and S. K. Das, "Security and discoverability of spread dynamics in cyberphysical networks," *IEEE Trans. Parallel Distrib. Syst.*, vol. 23, no. 9, pp. 1694–1707, Sept. 2012.

[18] M. Xue, W. Wang, and S. Roy, "Security concepts for the dynamics of autonomous vehicle networks," *Automatica*, vol. 50, no. 3, pp. 852–857, Mar. 2014.

[19] S. Roy, M. Xue, R. Dhal, et al., "Security of airborne network dynamics and algorithms: A graph-theoretic perspective," in *Proc. 1st ACM MobiHoc Workshop on Airborne Networks and Communications*, 2012, pp. 19–24.

[20] M. Xue and S. Roy, "Characterization of security levels for the dynamics of autonomous vehicle networks," in *Proc. Decision and Control (CDC)*, 2012, pp. 3916–3921.

[21] S. Roy and R. Dhal, "Situational awareness for dynamical network processes using incidental measurements," *IEEE J. Sel. Top. Signal Process.*, vol. 2, no. 9, pp. 304–316, Mar. 2015.

[22] R. Dhal and S. Roy, "Vulnerability of continuous-time network synchronization processes: A minimum energy perspective," in *Proc. Decision and Control (CDC)*, 2013, pp. 823–828.

[23] A. Cardenas, S. Amin, B. Sinopoli, et al., "Challenges for securing cyber physical systems," in *Proc. Workshop on Future Directions in Cyberphysical Systems Security*, DHS Vol. 5, No. 1 2009. Available at: https://ptolemy.berkeley.edu/projects/chess/pubs/601/cps-security-challenges.pdf.

[24] E. Ke Wang, Y. Ye, X. Xu, et al., "Security issues and challenges for cyber physical system," in *Proc. IEEE/ACM Int. Conf. Cyber, Physical and Social Computing (CPSCom)*, 2010, pp. 733–738.

[25] Q. Shafi, "Cyber physical systems security: A brief survey," in *Proc. 12th Int. Conf. Computational Science and Its Applications (ICCSA)*, 2012, pp. 156–150.

[26] Q. Zhu, C. Rieger, and T. Basar, "A hierarchical security architecture for cyber-physical systems," in *Proc. Int. Symp. Resilient Control Systems (ISRCS)*, 2011, pp. 15–20.

[27] X. Jin, A. Ray, and R. M. Edwards, "Integrated robust and resilient control of nuclear power plants for operational safety and high performance," *IEEE Trans. Nucl. Sci.*, vol. 57, no. 2, pp. 807–817, 2010.

[28] E. M. Atkins, "Cyber-physical aerospace: Challenges and future directions in transportation and exploration systems," in *Proc. 2006 National Science Foundation Workshop on Cyber-Physical Systems*, 2006. Available at: https://ptolemy.berkeley.edu/projects/chess/pubs/601/cps-security-challenges.pdf

[29] P. J. Antsaklis, "From hybrid to networked cyber-physical systems," in *Proc. American Control Conference (ACC)*, 2009, pp. 6–7.

[30] D. Shen, G. Chen, K. Pham, and E. Blasch, "A trust-based sensor allocation algorithm in cooperative space search problems," in *Proc. SPIE*, vol. 8044, p. 80440C, 2011.

[31] A. Banerjee, K. K. Venkatasubramanian, T. Mukherjee, S. Kumar, and S. Gupta, "Ensuring safety, security, and sustainability of mission-critical cyber-physical systems," *Proc. IEEE*, vol. 100, no. 1, pp. 283–299, Jan. 2012.

[32] National Aeronautics and Space Administration (NASA), "Space technology roadmap for TA11, modeling, simulation, information technology and processing roadmap," 2011. [Online]. Available: www.nasa.gov/pdf/501321main_TA11-MSITP-DRAFT-Nov2010-A1.pdf.

[33] National Aeronautics and Space Administration (NASA), "Cyber-Physical Systems Modeling and Analysis (CPSMA) Initiative," 2013. [Online]. Available: www.nasa.gov/centers/ames/cct/office/studies/cyber-physical_systems.html.

[34] D. Kuipers and M. Fabro, "Control systems cyber security: Defense in depth strategies." [Online]. Available: www.inl.gov/technicalpublications/Documents/3375141.pdf.

[35] D. Li, G. Chen, E. Blasch, and K. Pham, "Sensor attack avoidance: Linear quadratic game approach," in *Proc. Int. Conf. Information Fusion*, 2009, pp. 1131–1138.

[36] F. Pasqualetti, F. Dorfier, and F. Bullo, "Attack detection and identification in cyber-physical systems," *IEEE Trans. Autom. Control*, vol. 58, no. 11, pp. 2715–2729, Nov. 2013.

[37] A. A. Cárdenas, S. Amin, G. Schwartz, R. Dong, and S. Sastry, "A game theory model for electricity theft detection and privacy-aware control in AMI systems," in *Proc. 50th Annual Allerton Conference on Communication, Control, and Computing*, 2012, pp. 1830–1837.

[38] C. Y. T. Ma, N. S. V. Rao, and D. K. Y. Yau, "A game theoretic study of attack and defense in cyber-physical systems," in *Proc. IEEE Conf. Computer Communications (INFOCOM) Workshops*, 2011, pp. 708–713.

[39] Q. Zhu and T. Başar, "Robust and resilient control design for cyber-physical systems with an application to power systems," in *Proc. Robust and Resilient Control Design for Cyber-Physical Systems with an Application to Power Systems*, 2011, pp. 4066–4071.

[40] R. Mitchell and I. Chen, "A hierarchical performance model for intrusion detection in cyber-physical systems," in *Proc. IEEE Int. Conf. Wireless Communications and Networking (WCNC)*, 2011, pp. 2095–2100.

[41] W. Yu, X. Fu, E. Blasch, et al., "On effectiveness of hopping-based techniques for network forensic traceback," in *Proc. IEEE/ACIS Int. Conf. Software Engineering, Artificial Intelligence, Networking and Parallel/Distributed Computing (SNPD)*, 2013, pp. 101–106.

[42] W. Yu, X. Fu, E. Blasch, et al., "On effectiveness of hopping-based techniques for network forensic traceback," *IEEE Trans. Intl J. Netw. Distrib. Comput.*, vol. 1, no. 3, pp. 144–158, 2013.

[43] D. Zhang, L. Ge, W. Yu, et al., "On effective data aggregation techniques in host-based intrusion detection in MANET," *IEEE Trans Int. J. Security Netw.*, vol. 8, no. 4, pp. 179–193, 2013.

[44] Y. Liu, M. K. Reiter, and P. Ning, "False data injection attacks against state estimation in electric power grids," in *Proc. 16th ACM conference on Computer and Communications Security*, 2009, pp. 21–32.

[45] G. Chen, D. Shen, C. Kwan, et al., "Game theoretic approach to threat prediction and situation awareness," *IEEE Trans. Adv. Inform. Fusion*, vol. 2, no. 1, pp.1–14, Jun. 2007.

[46] "NIST & the Smart Grid," 2010. [Online] Available: www.nist.gov/smartgrid/nistandsmartgrid.cfm.

[47] IEEE, "IEEE P2030." [Online]. Available: http://grouper.ieee.org/groups/scc21/2030/2030_index.html.

[48] T. L. Crenshaw, E. Gunter, C. L. Robinson, L. Sha, and P. R. Kumar, "The simplex reference model: Limiting fault-propagation due to unreliable components in cyber-physical system architectures," in *Proc. Real-Time Systems Symp. (RTSS)*, 2007, pp. 400–412.

[49] A. A. Cárdenas, S. Amin, and S. Sastry, "Research challenges for the security of control systems," in *Proc. 3rd USENIX workshop on Hot Topics in Security (HotSec)*, 2008, pp. 1–6.

[50] NIST, "Guidelines for smart grid cyber security," 2010. [Online]. Available: http://csrc.nist.gov/publications/PubsNISTIRs.html#NIST-IR-7628.

[51] S. McLaughlin, D. Podkuiko, and P. McDaniel, "Energy theft in the advanced metering infrastructure," in *Proc. Critical Information Infrastructures Security*, 2010, pp. 176–187.

[52] Q. Yang, J. Yang, W. Yu, N. Zhang, and W. Zhao, "On a hierarchical false data injection attack on power system state estimation," in *Proc. IEEE Globe Communication (GLOBECOM)*, 2011, p. 1–5.

[53] K. Song, D. Seo, H. Park, H. Lee, and A. Perrig, "OMAP: One-way Memory Attestation Protocol for smart meters," in *Proc. 2011 9th IEEE International Symposium on Parallel and Distributed Processing with Applications Workshops (ISPAW)*, 2011, pp. 111–118.

[54] J. Lin, W. Yu, X. Yang, G. Xu, and W. Zhao, "On false data injection attacks against distributed energy routing in smart grid," in *Proc. ACM/IEEE Third Int. Conf. Cyber-Physical Systems (ICCPS)*, 2012, pp. 193–192.

[55] Q. Yang, L. Chang, and W. Yu, "On false data injection attacks against Kalman filtering in power system dynamic state estimation," in *Int. J. Security Commun. Netw.*, vol. 9, pp. 833–849, 2013.

[56] X. Chen, W. Yu, D. Griffith, N. Golmie, and G. Xu, "On cascading failures and countermeasures based on energy storage in the smart grid," in *Proc. ACM Research in Adaptive and Convergent Systems (RACS)*, 2014, pp. 291–296.

[57] M. Ardelt, C. Coester, and N. Kaempchen, "Highly automated driving on freeways in real traffic using a probabilistic framework," *IEEE Trans. Intell. Transp. Syst.*, vol. 13, pp. 1576–1585, 2012.

[58] Z. Bowu, X. Kai, C. Xiuzhen, H. Liusheng, and B. Rongfang, "Traffic clustering and online traffic prediction in vehicle networks: A social influence perspective," in *Proc. IEEE INFOCOM*, 2012, pp. 495–503.

[59] J. Hubaux, S. Capkun, and J. Luo, "The security and privacy of smart vehicles," in *IEEE Security Privacy Mag*, vol. 2, pp. 49–55, 2004.

[60] M. D. Dikaiakos, A. Florides, T. Nadeem, and L. Iftode, "Location-aware services over vehicular ad-hoc networks using car-to-car communication," *IEEE J. Sel. Areas Commun.*, vol. 25, no. 8, pp. 1590–1602, 2007.

[61] N. Ekedebe, H. Song, W. Yu, C. Lu, and Y. Wan, "Securing transportation cyber-physical systems," in *Securing Cyber Physical Systems*. Boca Raton, FL: CRC Press, pp. 163–196, 2014.

[62] N. Ekedebe, Z. Chen, G. Xu, C. Lu, and W. Yu, "On an efficient and effective intelligent transportation system (ITS) using field and simulation data," in *Proc. SPIE Defense, Security, and Sensing (DSS)*, pp. 91210B, 2014.

[63] E. Wang, Y. Ye, X. Xu, et al., "Security issues and challenges for cyber physical system," in *Proc. 2010 IEEE/ACM Int. Conf. Green Computing and Communications & Int. Conf. Cyber, Physical and Social Computing*, 2010, pp. 733–738.

[64] S. Ansari, S. G. Rajeev, and H. S. Chandrashekar, "Packet sniffing: A brief introduction," *IEEE Trans. Potentials*, vol. 21, no. 5, pp. 17–19, 2002.

[65] Y. Mo and B. Sinopoli, "False data injection attacks in control systems," in *Proc. 1st Workshop on Secure Control Systems*, 2010, pp. 1–6.

[66] R. Chabukswar, Y. Mo, and B. Sinopoli, "Detecting integrity attacks on SCADA systems," in *Proc. 18th IFAC World Congress*, 2011, pp. 11239–11244.

[67] A. Hamieh and J. Ben-Othman, "Detection of jamming attacks in wireless ad hoc networks using error distribution," in *Proc. IEEE Int. Conf. Communication (ICC)*, 2009, pp. 1–6.

[68] E. Guirguis, "On the effect of jamming attacks on cyber physical systems with the focus on target tracking applications." Ph.D. Dissertation, Texas State University, 2012.

[69] Y. Chen, W. Trappe, and R. P. Martin, "Detecting and localizing wireless spoofing attacks," in *Proc. Sensor, Mesh and Ad Hoc Communications and Networks*, 2007, pp. 193–202.

[70] P. Motta Pires and L. Oliveira, "Security aspects of SCADA and corporate network interconnection: An overview," in *Proc. Int. Conf. Dependability of Computer Systems*, 2006, pp. 127–134.

[71] P. A. S. Ralston, J. H. Graham, and J. L. Hieb, "Cyber security risk assessment for SCADA and DCS networks," *ISA Trans.*, vol. 46, no. 4, pp.583–594, Oct. 2007.

[72] S. Hong and M. Lee, "Challenges and direction toward secure communication in the SCADA system," in *Proc. 2010 8th Ann. Communication Networks and Services Research Conf.*, 2010, pp. 381–386.

[73] A. Teixeira, G. Dán, H. Sandberg, and K. H. Johansson, "A cyber security study of a SCADA energy management system: Stealthy deception attacks on the state estimator," in *Proc. Computing Research Repository*, 2010, pp. 11271–11277.

[74] B. Zhu, A. Joseph, and S. Sastry, "A taxonomy of cyber attacks on SCADA systems," in *Proc. 2011 Int. Conf. Internet of Things and 4th Int. Conf. Cyber, Physical and Social Computing*, 2011, pp. 380–388.

[75] Q. Yang, D. An, and W. Yu, "On time desynchronization attack against IEEE 1588 protocol in power grid systems," in *Proc. Energytech*, 2013, pp. 1–5.

[76] A. Wood and J. A. Stankovic, "Denial of service in sensor networks," *IEEE Trans. IEEE Comput.*, vol. 35, no. 10, pp. 54–62, 2002.

[77] X. Yang, J. Lin, P. Moulema, et al., "A novel en-route filtering scheme against false data injection attacks in cyber-physical networked systems," in *Proc. IEEE Int. Conf. Distributed Computing Systems (ICDCS)*, 2012, pp. 92–101.

[78] Trusted Computing Group. [Online]. Available: www.trustedcomputinggroup.org.

[79] Q. Yan, J. Han, Y. Li, R. H. Deng, and T. Li, "A software-based root-of-trust primitive on multicore platforms," in *Proc. 6th ACM Symp. on Information, Computer and Communications Security*, 2011, pp. 334–343.

[80] OSSEC. [Online]. Available: www.ossec.net.

[81] Advanced Intrusion Detection Environment (AIDE). [Online]. Available: http://sourceforge.net/projects/aide.

[82] D. Zhang, W. Yu, and R. Hardy, "A distributed network sensor based intrusion detection framework in enterprise networks," in *Proc. IEEE Military Communication Conference (MILCOM)*, 2011, pp. 1195–1200.

[83] R. Dhal, "Consequences of topological changes in cyber-physical networks: A system-theoretic perspective." Ph.D. Dissertation, Washington State University, 2014.

[84] W. Yu, D. Griffith, L. Ge, S. Bhattarai, and N. Golmie. "An integrated detection system against false data injection attacks in the smart grid," in *Proc. of Security and Communication Networks*, vol. 8, no. 2, pp. 91–109, 2015.

[85] M. M Farag, "Architectural enhancements to increase trust in cyber-physical systems containing untrusted software and hardware." Ph.D. Dissertation, Virginia Polytechnic Institute and State University, 2012.

[86] Springer Reference, "Roots of Trust," 2013. [Online]. Available: www.springerreference.com/docs/html/chapterdbid/317714.html.

[87] S. Choinyambuu, "A root of trust for measurement mitigating the lying endpoint problem of TNC," MSE Project Report, 2011.

[88] J. Heasman, "Implementing and detecting an ACPI BIOS rootkit," in *Proc. Black Hat Federal 368*, 2006. Available at: www.blackhat.com/presentations/bh-federal-06/BH-Fed-06-Heasman.pdf.

[89] H. Sirag, N. Bondugula, and R. Gupta, "Advanced persistent attacks: BIOS rootkit-Mebromi," 2011. Available at: www.scribd.com/document/186096697/BIOS-Rootkit-Mebromi.

[90] M. Christodorescu and S. Jha, "Static analysis of executables to detect malicious patterns," Wisconsin University-Madison Department of Computer Sciences, 2006. Available at: www.cs.cornell.edu/courses/cs711/2005fa/papers/cj-usenix03.pdf.

[91] J. Schiffman, T. Moyer, T. Jaeger, and P. McDaniel, "Network-based root of trust for installation," in *Proc. IEEE Security & Privacy (S&P)*, 2011, pp. 40–48.

[92] G. Shpantzer, "Implementing hardware roots of trust: The trusted platform module comes of age." SANS white paper, 2013. [Online]. Available: www.sans.org/readingroom/analysts-program/hardware-trust.

[93] J. Schiffman, T. Moyer, C. Shal, T. Jaeger, and P. McDaniel, "Justifying integrity using a virtual machine verifier," in *Proc. Computer Security Applications Conf.*, 2009, pp. 83–92.

[94] U. Kühn, K. Kursawe, S. Lucks, A. Sadeghi, and C. Stüble, "Secure data management in trusted computing," in *Proc. Cryptographic Hardware and Embedded Systems (CHES)*, 2005, pp. 324–338.

[95] W. Yu, S. Wei, D. Shen, et al., "On detection and visualization techniques for cyber security situation awareness," *Proc. SPIE*, vol. 8739, p. 87390R, 2013.

[96] M. Pipattanasomporn, H. Feroze, and S. Rahman. "Multi-agent systems in a distributed smart grid: design and implementation," in *Proc. Power Systems Conf. Expo.*, 2009, pp. 1–8.

[97] A. T. Al-Hammouri, L. Nordstrom, M. Chenine, et al., "Virtualization of synchronized phasor measurement units within real-time simulators for smart grid applications," in *Proc. Power and Energy Society General Meeting*, 2012, pp. 1–7.

[98] H. Georg, N. Dorsch, M. Putzke, and C. Wietfeld, "Performance evaluation of time-critical communication networks for smart grids based on IEC 61850," in *Proc. Computer Communications Workshops (INFOCOM WKSHPS)*, 2013, pp. 3417–3422.

[99] D. Zhang, S. Li, P. Zeng, and C. Zang, "Optimal microgrid control and power-flow study with different bidding policies by using Powerworld simulator," *IEEE Trans. Sustainable Energy*, vol. 5, no. 1, pp. 282–291, Jan. 2014.

[100] P. Forsyth and R. Kuffel, "Utility applications of a $RTDS_R$ simulator," in *Proc. Power Engineering Conf.*, 2007, pp. 112–117.

[101] Trilliant Incorporated, "Trilliant: A Trusted Solution Partner," 2009. [Online]. Available: www.trilliantinc.com/4-Rsrcs/-PDFs/TSB-TrustedPartner.pdf.

10 Security in the Cyber-Physical Electric Power Infrastructure

Bruce McMillin, Ravi Akella, Gerry Howser, Thoshitha Gamage, and Tom Roth

10.1 Abstract

The emerging smart electrical grid is a representative application of the growing interaction between computing, communication networks, and physical devices in modern system design. More widely referred to as cyber-physical systems (CPSs), the reliable and sound operation of these systems is of the utmost priority, given how every sector of society and the economy depend on them for day-to-day operation [1, 2].

The compound discrete computational and continuous physical nature of a CPS, however, provokes new system design challenges; not only do these have different semantics from each other, but the notion of system boundaries is blurred from purely cyber or purely physical systems. The embedded computers and communication networks therein govern both physical manifestations and computations, which in turn affect how these two major components interact with each other and the outside world [3]. As we shall see in this chapter, this lack of boundaries creates a host of new security and privacy vulnerabilities. Specifically, changes in the physical portion of the infrastructure are always observable and adversaries to the system can potentially derive sensitive system internal settings by observing these external system changes. This derived knowledge coupled with the semantic knowledge of the system can be used against the system. In terms of system security models centered around the classic CIA[1] triad [4], this is a confidentiality violation with serious system integrity and availability breach implications. Moreover, typical system security models are not necessarily adequately expressive in their ability to address these new breeds of vulnerabilities.

This chapter develops semantically integral CPS models that unify the computational and physical aspects of security in CPSs. These models are based on a special branch of computer security known as *information flow security*, which approaches confidentiality through restricting information moves between domain-segregated user groups within a system rather than strict access control as in traditional models; information flow security policies prevent information from flowing to users who are not authorized to receive it. However, preserving information flow security is a challenging task; information flow is inexorably intertwined between and among computing and physical components. Key to the approach is to integrate

[1] **C**onfidentiality, **i**ntegrity, and **a**vailability.

the information flow contained in the physical commodity with the information flow contained in the computing components. The models will (1) characterize confidentiality, (2) develop composable information flow architectures, (3) refine existing methods of information flow analysis to encompass multiple security domains, and (4) provide execution monitoring enforcement to preserve information flow security.

This chapter is organized as follows. Section 10.2 sets the stage for the problem; Section 10.3 discusses why existing approaches to security are inadequate; Section 10.4 discusses the inherent information flow present in a simple toy pipeline model; Section 10.5 shows how to reduce the information leakage through compensating physical events; and Section 10.6 uses the toy model from Section 10.4 to show how the same principles apply in a prototype distributed smart grid system. Specifically, Section 10.6 introduces two formal modeling techniques – the π-calculus and multiple security domain information flow – along with detailed analysis for both. Section 10.7 explains how this work is related to system theory; Section 10.8 is on future directions; and lastly Section 10.9 provides some challenge questions for the reader.

Readers interested in an introduction to information flow security within a CPS may read Sections 10.2–10.5. For those interested in distributed energy systems as a CPS, Section 10.6 describes an architecture. For the advanced reader, Section 10.6.1 delves deeply into process algebra and model logic models of CPSs.

10.2 Need and Challenges

A critical infrastructure consists of a multitude of interacting distributed components. The physical layer handles the movement of a certain physical element or a commodity such as water, power, or energy through a physical medium. The computing components, on the other hand, govern the movement of such elements using complex control algorithms, scheduling mechanisms, and protection systems. Critical infrastructure systems, as they advance, invariably blend the two. These include commodity transportation and distribution systems such as the national power distribution network, the national gas distribution network, factory automation, smart houses, emergency medical services, and intelligent highway systems. The sound operation of such systems is critical to both the economic and social stability of nations. In August 2003, an estimated 50 million people in the Midwest and Northeast portions of the United States and Ontario, Canada were affected by a power outage which in some parts of the United States lasted for four days [5]. It was estimated that the total cost in the United States was $4–10 billion, and in Canada there was a loss to gross domestic production of 0.7 percent for August. Although this event was triggered by a cascading failure in power lines loosely coupled with a cyber failure, rather than an intentional act of sabotage, such an event highlights the need for extensive research, methodologies, new tools, and new models of security to safeguard these critical infrastructures. The Stuxnet worm showed that carefully crafted exploits can hide from detection both on the computing domain and on the physical domain [6, 7].

Critical infrastructures such as advanced power grids (or "smart power grids") exemplify emerging specific security risks [8, 9], and their higher susceptibility to failures and attacks compared to their counterparts [10] as a result of increased computing and networking integration. Early incarnations of smart grids (e.g., smart metering systems) have been found to be vulnerable to sniffing attacks and malicious command injections [11, 12]. There has also been a growing appreciation of attacks on advanced cyber-physical electric power systems [13]. The Future Renewable Electric Energy Delivery and Management (FREEDM) Energy Research Center (ERC) [8] faces much greater security risks due to the increased complexity. Physical observation of smart meters, power transmission lines [14–16], power distribution lines, and usage patterns [17] can lead to breaches of overall confidentiality. By contrast, a great deal of existing critical infrastructure security work at the industrial and national lab level concentrates on protection of cyber assets [18–24]. However, to semantically integrate cyber-security with physical security through information flow requires a return to fundamental principles.

10.3 The Case for Information Flow Security

This chapter identifies two main approaches to system security: access control-based methods and information flow-based methods. Traditional security models such as HRU [25], Take-Grant [26], Bell-Lapadula (BLP) [27, 28], and Biba [29] take an access control-based approach to address the general security problem of assuring system security (the CIA triad) by restricting access to information and resources. However, there is growing evidence to suggest access control may not be the way forward in securing highly integrated cyber-physical infrastructure systems [30–37].

Information flow security suggests preventing unintended high-level (secure/private) domain information disclosures to the low-level (open/public) domain of systems. The critical argument here is that the physical portion of a CPS often presents a low-level observable output in space, time, and frequency that contains information about the rest of the system. A security policy may exactly define what low-level users are forbidden to know about the high-level domain, but the actual enforcement depends on sound implementation. Strict access control-based security mechanisms lack the expressiveness and the enforcement ability to prevent indirect security violations due to *implicit* and *explicit* information flows. Such security models can only impose special restrictions on information and resources but fail to prevent propagation [31] and cannot control how the data will be used after it has been read.

As an example, consider a process with a higher security clearance, which can easily indicate that it has accessed a sensitive piece of information to a lower security clearance process by setting a flag or making a variable visible/accessible to the lower security process. This is an indirect violation of the BLP *-property [27]; there is no direct "write-down" violation, yet the low-level process gets hold of sensitive information. Sound enforcement thus requires access be restricted only to processes guaranteed not to leak or improperly transmit sensitive information after reading [31].

Unfortunately, there is no provision in access control-based mechanisms to impose such a strong flow restriction [30–37]. Furthermore, there is implicit assumptions on trust involved here, which is not enforceable in any meaningful manner through access control.

The information flow security approach to this particular problem is to impose restrictions on the flow of information between partially ordered security clearance levels, rather than imposing restrictions on the subjects and objects involved. A violation in this model is the ability to deduce sensitive/confidential information at relatively inferior security clearances. In other words, information flows, as stated in the specific policy employed, should be prevented between processes that are not supposed to communicate [32, 38]. The sensitivity of information and the "communication" depends on how each process views (other processes) in the system.

Some of the fundamental information flow security properties include noninterference [39], noninference [40], and nondeducibility [41]. Noninference, and nondeducibility-based information security models are particularly attractive for CPSs, and noninterference is to a lesser extent. In essence, these properties define different levels of information deduction that could occur between domain-segregated – a high-level domain with a secret and a low-level domain attempting to acquire the secret – users. Noninterference [39] is consistent with most people's notion of security; high-level inputs cannot interfere with low-level outputs, making it impossible to identify and deduce those high-level actions. Low-level users of a noninterference secure system are unaware of any high-level activity. Noninference [40] secure systems prevent low-level users from determining whether observed actions are due to high-level or low-level activity. Nondeducibility [41] is the inability to deduce high-level actions based on low-level observations due to multiplicity of low-level observations; that is, every low-level observation has multiple possible high-level actions, which could have caused it (or vice versa).

10.4 Characterizing Confidentiality

To motivate the reader's understanding of access control-based models' inability to prevent implicit confidentiality violations, and to illustrate how information flow security-based models address these, consider the simplified conceptual model of a commodity distribution CPS (e.g., a gas-distribution network) depicted in Figure 10.1. Three intelligent flow controllers, each composed of a distributed grid intelligence (DGI), i.e., an embedded computing element, and a physical actuator (PA) interact with each other directly through communication links and indirectly through the underlying physical commodity flow. Here, the DGI is part of the cyber (computing) aspect, while the PA and the pipeline network is part of the physical aspect of the CPS.

Formally, DGIs can execute commands to the PAs: raise and lower flow, which raise and lower the flow of the commodity (gas) within the controlled pipeline. Whenever one of the DGIs exerts one of the possible commands on its respective section of the pipeline network, the corresponding physical repercussions will be felt on the

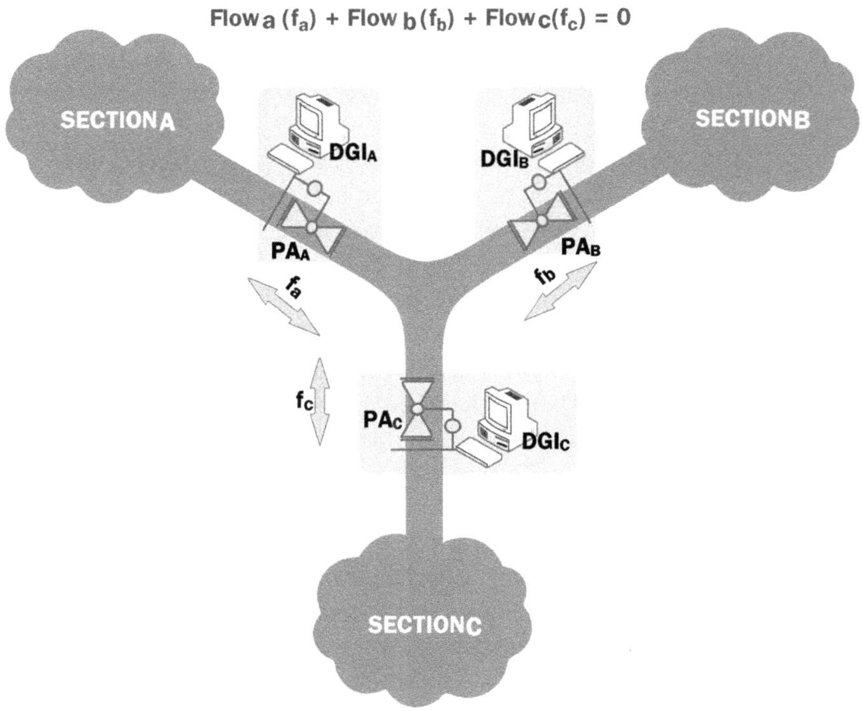

Figure 10.1 Pipeline network with three subnetworks (A, B, C) controlled by respective DGIs.

neighboring two pipeline sections; whenever one DGI makes a local flow change, the other DGIs can observe physical changes in the flow in their respective subnetworks. This relationship adheres to the flow conversation property, which can be expressed as an invariant as follows:

$$f_a + f_b + f_c = 0 \tag{10.1}$$

where f_a, f_b, and f_c represent the flow of the commodity (gas) in the pipes controlled at A, B, and C respectively. DGI_C may have a competitive reason[2] to keep its actions confidential from DGI_A. Let SD^G be the protection domain G and with high-level and low-level domains, $SD^{hl}, SD^{ll} \in G$. Thus, consider a case in which DGI_C increases its PA_C to $f_C + f'$, causing a flow increase in pipe $section_C$. Unless DGI_B reacts accordingly, DGI_C's actuation will result in a flow decrease in pipe $section_A$. The aggregated change adopted by the rest of the network Δf could be any of the following:

$$\Delta f = \{f'^{\uparrow} f_b\} \text{ or} \tag{10.2}$$

$$= \{f'^{\downarrow} f_a\} \text{ or} \tag{10.3}$$

$$= \{\alpha^{\uparrow} f_b \text{ and } \beta^{\downarrow} f_a : \mid \alpha - \beta \mid = f'\} \tag{10.4}$$

[2] The Russian–Ukrainian gas crisis is an example of this type of system partition [42].

where $\uparrow$ and $\downarrow$ indicate increase or decrease in flow, respectively. Let's suppose that DGI_C and DGI_A are owned by two rival companies eager to find out each others' flow information. Let's also consider DGI_C as an SD^{hl} process that initiates the first change. Due to the rivalry, this makes DGI_A an SD^{ll} process that should not know about DGI_C's actions. For the sake of this discussion, assume $DGI_B \in SD^{hl}$; DGI_B reacts to DGI_C's actions.

If DGI_B does nothing, then the flow change f is attributed to DGI_C, violating a strict notion of confidentiality by observation at DGI_A. DGI_A cannot observe DGI_C's cyber actions, but can deduce them from the physical observations on $section_A$. Even if DGI_B acts by allowing more flow through $section_B$, i.e., increasing its flow, but not in a timely manner, DGI_A can still deduce DGI_C's actions. For example, consider the following trace of global actions:

$$\phi = \{read_A(), write_C^{\uparrow}(f'), read_A(), write_B^{\uparrow}(f'), read_A()\} \tag{10.5}$$

This trace still violates confidentiality. Both scenarios can be shown, formally, to violate the property of noninterference [43]. From a strict access control point of view, DGI_C is not accessible to DGI_A. A confidentiality-based mechanism such as the BLP model would state that DGI_A should not read (actions of) DGI_C. Physical access to C's facility can also be restricted to anyone associated with DGI_A. Also, in terms of network security, DGI_C might already have instruments and software to prevent any form of network-oriented attack. Nevertheless, none of these security measures were able to prevent the confidential information leakage to DGI_A, who in this case is an adversary at a lower security domain. This shows why treating security in a disjoint fashion fails and why strict access control alone cannot prevent information flow security violations.

Relaxing this strict notion of confidentiality leads to two other possibilities. The obvious one is that the perceived change by DGI_C could be due to a usage change in C's subnetwork, B's subnetwork, or A's subnetwork. This follows the notion of noninference – the observed actions are consistent with low-level events. On the other hand, if the usage changes are considered high-level events, the system no longer satisfies noninference, but satisfies nondeducibility; the observation is consistent with any high-level event [44].

10.5 Compensation as a Security Measure

In the previous example, if DGI_C is required to lower its gas flow due to some reason (such as physical change or economic flow assignment), this intent first needs to be conveyed to DGI_B, if DGI_C wants to keep the secret from leaking to other competitive distributors like DGI_A. If DGI_C and DGI_B communicate via a (simplified) protocol:

```
DGI_C: send request to DGI_B (f')
DGI_C: delay by 1 message time
DGI_C: decrease flow by f'
```

```
DGI_B: receive request from DGI_C (f')
DGI_B: decrease flow by f'
```

when the cyber portion of the system synchronizes their actions in this manner, the resulting trace becomes:

$$\bar{\phi} = \{read_A(), write_C^{\uparrow}(f'), write_B^{\uparrow}(f'), read_A(), read_A()\} \tag{10.6}$$

which satisfies the property of noninterference. The argument presented here is that the two main aspects (the cyber and physical aspects) of a critical infrastructure need to be treated together to preserve and enforce information flow properties. The above analysis shows that considering one aspect alone is neither sufficient nor reasonable. By looking at (10.5) and (10.6) it can be seen that at the end of both traces, the value of f_A has returned to the original value as if nothing has happened. This was possible because the effect of input $write_C^{\uparrow}(f')$ on the pipeline has been compensated by the effect of $write_B^{\uparrow}(f')$. Furthermore, both of these commands were issued by members of the same security domain with respect to DGI_A. In general, though, noninterference is too strict for many infrastructures, and the opportunities to completely hide high-level actions do not often occur. The less restrictive models of noninference and nondeducibility can be employed if sophisticated semantics of the problem are taken into account.

10.6 Confidentiality in Electric Smart Grid Systems

In CPSs, several privacy and confidentiality issues emerge as an unintended consequence of coupling cyber systems with physical systems. In this section, we explore such issues in an electric smart grid, where the participants in a neighborhood can share energy generated from renewable sources installed at their respective houses.

In a simple example setting of such a system, Fred and Barney share their renewable energy generation to minimize the net energy drawn from the electric utility, as shown in Figure 10.2. Fred shares his "excess" wind energy with Barney on a windy day, and Barney shares his "excess" solar energy with Fred on a sunny day. The excess energy generated after their needs are satisfied is sold to the utility for a profit or to receive energy credits. The fairness of energy sharing in this setting can be violated if one of them gets greedy. For example, Fred can install a battery to store the generated wind energy at his house and can sell back to the utility at a higher price per kilowatt hour during peak hours when the demand pricing applies. On a windy day, Barney can deduce that he is not receiving any power from Fred's house as he should in a fair system. Barney's observations include the fact that it is a windy day with ample scope for wind energy generation (which is also asserted by physically observing the spinning of Fred's wind turbine), and the readings he obtains from Fred's power line and the utility power line coming into his home using a device such as a multimeter. Since Barney is unaware of the battery Fred is using to store the generated energy, he presumes that Fred is consuming all of what is being produced at his house. Fred's actions are therefore completely hidden from Barney. Barney can use an additional observation point, Fred's power line to the utility, to deduce that he is not being fair.

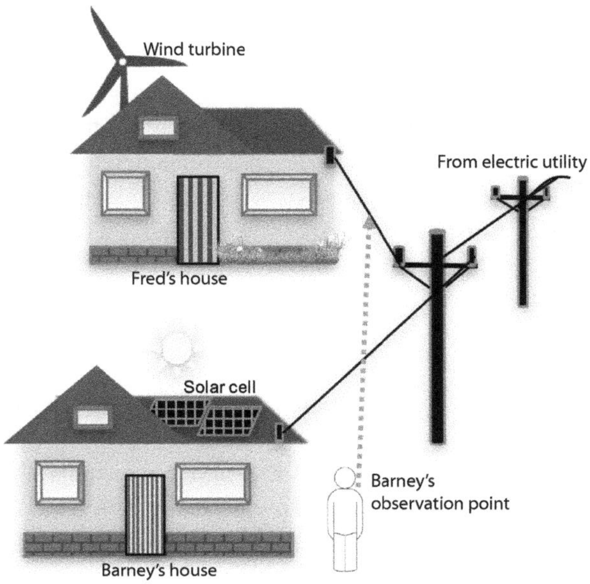

Figure 10.2 A simple case of information flow between two neighbors sharing renewable energy.

In a similar manner, Fred can discharge his battery to the electric utility when he observes that Barney is not drawing power from the grid. If the wind turbine is not spinning at the time Barney makes his observations and Fred continues to discharge his battery to the utility grid, Barney can confirm that Fred is being dishonest. The same is not true if Barney observes that the wind turbine is spinning. In this way, unintended information flows between the two, emerging from their physical observations of one another, leading to confidentiality, integrity, and privacy violations.

In a larger realization of the example of Fred and Barney, such as the FREEDM system [45], a multitude of complex cyber and physical interactions pose a dramatic challenge for determining information flow. FREEDM is an ERC funded by the National Science Foundation (NSF). The FREEDM smart grid optimizes the utilization of renewable energy generation and storage resources, to be integrated with the existing legacy grid. This is achieved through a DGI that consists of cyber processes that perform distributed computation to efficiently manage physical system resources.

Such distributed systems are of interest for multiple reasons. Distributed systems are scalable, and, unlike the current power grid which has inherently centralized generation (such as coal, hydro, or nuclear plants), renewable resources are inherently distributed and customer-owned (such as in the example in Figure 10.2). Also inherent to renewable resources is the problem of intermittency – suddenly wind generation may stop if the wind becomes calm and solar may stop if clouds appear. Due to this unpredictable intermittency, the system must be more reactive and less proactive. In complex systems such as the FREEDM smart grid, such factors add to the already challenging information and confidentiality requirements, making the analysis of information flow properties harder. As such, a cyber-physical solution to smart grid management is required. The effects of power migration at one node can divulge usage patterns (as in the pipeline model) to users observing their gateway power at other nodes, perhaps violating confidentiality. If a user has access to the state of their DGI, further information can be obtained. Unrestricted information flow can potentially be used against the system for economic gains. Under anticipated cap-and-trade schemes, generators of renewable energy may withhold power to sell at a premium.

The analysis of renewable energy systems in this chapter is limited to security issues caused by real flows. Power systems also contain significant dynamic responses that increase their vulnerability to information disclosure and their vulnerability to integrity attacks. Even in normal operation, additional governance is needed within the CPS to maintain system stability. Executable invariants [46] are a way, within a distributed energy system, to govern cyber behavior. Our related work shows how these invariants can be used to discover resources that have been compromised [47, 48]. Section 10.8 briefly discusses how dynamic response can be used in compensation; however, much work needs to be done in the area of security under system dynamic responses and is largely beyond the scope of this chapter.

The architecture of the FREEDM system extends the toy example of Section 10.4 into a succinct system, as shown in Figure 10.3. FREEDM forms a micro-grid of energy storage (DESD), energy resources (DRER), and LOADs to share power for the good of the entire system. FREEDM is envisioned as an architecture for future "smart distribution" systems. As in the pipeline system, intelligent flow controllers (nodes) contain physical actuators (solid state transformers [SSTs]) that control power flow to and from a shared electrical bus, under direction of cooperating DGIs.

Each node is potentially owned and located in a house or business. Within each node, the DGI processes compute a power cost and use a drafting process [49]. Drafting is a receiver-initiated load-balancing procedure; if a node (draft) has available power-generation capacity, it solicits bids from loaded nodes (source). Each DGI

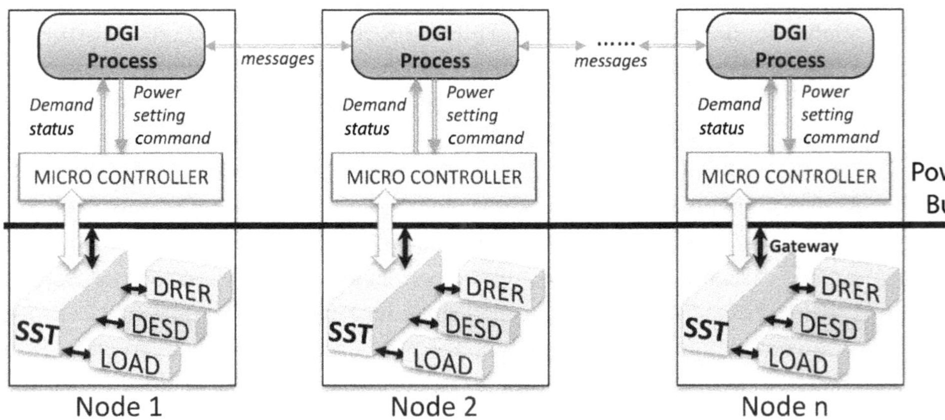

Figure 10.3 FREEDM power management architecture.

Algorithm 10.1: DGI power management algorithm

```
{Variables: R:Renewable, L:Load, S:Storage, G:Gateway Power, ES: Energy State}
Get Values: R/L/S/G
ComputeEnergyState (ES⇐{Supply, Demand, Normal})
if (ES←Demand) then
   Broadcast ("Demand")
   if Recv ("Draft Request") then
      Send ("Draft Response")
   end if
   if Recv ("Draft Select") then
      Send ("Accept")
      Set R/L/S/G
   end if
end if
if (ES←Supply) then
   Broadcast "Draft Request" to Demand nodes
   if Recv("Draft Response") then
      Send ("Draft Select")
   end if
   if Recv ("Accept") then
      Set R/L/S/G
   end if
end if
```

maintains a (potentially out-of-date) load table and the DGI makes its decisions as in Algorithm 10.1 for power management. Other work [46] has used this same model in a traditional power engineering framework to govern power system security and to do the energy management computation using a consensus approach [50].

The energy demand state of each node on the distribution grid is determined by computing the actual load at the SST, referred to as the gateway. The gateway value represents the net draw or contribution of power over the power bus shared by the nodes on the distribution grid. At any given time, a node can supply power or is in demand of power based on the *gateway* value. Any change in the load state is communicated with other nodes over a communication network of DGIs. The power-balancing algorithm at a supply node factors in the demand costs from each demand node, and a forecast of local power requirement in making an optimal selection of node(s) it can supply to. This selection process can get more sophisticated by including other optimization metrics such as market factors and duration of the contract. The actual step of power "migration" involves the supply node setting its gateway value such that the excess power goes back to the shared power bus. The node that was in the demand state receives this power over the shared power line, changing the value of the gateway at its SST. This power migration process continues in multiple rounds until the demand node changes its state to normal or the time-based contract between the two nodes expires.

Intuitively, the power migration step can divulge confidential information of the supply and demand nodes. One can correspond the cyber messages (including advertised demand states) to the physical observations on the shared power bus for deducing

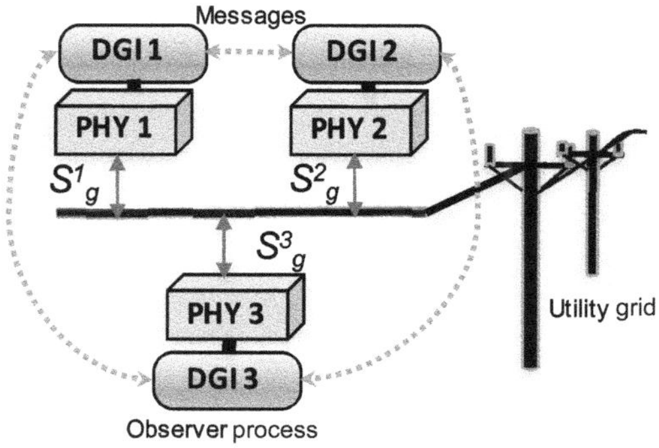

Figure 10.4 An observer process interacting with the FREEDM system (process PHY is an abstraction of the combination of SST, DESD, DRER, and LOAD in Figure 10.3).

otherwise confidential cyber-physical activity, as in the case of the observer process in Figure 10.4. Much as in the example of the pipeline system or in the Fred and Barney example, an observer who is not participating in the power migration process can also deduce what is occurring within the power system by observing the change in the power flow values. The confidentiality and privacy requirement is to make this system, especially the power migration step, nondeducible to an observer who may or may not be part of the distribution grid.

While not covered here, similar concepts exist in electric power transmission systems. The work in [51] explores information leakage for advanced power electronics in the form of flexible AC transmission (FACTS) devices in which actions on the power network by the power electronics provide deducible information about the device settings. Work using the methods of this chapter [52] showed how to hide these events within certain power system topologies.

10.6.1 Formal Models of Composition of Cyber and Physical Processes

The analysis, until now, has been manual and informal. Mechanical specification and verification of functional properties of cyber systems (safety, liveness) is well known. Some work is crossing into the physical world. For example, Platzer [53] integrated process algebraic theory with hybrid systems theory for functional correctness in CPSs. However, a unified framework to model and verify distributed CPSs is required. The challenge is to be able to represent the diverse cyber and physical aspects of a CPS on a uniform semantic basis. Earlier, Wolthusen et al. [54] used π-calculus to detect security attacks in a modeled process control system. Both the adversary and the system in their work were modeled as interacting π-processes. Our work uses the π-calculus to model distributed CPSs and to verify confidentiality rather than functional correctness. In the later sections, we show our approach using π-calculus to model complex CPSs and use behavioral equivalence of π-processes to verify information flow properties within them.

10.6.2 The π-calculus

The π-calculus [55] is a process algebra, suitable for formal modeling of concurrent systems in which communication takes place over dynamic channels. It also offers formal methods to verify system specification using process equivalence testing based on various forms of bisimulation. The process definitions in π-calculus utilize a set of operators with semantics to represent the logic of a concurrent and distributed system, as shown in Table 10.1. The communication channels are labeled from a set of *names* (x, y, z, etc.) and a set of *variables* (m, n, p, q, r, etc.), which can be used to label terms such as messages that are used to define a process. A null process is an atomic process in which no action takes place, typically used to represent a halted process; hence **0**. A non-deterministic choice operator ("+") between two processes indicates that either of them or both of them may be performed. The concurrency between two processes P and Q is represented using the parallel composition operator. Communication is represented using the message output and input patterns shown. The expression, $\bar{x} < n >$.P is also known as output prefixing, meaning the name "n" is first output on channel x before continuing as process P. Similarly, $x(m)$.P is input prefixing in which the process P is waiting to receive the name "m" over channel x before executing its process definition. The names are "bound" to a channel if they are input/output prefixed. Otherwise, they are "free" names.

Replication semantic !P is used to represent a process that creates a new copy of itself and runs in parallel, mainly used to model network daemon processes. Creation of a new name in the process P is captured by the name restriction operator (as in νn.P). The conditional operator makes sure the process moves forward only if the terms x and y match, else it is blocked. For representing dynamic channels, the representation $\{M/x\}$ is used to replace the occurrence of channel x with a new term M in process P. The internal actions within a process that occur but are not observable outside the scope of the process are called silent actions, represented by τ.

This simple and powerful formalism to represent dynamic and concurrent aspects of the processes make it an ideal choice for modeling CPSs. The *pi* specification of CPSs includes the cyber and physical processes, and their interactions modeled using the above formalism. To capture the continuous aspects of physical processes (such as gas flow and power flow), the physical laws guiding them will be encoded as logical

Table 10.1 π-calculus operational semantics

0	Null process
P+Q	Non-deterministic choice
P|Q	Parallel composition
$\bar{x} < n >$.P	Message output
$x(m)$.P	Message input
!P	Replication
νn.P	Name restriction
if $x = y$ then	Conditional operator
$\{M/x\}$	Relabeling

events that can be integrated into the specification. This latter portion requires some semantic thought: What are the significant events within the system? If we examine the pipeline flow problem of Section 10.4, the number of events is rather restricted, only consisting of increase, decrease, or maintain flow. As such, the complexity of this mapping is not high, and certainly less than a discretization of a continuous flow.

10.6.3 Information Flow Properties for CPSs

In process algebras like the calculus of communicating systems (CCS) [56] and π-calculus, security properties are defined by exploiting behavioral equivalence between the processes, as observed by the low-level observer. Noninterference was previously studied in bisimulation semantics of CCS as a nondeducibility property in [57] and in the typed π-calculus framework by Crafa et al. [58]. To relate a complex cyber algorithm with physical dynamics, the nondeducibility property in π-calculus based on the behavioral equivalence is reduction closed barbed congruence [59]. In the remainder of this section, we introduce some definitions that are used in the specification and verification proofs of the FREEDM system.

CONTEXT A *context*, C, is a subprocess defined in π-calculus to limit the environment within the system with which an observer process interacts. Formally, it can be defined as:

$$C :: [.] \mid \pi.C[.] + M \mid \nu aC[.] \mid C[.]|\mathrm{P} \mid !C[.] \tag{10.7}$$

where π is a process term formed from the grammar in π-calculus, and M and P are processes within the system.

In this definition, the "[.]" represents a placeholder for an observer process that may interact with each term in the specification. The notion of context is central to our proofs to abstract only parts of the system which are exposed to an observer. We now move on to various definitions of process equivalence that differ only in the treatment of internal silent actions (τ actions). In the physical world, such actions may not be observable, but the fact that they have occurred may divulge additional information. If a process P can perform an input action with subject μ, disregarding any τ actions, we denote it as $P \downarrow_{\mu}$. For a process Q, if τ actions are taken into account for every action with subject μ, we denote it as $Q \Downarrow_{\mu}$. This means that we want a stronger form of the definition of process Q that might perform several internal actions before and after the μ action ($Q \xrightarrow{\tau*}\xrightarrow{\mu}\xrightarrow{\tau*}$).

DEFINITION 10.1 (*Barbed bisimilarity.*) Two processes P and Q are barbed bisimilar ($P\dot{\approx}Q$), if they satisfy the following criteria:

1. $P \downarrow_{\mu}$ implies $Q \Downarrow_{\mu}$ and vice-versa; and
2. $P \xrightarrow{\tau} P'$ implies $Q \xrightarrow{\tau} Q'$ and vice-versa,

where P' and Q' are the respective new states of P and Q after performing the τ action.

DEFINITION 10.2 (*Reduction closed barbed congruence.*) Two processes, P and Q are reduction closed barbed congruent ($P \dot{\simeq}^c Q$), if

1. they are barbed bisimilar, i.e., $P \dot{\approx} Q$ and
2. $C[P] \dot{\approx} C[Q]$, for every context C.

Based on the above notion of equivalences in π-calculus, we define a basic formulation of nondeducibility using reduction closed barbed congruence. As introduced earlier, bound names (bn) are not accessible outside the scope of process they are input/output prefixed to. For example, in the process $\nu n.P$, the name n is bound.

DEFINITION 10.3 (*Basic π-nondeducibility.*) A process P satisfies basic π-nondeducibility (basic π-ND) with respect to system S, if $P \dot{\simeq}^c P|\mathcal{H}$, where $\mathcal{H}$ is a high-level process composed of the names $\in$$\{$bn(S)-bn(P)$\}$.

In simple words, an observer process P in any context of process S cannot distinguish between the normal operational state of the system S, and the state of the system when composed with a high-level process. An example of such a high-level process is the power migration step in the FREEDM system, which is a result of private interaction between corresponding DGIs. Thus, the observer should not be able to deduce any information about the bound names in the high-level process, which are used to logically represent the private interaction between processes other than P. In the next section, basic π-ND is used to analyze information leakage to an observer in various scenarios of the FREEDM system. This approach can be applied to other CPS infrastructures to ensure confidentiality of cyber-physical interactions. It is also possible to extend this approach to define other information flow properties similar to basic π-ND.

10.6.4 Modeling of FREEDM Using the π-calculus

In π-calculus, the communication structure among existing processes can change over time. Dynamic communication (link mobility) among peers can be represented using two features: (1) restriction of a message to be sent or received over a communication channel; and (2) scope extrusion. Scope extrusion allows the abstraction of the dynamically changing scope of the communication domain, by sending channel names as messages to processes that did not have access to those channels before. The π-calculus offers other useful features such as abstractions to model adversarial behavior, and different notions of process equivalences to define different notions of security. These features are applied in the Future Renewable Electric Energy Delivery and Managements (FREEDMs) system specification presented in this section.

The π-calculus was mainly used to represent two aspects brought by the dynamic nature of communication in FREEDM: (1) the private communication among peers based on the demand state; and (2) the ability to distinguish between messages with the same name originating from different peers based on channels of communication. In Figure 10.5, IEM3 which was in $supply$ initially, is involved in private communication

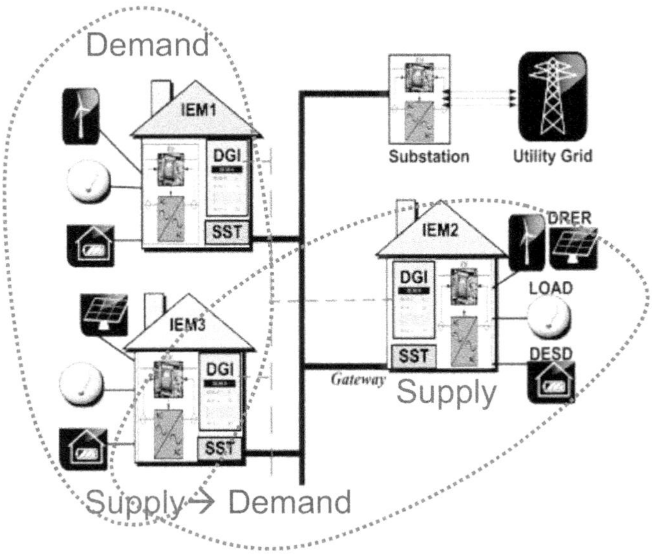

Figure 10.5 Need for scope extrusion in FREEDM.

with the *demand* node, IEM1. However, IEM3 transitions from a supply state to a demand state, thereby establishing communication with a new supply node, IEM2. Likewise, the communication structure changes based on the demand status of the nodes. This needs to be taken into account in the process specification of the system.

In [60], the authors used π-calculus to model distributed processes running peer-to-peer algorithms and verified their functional correctness. The interested reader is encouraged to consult this paper for an example of the methodology before proceeding. We apply a similar approach for modeling a DGI process of the FREEDM system, as shown in Figure 10.6. Communication among various π processes takes place over dedicated channels that originate or terminate at respective input/output ports. For example, the demand state of the DGI is advertised to the peer DGIs over dedicated ports e_1, e_2, e_3 for normal, supply, and demand messages, respectively. The messages used in the drafting method originate from ports m_1, m_2, m_3, m_4. Similarly, the communication with local devices ((house load (LOAD), SST, renewable, and storage) have their corresponding ports labeled s_l, s_g, s_r, s_s, respectively. DGIs have universally unique identifiers (UUIDs) that are used to append to every message. The *state* port is a state variable used only by the DGI process for internal reference. Similarly, the SST has its own port s_g to keep track of the net power flow entering or leaving this SST. Formal π-calculus specifications of various cyber and physical processes in the FREEDM system are laid out in (10.8)–(10.14).

The DGI process definition in (10.8) follows the power management algorithm of Algorithm 10.1. A "$\rightarrow$" symbol is used over names in the process definitions to represent a vector of values sent/received over the corresponding port. The modeling of dynamic communication triggered by the change in demand state is captured in the

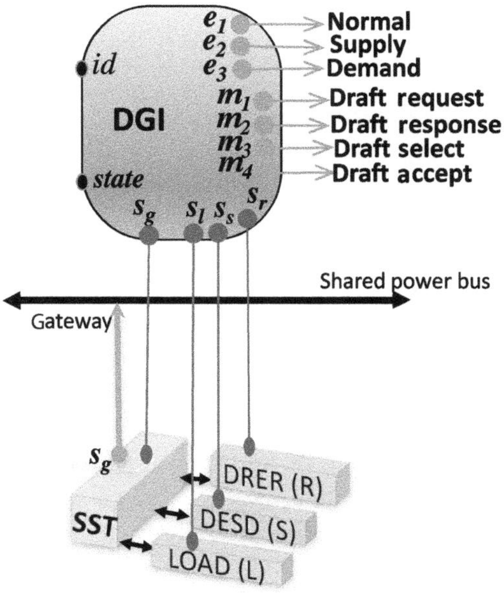

Figure 10.6 π-characterization of a DGI in FREEDM.

process definition using a scope extrusion mechanism (as in $\{^{SND}/_{Demand}\}$). Ω represents a power migration step between two drafting nodes (in supply and demand states respectively), initiated after successful Distributed Grid Intelligence (DGI) negotiations, as defined in (10.9). The interaction among the physical processes at each node (including Solid State Transformer (SST), DRERs, DESDs and LOAD) is reduced to a single process, PHY, that sends or receives a vector of corresponding power settings, as in (10.10).

$$\begin{aligned} & DGI(id^x,\vec{m},\vec{e},\vec{s^x},\vec{state^x}) \triangleq \\ & \quad s^x(\vec{s^x}).\tau.\bar{state^x} < \vec{state} > .DGI(id^x,\vec{m},\vec{e},\vec{s^x},\vec{state^x}) + e^x(\vec{e^y}).\tau. \\ & \quad DGI(id^x,\vec{m},\vec{e},\vec{s^x},\vec{state^x}) + state^x(SND).if\ SND = Supply \\ & \quad then\ \{^{SND}/_{Supply}\}.\bar{m}_1^d < m_1^x > .\Omega(id^x,id^d) + \ state^x(SND). \\ & \quad if\ SND = Demand\ then\ \{^{SND}/_{Demand}\}.\bar{e^x} < e_3^{\vec{y}} > .DGI(id^x,\vec{m}, \\ & \quad \vec{e},\vec{s^x},\vec{state^x}) \ + \ state^x(Demand).m_1^x(m_1^s).\bar{m}_2^s < m_2^x > .\Omega(id^x,id^s) \end{aligned} \tag{10.8}$$

$$\begin{aligned} & \Omega(id^x,id^y) \triangleq \\ & \quad state^x(SND).if\ SND = Supply\ then\ \nu C_{xy}(m_2^x(m_2^y).\bar{m}_3^y < m_3^x > . \\ & \quad m_4^x(m_4^y)).\tau.\bar{s^x} < \vec{s^x} > .0 + state^x(SND). \\ & \quad if\ SND = Demand\ then\ \nu C_{xy}(m_3^x(m_3^y).\bar{m}_4^x < m_4^y >).\tau.\bar{s^x} < \vec{s^x} > .0 \end{aligned} \tag{10.9}$$

$$\begin{aligned} & PHY(id^x,\vec{s^x}) \triangleq \\ & \quad \tau.s^x(\vec{s^x}).\bar{bus} < s_g^x > .PHY(id^x,\vec{s^x}) + \bar{s^x} < \vec{s^x} > .\tau.PHY(id^x,\vec{s^x}) \end{aligned} \tag{10.10}$$

$$INV(\vec{\delta},3) \triangleq$$
$$bus(s_g^1).bus(s_g^2).bus(s_g^3).grid(u).INV(\vec{\delta},3) + \bar{grid} < u > . \quad (10.11)$$
$$\bar{bus} < s_g^1 + \delta^1 > .\bar{bus} < s_g^2 + \delta^2 > .\bar{bus} < s_g^3 + \delta^3 > .INV(\vec{\delta},3)$$

$$INV(\vec{\delta},n) \triangleq$$
$$\prod_{i \in n} bus(s_g^i).grid(u).INV(\vec{\delta},n) + \bar{grid} < u > . \prod_{i \in n} \bar{bus}(s_g^i + \delta^i). \quad (10.12)$$
$$INV(\vec{\delta},n), where\ |n| = \#Nodes.$$

The processes DGI, Ω, and PHY defined thus far do not complete the system specification. The invariance of power flow on the shared power bus that interconnects multiple SSTs is required as this is a property of the underlying power system and hence cannot be violated. It is defined by Kirchhoff's current laws for real power flow; the sum of the currents leaving or entering an SST at its gateway (s_g^i) must equal those of the connecting buses. The 3-SST base case shown in (10.11) captures this idea by specifying that if a change in power flow (by an amount, δ^1) occurs at any SST due to a draw from the utility grid, it is followed by compensatory events (δ^2, δ^3) at the other SSTs to maintain the physical invariant. The invariant must be expressed in the model for each SST connection. For an n node system, the invariant acting on the bus can be developed inductively as in (10.12). The overall system specification can be achieved by composing the invariant process with the composition of cyber and physical processes, as shown in (10.13) and (10.14):

$$Node(id^x,\vec{m},\vec{e},\vec{s^x},\vec{state^x}) \triangleq$$
$$DGI(id^x,\vec{m},\vec{e},\vec{s^x},\vec{state^x})|PHY(id^x,\vec{s^x}) \quad (10.13)$$

$$System, S \triangleq$$
$$Node(id^1,\vec{m},\vec{e},\vec{s^1},\vec{state^1})|Node(id^2,\vec{m},\vec{e},\vec{s^2},\vec{state^2})|\ldots\ldots| \quad (10.14)$$
$$Node(id^n,\vec{m},\vec{e},\vec{s^n},\vec{state^n})|INV(\vec{\delta},n)$$

10.6.5 Verification of Information Flow in FREEDM Using π-calculus

In this section, a comprehensive analysis of operations in various configurations of the FREEDM system is presented. An observer process that participates legitimately in the system is used to verify information flow leakage during critical operations. A configuration of the FREEDM system satisfies basic π-ND if the observer process in any context cannot distinguish between the system states that differ only in the occurrence of a power migration step. Hence, verification of the equivalence $Node3 \dot{\simeq}^c Node3|H$ will be performed.

Observer (Context) in the System without DGI

An observer with no DGI cannot send or receive cyber messages within the system. Conceptually, Theorem 10.1 establishes that such an observer in a basic power system cannot deduce individual power settings.

THEOREM 10.1 The system without DGI is inherently secure with respect to a low-level observer.

Proof. To prove that $Node3 \dot{\simeq}^{c} Node3|H$, it is required to establish the following:

- $Node3 \dot{\approx} Node3|H$
- $C[Node3] \dot{\approx} C[Node3|H]$, for the context C, defined in (10.15).

$$C :: [.]|S \tag{10.15}$$

The high-level process in this case is the sequence of settings on the bus, made by nodes other than the observer. Without DGI, the system with two nodes functions as:

$$\begin{aligned} S' \triangleq Node(id^1, \vec{s^1})|Node(id^2, \vec{s^2})|INV(\vec{\delta}, 3) := (\tau.\bar{bus} < s_g^1 > .S' + \tau. \\ bus(s_g^1).S')|(\tau.\bar{bus} < s_g^2 > .S' + \tau.bus(s_g^2).S')|INV(\vec{\delta}, 3) \end{aligned} \tag{10.16}$$

For $S' \xrightarrow{\bar{bus}<s_g^1+\delta^1>} \xrightarrow{\bar{bus}<s_g^2+\delta^2>} \hat{S}'$, and $Node3 \downarrow_{\bar{bus}<s_g^1>, \bar{bus}<s_g^2>}$ is $Null$. Similarly, $Node3|H := Node3|(\bar{bus} < s_g^1 + \delta^1 > |\bar{bus} < s_g^2 + \delta^2 >) \Downarrow_{\bar{bus}<s_g^1>, \bar{bus}<s_g^2>}$ is $Null$. Inserting the observer node in the context of (10.15), the following are indistinguishable:

$C[Node3] := Node3|(\tau.\tau.grid(u).S' + \tau.bus(s_g^1).S')|(\tau.\tau.grid(u).S' + \tau.bus(s_g^2).S')|\tau.\tau.bus(s_g^3).grid(u).INV(\vec{\delta}, 3)$

$C[Node3|H] := Node3|(\tau.\tau.grid(u).S' + \tau.bus(s_g^1 + \delta^1).S')|(\tau.\tau.grid(u).S' + \tau.bus(s_g^2 + \delta^2).S')|\tau.\tau.bus(s_g^3).grid(u').INV(\vec{\delta}, 3)$

Therefore, $C[Node3] \downarrow_{\bar{bus}<s_g^1>, \bar{bus}<s_g^2>} \dot{\approx} C[Node3|H] \Downarrow_{\bar{bus}<s_g^1+\delta^1>, \bar{bus}<s_g^2+\delta^2>}$ and hence the conclusion, $Node3 \dot{\simeq}^{c} Node3|H$. $Node3$ does not observe a change in gateway value s_g^3 due to the maintenance of the invariant on the bus, such that $s_g^1 + s_g^2 + s_g^3 + u = (s_g^1 + \delta^1) + (s_g^2 + \delta^2) + s_g^3 + u'$. □

We now consider the configuration in which two nodes equipped with DGI are involved in a power migration step. An observer process with DGI can have different observations based on its advertised state (*supply*, *demand*); nonetheless, basic π-ND is violated in both cases.

Observer (Context) in Supply State in the System with DGI

The observer process in the supply state is assumed to function in the same way as other supply nodes in the system with the exception that it can send a fake draft response to a demand node to avoid participating in the actual power migration step. The specification of such a process, $Node3_s$, is defined in (10.17). The interactions between the other two nodes in the system, S'', is captured in (10.18).

$$\begin{aligned} Node3_s \triangleq Node(id^3, \vec{m}, \vec{e}, \vec{s^3}, \vec{state^3}) := \tau.\tau.\bar{bus} < s_g^3 > .\tau.\bar{state^3} \\ < \vec{state^3} > .Node3_s + e^3(\vec{e^2}).\tau.Node3_s + state^3(SND). \\ \{^{SND}/_{Supply}\}.\bar{m}_1^2 < m_1^3 > .Node3_s + state^3(Supply).m_2^3 \\ (m_2^2).Node3_s \end{aligned} \tag{10.17}$$

$$S'' \triangleq Node(id^1, \vec{m}, \vec{e}, \vec{s^1}, \vec{state^1}) | Node(id^2, \vec{m}, \vec{e}, \vec{s^2}, \vec{state^2}) | INV(\vec{\delta}, 3)$$
$$:= \tau.\tau.\bar{bus} < s_g^1 > .\tau.\bar{state^1} < \vec{state^1} > .S'' + \tau.\tau.\bar{bus} < s_g^2 > .\tau.$$
$$\bar{state^2} < \vec{state^2} > .S'' + e^1(\vec{e^2}).\tau.S'' + state^1(Supply).state^2 \quad (10.18)$$
$$(Demand).\tau.\tau.S'' + state^2(Demand).\bar{e^2} < \vec{e_3^1} > .S'' + e^2(\vec{e^1}).\tau.$$
$$S''|INV(\vec{\delta}, 3)$$

When composed with a high-level process performing the power migration step, S'' transforms into a new process, as shown in (10.19). In Theorem 10.2, we show how the observer, $Node3_s$, deduces the power migration between the two nodes in S'':

$$S''|\Omega(id^1, id^2) \triangleq Node(id^1, \vec{m}, \vec{e}, \vec{s^1}, \vec{state^1}) | Node(id^2, \vec{m}, \vec{e}, \vec{s^2}, \vec{state^2}) |$$
$$\Omega(id^1, id^2) | INV(\vec{\delta}, 3) := \tau.\tau.\bar{bus} < s_g^1 > .\tau.\bar{state^1} < \vec{state^1} > .S''$$
$$+ \tau.\tau.\bar{bus} < s_g^2 > .\tau.\bar{state^2} < \vec{state^2} > .S'' + e^1(\vec{e^2}).\tau.S'' +$$
$$state^1(Supply).state^2(Demand).\tau.\tau.S'' + state^2(Demand). \quad (10.19)$$
$$\bar{e^2} < \vec{e_3^1} > .S'' + e^2(\vec{e^1}).\tau.S'' + \nu C_{12}(\tau.\tau.\tau.\tau.\tau.\tau.\tau.\tau).$$
$$(\tau.\tau.\bar{bus} < s_g^1 > + \tau.\tau.\bar{bus} < s_g^2 >).S''$$

THEOREM 10.2 Confidentiality in the three-node system is compromised due to a low-level observer in the supply state.

Proof. Following Definition 10.6.3, $Node3_s \dot{\simeq}^c Node3_s|\Omega(id^1, id^2)$[3] if the following conditions are satisfied:

1. $Node3_s \dot{\approx} Node3_s|\Omega^{12}$
2. $C[Node3_s] \dot{\approx} C[Node3_s|\Omega^{12}]$, for the context C, defined in (10.15).

From the process definition of Ω^{12}, a secure communication channel is established between nodes 1 and 2, thereby hiding the private communication from an external observer. This means that $Node3_s$ cannot deduce if the drafting process between the other two nodes has been initiated, also possibly by knowing their demand states. As $Node3_s$ does not participate in the power migration process, Ω^{12}, it may seem that $Node3_s \dot{\approx} Node3_s|\Omega^{12}$. However, in the other contexts of the system that the observer is placed in, as shown in (10.20) and (10.21), the physical observability on the shared power bus coupled with the observer's knowledge of the demand states of the other two nodes reveal critical information about power migration between them. Formally, this is captured in (10.20), as $C[Node3_s] \downarrow_{\Omega^{12}}$ results in $\{e^2(\vec{e^3})\}$. Also, in (10.21), $C[Node3_s|\Omega^{12}] \Downarrow_{\Omega^{12}}$ yields the observable set $\{e^2(\vec{e^3}), bus(s_{g'}^3)\}$. The event $bus(s_{g'}^3)$ was observable due to the physical invariant on the shared power bus, $bus(s_g^1)$. $bus(s_g^2)$. $bus(s_g^3)$:

[3] $\Omega(id^1, id^2)$ will be used as Ω^{12} for simplicity.

$$
\begin{aligned}
C[Node3_s] &\triangleq S''|Node3_s := \tau.\tau.\bar{bus} < s_g^1 > .\tau.\bar{state}^1 < \vec{state}^1 > .C+ \\
&\tau.\tau.\bar{bus} < s_g^2 > .\tau.\bar{state}^2 < \vec{state}^2 > .C \ + e^1(\vec{e^2}).\tau.C + state^1 \\
&(Supply).state^2(Demand).\tau.\tau.C \ + state^2(Demand).\bar{e^2} < \vec{e^1_3}, \vec{e^3_3} > \\
&.C \ + e^2(\vec{e^1}).\tau.C + \tau.\tau.\bar{bus} < s_g^3 > .\tau.\bar{state}^3 < \vec{state}^3 > .C \\
&e^3(\vec{e^2}).\tau.C
\end{aligned} \tag{10.20}
$$

$$
\begin{aligned}
C[Node3_s|\Omega^{12}] &\triangleq S''|Node3_s|\Omega^{12} := \tau.\tau.\tau.\tau.C + e^1(\vec{e^2}).\tau.C \ + state^2 \\
&(Demand).\bar{e^2} < \vec{e^1_3}, \vec{e^3_3} > .C \ + \tau.\tau.\bar{bus} < s_{g'}^3 > .\tau.\bar{state}^3 \\
&< \vec{state}^3 > .C \ + e^3(\vec{e^2}).\tau.C
\end{aligned} \tag{10.21}
$$

Therefore, the observer is able to distinguish the case of a power migration between nodes 1 and 2 and non-occurrence of power migration, by determining that node 2 has recently been in the demand state. Hence, $Node3_s \not\dot{\approx}^C Node3_s \, |\Omega(id^1, id^2)$. □

Observer (Context) in Demand State in the System with DGI

The case in which the observer is in the demand state is similar to the case in which the observer is in the supply state. However, the evaluation context consists of the observer node in the demand state, $Node3_d$.

$$
\begin{aligned}
Node3_d &\triangleq Node(id^3, \vec{m}, \vec{e}, \vec{s^3}, \vec{state}^3) := \tau.\tau.\bar{bus} < s_g^3 > .\tau.\bar{state}^3 \\
&< \vec{state}^3 > .Node3_d \ + e^3(\vec{e^2}).\tau.Node3_d \ + state^3(SND). \\
&\{^{SND}/_{Demand}\}.\bar{e^3} < \vec{e^1_3} > .Node3_d \ + state^3(Demand). \\
&m_1^3(m_1^1).\bar{m}_2^1 < m_2^3 > .Node3_d
\end{aligned} \tag{10.22}
$$

THEOREM 10.3 The system is not secure with respect to a low-level observer in the demand state.

Proof. For the same reason as mentioned in Theorem 10.2, proving $Node3_d \approx Node3_d|\Omega^{12}$ is trivial, because $Node3_d$ does not interact with the process Ω^{12}:

$$
\begin{aligned}
C[Node3_d] &\triangleq S''|Node3_d := \tau.\tau.\bar{bus} < s_g^1 > .\tau.\bar{state}^1 < \vec{state}^1 > .C + \\
&\tau.\tau.\bar{bus} < s_g^2 > .\tau.\bar{state}^2 < \vec{state}^2 > .C \ + e^1(\vec{e^2}).\tau.C \ + state^1 \\
&(Supply).state^2(Demand).\tau.\tau.C \ + state^2(Demand).\bar{e^2} < \vec{e^1_3}, \vec{e^3_3} > \\
&.C \ + e^2(\vec{e^1}).\tau.C + \tau.\tau.\bar{bus} < s_g^3 > .\tau.\bar{state}^3 < \vec{state}^3 > .C + \\
&e^3(\vec{e^2}).\tau.C
\end{aligned} \tag{10.23}
$$

$$
\begin{aligned}
C[Node3_d|\Omega^{12}] &\triangleq S''|Node3_d|\Omega^{12} := \tau.\tau.\tau.\tau.C + e^1(\vec{e^2}).\tau.C \ + state^2 \\
&(Demand).\bar{e^2} < \vec{e^1_3}, \vec{e^3_3} > .C \ + \tau.\tau.\bar{bus} < s_{g'}^3 > .\tau.\bar{state}^3 \\
&< \vec{state}^3 > .C \ + e^3(\vec{e^2}).\tau.C
\end{aligned} \tag{10.24}
$$

$C[Node3_d] \downarrow_{\Omega^{12}}$ is { $e^2(\vec{e^3})$, $m_1^3(m_1^1)$} while $C[Node3_d|\Omega^{12}] \Downarrow_{\Omega^{12}}$ is { $e^2(\vec{e^3})$, $m_1^3(m_1^1)$, $bus(s_{g'}^3)$}. The context $C[Node3_d|\Omega^{12}] \Downarrow_{\Omega^{12}}$ captures the notion that $bus(s_g^1)$. $bus(s_g^2).bus(s_g^3)$ takes place on the bus leading to an observable event $bus(s_{g'}^3)$. The observer is able to distinguish the case of a power migration between nodes 1 and 2 and non-occurrence of power migration, by determining that node 1 is in the supply state and node 2 is in the demand state. Clearly, $Node3_d \not\dot{\simeq}^c Node3_d|\Omega(id^1, id^2)$. □

Making the FREEDM System π-ND-secure

In the previous sections, it was proved that the FREEDM system is not π-ND when there are three nodes in the system. The observer process was able to correlate the observable events due to the physical invariant to the cyber messages involving demand states of the nodes in the system. In a larger deployment, however, multiple power migrations can occur between any pair of nodes, preserving confidentiality. This fact is verified in Theorem 10.4.

THEOREM 10.4 In the FREEDM system with five nodes (Figure 10.7), confidentiality is preserved with respect to a low-level observer in the supply state when more than one power migration occurs in parallel.

Proof. For five nodes, (10.18) can be rewritten with the new invariant, $INV(\vec{\delta}, 5)$ instead of $INV(\vec{\delta}, 3)$. This proof is presented for a case with two power migrations occurring in parallel; however, it can be extended to a case with more. Following the definition of π-ND, we compute the trace $C[Node3_s] \downarrow_{\Omega^{12}}$, which yields the set $\{e^2(\vec{e^1}), e^2(\vec{e^4}), e^3(\vec{e^2}), e^3(\vec{e^5})\}$. Similarly, $C\ [Node3_s|\Omega^{12}] \Downarrow_{\Omega^{12}}$ equates to the set $\{e^2(\vec{e^1}), e^2(\vec{e^4}), e^3(\vec{e^2}), e^3(\vec{e^5}), bus(s_{g'}^3)\}$. From the result traces, the observer cannot infer the event preceding the observation due to the physical invariant, $bus(s_{g'}^3)$. It could be due to Ω^{12} or Ω^{34} or a combination of both. Therefore, $Node3_s \dot{\simeq}^c Node3_s$ $|\Omega^{12}$ as well as $Node3_s \dot{\simeq}^c Node3_s\ |\Omega^{45}$. □

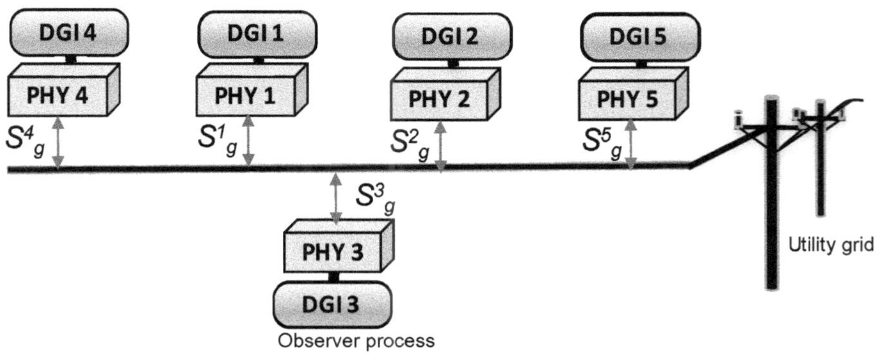

Figure 10.7 A five-node FREEDM system including an observer.

Manual proofs are tedious. Process algebraic equivalence checkers like the Mobility Workbench (MWB) [61] and ProVerif [62] are employed to automate the verification process. These tools accept the system specification in terms of π-calculus syntax and verify certain equivalence properties. The MWB is a tool developed for polyadic π-calculus and verifies observational equivalence between processes. The π-ND property used in this analysis, however, uses a finer refinement of observational equivalence, the *reduction closed barbed congruence* (RCBC) and hence the capability of this tool is limited. However, it can be proved that the RCBC relation does not apply to two processes by proving that the two processes are not weakly observational equivalent. ProVerif, on the other hand, is more robust because it accepts the applied π-calculus specification of the system and provides in-built commands to verify a number of secrecy properties that are of interest to CPS analyses. Various analyses of the FREEDM specification are carried out using each of these tools. It is to be noted that the FREEDM system only serves as a test example of a complex distributed CPS, and the modeling shown in the following provides the reader with how this analysis approach can be extended to other CPSs.

10.6.6 Mobility Workbench

The MWB is an automated tool for manipulating and analyzing mobile concurrent systems specified in π-calculus. It proves that something is barbed bisimilar (BB, $\dot{\approx}$) but not RCBC ($\dot{\simeq}^c$). That is, if two processes P, Q are such that P $\dot{\approx}$Q , they need not be P $\dot{\simeq}^c$Q. However, the opposite is true. Therefore, RCBC is a finer refinement of BB. To verify the weak or observational equivalence, MWB is used as follows:

weq *Process1 Process2*

The MWB encoding for the FREEDM system has been previously published [63]. The results obtained for a three-node and five-node system are summarized in Table 10.2.

Table 10.2 Basic π-ND results for observer in supply and demand states using MWB

			MWB	
Process 1	**Process 2**	**Comment**	$\dot{\approx}$	**Complexity**
S: Node1 | Node2	S|Ω^{12} |$Node3_S$	Power migration between two nodes with respect to an observer in the supply state	x	28.48 mins
S: Node1 | Node2	S|Ω^{12} |$Node3_D$	Power migration between two nodes with respect to an observer in the demand state	x	29.12 mins
S': Node1 |..| Node5	S'|Ω^{12}|Ω^{45} |$Node3_S$	Two consecutive power migrations in a five node system with respect to an observer in the supply state	✓	49.3 mins
S': Node1 |..| Node5	S'|Ω^{12}|Ω^{45} |$Node3_D$	Two consecutive power migrations in a five node system with respect to an observer in the demand state	✓	50.1 mins

10.6.7 ProVerif

ProVerif [64] is a tool primarily intended to verify cryptographic protocols. It accepts as input the protocol specification in applied π-calculus [65] and verifies claimed security properties. The tool helps uncover potential security violations in the implementation of cryptographic protocols in the presence of an attacker who may have complete control over the communication channels. An attacker with such a capability, often referred to as a "Dolev-Yao" [66] attacker, can read, modify, and inject messages into the system. In this work, this tool was used in a different way to reveal information flow violations in CPS environments by abstracting away the cryptographic primitives defined in the tool language. The verification of π-ND using ProVerif is inspired by the privacy verification of electronic voting protocol, FOO92 in [67].

π-ND

The encoding of the FREEDM system in ProVerif is shown in Figures 10.8–10.12. The observer is able to break the confidentiality of power migration among the nodes by distinguishing observer contexts in supply and demand states as above. Additional properties that can be verified using ProVerif are given in Figure 10.13.

Strong Secrecy

Strong secrecy is preserved when the attacker is unable to distinguish when the secret changes. With respect to the FREEDM system, the attacker should not distinguish between the cases when a supply node migrates a units of power in a single power migration step and b units of power. To perform such an attack, the attacker is assumed to detect power withdrawal from the utility and read the event of a spike in the power on the shared power bus. Strong secrecy can be verified in ProVerif using the **noninterf** keyword, as shown in Figure 10.13. It can be argued that strong secrecy is preserved in the FREEDM system since the migrations currently take place in a fixed quantum of power. This property is particularly useful to analyze the effect of partial information revealed to the attacker. For example, in the case of a series of

```
free g1,g2:G [private].
free g3:G.
free delta1, delta2:netValue [private].
```

Figure 10.8 Defining variables and names to initialize the FREEDM ProVerif script.

```
(* Invariant process*)
fun Inv(G):netValue.
reduc forall g:G, value:netValue;
      sum(Inv(g), value)= value.
```

Figure 10.9 ProVerif process defining the physical invariant of flow.

```
(* DGI process1 in Supply state *)
let DGI_Supply =
        out(c, Status);
        (*Request sent to only demand nodes*)
        out(c, Request);
        in(ch, x_Response:bitstring);
        if x_Response = Response then
        out(ch, choice[Select, Reject]);
        in(ch, x_Accept:bitstring);
        if x_Accept = Accept then
        out(c_phy, Inv(g1));
        let y=sum(Inv(g1), delta1) in
        out(ch, Set).
```

Figure 10.10 ProVerif process for a DGI node in the supply state.

```
(* DGI process2 in Demand state *)
let DGI_Demand =
        out(c,Status);
        in(c,x_Request:bitstring);
        if x_Request=Request then
        out(ch, Response);
        in(ch, x_Select:bitstring);
        if x_Select=Select then
        out(ch, Accept);
        out(c_phy, Inv(g2));
        let y=sum(Inv(g2), delta2) in
        out(ch, Set).
```

Figure 10.11 ProVerif process for a DGI node in the demand state.

```
(* The main process defining the system *)
process (!(DGI_Supply) | !(DGI_Demand) )
```

Figure 10.12 ProVerif process defining FREEDM.

power migrations between a supply–demand pair involving migratable quantum of non-uniform magnitude, the adversary can distinguish each of the series indicating that he obtains partial information. However, in a FREEDM system with five or more nodes, strong secrecy is always preserved unless there is a unique pair of nodes in supply and demand states, respectively.

Weak Secrecy

A weak secret refers to the information that the attacker may guess through passive or active observation of the system. A common example of weak secrets are

Table 10.3 Results of verification with ProVerif

		Three-node system		Five-node system	
Property verified	**Process/name**	**Result**	**Complexity**	**Result**	**Complexity**
Basic π-ND	DGI_Supply, DGI_Demand	x	1.28 s	✓	1.45 s
Strong secret	g1, g2	✓	1.14 s	✓	1.14 s
Weak secret	delta1, delta2	✓	1.14 s	✓	1.14 s

```
query attacker(delta2).
weaksecret delta1.
noninterf g1, g2.
```

Figure 10.13 Using ProVerif secrecy features on FREEDM.

human-memorable passwords used in some protocols, which are often values with low information entropy. The attacker may enumerate all the possible values and end up with the exact value though repeated trial. In ProVerif, any name can be verified to determine if it is a weak secret, using the **weaksecret** keyword as shown in Figure 10.13.

The time taken for the verification of all the properties is about 1.14 seconds and about 1.45 seconds for weak equivalence on a computer with a 2.4 GHz Intel Core 2 Duo processor and 2 GB memory running Mac OS X 10.5. This proves that ProVerif is far more efficient compared to the MWB. The results with ProVerif are summarized in Table 10.3.

10.7 Refine Existing Methods of Information Flow Analysis to Encompass Multiple Security Domains

Most security models partition the system into a high domain and a low domain. This is a simplification of the multiple levels of Bell–Lapadula. If the system consists of more than two security levels, the assumption is that the progressively higher security levels are contained within the next lower security level. Indeed, this model has served us well since the advent of the medieval castle, with its outer walls and interior keep (Figure 10.14) [68]. Unfortunately, this model does not work well for modern CPSs because the security levels often overlap and are rarely entirely contained as in the case of the medieval castle and keep system (Figure 10.15).

We have good security tools, but CPSs are more complicated than the computer systems these tools were designed to model. Information flow security is important and in CPSs we are not able to limit our view to denying information flows from one security domain to another. The physical nature of the system leaks information,

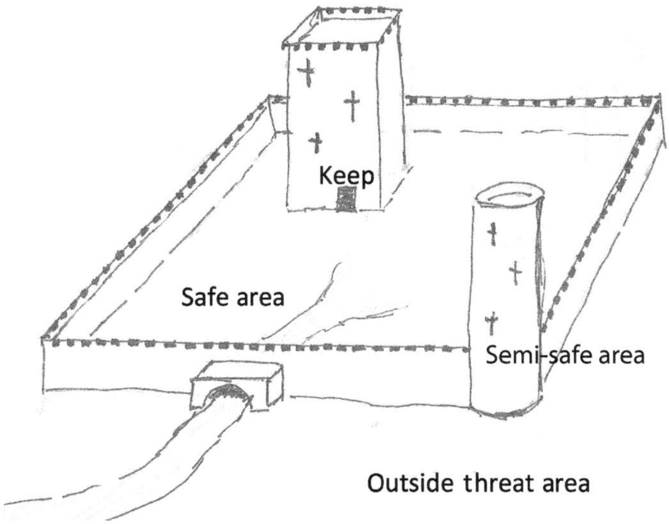

Figure 10.14 A medieval castle model of security.

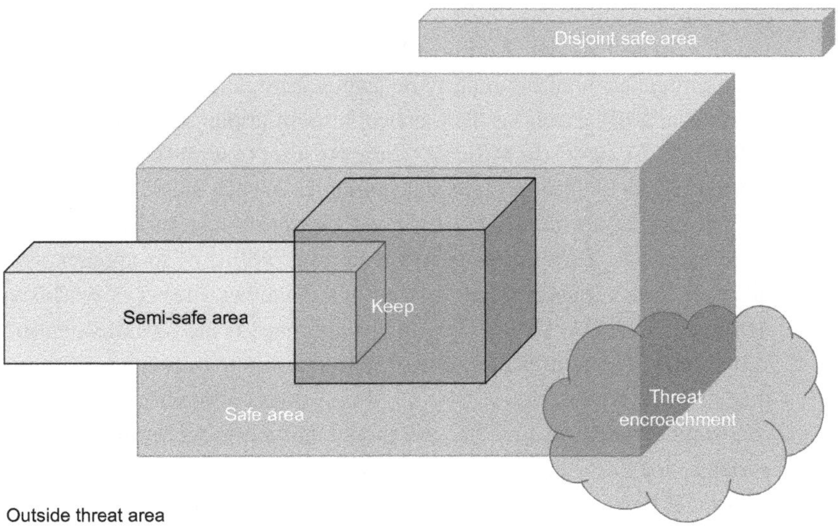

Figure 10.15 Problematic overlapping security domains.

and novel attacks – such as Stuxnet [69] – have shown the limitations of viewing threats as only attempts to steal information. CPS information flow models must now explore information flow across security domain boundaries as being critical to the safe operation of the system.

Clearly, access control models become irrelevant within this context. In the overlapping domains of a CPS, both cyber and physical entities share portions of the same security domain, so requesting access does nothing to track the potential information flow. Likewise for noninterference, no rigorous partition can be formed across all information flows. As such, as in previous sections, we turn to nondeducibility.

There are two ways to define nondeducibility: trace-based and modal frame-based. The trace-based version of nondeducibility [70, 71] holds if what is observed at a low level is consistent with any number of high-level actions and therefore does not leak any information to the low level. In a modal framework, a system is defined as a set of worlds W, where a world $w \in W$ represents a single execution of the system. Let two information functions $f_1(w)$ and $f_2(w)$ define a view over each world. A system is secure with respect to the modal definition if:

$$ND(ES) : (\forall z : f_2^{-1}(z) \neq \emptyset)(\forall w \in W)(\exists w' \in W : f_1(w) = f_1(w') \wedge f_2(w') = z) \quad (10.25)$$

where $f_1(w)$ and $f_2(w)$ define high- and low-level security domains. For each observation z, over every command sequence w, there exists some execution w' where the command sequence w and the observation z coexist. This means that it is possible to pair each observation with each command sequence and still have a valid system execution. In this modal framework, the state of ND(ES) can be evaluated as execution progresses. As such, any analysis is not limited to past behavior of the system, as in the trace-based model. Additionally, $f_1(w)$ and $f_2(w)$ are actually partitions of the system; the notion of high- and low-level security domains is arbitrary.

Howser [72] extended this concept to multiple security domains through the MSDND(ES) model. In MSDND, there are a set of worlds, $\{W\}$, consisting of distinct worlds, $w_0, w_1, \ldots, w_n$ with m state variables in each world, $s_1, s_2, \ldots, s_m$.

The worlds are connected by a set of transitions, $\{wRw'\}$. Changing any state variable causes a transition from the current world, w, to another world, w', where all other state variables retain their values. Together, the set of worlds and transitions define a frame, $\mathfrak{F} = \{W, R\}$. Key to this approach is the idea of valuation functions. A valuation $V^i_{s_x}(w)$ returns the value of state variable s_x as seen by an entity i in world w; that is, $V^i_{s_x}(w) = (s_x : s_x = \bot \vee \top)$. If no valuation function exists to return the value of a state variable, say s_i, then the model can never determine the value of that state variable nor the value of any logical expression dependent upon that state variable. This latter concept becomes important when evaluating CPSs.

To see how valuation functions work, consider a generic case of a single security domain SD^i with two valuations, $V^i_x(w)$ and $V^i_y(w)$. There exists some world with a pair of states, which may or may not be in the same security domain, where one must be true and the other false (exclusive OR), but an entity i has no valuation function for those states. Because there is no valuation function for the states in security domain SD^i, it cannot be determined which state is true and which is false:

$$\begin{aligned}\text{MSDND(ES)} = \quad & \exists w \in W : w \vdash \Box\left[(s_x \vee s_y) \wedge \neg(s_x \wedge s_y)\right] \\ & \wedge \left[w \vDash (\nexists V^i_x(w) \wedge \nexists V^i_y(w))\right]\end{aligned}$$

In the special case where s_x is $\varphi = \top$ and s_y is $\neg\varphi = \top$, MSDND(ES) reduces to:

Table 10.4 Process control system security domains

Domain	Valuation	
SD^0	V_φ^0	Paint separator
SD^1	V_φ^1	Virus
SD^2	V_φ^2	Controller
SD^3	V_φ^3	Monitor
SD^4	V_φ^4	Human operator

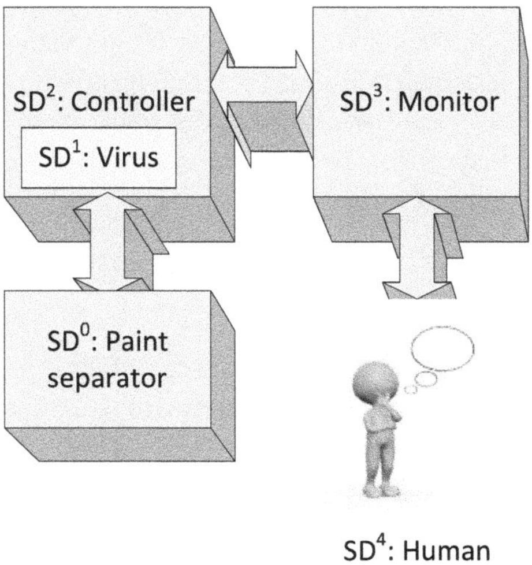

Figure 10.16 Abstract plant model with security domains.

$$\text{MSDND(ES)} = \exists w \in W : w \vdash \Box\left[\varphi \otimes \neg\varphi\right] \wedge \left[w \models (\nexists V_\varphi^i(w))\right]$$

To see why multiple security domain models are critical for CPS, consider the simple plant control model of Figure 10.16. A physical process makes paint under the control of a controller. There is a monitor that reads what the controller senses from the physical process and displays the result to a human. Consider a portion of the controller to be a program that is not functioning according to what is anticipated. This could be due to many reasons, ranging from software bugs to an actual virus. As such there are five separate security domains as defined in Table 10.4. In this model, there are not only multiple security domains, but the notion of "high" and "low" security is not relevant. Let φ be "Paint is being separated properly." Either φ or $\neg\varphi$ must be true at all times. Under normal conditions, the controller/monitor system will monitor the paint separator and make adjustments to ensure that φ is *true*.

If the system is operating correctly (without the virus present), we expect that all operations done by the controller are successfully carried out and reported back to the monitor, deducibly. In other words, every action in a domain is uniquely identifiable in another domain; the paint separation process is reported correctly to the operator, operator comments are carried out by the controller to the paint separator, etc.

The key to showing the integrity of such a system is to show that the desired information flow cannot be disrupted. By showing that the system is not MSDND secure, one shows that every observation and command is uniquely attributable to its corresponding command and observation. As a starting step, we must establish that the paint separator correctly reports its status and dutifully follows commands sent to it from within the controller.

In the following, information transfer is represented by $I_{i,j}\varphi$, which clearly states how i gains knowledge of φ. The information transfer operator inherently assumes j will not lie to i; however, this restriction allows liars to lie to trusting agents. It is key to realize the information is transferred directly to an agent who has no direct way to evaluate whether or not j is a liar. As such, the status of the paint is not MSDND(ES) secure at the paint separator. To see why, inherently the physical paint separator correctly reports the status of its paint.

THEOREM 10.5 The paint status is not Multiple Security Domains Nondeducibility (MSDND)(ES) secure at the paint separator.

Proof. To clarify notation, let s_1 be the state in which the centrifuge is within nominal bounds ($\varphi = \top$), and let s_2 be the state in which the centrifuge is not operating within nominal bounds ($\varphi = \bot$ and $\neg\varphi = \top$).

To show MSDND(ES), we must find a world w such that:

$$\begin{aligned}\text{MSDND(ES)}= \quad & \exists w \in W : w \vdash \Box[(s_1 \vee s_2) \wedge \neg(s_1 \wedge s_2)] \\ & \wedge \left[w \models (\nexists V^i_{s_1}(w) \wedge \nexists V^i_{s_2}(w))\right]\end{aligned}$$

It is obvious that $(s_1 \oplus s_2) = true$, so the first condition for MSDND(ES) is met. However, the separator directly measures the paint and therefore both $V^0_{s_1}(w)$ and $V^0_{s_2}(w)$ are correctly evaluated for any w and the conditions for MSDND(ES) are not met. □

The virus, to cause the maximum amount of disruption, can completely block the information flow from the paint separator to the controller. However, this is a trivial case as it is easily detected via timeouts on readings from the paint separator. Instead, the analysis focuses on the more insidious case in which the virus fabricates readings to create false information flow. These fabricated readings cause an observation at the monitor to be consistent with multiple possibilities within the physical system, essentially making the system nondeducible from the perspective of the human operator.

THEOREM 10.6 Formally, this requires that the system is MSDND secure for SD^2 so that any entity i within SD^2, SD^3, and SD^4 will believe all is well or φ.

Proof. By definition $(\varphi \oplus \neg\varphi) = true$, so the first condition for MSDND(ES) is met. If φ cannot be correctly evaluated in SD^2, then both conditions are met.

The argument for this goes as follows.

[Paint status is normal and $\varphi = true$]

1.	φ	Paint status normal
2.	$w \models V_\varphi^0(w) = true$	Definition of $w \models V_\varphi^0(w)$
3.	$I_{1,0}\varphi$	Paint separator reports to virus
4.	$I_{2,1}\varphi$	**Virus *always* reports "all is fine"**
5.	$w \models V_\varphi^2(w) = true$	$V_\varphi^2(w)$ **always** returns *true*

[Paint Status is not normal and $\neg\varphi = true$]

1.	$\neg\varphi$	Paint status is not normal
2.	$w \models V_\varphi^0(w) = false$	Definition of $w \models V_\varphi^0(w)$
3.	$I_{1,0}\neg\varphi$	Sensor reports problem to virus
4.	$I_{2,1}\varphi$	**Virus *always* reports "all is fine"**
5.	$w \models V_\varphi^2(w) = true$	$V_\varphi^2(w)$ **always** returns *true*

Since $I_{2,1}$ is the same in all cases, the controller, monitor, and human all accept the reading as normal operation (φ). Therefore, unknown to entities in SD^2, SD^3, and SD^4, $V_\varphi^2(w)$ and $V_{\neg\varphi}^2(w)$ cannot be evaluated. We now have all the requirements to conclude that φ is MSDND(ES) secure from SD^2. □

Howser [73] developed the above model to show formally how the Stuxnet virus operated.

10.8 Execution Monitoring Enforcement to Preserve Information Flow Security

In general, satisfaction of information flow properties is not enforceable at system runtime [74]. This rather disheartening result would tend to preclude any sort of security mechanism. In practice, though, there is a strong desire for runtime monitoring to add security capability to design-time security enforcement; an operating environment may undergo unexpected changes not anticipated in the specification of a security policy. Conceptually, runtime enforcement is monitoring a system during its execution to identify whether it deviates from the set of pre-identified desired behaviors and to take appropriate measures. This is also known as execution monitoring (EM) enforceability. Execution monitors work by monitoring the computational steps of untrusted programs and intervening whenever execution is about to violate the security policy being enforced [75]. Each of the intended behaviors (or forbidden behaviors) of the system are in fact security invariants, which can be evaluated based on a certain *security predicate*; a predicate acts as a decision engine. In order to facilitate this process, system executions are modeled as state sequences; transitions between states capture the changes to the state variables or specific events.

Some of the premier work of formally defining concurrent system executions was done by [76], where these authors introduced the concept of *property*. By definition, a property is a set of infinite sequences of states, also known as *executions*. Lamport [77] categorized system properties into two major classes: *safety* and *liveness*. Schneider argued that only safety properties can be EM enforced [74] and that EM-enforceable mechanisms terminate any execution which is about to violate the corresponding security predicate. The initial argument was that every system property is either a safety property, a liveness property, or the intersection [76]. McLean, however, argued that possibilistic properties such as information flow security properties are not preserved under refinement and they are neither safety nor liveness properties. This is simply because they are not execution sets but sets of execution sets [78, 79].

In an executing system, two questions are critical: (1) What system behaviors are runtime enforceable? (2) What are the appropriate measures to take once a violation is detected? Information flow security policy enforcement is twofold: compile time enforcement and runtime enforcement. Mechanisms implemented under these two types attempt to mitigate both implicit and explicit information flow violations. Explicit information flow violations occur when confidential information is directly (and possibly intentionally) passed to the public domain. Implicit information flows are comparatively passive in nature where (potentially) unintentional flow violations are a result of the structure and/or the interactions within the systems such as in cyber-physical systems (CPSs). Some of the recent approaches to enforce information flow security are secure-type systems [30, 31, 80], mechanisms based on petri nets [81–83], process algebra and program logic [37, 84–91], automata and predicates [35, 74, 92, 93], computation slicing [94–96], and bisimulation [89, 97–99]. For the most part, these are static and compile time techniques. Recent work has made progress in EM enforcement of non-interference for purely cyber systems through static analysis of typed systems [100–103].

Schneider's findings established the earliest threshold on EM enforceable security policies by stating that only safety properties can be enforced using a monitoring mechanism [74]. The monitor in this case, a büchi-like state automata called the *security automata*, enforces a certain security policy by terminating the target when a security violation is detected in an execution. The safety requirement, unfortunately, precludes many interesting security properties from being EM enforced. Notably, flow-based security properties are not safety properties [78, 104] and are not enforceable using Schneider's security automata. These properties are defined over sets of executions [78] rather than over sets of execution sets. Thus, the decision to terminate an execution cannot be purely based on a detected violation of a single execution. This fundamental issue has led to two alternative classes of resolutions, extending the capability of the security automata using an edit automata [35, 75, 92, 93, 105–107] or developing monitors that can enforce nonsafety properties [105, 108, 109].

In monitoring, an invalid execution $\bar{\sigma}$, from a safety property perspective, has some finite prefix $\bar{\sigma}_i$ which invalidates the security predicate, and cannot be corrected by any additional β, i.e., $\neg\hat{\wp}(\bar{\sigma}_i)$ and $\forall\beta \in \Sigma^{\infty} : \neg\hat{\wp}(\bar{\sigma}_i\beta)$. However, an edit automata can inject finite-length executions at point $\bar{\sigma}_i$ in order to rectify the violation. By definition, a "corrected" execution of this form is not a safety property. Ligatti et al. characterized such correctable properties as *infinite renewal properties* [105, 108, 109]. Thus, an infinite-length execution with infinitely many invalid prefixes still satisfies renewal property as long as it has infinitely many valid prefixes. As a result, every safety property is a renewable property with an infinitely long valid prefix, and some, not all, liveness properties are also renewable properties [109].

The above characterization of nonsafety properties as renewable properties allows them to be enforced using edit automata. An edit automata effectively can expand an invalid execution by injecting corrections. The challenge in Cyber-Physical Systems (CPSs) is to develop an edit automata that embraces the physical system's semantics, including those in the frequency and time domains and enforces nonsafety properties within the semantic context of CPSs.

Returning to the example of the pipeline,

$$\bar{\phi} = \{read_A(), write_C^{\uparrow}(f'), write_B^{\uparrow}(f'), read_A(), read_A()\} \tag{10.26}$$

an edit automata corrects the event $write_C^{\uparrow}(f')$ by inserting $write_B^{\uparrow}(f')$, forming a *compensating couple*.

A *compensating couple* is defined as two consecutive compensating state transition commands both issued by subjects in the same protection domain. Compensation for database systems is well understood and recent work has formalized compensation for complex synchronization [110]. In compensation, suppose a certain system is information flow secure up to some stable state q_k. Let's define the trace leading to q_k as $\phi[\ldots k]$. Instead of considering a single action that results in a state transition, consider a compensating couple $< \phi_1^c, \phi_2^c >$ both issued by processes in the protection domain SD^G. The trace ϕ is extended as $\phi[\ldots k+1] = \phi[\ldots k] + \phi_1^c + \phi_2^c$. Here, the superscript c is used to denote actions resulting in some change in the system; *read* commands are excluded from c.

If every extension to ϕ is considered in the above manner, the resulting trace is guaranteed to maintain a wide variety of information flow security properties. Technically, this can be stated as follows:

$$\forall k,\ \rho(\phi[\ldots k], SD^{\bar{G}}) = \rho(\pi(\phi[\ldots k], SD^G), SD^{\bar{G}})$$

which preserves the definition of noninterference. Here, $\bar{G}$ is the complement of G. This also shows that processes in $\bar{G}$ are unaware of any event occurrences in G which preserves *noninference* security. In a noninference secure system, low-level subjects are left in doubt as to whether or not any high-level event(s) has occurred. Event compensation, by definition, only allows selected traces, i.e., compensating actions, to make progress. Noncompensating traces need to be rejected by an enforcement mechanism. Further, any compensated trace is still a valid trace even after purging all high-level events.

In a purely cyber system, a compensating couple would represent *event undo*, and in a pure physical world it is an *event complement*. For event compensation to work, two (or more) processes within the same security clearance first need to establish a secure connection and exchange information regarding an event undo. Once an agreement

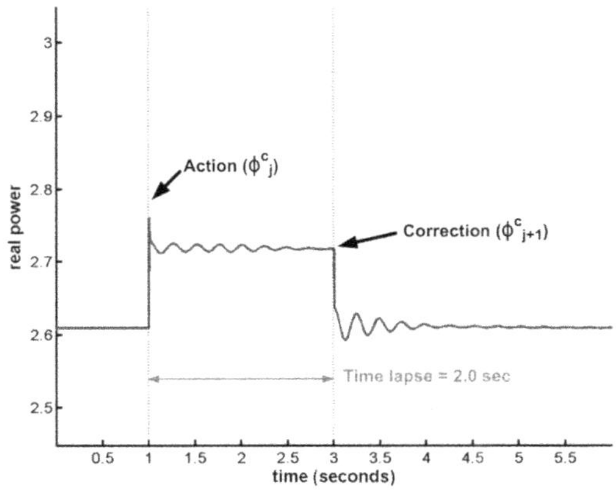

Figure 10.17 A short discrete vulnerable period.

is reached, both parties ensure that a compensating couple is executed in a timely manner. This will ensure that the processes outside the protection domain are kept unaware of the changes. At most, a third party may be able to observe a sudden pulse as conceptually shown in Figure 10.17, but these kind of discrete pulses can be a result of natural causes as well.[4] Conceptually, the period of vulnerability needs to be minimized as much as possible by a pre-compensation agreement and a commitment for action. In reality, in a CPS, the period of vulnerability is tied to the timing and frequency requirements of the entire infrastructure. Preliminary work showed how noninterference within the frequency domain for a cyber-physical power system could be verified in terms of the system actuation and (Nyquist) sampling rate [111].

In a CPS, event compensation does make sense because commands that affect the equilibrium of the system (write operations) need to be followed up with other events to maintain the system invariant. For example, if DGI_C is required to lower its gas flow due to some unavoidable reason, this intent first needs to be conveyed to DGI_B if DGI_C wants to keep the secret from leaking to other competitive distributors like DGI_A. Compensating couples help maintain the invariant while ensuring unauthorized information dissemination is prevented outside the protection domain of execution.

However, since event compensation is done for traces, the underlying execution still suffers from periodic vulnerability. For example, after the first event of a compensating couple is executed, the system reaches a vulnerable state (see Figure 10.17). At this point, the security automata steps in and enumerates through all possible next-state transition actions to decide the matching compensating action (similar to what is shown in Δf of (10.4). On top of this matching command search is the actual cyber communication and the agreement for compensation occurring between the communicating parties.

The compensating couple can also be viewed as the request–reply message passing of Section 10.4. Using the communication network, $DGI_i \in SD^G$ makes a *request*

[4] An air bubble or a clog in the pipeline, lighting striking a power transmission line etc.

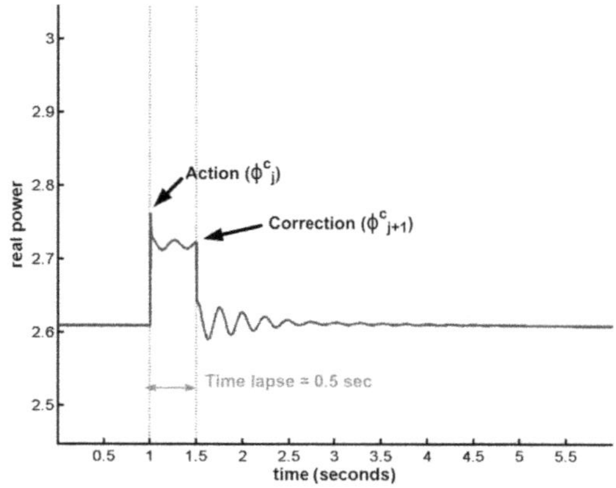

Figure 10.18 Compensating couple in smart grid.

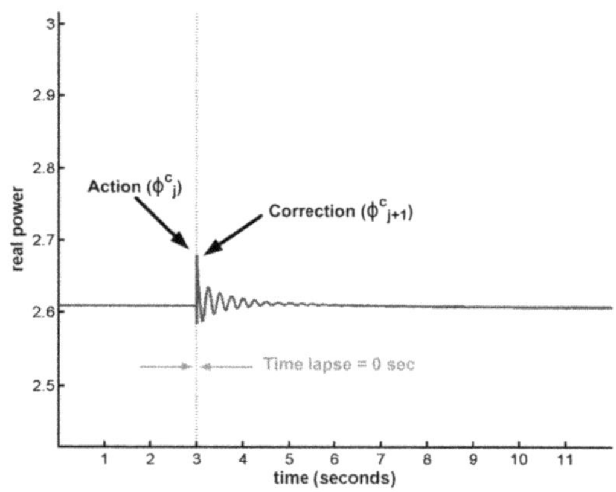

Figure 10.19 Compensating couple with overlapping frequency response.

to $DGI_j \in SD^G$ that it intends to execute a state transition command. DGI_j could decide either to acknowledge (reply) or ignore this request. Once an agreement has been made, DGI_i issues the first command of the compensating couple, which will be immediately followed by DGI_j's matching command. If ignored, the originating party could wait for a certain predefined response time and make another attempt.

The compensating couple only makes sense in the semantic context of CPSs and cannot really be arbitrarily extended. Rapid changes can cause system instability in the frequency domain and periodic changes can divulge confidentiality [15]. Figure 10.18 depicts compensation in time under an electric power system. The first event, ϕ_1^c results from a programmed recovery action by DGI_3 to correct a contingency in the power system [112]. The physical response of the system is modeled using a load–flow power systems calculation. As the time between the event (action by DGI_3) and the compensating event ϕ_2^c (action by DGI_7) shrinks (Figure 10.19), so does the vulnerable period until the event and its compensation are hidden by their overlapping frequency response.[5]

For nondeducibility security, the two actions must have frequency response characteristics that are indistinguishable in some sense, and also indistinguishable from the commonly occurring disturbances on the system. For example, if the natural frequency response to each action is known, ϕ_1^c could include both an algebraic compensation and an obfuscating dynamic compensation that is similar to ϕ_2^c, and vice versa. Then an independent observer would be unable to determine which system response is due to which event. Similarly, ϕ_1^c could include frequency modulation that spreads the spectrum of the system's dynamic response; some element of randomness could be added to push the system's response below the noise floor of the observer's measurement capabilities. This technique is frequently used in power conversion and in communication systems to meet regulatory requirements on electromagnetic noise emissions. Thus, the information flow must also be characterized in terms of the number and position of observers. Returning again to the example in Figure 10.19, the compound effect of compensation, while it can be hidden from some observers, inherently there will or could be other observers in the system that are capable of deducing the specific underlying actions [52]. Ideally, this analysis can be made compositional by developing fundamental "connectors" such as the structures of Figure 10.1 to assemble CPS architectures. With the understanding of common characteristics of CPSs, we envision that we will be able to develop other such mechanisms that can be used to safeguard confidentiality (in terms of information flow security) within CPSs in addition to the event compensation idea. Recent work has explored algorithms for locating and executing compensating couples in regularly structured networks [52] with complexities in the size of the physical interconnection graph.

10.9 The Way Forward

This chapter has painted a picture of CPS security as something far beyond just cybersecurity. To be effective, cyber-physical security must embrace both the cyber and the physical semantics of a CPS. Information flow is the unifying factor; information is present both as discrete information with the cyber and networking world, as well as observational information from the physical world. Traditional access control models are inadequate to describe CPS security as the security domains overlap. Moreover, traditional notions of high and low security levels are inadequate to describe the interaction among CPS security domains.

[5] Unfortunately, it still may be possible to recover the original convolution of compensating actions using Fourier analysis. In an interconnected physical system, steady-state relationships are algebraic and therefore the inverse of an action can be found to determine a compensating couple. However, dynamic relationships are governed by differential equations so that exact compensation is generally not possible in a dynamic sense. Each action will excite different portions of the physical system with different natural frequencies.

A critical step is to be able to reason about both the physical and cyber semantics in a uniform representation. Formal models of process algebra can represent both aspects in a way that can be reasoned about. Additionally, common semantics of cyber and physical can develop coordinated compensating events. Key to all is understanding information, rather than access.

Acknowledgment

The authors acknowledge the support of the Future Renewable Electric Energy Delivery and Management Center, a National Science Foundation supported Engineering Research Center, under grant NSF EEC-081212, the Missouri S&T Intelligent Systems Center, and the Missouri S&T Energy Research and Development Center.

References

[1] "IFIP WG 11.10 on critical infrastructure protection." [Online]. Available: www.ifip1110.org/Mission.

[2] "Department of Homeland Security Critical Infrastructure Program." [Online]. Available: www.dhs.gov/files/programs/gc_1189168948944.shtm.

[3] E. A. Lee, "Position paper: Cyber-physical systems – Are computing foundations adequate?" in *NSF Workshop On Cyber-Physical Systems: Research Motivation, Techniques and Roadmap*, Austin, TX, October 16–17, 2006.

[4] M. Bishop, *Computer Security: Art and Science*. Boston, MA: Addison-Wesley Longman, 2002.

[5] U.S. and C.P.S.O.T. Force, "Final report on the August 14th blackout in the United States and Canada," Department of Energy, Tech. Rep., 2004.

[6] K. Zetter, "Iran: Computer malware sabotaged uranium centrifuges," *Wired Magazine*, Nov. 2010.

[7] P. Kerr, J. Rollins, and C. Theohary, "The Stuxnet computer worm: Harbinger of an emerging warfare capability," Congressional Research Service Report for Congress, Dec. 2010.

[8] A. Huang, "Renewable energy system research and education at the NSF FREEDM systems center," in *Power & Energy Society General Meeting, 2009*, July 26–30, 2009, pp. 1–6.

[9] S. Massoud Amin and B. Wollenberg, "Toward a smart grid: Power delivery for the 21st century," *Power Energy Mag.*, vol. 3, no. 5, pp. 34–41, Sept.–Oct. 2005.

[10] A. Faza, S. Sedigh, and B. McMillin, "Reliability analysis for the advanced electric power grid: From cyber control and communication to physical manifestations of failure," in *Proc. Int'l Conf. Computer Safety, Reliability and Security (SAFECOMP' 09)*, Sept. 2009, pp. 257–269.

[11] K. Zetter, "Security pros question deployment of smart meters," *Wired Magazine*, Apr. 2010.

[12] C. A. Gunter, W. H. Sanders, D. E. Bakken, "Trustworthy cyber-infrastructure for power," in *TAS 2006 Workshop on Research Directions for Security and Networking*

in Critical Real-Time and Embedded Systems (CRTES '06), San Jose, CA, USA, April 4, 2006.

[13] Z. Qin, Q. Li, and M.-C. Chuah, "Defending against unidentifiable attacks in electric power grids," *IEEE Trans. Parallel Distrib. Syst.*, vol. 24, no. 10, pp. 1961–1971, 2013.

[14] H. Tang and B. McMillin, "Analysis of the security of information flow in the Advanced Electric Power Grid using Flexible Alternating Current Transmission System (FACTS)," in *Critical Infrastructure Protection*, E. Goetz and S. Shenoi, Eds. New York: Springer, 2008, pp. 43–56.

[15] H. Tang and B. McMillin, "Security property violation in CPS through timing," in *Proc. 1st Workshop on Cyber-Physical Systems*. Piscataway, NJ: IEEE Computer Society Press, 2008.

[16] D. Ditch and B. McMillin, "The security implication of multiple observers in a distributed system," in *Computer Software and Applications Conf., 2009*, vol. 2, Jul. 2009, pp. 341–346.

[17] V. Srinivasan, J. Stankovic, and K. Whitehouse, "Protecting your daily in-home activity information from a wireless snooping attack," in *UbiComp '08: Proc. 10th Int'l Conf. Ubiquitous Computing*, 2008, pp. 202–211.

[18] C. Davis, J. Tate, H. Okhravi, et al., "SCADA cyber security testbed development," in *2006 38th North American Power Symp.*, Sept. 2006, pp. 483–488.

[19] D.-J. Kang, J.-J. Lee, S.-J. Kim, and J.-H. Park, "Analysis on cyber threats to SCADA systems," Tech. Rep., Oct. 2009.

[20] J. Stamp, J. Dillinger, W. Young, and J. DePoy, "Common vulnerabilities in critical infrastructure control systems," Sandia National Laboratories, Tech. Rep., 2003.

[21] A. McIntyre, B. Becker, and R. Halbgewachs, "Security metrics for process control systems," Sandia National Laboratories, Tech. Rep.,.2007.

[22] C.-C. Liu, C.-W. Ten, and M. Govindarasu, "Cybersecurity of SCADA Systems: Vulnerability assessment and mitigation," in *2009 IEEE/PES Power Systems Conf. Expo.*, Mar. 2009, pp. 1–3.

[23] C.-W. Ten, C.-C. Liu, and M. Govindarasu, "Vulnerability assessment of cybersecurity for SCADA systems using attack trees," in *2007 IEEE Power Engineering Society General Meeting*, Jun. 2007, pp. 1–8.

[24] E. Johansson, T. Sommestad, and M. Ekstedt, "Issues of cyber security in SCADA-systems: On the importance of awareness," in *20th Int'l Conf. Exhibition on Electricity Distribution*, Jun. 2009, pp. 1–4.

[25] M. A. Harrison, W. L. Ruzzo, and J. D. Ullman, "Protection in operating systems," *Commun. ACM*, vol. 19, no. 8, pp. 461–471, 1976.

[26] R. J. Lipton and L. Snyder, "A linear time algorithm for deciding subject security," *J. ACM*, vol. 24, no. 3, pp. 455–464, 1977.

[27] D. Bell and L. LaPadula, "Secure computer systems: Mathematical foundations," MITRE Corporation, Tech. Rep., Mar. 1973.

[28] D. E. Bell and L. J. Lapadula, "Secure computer system: Unified exposition and multics interpretation," MITRE Corporation, Tech. Rep., 1976.

[29] K. J. Biba, "Integrity considerations for secure computer systems," MITRE Corporation, Tech. Rep., Apr. 1977.

[30] A. Myers and B. Liskov, "Protecting privacy in a decentralized environment," in *Proc. DARPA Information Survivability Conf. Expo.*, vol. 1, 2000, pp. 266–277.

[31] A. Sabelfeld and A. Myers, "Language-based information-flow security," in *IEEE J. Sel. Areas Commun.*, vol. 21, no. 1, 2003, pp. 5–19.

[32] J. Wittbold and D. Johnson, "Information flow in nondeterministic systems," in *Proc. 1990 IEEE Computer Society Symposium on Research in Security and Privacy*, May 1990, pp. 144–161.

[33] J. McLean, "Security models and information flow," *Proc. 1990 IEEE Computer Society Symposium on Research in Security and Privacy*, May 1990, pp. 180–187.

[34] J. A. Goguen and J. Meseguer, "Unwinding and inference control," *Proc. IEEE Computer Society Symposium on Research in Security and Privacy*, 1984, p. 75.

[35] N. Nagatou and T. Watanabe, "Run-time detection of covert channels," *in ARES '06: Proc. 1st Int'l Conf. Availability, Reliability and Security*, 2006, pp. 577–584.

[36] S. McCamant and M. D. Ernst, "Quantitative information flow as network flow capacity," in *PLDI '08: Proc. 2008 ACM SIGPLAN Conf. Programming Language Design and Implementation*, 2008, pp. 193–205.

[37] C. Bryce, J. Banatre, and D. Le Metayer, "An approach to information security in distributed systems," in *Proc. Fifth IEEE Computer Society Workshop on Future Trends of Distributed Computing Systems*, Aug. 1995, pp. 384–394.

[38] T. Fine, J. Haigh, R. O'Brien, and D. Toups, "Noninterference and unwinding for lock," in *Proc. Computer Security Foundations Workshop II*, Jun. 1989, pp. 22–28.

[39] J. Goguen and J. Meseguer, "Security policies and security models," in *Proc. 1982 IEEE Symp. Security and Privacy*, 1982, p. 11

[40] C. O'Halloran, "A calculus of information flow," in *Proc. European Symp. Research in Computer Security*, 1990, pp. 147–159.

[41] D. Sutherland, "A model of information," *in Proc. 9th National Computer Security Conference*, Sept. 1986, pp. 175–183.

[42] A. Kovacevic, *The Impact of the Russia–Ukraine Gas Crisis in South-Eastern Europe*. Oxford: Oxford Institute for Energy Studies, 2009.

[43] T. T. Gamage and B. M. McMillin, "EM enforcing information flow properties using compensating events," in *Proc. 42nd Hawaii International Conference on Systems Science*, Jan. 2009, pp. 1–7.

[44] R. Akella and B. McMillin, "Model-checking BNDC properties in cyber-physical systems," in *Computer Software and Applications Conf., 2009*, Jul. 2009, pp. 660–663.

[45] A. Huang, "Renewable energy system research and education at the NSF FREEDM systems center," in *Power & Energy Society General Meeting, 2009*, Jul. 2009, pp. 1–6.

[46] T. Paul, J. Kimball, M. Zawodniok, et al., "Unified invariants for cyber-physical switched system stability," *IEEE Trans. Smart Grid*, vol. 5, no. 1, pp. 112–120, Jan. 2014.

[47] T. Roth and B. McMillin, "Breaking nondeducible attacks on the smart grid," in *7th CRITIS Conf. Critical Information Infrastructures Security*, September 17–18, 2012, pp. 80–91.

[48] T. Roth and B. McMillin, "Physical attestation of cyber processes in the smart grid," in *Critical Information Infrastructures Security*. E. Luiijf and P. Hartel, Eds. New York: Springer, 2013, pp. 96–107.

[49] L. M. Ni, C.-W. Xu, and T. B. Gendreau, "A distributed drafting algorithm for load balancing," *IEEE Trans. Software Eng.*, vol. 11, pp. 1153–1161, 1985.

[50] Z. Zhang and M.-Y. Chow, "Incremental cost consensus algorithm in a smart grid environment," in *Power and Energy Society General Meeting, 2011*, Jul. 2011, pp. 1–6.

[51] R. Akella, H. Tang, and B. McMillin, "Analysis of information flow security in cyberphysical systems," *Int. J. Crit. Infrastruct. Prot.*, vol. 3–4, pp. 157–173, Dec. 2010.

[52] T. Gamage, T. Roth, B. McMillin, and M. Crow, "Mitigating event confidentiality violations in smart grids: An information flow security-based approach," *IEEE Trans. Smart Grid*, vol. 4, no. 3, pp. 1227–1234, Sept. 2013.

[53] A. Platzer, "Integrative challenges of cyber-physical systems," in *Workshop on Usable Verification*, November 2010.

[54] T. McEvoy and S. Wolthusen, "A formal adversary capability model for SCADA environments," in *Critical Information Infrastructures Security*, C. Xenakis and S. Wolthusen, Eds. Berlin: Springer, 2011, pp. 93–103.

[55] R. Milner, J. Parrow, and D. Walker, "A calculus of mobile processes," *Inf. Comput.*, vol. 100, no. 1, pp. 1–40, 1992.

[56] R. Milner, *Communication and Concurrency*. Upper Saddle River, NJ: Prentice Hall, 1989.

[57] R. Focardi and R. Gorrieri, "A classification of security properties for process algebras," *Int J Inf Comput. Secur.*, vol. 3, no. 1, pp. 5–33, 1994/1995.

[58] S. Crafa and S. Rossi, "A theory of noninterference for the π-calculus," in *Trustworthy Global Computing*, R. De Nicola and D. Sangiorgi, Eds. Berlin: Springer, 2005, pp. 2–18.

[59] D. Sangiorgi and D.Walker, The π-calculus: A Theory of Mobile Processes. Cambridge: Cambridge University Press, 2001.

[60] R. Bakhshi and D. Gurov, "Verification of peer-to-peer algorithms: A case study," *Electron. Notes Theor. Comput. Sci.*, vol. 181, pp. 35–47, Jun. 2007.

[61] B. Victor and F. Moller, "The mobility workbench: A tool for the pi-calculus," in *Proc. 6th Int'l Conf. Computer Aided Verification (CAV)*, 1994, pp. 428–440.

[62] B. Blanchet, M. Abadi, and C. Fournet, "Automated verification of selected equivalences for security protocols," in *Logic in Computer Science, 2005*, Jun. 2005, pp. 331–340.

[63] R. Akella, F. Meng, D. Ditch, B. McMillin, and M. Crow, "Distributed power balancing for the FREEDM system," in *1st IEEE Int'l Conf. Smart Grid Communications (SmartGridComm), 2010*, Oct. 2010, pp. 7–12.

[64] "ProVerif: Cryptographic protocol verifier in the formal model." [Online]. Available: http://prosecco.gforge.inria.fr/personal/bblanche/proverif.

[65] M. Abadi and C. Fournet, "Mobile values, new names, and secure communication," in *Proc. 28th ACM SIGPLAN-SIGACT Symp. on Principles of Programming Languages*, 2001, pp. 104–115.

[66] D. Dolev and A. Yao, "On the security of public key protocols," *IEEE Trans. Inf. Theory*, vol. 29, no. 2, pp. 198–208, Mar. 1983.

[67] S. Kremer and M. Ryan, "Analysis of an electronic voting protocol in the applied pi calculus," in *Proc. 14th European conference on Programming Languages and Systems*. 2005, pp. 186–200.

[68] G. Jansons, "Drawing by G. Jansons 'Reconstruction of Turaida Castle, 16th Century.'" [Online]. Available: www.turaida-muzejs.lv/exhibitions/the-turaida-medieval-castle/development-of-castle-construction.

[69] R. Langner, "Stuxnet: Dissecting a cyberwarfare weapon," *IEEE Secur Privacy*, vol. 9, no. 3, pp. 49–51, 2011.

[70] J. McLean, "Security models and information flow," in *Proc. Research in Security and Privacy,* 1990, May 1990, pp. 180–187.

[71] M. Bishop, *Computer Security: Art and Science*. Boston, MA: Addison-Wesley, 2003.

[72] G. Howser and B. McMillin, "A multiple security domain model of a drive-by-wire system," in *IEEE 37th Ann. Computer Software and Applications Conf.*, 2013, pp. 369–374.

[73] G. Howser and B. McMillin, "A modal model of Stuxnet attacks on cyber-physical systems: A matter of trust," in *Proc. 8th Int'l Conf. Software Security and Reliability (SERE)*, 2014, pp. 225–234.

[74] F. B. Schneider, "Enforceable security policies," *ACM Trans. Inf. Syst. Secur.*, vol. 3, no. 1, pp. 30–50, 2000.

[75] K. W. Hamlen, G. Morrisett, and F. B. Schneider, "Computability classes for enforcement mechanisms," *ACM Trans. Program. Lang. Syst.*, vol. 28, no. 1, pp. 175–205, 2006.

[76] B. Alpern and F. B. Schneider, "Defining liveness," Cornell University, Tech. Rep., 1984.

[77] L. Lamport, "Proving the correctness of multiprocess programs," *IEEE Trans. Software Eng.*, vol. SE-3, no. 2, pp. 125–143, Mar. 1977.

[78] J. McLean, "A general theory of composition for a class of 'possibilistic' properties," *IEEE Trans. Software Eng.*, vol. 22, no. 1, pp. 53–67, 1996.

[79] D. M. Volpano, "Safety versus secrecy," in *SAS '99: Proc. 6th Int'l Symp. Static Analysis*, 1999, pp. 303–311.

[80] S. Zdancewic and A. C. Myers, "Secure information flow and CPS," in *ESOP '01: Proc. 10th European Symp. Programming Languages and Systems*, 2001, pp. 46–61.

[81] W. Li, R. Wu, and H. Huang, "Colored petri nets based modeling of information flow security," in *2nd International Workshop on Knowledge Discovery and Data Mining,* Jan. 2009, pp. 681–684.

[82] V. Varadharajan, "Petri net based modelling of information flow security requirements," in *Proc. Computer Security Foundations Workshop III*, Jun. 1990, pp. 51–61.

[83] V. Varadharajan, "Hook-up property for information flow secure nets," in *Proc. Computer Security Foundations Workshop IV*, Jun. 1991, pp. 154–175.

[84] P. Ryan and S. Schneider, "Process algebra and non-interference," in *Proc. 12th IEEE Computer Security Foundations Workshop*, 1999, pp. 214–227.

[85] G. Lowe, "Quantifying information flow," in *IEEE Computer Security Foundations Workshop*, 2002, p. 18.

[86] L. Beringer and M. Hofmann, "Secure information flow and program logics," in *Proc. 20th IEEE Computer Security Foundations* Symp., Jul. 2007, pp. 233–248.

[87] W. Hunt, R. Krug, S. Ray, and W. Young, "Mechanized information flow analysis through inductive assertions," in *2008 Formal Methods in Computer-Aided Design*, Nov. 2008, pp. 1–4.

[88] K. O'Neill, M. Clarkson, and S. Chong, "Information-flow security for interactive programs," in *19th IEEE Computer Security Foundations Workshop (CSFW'06)*, 2006, pp. 12–201.

[89] F. Pottier, "A simple view of type-secure information flow in the π-calculus," in *Proc. 15th IEEE Computer Security Foundations Workshop, CSFW-15*, 2002, pp. 320–330.

[90] F. Martinelli and I. Matteucci, "An approach for the specification, verification and synthesis of secure systems," *Electron. Notes Theor. Comput. Sci.*, vol. 168, pp. 29–43, 2007.

[91] F. Martinell and I. Matteucci, "Through modeling to synthesis of security automata," *Electron. Notes Theor. Comput. Sci.*, vol. 179, pp. 31–46, 2007.

[92] H. Mantel, "Possibilistic definitions of security: An assembly kit," in *Proc. 13th IEEE Computer Security Foundations Workshop, CSFW-13*, 2000, pp. 185–199.

[93] H. Mantel, "Unwinding possibilistic security properties," in ESORICS '00: *Proc. 6th European Symp. Research in Computer Security*, 2000, pp. 238–254.

[94] W. Masri, A. Podgurski, and D. Leon, "Detecting and debugging insecure information flows," in *15th Int. Symp. Software Reliability Engineering*, Nov. 2004, pp. 198–209.

[95] A. Sen and V. Garg, "Formal verification of simulation traces using computation slicing," *IEEE Trans. Comput.*, vol. 56, no. 4, pp. 511–527, Apr. 2007.

[96] N. Mittal, A. Sen, and V. Garg, "Solving computation slicing using predicate detection," *IEEE Trans. Parallel Distrib. Syst.*, vol. 18, no. 12, pp. 1700–1713, Dec. 2007.

[97] R. v. d. Meyden and C. Zhang, "Information flow in systems with schedulers," in *Proc. 21st IEEE Computer Security Foundations Symp.*, Jun. 2008, pp. 301–312.

[98] R. Focardi, R. Gorrieri, and F. Martinelli, "Real-time information flow analysis," *IEEE J. Sel. Areas Commun.,* vol. 21, no. 1, pp. 20–35, Jan. 2003.

[99] G. Smith, "Probabilistic noninterference through weak probabilistic bisimulation," in *Proc. 16th IEEE Computer Security Foundations Workshop*, Jun.–Jul. 2003, pp. 3–13.

[100] G. L. Guernic, "Automaton-based confidentiality monitoring of concurrent programs," in *IEEE Computer Society, CSF*, 2007, pp. 218–232.

[101] P. Shroff, S. F. Smith, and M. Thober, "Dynamic dependency monitoring to secure information flow," in *IEEE Computer Society, CSF*, 2007, pp. 203–217.

[102] T. H. Austin and C. Flanagan, "Efficient purely-dynamic information flow analysis," in *PLAS*, 2009, pp. 113–124.

[103] V. N. Venkatakrishnan, W. Xu, D. C. DuVarney, and R. Sekar, "Provably correct runtime enforcement of non-interference properties," in *ICICS*, 2006, pp. 332–351.

[104] A. Zakinthinos and E. S. Lee, "A general theory of security properties," in *SP '97: Proc. 1997 IEEE Symp. Security and Privacy*, 1997, p. 94.

[105] J. Ligatti, L. Bauer, and D. Walker, "Edit automata: Enforcement mechanisms for run-time security policies," *Int J. Inf. Secur.*, vol. 4, no. 1–2, pp. 2–16, Feb. 2005.

[106] L. Bauer, J. Ligatti, and D. Walker, "More enforceable security policies," in *Proc. FLoC'02 workshop on Foundations of Computer Security,* July 2002, pp. 95–104.

[107] D. Beauquier, J. Cohen, and R. Lanotte, "Security policies enforcement using finite edit automata," *Electron. Notes Theor. Comput. Sci.*, vol. 229, no. 3, pp. 19–35, 2009.

[108] J. A. Ligatti, "Policy enforcement via program monitoring," Ph.D. dissertation, Princeton University, 2006.

[109] J. Ligatti, L. Bauer, and D. Walker, "Run-time enforcement of nonsafety policies," *ACM Trans. Inf. Syst. Secur.*, vol. 12, no. 3, Jan. 2009.

[110] C. Colombo, G. J. Pace, and P. Abela, "Compensation-aware runtime monitoring," in *Proc. 1st Int'l Conf. Runtime Verification*, 2010, pp. 214–228.

[111] Y. Sun, B. McMillin, X. F. Liu, and D. Cape, "Verifying noninterference in a cyber-physical system: The advanced electric power grid," in *Proc. 7th Int'l Conf. Quality Software (QSIC)*, Oct. 2007, pp. 363–369.

[112] A. Armbruster, M. Gosnell, B. McMillin, and M. Crow, "The maximum flow algorithm applied to the placement and distributed steady-state control of FACTS devices," in *Proc. 37th Annual North American Power Symp. (NAPS)*, Oct. 2005, pp. 77–83.

11 Bio-inspired Framework for Allocation of Protection Resources in Cyber-Physical Networks

Victor M. Preciado, Michael Zargham, Chinwendu Enyioha, Cameron Nowzari, Shuo Han, Masaki Ogura, Ali Jadbabaie, and George Pappas

11.1 Introduction

Understanding spreading processes in complex networks and designing control strategies to contain them are relevant problems in many different settings, such as epidemiology and public health [1], computer viruses [2], information propagation in social networks [3], or security of cyber-physical networks [4]. In this chapter, we describe a bio-inspired framework for optimal allocation of resources to prevent spreading processes in complex cyber-physical networks. Our motivation is recent advancement on the problem of containing epidemics in human contact networks. The most popular dynamic epidemic model is the susceptible–infected–susceptible (SIS) model [5, 6]. In this model, a given population is divided into two compartments. The first compartment, called "susceptible" (S), contains individuals who are healthy, but susceptible to becoming infected. The second compartment is called "infected" (I) and contains individuals who are infected and able to recover from the disease. Individuals can transition from S to I as they become infected, and from I to S as they recover. In addition to the SIS model, there are many other models able to model more realistic spreading processes. This is often done by adding extra compartments representing a variety of disease stages. There are many works that analyze variations of the SIS model, such as extensions to higher number of disease states [7–13], or explicit modeling of birth and mortality rates [14, 15]. Stability results are obtained in [15–17] using Lyapunov analysis, or in [18] using Volterra integral models.

In the literature, we find several approaches to model spreading mechanisms in arbitrary contact networks. The analysis of this question in arbitrary (undirected) contact networks was first studied by Wang et al. [19] for a SIS discrete-time model. In [20], Ganesh et al. studied the epidemic threshold in a continuous-time SIS spreading process. In both continuous- and discrete-time models, there is a close connection between the speed of the spreading and the spectral radius of the network (i.e., the largest eigenvalue of its adjacency matrix) [21]. Designing strategies to contain spreading processes in networks is a central problem in public health and network security. In this context, the following question is of particular interest: Given a contact network (possibly weighted and/or directed) and resources that provide partial protection (e.g., vaccines and/or antidotes), how should one distribute these resources throughout the network in a cost-optimal manner to contain the spread? This question has been addressed in several papers. Cohen et al. [22] proposed a heuristic vaccination strategy called *acquaintance immunization policy* and proved it to be much more efficient than random vaccine allocation. In [23], Borgs et al. studied theoretical limits in the control of spreads in undirected networks with a non-homogeneous distribution of antidotes. Chung et al. [24] studied a heuristic immunization strategy based on the PageRank vector of the contact graph. In the control systems literature, Wan et al. proposed in [25, 26] a method to design control strategies by allocating heterogeneous resources in undirected networks. In [27], the authors present a spectral analysis of proximity random graphs with applications to virus spread. In [28], the authors study the problem of minimizing the level of infection in an undirected network using corrective resources within a given budget. In [29], a linear-fractional optimization program was proposed to compute the optimal investment on disease awareness over the nodes of a social network to contain a spreading process. In particular, we will cover in detail the work in [30–33], where the authors developed a convex formulation to find the optimal allocation of protective resources in a network. An analysis of greedy control strategies and worst-case conditions was presented in [34]. Recent extensions include the analysis of more general epidemic models [35], competing diseases [36–38], time-switching networks [39–41], and non-Poissonian spreading and recovery rates [42, 43]. A novel data-driven optimization framework has also been recently proposed by Han et al. in [44]. A distributed framework for optimal allocation of resources has also been proposed in [45]. A novel analysis of epidemic models in arbitrary graphs using tools from positive systems can be found in [46].

In this chapter, we describe an optimization-based framework to find the optimal allocation of protection resources in *weighted and directed* networks of *nonidentical* agents in polynomial time. In our study, we consider two types of containment resources:

- *Preventive* resources able to protect (or "immunize") nodes against the spreading (such as vaccines in a viral infection process). These types of resources are allocated in nodes and/or edges of the network before the spread has reached them, so that this element is protected from the spread. The effect of this resource is to reduce the rate at which the spread can reach this element.
- *Corrective* resources able to neutralize the spreading after it has reached a node (such as antidotes in a viral infection). Notice that, in contrast with preventive resources, corrective resources are used after the spread has reached a node in the network. The effect of this type of resource is to increase the rate of recovery of an element after the spread has reached it.

In the framework herein presented, we assume there are costs associated with these resources and study the problem of finding the cost-optimal distribution of resources throughout the network to contain the spreading. The aforementioned protection resources have an associated cost that depends on the level of protection achieved by the resource. For example, the larger the investment in vaccines and antidotes, the higher the level of protection achieved by the population in which the resources have been distributed. One of the main questions in epidemiology and public health is to find the optimal allocation of preventive and corrective resources to contain an epidemic outbreak in a cost-optimal manner. An identical question can be asked in the context of designing protection strategies for other cyber-physical networks, motivating the main problem covered in this chapter:

PROBLEM 11.1 Find the cost-optimal allocation of preventive and corrective resources to protect a cyber-physical network against spreading processes.

In the field of systems reliability, there is a well-developed theory of preventive and corrective maintenance for single components or machines, but there is a lack of a theoretical framework to analyze large-scale interdependent systems [47]. The state-of-the-art in the reliability analysis of networked systems is mostly based on Markov models [47–50]. These models usually suffer from scalability issues, since the state space grows exponentially fast with the number of components under consideration. Similar Markov models have also been proposed in the analysis of disease spreading in networked populations. A rich and growing literature is arising in this context, proposing a variety of approaches to find efficient allocation of protection resources to contain an epidemic outbreak. In a series of papers, Preciado et al. developed a mathematical framework, based on dynamic systems theory and convex optimization, to find the optimal distribution of protection resources in a complex network [30, 51–53]. In particular, they showed that it is possible to *find the cost-optimal distribution of vaccines and antidotes in a (possibly weighted and directed) social network of nonidentical nodes in polynomial time* using geometric programming [31]. This framework has also been extended to find the *allocation of traffic-control resources* to find the cost-optimal traffic profile in a transportation network to contain the spread of a disease among cities [32].

11.2 Mathematical Framework

We introduce notation and preliminary results needed in our derivations. In the rest of the chapter, we denote by $\mathbb{R}^n_+$ (respectively, $\mathbb{R}^n_{++}$) the set of n-dimensional vectors with nonnegative (respectively, positive) entries. We denote vectors using boldface letters and matrices using capital letters. I denotes the identity matrix and $\mathbf{1}$ the vector of all 1s. $\Re(z)$ denotes the real part of $z \in \mathbb{C}$.

11.2.1 Graph Theory

A *weighted, directed* graph (also called a digraph) is defined as the triad $\mathcal{G} \triangleq (\mathcal{V}, \mathcal{E}, \mathcal{W})$, where (1) $\mathcal{V} \triangleq \{v_1, \ldots, v_n\}$ is a set of n nodes, (2) $\mathcal{E} \subseteq \mathcal{V} \times \mathcal{V}$ is a set of ordered pairs of nodes called directed edges, and (3) the function $\mathcal{W} : \mathcal{E} \to \mathbb{R}_{++}$ associates *positive* real weights to the edges in $\mathcal{E}$. By convention, we say that (v_j, v_i) is an edge from v_j pointing toward v_i. We define the in-neighborhood of node v_i as $\mathcal{N}_i^{in} \triangleq$

$\{j : (v_j, v_i) \in \mathcal{E}\}$, i.e., the set of nodes with edges pointing toward v_i. We define the weighted *in-degree* (resp., out-degree) of node v_i as $\deg_{in}(v_i) \triangleq \sum_{j \in \mathcal{N}_i^{in}} \mathcal{W}((v_j, v_i))$ (resp., $\deg_{out}(v_i) \triangleq \sum_{j \in \mathcal{N}_i^{out}} \mathcal{W}((v_j, v_i))$); in other words, the weighted degrees are the sum of the edge weights attached to a node.

The *adjacency matrix* of a weighted, directed graph $\mathcal{G}$, denoted by $A_{\mathcal{G}} = [a_{ij}]$, is an $n \times n$ matrix defined entry-wise as $a_{ij} = \mathcal{W}((v_j, v_i))$ if edge $(v_j, v_i) \in \mathcal{E}$, and $a_{ij} = 0$ otherwise [54]. Given an $n \times n$ matrix M, we denote by $\mathbf{v}_1(M), \ldots, \mathbf{v}_n(M)$ and $\lambda_1(M), \ldots, \lambda_n(M)$ the set of eigenvectors and corresponding eigenvalues of M, respectively, where we order them in decreasing order of their real parts, i.e., $\Re(\lambda_1) \geq \Re(\lambda_2) \geq \ldots \geq \Re(\lambda_n)$. We call $\lambda_1(M)$ and $\mathbf{v}_1(M)$ the dominant eigenvalue and eigenvector of M. The spectral radius of M, denoted by $\rho(M)$, is the maximum modulus of an eigenvalue of M.

In this chapter, we only consider graphs with positively weighted edges; hence, the adjacency matrix of a graph is always nonnegative. Conversely, given an $n \times n$ nonnegative matrix A, we can associate a directed graph $\mathcal{G}_A$ such that A is the adjacency matrix of $\mathcal{G}_A$. Finally, a nonnegative matrix A is *irreducible* if and only if its associated graph $\mathcal{G}_A$ is strongly connected.

In our derivations, we use the Perron–Frobenius lemma, from the theory of nonnegative matrices [55]:

LEMMA 11.1 (Perron–Frobenius) Let M be a nonnegative, irreducible matrix. Then, the following statements about its spectral radius, $\rho(M)$, hold:

(a) $\rho(M) > 0$ is a simple eigenvalue of M,
(b) $M\mathbf{u} = \rho(M)\mathbf{u}$, for some $\mathbf{u} \in \mathbb{R}^n_{++}$, and
(c) $\rho(M) = \inf\{\lambda \in \mathbb{R} : M\mathbf{u} \leq \lambda\mathbf{u} \text{ for } \mathbf{u} \in \mathbb{R}^n_{++}\}$.

REMARK 11.1 Since a matrix M is *irreducible* if and only if its associated digraph $\mathcal{G}_M$ is strongly connected, the above lemma also holds for the spectral radius of the adjacency matrix of any (positively) weighted, strongly connected digraph.

COROLLARY 11.1 Let M be a nonnegative, irreducible matrix. Then, its eigenvalue with the largest real part, $\lambda_1(M)$, is real, simple, and equal to the spectral radius $\rho(M) > 0$.

11.2.2 Stochastic Spreading Model in Arbitrary Networks

We formulate the simplest version of the problem under consideration using a generalization of the SIS model, popularly used to model spreading dynamics in networks, such as the propagation of diseases in a networked population [21, 56, 57] or malware in a computer network [58–61]. This generalization of the SIS model, called the *heterogeneous networked SIS model* (HeNeSIS), is a continuous-time networked Markov process in which each node in the network can be in one of two possible states, namely susceptible or infected. In the context of systems reliability, each node in the networked Markov process represents a component in a networked infrastructure, and the *susceptible* and *infected* states correspond to *operational* and *faulty* states of these components, respectively. Over time, each node $v_i \in \mathcal{V}$ in the networked Markov

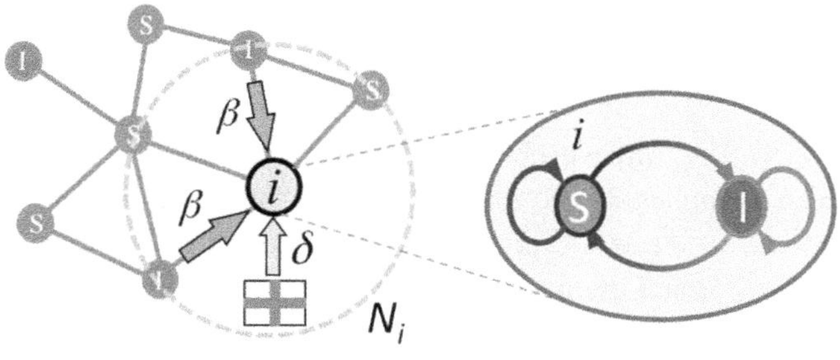

Figure 11.1 Networked Markov process with two states per node, corresponding to the HeNeSIS spreading model. Infected (resp., susceptible) nodes contain the letter "I" (resp., "S") inside.

process can change its state according to a stochastic process parametrized by (1) the edge propagation rate β_{ij}, and (2) its node recovery rate δ_i. In what follows, we shall describe the dynamics of the HeNeSIS model (Figure 11.1).

The dynamics of the HeNeSIS model can be described as follows. The state of node v_i at time $t \geq 0$ is a binary random variable $X_i(t) \in \{0, 1\}$. The state $X_i(t) = 0$ (resp., $X_i(t) = 1$) indicates that node v_i is in the susceptible (resp., infected) state. We define the vector of states as $X(t) = (X_1(t), \ldots, X_n(t))^T$. The state of a node can experience two possible stochastic transitions:

1. Assume node v_i is in the susceptible state at time t. This node can switch to the infected state during the time interval $[t, t + \Delta t)$ with a probability that depends on: (1) the propagation rates $\{\beta_{ij}, \text{for } j \in \mathcal{N}_i^{in}\}$, and (2) the states of its in-neighbors $\{X_j(t), \text{for } j \in \mathcal{N}_i^{in}\}$. Formally, the probability of this transition is given by

$$\Pr(X_i(t + \Delta t) = 1 | X_i(t) = 0, X(t)) = \sum_{j \in \mathcal{N}_i^{in}} \beta_{ij} X_j(t) \Delta t + o(\Delta t) \tag{11.1}$$

where $\Delta t > 0$ is considered an asymptotically small time interval.

2. Assuming node v_i is infected, the probability of v_i recovering back to the susceptible state in the time interval $[t, t + \Delta t)$ is given by

$$\Pr(X_i(t + \Delta t) = 0 | X_i(t) = 1, X(t)) = \delta_i \Delta t + o(\Delta t) \tag{11.2}$$

where $\delta_i > 0$ is the curing rate of node v_i.

In the context of failure propagation in networked infrastructure, β_{ij} represents the Poisson rate at which a failure in the element located at node v_j propagates to the element in node v_i. Similarly, δ_i represents the Poisson rate at which a fault at component v_i is cleared. This HeNeSIS model is therefore a continuous-time Markov process with 2^n states in the limit $\Delta t \to 0^+$. Unfortunately, the exponentially increasing state space makes this model hard to analyze for large-scale networks. Using the Kolmogorov forward equations and a mean-field approach [21], one can approximate the dynamics of the spreading process using a system of n ordinary differential equations, as follows. Let us define $p_i(t) \triangleq \Pr(X_i(t) = 1) = E(X_i(t))$, i.e., the probability of node v_i being infected (or faulty) at time t. Hence, the Markov differential equation [62] for the state $X_i(t) = 1$ is:

$$\frac{dp_i(t)}{dt} = (1 - p_i(t)) \sum_{j=1}^{n} \beta_{ij} p_j(t) - \delta_i p_i(t) \tag{11.3}$$

Considering $i = 1, \ldots, n$, we obtain a system of nonlinear differential equations with a complex dynamics. In the following, we derive a sufficient condition for the spreading process to die out exponentially fast. Let us define the vector $\mathbf{p}(t) \triangleq (p_1(t), \ldots, p_n(t))^T$, and the matrices $B_{\mathcal{G}} \triangleq [\beta_{ij}]$, $D \triangleq \operatorname{diag}(\delta_i)$. Notice that $B_{\mathcal{G}}$ is the weighted adjacency matrix of a weighted, directed graph with edge-weight function $\mathcal{W}(v_j \to v_i) = \beta_{ij}$; in other words, the weights of the directed link from v_j to v_i is β_{ij}. The ordinary differential equation (ODE) under consideration presents an equilibrium point at $\mathbf{p}^* = 0$, called the disease-free (or fault-free) equilibrium. A stability analysis of this ODE around the equilibrium provides the following stability result [30]:

PROPOSITION 11.1 Consider the nonlinear HeNeSIS model in (11.3) and assume $\beta_{ij}, \delta_i > 0$. Then, if the eigenvalue with the largest real part of $B_{\mathcal{G}} - D$ satisfies

$$\Re\left[\lambda_1(B_{\mathcal{G}} - D)\right] \leq -\varepsilon \tag{11.4}$$

for some $\varepsilon > 0$, the disease-free equilibrium ($\mathbf{p}^* = \mathbf{0}$) is globally exponentially stable, i.e., $\|\mathbf{p}(t)\| \leq \|\mathbf{p}(0)\| K \exp(-\varepsilon t)$, for some $K > 0$.

11.3 A Quasiconvex Framework for Optimal Resource Allocation

Assume that the fault propagation and recovery rates, β_{ij} and δ_i, are adjustable by allocating protection resources on the edges and nodes of the networked Markov process. We consider two types of protection resources: (1) preventive resources (e.g., vaccinations in the case of disease spreading), and (2) corrective resources (e.g., antidotes). We assume that the propagation rate β_{ij} can be reduced using preventive resources. Also, allocating corrective resources at node v_i increases the recovery rate δ_i. We assume that we are able to simultaneously modify the fault propagation and recovery rates of v_i within feasible intervals $0 < \underline{\beta}_{ij} \leq \beta_{ij} \leq \bar{\beta}_{ij}$ and $0 < \underline{\delta}_i \leq \delta_i \leq \bar{\delta}_i < \Delta$, where Δ is a uniform upper bound in the achievable recovery rate, which is assumed to be known a priori. The particular values of β_{ij} and δ_i depend on the amount of preventive and corrective resources allocated at node v_i. We consider that protection resources have an associated cost. We define two cost functions, the prevention (or vaccination) cost function $f_{ij}(\beta_{ij})$ and the correction (or antidote) cost function $g_i(\delta_i)$, that account for the cost of tuning the fault propagation and recovery rates to $\beta_{ij} \in \left[\underline{\beta}_{ij}, \bar{\beta}_{ij}\right]$ and $\delta_i \in \left[\underline{\delta}_i, \bar{\delta}_i\right]$, respectively.

In this context of protection design, one can study a type of resource allocation problem, called the *budget-constrained* allocation problem. In the *budget-constrained* problem we are assigned a total budget C to invest in protection resources and we need to find the best allocation of preventive and/or corrective resources to maximize a measure of the network resilience. In [30] and [31], the authors proposed a measure of the network resilience based on the norm of the vector of probabilities of fault probabilities, $\|\mathbf{p}(t)\|$. In particular, the exponential rate of decay of such a vector is a measure of the ability of the networked infrastructure to recover from random failures. In other words, assuming that we are able to control the system to satisfy the condition $\|\mathbf{p}(t)\| \leq \|\mathbf{p}(0)\| K \exp(-\varepsilon t)$, the exponential decay rate ε measures the ability of the networked infrastructure to "self-heal" from random contingencies.

Based on Proposition 11.1, the decay rate of an epidemic outbreak is determined by ε in (11.4). Thus, given a budget C, the budget-constrained allocation problem is formulated as follows:

PROBLEM 11.2 (*Budget-constrained allocation*) Given the elements (1) a directed network $\mathcal{G} = (\mathcal{V}, \mathcal{E})$ representing failure dependencies between components in a networked infrastructure; (2) a set of cost functions $f_{ij}(\beta_{ij}), g_i(\delta_i)$, (3) bounds on the fault propagation and recovery rates $0 < \underline{\beta}_{ij} \leq \beta_{ij} \leq \overline{\beta}_{ij}$ and $0 < \underline{\delta}_i \leq \delta_i \leq \overline{\delta}_i$, and (4) a total budget C, find the cost-optimal distribution of (preventive and corrective) protection resources to maximize the exponential decay rate ε.

Based on Proposition 11.1, we can state this problem as the following optimization program:

$$\underset{\varepsilon, \{\beta_{ij}\}_{(j,i)\in\mathcal{E}}, \{\delta_i\}_{i=1}^n}{\text{maximize}} \quad \varepsilon \tag{11.5}$$

$$\text{subject to } \Re\left[\lambda_1 (B_{\mathcal{G}} - D)\right] \leq -\varepsilon \tag{11.6}$$

$$\sum_{(j,i)\in\mathcal{E}} f_{ij}(\beta_{ij}) + \sum_{i\in\mathcal{V}} g_i(\delta_i) \leq C \tag{11.7}$$

$$\underline{\beta}_{ij} \leq \beta_{ij} \leq \overline{\beta}_{ij}, \ (j,i) \in \mathcal{E}; \ \underline{\delta}_i \leq \delta_i \leq \overline{\delta}_i, \ i \in \mathcal{V} \tag{11.8}$$

where (11.7) is the budget constraint.

In the following section, we propose an approach to find the optimal budget-constraint allocation in polynomial time for weighted and directed contact networks, under certain convexity assumptions on the cost functions f_{ij} and g_i.

11.3.1 A Geometric Programming Approach

We propose a convex formulation to solve the budget constraint in weighted, directed networks using *geometric programming* (GP) [63]. Geometric programs are a type of quasiconvex optimization problem that can be easily transformed into convex programs and solved in polynomial time. We start our exposition by briefly reviewing some concepts used in our formulation. Let $x_1, \ldots, x_n > 0$ denote n decision variables and define $\mathbf{x} \triangleq (x_1, \ldots, x_n) \in \mathbb{R}^n_{++}$. In the context of GP, a *monomial* $h(\boldsymbol{x})$ is defined as a real-valued function of the form $h(\mathbf{x}) \triangleq d x_1^{a_1} x_2^{a_2} \ldots x_n^{a_n}$ with $d > 0$ and $a_i \in \mathbb{R}$. A *posynomial* function $q(\mathbf{x})$ is defined as a sum of monomials, i.e., $q(\mathbf{x}) \triangleq \sum_{k=1}^K c_k x_1^{a_{1k}} x_2^{a_{2k}} \ldots x_n^{a_{nk}}$, where $c_k > 0$. Posynomials are closed under addition, multiplication, and nonnegative scaling. A posynomial can be divided by a monomial, with the result a posynomial.

A GP is an optimization problem of the form (see [64] for a comprehensive treatment):

$$\begin{aligned} &\text{minimize } f(\mathbf{x}) \\ &\text{subject to } q_i(\mathbf{x}) \le 1,\ i = 1, \dots, m \\ &\qquad\qquad h_i(\mathbf{x}) = 1,\ i = 1, \dots, p \end{aligned} \tag{11.9}$$

where q_i are posynomial functions, h_i are monomials, and f is a convex function in log-scale.[1] A GP is a quasiconvex optimization problem [63] that can be transformed to a convex problem. This conversion is based on the logarithmic change of variables $y_i = \log x_i$, and a logarithmic transformation of the objective and constraint functions (see [64] for details on this transformation). After this transformation, the GP in (11.9) takes the form

$$\begin{aligned} &\text{minimize } F(\mathbf{y}) \\ &\text{subject to } Q_i(\mathbf{y}) \le 0,\ i = 1, \dots, m \\ &\qquad\qquad \mathbf{b}_i^T \mathbf{y} + \log d_i = 0,\ i = 1, \dots, p \end{aligned} \tag{11.10}$$

where $Q_i(\mathbf{y}) \triangleq \log q_i(\exp \mathbf{y})$ and $F(\mathbf{y}) \triangleq \log f(\exp \mathbf{y})$. Also, assuming that $h_i(\mathbf{x}) \triangleq d_i x_1^{b_{1,i}} x_2^{b_{2,i}} \dots x_n^{b_{n,i}}$, we obtain the equality constraint above, with $\mathbf{b}_i \triangleq (b_{1,i}, \dots, b_{n,i})$, after the logarithmic change of variables. Notice that, since $f(\mathbf{x})$ is convex in log-scale, $F(\mathbf{y})$ is a convex function. Also, since q_i is a posynomial (therefore, convex in log-scale), Q_i is also a convex function. In conclusion, (11.10) is a convex optimization problem in standard form and can be efficiently solved in polynomial time [63].

To solve Problem 11.2 using GP, it is convenient to define the "complementary" recovery rate $\widehat{\delta}_i \triangleq \Delta - \delta_i$. We can also define a "complementary" recovery cost function as $\widehat{g}_i(\widehat{\delta}_i) \triangleq g_i(\delta_i) = g_i(\Delta - \widehat{\delta}_i)$; in other words, instead of defining the recovery cost in terms of the recovery rate, δ_i, we define it in terms of its complementary value, $\widehat{\delta}_i$. Hence, Problem 11.2 can be formulated as a GP if the cost functions $f_{ij}(\beta_{ij})$ and $\widehat{g}_i(\widehat{\delta}_i)$ are posynomials (see [64, section 8], for a treatment about the modeling abilities of monomials and posynomials). Therefore, the total cost function $\sum_{(j,i)\in\mathcal{E}} f_{ij}(\beta_{ij}) + \sum_{i\in\mathcal{V}} \widehat{g}_i(\widehat{\delta}_i)$ is also a posynomial. In [31], Problem 11.2 is transformed into a GP using results from the theory of nonnegative matrices and the Perron–Frobenius lemma. The resulting formulation is described below [31]:

THEOREM 11.1 Consider the following elements:

1. A directed graph $\mathcal{G} = (\mathcal{V}, \mathcal{E})$ representing failure dependencies in a networked infrastructure.
2. Posynomial cost functions $\{f_{ij}(\beta_{ij})\}_{(j,i)\in\mathcal{E}}$ and $\{\widehat{g}_i(\widehat{\delta}_i)\}_{i\in\mathcal{V}}$.
3. Bounds on the failure propagation and recovery rates $0 < \underline{\beta}_{ij} \le \beta_{ij} \le \overline{\beta}_{ij}$ and $0 < \underline{\delta}_i \le \delta_i \le \overline{\delta}_i < \Delta$.
4. A maximum budget C to invest in protection resources.

[1] Geometric programs in standard form are usually formulated assuming $f(\mathbf{x})$ is a posynomial. In our formulation, we assume that $f(\mathbf{x})$ is in the broader class of convex functions in logarithmic scale.

Then, the optimal allocation of protection resources on edge (v_j, v_i) is given by $f_{ij}\left(\beta_{ij}^*\right)$ and the optimal allocation of recovery resources at node v_i is $g_i\left(\Delta - \widehat{\delta}_i^*\right)$, where $\beta_{ij}^*, \widehat{\delta}_i^*$ *are the optimal solution of the following GP*:

$$\underset{\widehat{\lambda}, \{\beta_{ij}\}_{(j,i)\in\mathcal{E}}, \{u_i, \widehat{\delta}_i\}_{i\in\mathcal{V}}}{\text{minimize}} \quad \widehat{\lambda} \tag{11.11}$$

$$\text{subject to} \sum_{j=1}^{n} \beta_{ij} u_j + \widehat{\delta}_i u_i \leq \widehat{\lambda} u_i \tag{11.12}$$

$$\sum_{(j,i)\in\mathcal{E}} f_{ij}\left(\beta_{ij}\right) + \sum_{i\in\mathcal{V}} \widehat{g}_i\left(\widehat{\delta}_i\right) \leq C \tag{11.13}$$

$$\underline{\beta}_{ij} \leq \beta_{ij} \leq \overline{\beta}_{ij}, \ (j,i) \in \mathcal{E}; \ \Delta - \overline{\delta}_i \leq \widehat{\delta}_i \leq \Delta - \underline{\delta}_i, \ i \in \mathcal{V} \tag{11.14}$$

It is easy to verify that the above formulation is a GP; hence, it can be efficiently transformed into a convex optimization program and solved in polynomial time [31]. The tools presented are illustrated with a numerical simulation involving the worldwide air transportation network.

11.3.2 Controlling Epidemic Outbreaks in a Transportation Network

We apply the above results to the design of a cost-optimal protection strategy against epidemic outbreaks that propagate through the air transportation network [65]. We analyze real data from the worldwide air transportation network and find the optimal distribution of vaccines and antidotes to prevent the viral spread of an epidemic outbreak. We consider the budget-constrained problems in our simulations. We limit our analysis to an air transportation network spanning the major airports in the world; in particular, we consider only airports having an incoming traffic greater than 10 million passengers per year (MPPY). There are 56 such airports worldwide and they are connected via 1843 direct flights, which we represent as directed edges in a graph. To each directed edge (i, j), we assign a “contact” weight, a_{ji}, equal to the number of passengers taking that flight throughout the year[2] (in MPPY units).

In this problem, we assume that allocating preventive resources (e.g., vaccines) at a particular airport reduces the propagation rate of all the incoming links in proportion to the incoming traffic. In other words, we assume that $\beta_{ij} = \beta_i a_{ij}$, where a_{ij} is the number of passengers per year (in MPPY) that travel from airport v_j to airport v_i, and β_i is a scaling factor that depends on the destination airport only. In our simulations, we consider the following cost functions $f_{ij}\left(\beta_{ij}\right) = f\left(\beta_i\right)$ and $g_i\left(\delta_i\right) = g\left(\delta_i\right)$, where f and g are the following functions (plotted in Figure 11.2):

[2] Although we could have chosen other functions of the traffic to design these contact weights, we illustrate our framework using this simple set of weights. Using different, possibly nonlinear functions, to generate these weights does not influence the tractability of our framework.

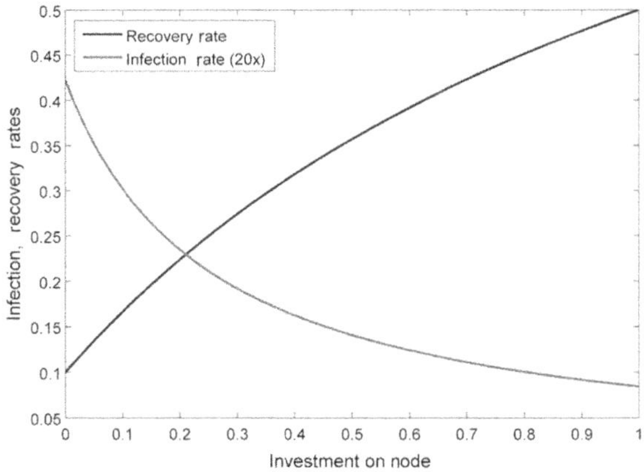

Figure 11.2 Propagation rate (decreasing, convex function, and multiplied by 20 to improve visualization) and recovery rate (increasing, concave function) achieved at node v_i after an investment in protection (in abscissas) is made on that node.

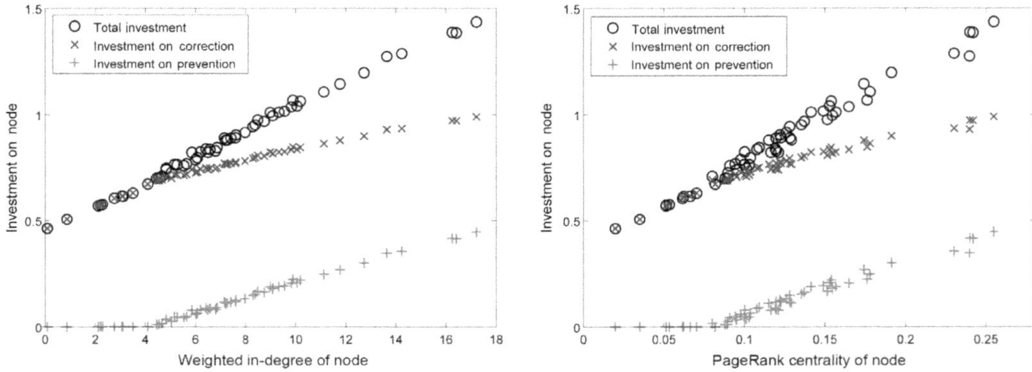

Figure 11.3 Results from the budget-constrained allocation problem. We present scatter plots representing the investments in different types of protection per node versus the in-degrees (left figure) and the PageRank centralities (right figure).

$$f_i\left(\beta_i\right) = \frac{\beta_i^{-1} - \bar{\beta}_i^{-1}}{\underline{\beta}_i^{-1} - \bar{\beta}_i^{-1}}, \quad g_i\left(\delta_i\right) = \frac{\left(1-\delta_i\right)^{-1} - \left(1-\underline{\delta}_i\right)^{-1}}{\left(1-\bar{\delta}_i\right)^{-1} - \left(1-\underline{\delta}_i\right)^{-1}} \tag{11.15}$$

Notice that as we increase the amount invested in vaccines, the propagation rate of that node is reduced from $\bar{\beta}_i$ to $\underline{\beta}_i$. Similarly, as we increase the amount invested in antidotes at node v_i, the recovery rate grows from $\underline{\delta}_i$ to $\bar{\delta}_i$. Notice that both cost functions present diminishing marginal benefits on investment.

Using the air transportation network and the cost functions specified above, we solve the budget-constrained allocation problem using the geometric programs in Theorem 11.1. In the left and right subplots in Figure 11.3, we present scatter plots comparing the distribution of resources over each airport (y-axis) versus the in-degrees

(x-axis in the left plot) and the PageRank[3] centralities (x-axis in the right plot) of the nodes in the network [66]. In both scatter plots, the ordinates represent investments on prevention (+'s), correction (x's), and total investment (the sum of prevention and correction investments, in black circles) for each airport under consideration. We observe a nontrivial pattern in the allocation of investments for protections. In particular, for airports with incoming traffic less than 4 MPPY, only corrective resources are needed. Airports with incoming traffic over 4 MPPY receive both preventive and corrective resources. In the right subplot in Figure 11.3, we include a scatter plot of the amount invested on prevention and correction for each airport versus its PageRank centrality in the transportation network. We observe that there is a strong correlation between the network centrality measures and the level of investment per node. In particular, there is an almost affine relationship between the total level of investment (black circles in Figure 11.3, left) and the incoming traffic of an airport. Furthermore, there is a clear piecewise linear affine relationship between the levels of investment in prevention and correction (Figure 11.3, left). Similar relationships also hold when comparing the levels of investment versus the PageRank centralities in the airport network (Figure 11.3, right).

Notice that the above distribution of protection resources corresponds to the particular cost functions chosen for our simulations. Changes in these cost functions allow us to observe interesting phenomena in the optimal distribution of protection resources, such as airports with a zero protection assignment at optimality, or a distribution of resources with a negative correlation with centrality measures. For example, it is possible to build cases in which nodes with low centrality (e.g., nodes with low incoming traffic and PageRank) are assigned at optimality a higher level of protection than more central nodes [67].

11.4 Toward a General Framework for Network Protection

The framework presented in this chapter has been recently extended in several directions. In what follows, we briefly describe the following extensions: (1) a generalized framework to cover more realistic epidemic models (beyond SIS); (2) a novel data-driven framework able to handle network uncertainties; and (3) an analysis tool that allows us to study non-Poissonian transmission and recovery rates.

11.4.1 Generalized Epidemic Models

Nowzari et al. [35, 68] recently studied a model of spreading, called the generalized susceptible–exposed–infected–vigilant (G-SEIV) model (Figure 11.4), that generalizes many of the models in the literature, including SIS, SIR, SIRS, SEIR, SEIV, SEIS, and SIV [14, 69]. This model has two infectious states, *infected* (*I*) and *exposed* (*E*),

[3] The PageRank vector $\mathbf{r}$, before normalization, can be computed as $\mathbf{r} = (I - \alpha A_{\mathcal{G}} \text{diag}(1/\deg_{out}(v_i)))^{-1}\mathbf{1}$, where $\mathbf{1}$ is the vector of all 1s and α is typically chosen to be 0.85.

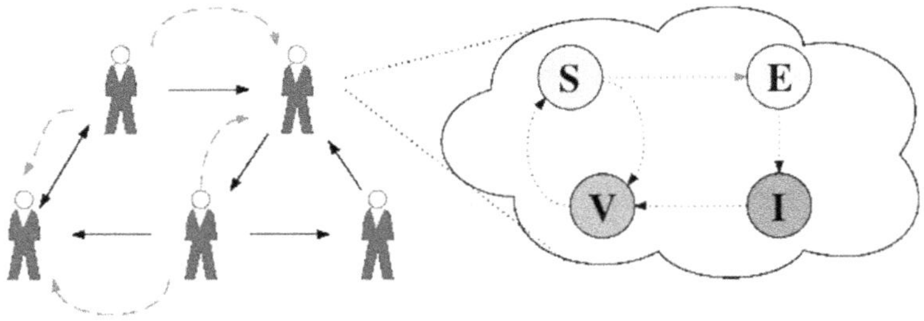

Figure 11.4 G-SEIV model in a network of individuals. ©2017 IEEE. Reprinted, with permission, from [35].

that allow us to model human behavioral changes. An individual is in the exposed state if she is infected and contagious, but not yet aware that she is sick (i.e., in an asymptomatic incubation period). Individuals in the infected state are infected and aware of the disease, which induces a different behavior. For instance, a person knowingly infected with a disease may have much less contact with others, yielding less chance of spreading the infection. The dynamics of this model is described below. The G-SEIV model also includes a *vigilant* (V) state, which represents healthy individuals being aware of the disease being spread. Hence, individuals in the vigilant state are more careful in their social contacts and less likely to be infected.

Let us denote by $[S_i(t), E_i(t), I_i(t), V_i(t)]^T$ the probability vector associated with node i being in each one of these states: susceptible, exposed, infected, or vigilant, respectively. Using a mean-field approximation, the dynamics of the G-SEIV model can be described as:

$$\begin{aligned}
\dot{S}_i(t) &= \gamma_i V_i(t) - \theta_i S_i(t) - S_i(t)\left(\sum_{j\in\mathcal{N}_i^{\text{in}}} \beta_i^E E_j(t) + \beta_i^I I_j(t)\right)\\
\dot{E}_i(t) &= S_i(t)\left(\sum_{j\in\mathcal{N}_i^{\text{in}}} \beta_i^E E_j(t) + \beta_i^I I_j(t)\right) - \varepsilon_i E_i(t)\\
\dot{I}_i(t) &= \varepsilon_i E_i(t) - \delta_i I_i(t)\\
\dot{V}_i(t) &= \delta_i I_i(t) + \theta_i S_i(t) - \gamma_i V_i(t)
\end{aligned} \tag{11.16}$$

Using nonlinear analysis techniques, Nowzari et al. derived the following necessary and sufficient condition for the disease to die out exponentially fast:

THEOREM 11.2 (*Conditions for stability of disease-free equilibrium*) The disease-free equilibrium of the G-SEIV model is globally exponentially stable if and only if the matrix

$$Q = \begin{bmatrix} TB^E A_{\mathcal{G}} - E & TB^I A_{\mathcal{G}} \\ E & -D \end{bmatrix} \tag{11.17}$$

is Hurwitz, where $B^E = \text{diag}\left(\beta^E\right), B^I = \text{diag}\left(\beta^I\right), D = \text{diag}\left(\delta\right), E = \text{diag}\left(\varepsilon\right)$, $T = \text{diag}\left(\frac{\gamma}{\theta+\gamma}\right)$.

The above result can be used to mitigate, or eliminate completely, the spreading of the disease. In [35], the authors considered three types of resources that are available to control the disease: corrective resources (e.g., antidotes), preventive resources (e.g., vaccines), and preemptive resources (e.g., awareness campaigns and/or limiting traffic). Under mild conditions on the cost functions of these resources, the authors were able to bound the rate of spreading of the undesired disease.

11.4.2 Data-Driven Allocation

Although current vaccination strategies assume full knowledge about the network structure and spreading rates, in most practical applications this information is only partially known. To elaborate on this point, let us consider the following setup. Assume that each node in the network represents subpopulations (e.g., city districts) connected by edges that are determined by commuting patterns between districts. In practice, one can use traffic information and geographical proximity to infer the existence of an edge connecting districts. For example, in Figure 11.5, we represent such a network for those districts in West Africa affected by the 2014 Ebola outbreak. On the other hand, it is very challenging to use this information to estimate the contact rates between subpopulations. Inspired by this example, we considered in [44] a networked SIS model taking place in a contact network of unknown contact rates. To extract information about these unknown rates, we assumed that we have access to time series describing the evolution of the spreading process observed from a collection of sensor nodes during a finite time interval.

In contrast to current network identification heuristics, in which a single network is identified to explain the observed data, the authors in [44] developed a robust optimization framework in which an uncertainty set containing all networks that are coherent with empirical observations is defined. This characterization of the uncertainty set of networks is tractable in the context of *conic geometric programming*, recently proposed by Chandrasekaran and Shah [70]. In this context, the authors were able to efficiently find the optimal allocation of resources to control the worst-case spread that can take place in the uncertainty set of networks. In order to extract information about the contact rates, the authors considered two different sources of information that are usually available in epidemiological problems. These sources can be classified as (1) *prior information* about the network topology and parameters of the disease, and (2) *empirical observations* about the spreading dynamics. In particular, one can consider the following pieces of prior information:

1. Assume that the sparsity pattern of the contact matrix $B_{\mathcal{G}}$ is given, although its entries are unknown. This piece of information may be inferred from geographical proximity, commuting patterns, or the presence of transportation links connecting subpopulations.

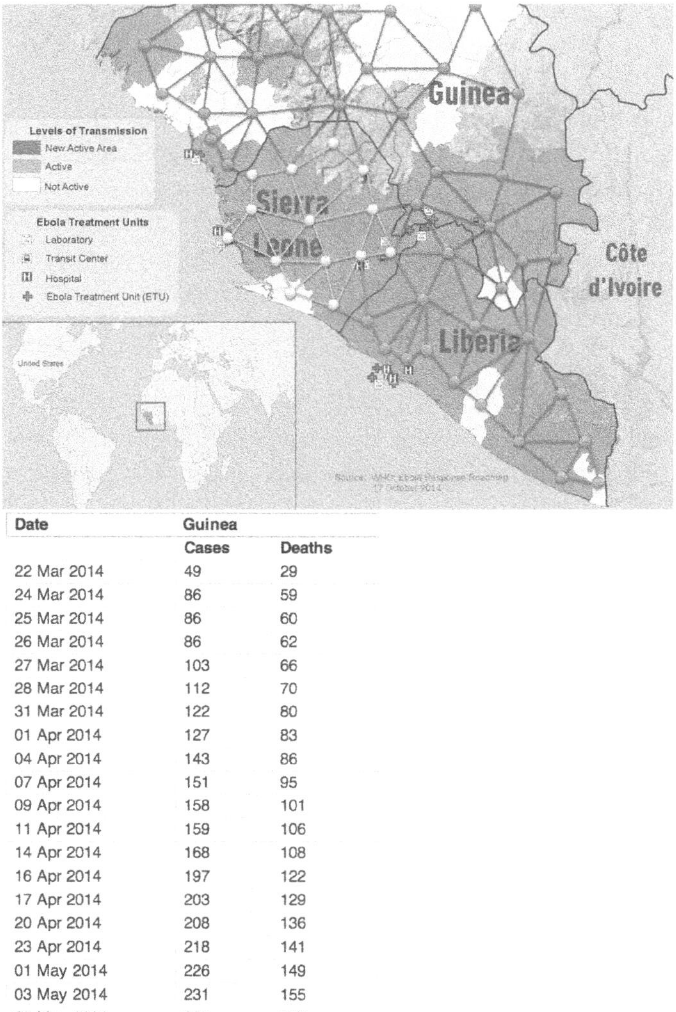

Date	Guinea	
	Cases	Deaths
22 Mar 2014	49	29
24 Mar 2014	86	59
25 Mar 2014	86	60
26 Mar 2014	86	62
27 Mar 2014	103	66
28 Mar 2014	112	70
31 Mar 2014	122	80
01 Apr 2014	127	83
04 Apr 2014	143	86
07 Apr 2014	151	95
09 Apr 2014	158	101
11 Apr 2014	159	106
14 Apr 2014	168	108
16 Apr 2014	197	122
17 Apr 2014	203	129
20 Apr 2014	208	136
23 Apr 2014	218	141
01 May 2014	226	149
03 May 2014	231	155

Figure 11.5 Network of districts in West Africa affected by the 2014 Ebola outbreak. Courtesy of the World Health Organization.

2. Assume that upper and lower bounds on the spreading rates associated to each edge, i.e., $\beta_{ij} \in \left[\underline{\beta}_{ij}, \overline{\beta}_{ij}\right]$, for all $(i, j) \in \mathcal{E}$, are available. This could be inferred from traffic densities and subpopulation sizes.
3. In practice, each district contains a large number of individuals. Therefore, one can use the average recovery rate in the absence of vaccination as an estimation of the nodal recovery rate. We denote this "natural" recovery rate by δ_i^0, and assume it to be known.

Apart from these pieces of prior information, the authors in [44] also assumed that they had access to partial observations about the evolution of the spread over a finite time interval. In particular, assume that we observe the dynamics of the disease for $t \in [0, T]$ from a collection of sensor nodes $\mathcal{V}_S \subseteq \mathcal{V}$. Based on these pieces of information, one can define an uncertainty set that contains all contact matrices $B_\mathcal{G}$ consistent with both empirical observations and prior knowledge. This set contains those contact matrices $B_\mathcal{G}$ such that the transmission rates $\{\beta_{ij}\}$ are consistent with the disease dynamics.

In order to eradicate the disease at the fastest rate possible, the authors in [44] considered the following control problem:

PROBLEM 11.3 (*Data-driven optimal allocation*) Assume the following pieces of information about a viral spread are given:

1. prior information about the state matrix (as described in P1–P3);
2. a finite (and possibly sparse) data series representing partial evolution of the spread over a set of sensor nodes $\mathcal{V}_S \subseteq \mathcal{V}$ during the time interval $t \in [T]$ (i.e., $\mathcal{D}$ in (11.1));
3. a set of vaccine cost functions g_i for all $i \in \mathcal{V}_C$, and a range of feasible recovery rates $\left[\underline{\delta}_i^c, \overline{\delta}_i^c\right]$ such that $1 - \delta_i^0 = \overline{\delta}_i^c \geq \delta_i^c \geq \underline{\delta}_i^c > 0$;
4. a fixed budget $C > 0$ to be allocated throughout a set of control nodes in $\mathcal{V}_C \subseteq \mathcal{V}$, so that $\sum_{i \in \mathcal{V}_C} g_i(\delta_i^c) \leq C$.

Find the cost-constrained allocation of control resources to eradicate the disease at the fastest possible exponential rate, measured as $\rho(M(B_\mathcal{G}, \mathbf{d}^c))$, over the uncertainty set $\Delta_{B_\mathcal{G}}$ of contact matrices coherent with prior knowledge and the observations in $\mathcal{D}$.

From the perspective of optimization, Problem 11.3 is equivalent to finding the optimal allocation of resources to minimize the worst-case (i.e., maximum possible) decay rate $\rho(M(B_\mathcal{G}, \mathbf{d}^c))$ for all $B_\mathcal{G} \in \Delta_{B_\mathcal{G}}$. In Han et al. [44], a robust optimization framework was developed to solve this problem, even in the presence of sparse observations.

11.4.3 Non-Poissonian Rates

The vast majority of spreading models over networks assume exponentially distributed transmission and recovery rates. In contrast, empirical observations indicate that most real-world spreading processes do not satisfy this assumption [71–73]. For example, the transmission rates of human immunodeficiency viruses present a distribution far from exponential [74]. In the context of online social networks, empirical studies show that the rate of spreading of information follows (approximately) a log-normal distribution [75, 76].

There are only a few results available for analyzing spreading processes over networks with non-exponential transmission and recovery rates. The experimental study in [77] confirmed the dramatic effect that non-exponential rates can have on the speed of spreading, as well as on the epidemic threshold. In [78], an analytically solvable (although rather simplistic) model of spreading with non-exponential rates was proposed. An approximate criterion for epidemic eradication over graphs with general transmission and recovery times based on asymptotic approximations was proposed in [79].

In recent work [42, 43], the authors propose an alternative approach to analyze general transmission and recovery rates using phase-type distributions. In particular, they derive conditions for disease eradication using transmission and recovery times that follow phase-type distributions (see, e.g., [80]). The class of phase-type distributions is dense in the space of positive-valued distributions [81], hence it can be used to theoretically analyze arbitrary transmission and recovery rates. Furthermore, there are efficient algorithms to compute the parameters of phase-type distributions to approximate any given distribution [80]. The key tool in this analysis is a vectorial representation proposed in [82], which can be used to represent phase-type distributions.

11.5 Comparisons with Common Heuristics

Usual approaches to distribute protection resources in a network of agents susceptible to cascade failures are heuristics based on network centrality measures [66]. As in the optimal framework presented in the previous section, much of the literature uses a bio-inspired epidemic model when studying harmful process with the ability to spread between interconnected agents. The main idea behind heuristic protection strategies is to rank agents according to different measures of importance based on their location in the network and greedily distribute protection resources based on each agent's rank. For example, Cohen et al. [22] proposed a simple protection strategy called *acquaintance immunization policy* in which the most connected node of a randomly selected node is given protective resources. This strategy was proved to be much more efficient than random allocation of protective resources. Hayashi et al. [83] proposed a simple heuristic called *targeted immunization*, consisting of greedily choosing nodes with the highest degree (number of connections) in scale-free graphs. Chung et al. [24] studied a greedy heuristic protection strategy based on the PageRank vector of the contact graph. Tong et al. [84] and Giakkoupis et al. [85] proposed greedy heuristics based on protecting those agents that induce the highest drop in the dominant eigenvalue of the contact graph. Recently, Prakash et al. [86] proposed several greedy heuristics to contain harmful cascades in directed networks when nodes can be partially protected (instead of completely removed, as assumed in previous work). These heuristics, as those in [83, 84], are based on eigenvalue perturbation analysis.

The heuristic methods in the literature are designed for a single resource type, predominantly the protective resources. A simplified variant of the budget-constrained allocation problem is presented with only protection-type resources in order to compare the optimal solution with heuristic solutions.

PROBLEM 11.4 The *Network Protection Problem* is given by

$$\max_{\beta,\varepsilon} \varepsilon$$

$$s.t.\ \Re[\lambda_1(BA-\delta I)] \leq -\varepsilon$$

$$\sum_{i=1}^{n} f(\beta_i) \leq C$$
$$\underline{\beta} \leq \beta_i \leq \bar{\beta} \qquad \forall i \in V.$$

11.5.1 Greedy, Centrality-Based Strategies

DEFINITION 11.1 Extract the *effective objective* in Problem 11.4 which is induced by the epigraph form. Define

$$\varepsilon(\beta) = -\Re[\lambda_1(BA - \delta I)] \tag{11.18}$$

where $B = \text{diag}\ (\beta)$ for any feasible resource allocation β.

Monotonicity and continuity of the function $\varepsilon(\beta)$ guarantee that fixing any feasible β and maximizing over ε always causes the constraint $\Re[\lambda_1(BA - \delta I)] \leq -\varepsilon$ to become tight. At the optimal point (β^*, ε^*) of Problem 11.4 satisfies

$$\varepsilon^* = -\Re[\lambda_1(\text{diag}(\beta^*)A - \delta I)].$$

When solving the resource allocation β, $\varepsilon(\beta)$ is treated as the effective objective in Problem 11.4.

DEFINITION 11.2 Define the *efficiency* of a feasible resource allocation β as

$$Q(\beta) = \frac{\varepsilon(\beta) - \varepsilon(\bar{\beta})}{\varepsilon(\beta^*) - \varepsilon(\bar{\beta})} \in [0, 1] \tag{11.19}$$

where β^* is a resource allocation achieving the maximum in Problem 11.4.

The effective objective $\varepsilon(\beta)$ and the costs functions $f(\beta_i)$ are monotonically non-increasing in the resource allocations β_i at each node, therefore $\bar{\beta}$ trivially achieves the minimum over the set of feasible resource allocations β.

DEFINITION 11.3 Let v be a centrality vector. Given a budget sufficient to completely protect k nodes: $C = kf(\underline{\beta})$, the greedy protection strategy $\hat{\beta}_v$ is to completely protect k nodes with the highest values in v. Define the protection fraction: $r = k/N$ where $N = n + m$ is the the total number of nodes.

Common centrality measures used for heuristics are degree and eigenvector centrality, [83]. PageRank centrality is used in place of eigenvector centrality in the case of general digraphs, [24]. While PageRank depends on a parameter α, we drop the α from our notation because our results hold for the whole family of PageRank vectors generated by nontrivial choices of $\alpha \in (0, 1)$.

THEOREM 11.3 Given the *Network Protection Problem* defined in Problem 11.4, with budget C, there exists a network G satisfying

$$Q(\hat{\beta}_{DEG}) = Q(\hat{\beta}_{PR}) = 0$$

where $r \in (0, 1)$ is the fraction of nodes that can be fully protected.

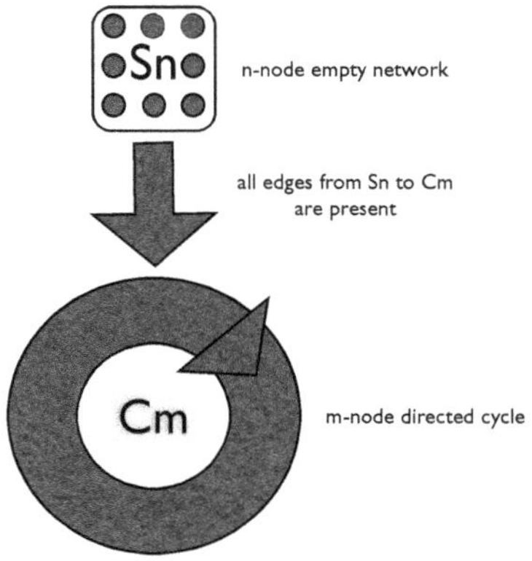

Figure 11.6 We construct the network G to prove Theorem 11.3.

Theorem 11.3 is based on a worst-case graph construction, as shown in Figure 11.6. Define the subgraph C_m as an m node directed cycle, S_n as an n node empty network and there are edges from all nodes $i \in S_n$ to all nodes $j \in C_m$. Formally, the edge set is given by

$$(i,j) \in \mathcal{E} \text{ if any of } \begin{cases} i \in S_n, j \in C_m \\ i, j = i+1 \in C_m \\ i = m+n, j = n+1 \in C_m \end{cases} \tag{11.20}$$

and in all other cases $(i,j) \notin \mathcal{E}$. All weights are given by $\mathcal{W}(i,j) = 1$ for $(i,j) \in \mathcal{E}$. Given a vaccination fraction r, the size of the subgraphs C_m and S_n must satisfy $m > n+2$ and $rN < n$ in order to generate a network for which greed heuristics have zero efficiency. Such m and n exist for any $r \in (0,1)$, but the networks required become very large as $r \to 1$.

The weakly connected network G results in a spreading process dominated by the nodes in C_m, even though nodes in S_n have larger centralities. A generalization of Theorem 11.3, which builds a strongly connected network with arbitrarily small efficiency, can be found in [67].

REMARK 11.2 The proof of Theorem 11.3 makes use of a constructive example for the centrality measures which identify nodes which are the most likely to fail: (a) out-degree and (b) PageRank with a random walk defined as moving up the edges. If one uses centrality measures which identify nodes which would be the most potent seeds such as (c) in-degree or (d) PageRank computed using a random walk that flows down the edges, one can construct an alternative G by simply reversing the direction of the edges from S_n to C_m. Using this alternative network, one can reproduce Theorem 11.3 for (c) and (d).

11.5.2 Greedy Heuristics and Workstation Protection

Consider a simple application in which such a worst-case network might arise naturally: nodes are computers belonging to individuals in a work environment. Edges indicate access to files on another person's computer.

Each workstation in C_m is an element in the cyber layer, paired with one or more plants in the physical layer. Workstations in S_n exist only in the cyber layer and belong to a group of administrators who can access files on all workstations in C_m.

Workers have limited access to each other's files, but do not have access to files on the administrator's computers. A virus may spread when an uninfected computer accesses an infected computer. It is assumed that an infected workstation cannot adequately control its associated plant, which leads to physical layer failures. Protection resources take the form of antivirus software with updates on a variable time interval, software updated more frequently providing a smaller infection rate β, but updates incurring a greater cost $f(\beta)$. The cost function

$$f(\beta_i) = \frac{\underline{\beta}(\frac{\bar{\beta}}{\beta_i} - 1)}{\bar{\beta} - \underline{\beta}} \tag{11.21}$$

is chosen to satisfy $f(\bar{\beta}) = 0$, $f(\underline{\beta}) = 1$ and $f(\beta) \propto 1/\beta$. This allows us to choose capacity C equal to the number of nodes to which we wish to be able to allocate maximum protection. In our example, the infection rate with outdated antivirus software is $\bar{\beta} = 0.5$ while the maximum update rate achieves an infection rate of $\underline{\beta} = 0.01$. Choosing a budget of $C = 3$ for a network with $n = 3$ and $m = 6$ (such as in G or A shown in Figure 11.7), the fraction of nodes that can be maximally protected is $r = 1/3$. An infected machine has recovery rate $\delta = 0.3$, based on curative resources in the form of IT staff, which are uniformly available.

In the example, four heuristic algorithms based on greedily allocating resources with respect to centrality measures are considered. The centrality measures are out-degree, total degree, PageRank with $\alpha = 0.1$ and symmetrized PageRank with $\alpha = 0.1$. Symmetrized PageRank is computed by allowing the random walk move over a directed edge in either direction. The worst-case networks are products of extreme asymmetry between C_m and S_n; the symmetric centrality measure shows that even symmetric centrality measures don't overcome the potential for arbitrarily poor behavior. In Figure 11.8 the top table shows all of the centrality vectors for the example problem in the networks A and G. The network G is the network constructed in our analytical proofs. The network A is an example of a less structured employee collaboration network which we include to demonstrate two points: (1) our

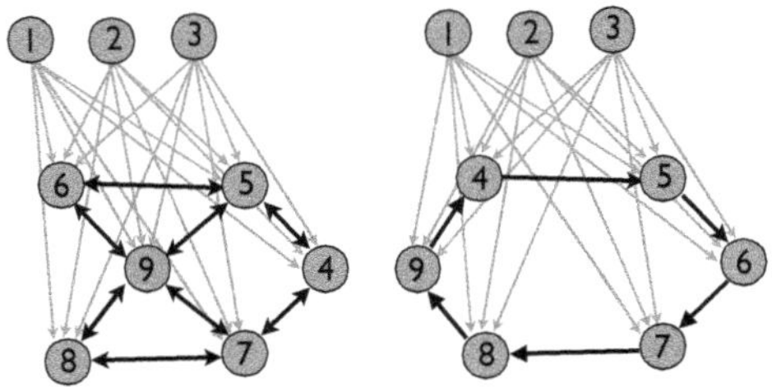

Figure 11.7 Network G with vertices $S_3 = \{1, 2, 3\}$ and $C_6 = \{4, 5, \ldots, 9\}$ satisfies the conditions for the counter-example network defined in Theorem 11.3. In Network A, the subgraph on C_6 is relaxed to be less structured for demonstration purposes.

		Network A (Cm = Undirected Network)				Network G (Cm= Directed Cycle)			
		Out Degree Centrality	Total Degree Centrality	Page Rank Centrality	Sym Page Rank Centrality	Out Degree Centrality	Total Degree Centrality	Page Rank Centrality	Sym Page Rank Centrality
Centrality Measures	node 1 (Sn)	6	6	0.289	0.093	6	6	0.299	0.122
	node 2 (Sn)	6	6	0.289	0.093	6	6	0.299	0.122
	node 3 (Sn)	6	6	0.289	0.093	6	6	0.299	0.122
	node 4 (Cm)	2	7	0.020	0.105	1	5	0.017	0.106
	node 5 (Cm)	3	9	0.023	0.128	1	5	0.017	0.106
	node 6 (Cm)	2	7	0.020	0.104	1	5	0.017	0.106
	node 7 (Cm)	3	9	0.023	0.128	1	5	0.017	0.106
	node 8 (Cm)	2	7	0.020	0.104	1	5	0.017	0.106
	node 9 (Cm)	4	11	0.026	0.151	1	5	0.017	0.106
Legend	Nodes selected for protection								

		Network A (Cm = Undirected Network)						Network G (Cm= Directed Cycle)					
		No Allocation	Out Degree Allocation	Total Degree Allocation	Page Rank Allocation	Sym Page Rank Allocation	Optimal Allocation	No Allocation	Out Degree Allocation	Total Degree Allocation	Page Rank Allocation	Sym Page Rank Allocation	Optimal Allocation
strategy profile β	node 1 (Sn)	0.5	0.01	0.5	0.01	0.5	0.5	0.5	0.01	0.01	0.01	0.01	0.5
	node 2 (Sn)	0.5	0.01	0.5	0.01	0.5	0.5	0.5	0.01	0.01	0.01	0.01	0.5
	node 3 (Sn)	0.5	0.01	0.5	0.01	0.5	0.5	0.5	0.01	0.01	0.01	0.01	0.5
	node 4 (Cm)	0.5	0.5	0.5	0.5	0.5	0.0261	0.5	0.5	0.5	0.5	0.5	0.0196
	node 5 (Cm)	0.5	0.5	0.01	0.5	0.01	0.0174	0.5	0.5	0.5	0.5	0.5	0.0196
	node 6 (Cm)	0.5	0.5	0.5	0.5	0.5	0.0261	0.5	0.5	0.5	0.5	0.5	0.0196
	node 7 (Cm)	0.5	0.5	0.01	0.5	0.01	0.0174	0.5	0.5	0.5	0.5	0.5	0.0196
	node 8 (Cm)	0.5	0.5	0.5	0.5	0.5	0.0261	0.5	0.5	0.5	0.5	0.5	0.0196
	node 9 (Cm)	0.5	0.5	0.01	0.5	0.01	0.0131	0.5	0.5	0.5	0.5	0.5	0.0196
convergence rate ε	-	-1.11	-1.11	0.152	-1.11	0.152	0.248	-0.2	-0.2	-0.2	-0.2	-0.2	0.28
Efficiency	-	0	0	0.929	0	0.929	1	0	0	0	0	0	1
Legend	unstable rate	stable rate		optimal allocation									

Figure 11.8 (Top) A variety of centrality measures are used as the basis for greedy algorithms; these measures are reported for networks A and G. (Bottom) The allocation strategies tested are detailed, their exponential convergence rate bounds ε, and their efficiencies are reported for comparison purposes. For the case of the counter-example network G, none of the greedy type algorithms yield a stable convergence rate.

constructed network G is not unique and (2) symmetrizing heuristics are less fragile than heuristics that respect edge direction.

In G and A the out-degree and PageRank heuristics allocate all resources to the admins, S_n. This is ineffective because even though the admins are the most likely to become infected, the worker group C_m cannot access their files and become infected. Furthermore, the failure of the admin workstation does not lead directly to physical layer failures. Figure 11.8 (bottom) shows the infection rate profiles generated by the various heuristics and the optimal solution. A strategy is ineffective if the convergence rate epsilon is negative because this corresponds to unstable dynamics, where the computer virus is spreading faster than the IT staff can repair workstations. The result is a complete failure to consistently control any of the plants in the physical layer of the system.

11.6 Conclusions

We have studied the problem of allocating protection resources in weighted, directed networks to contain spreading processes, such as the propagation of viruses in computer networks, cascading failures in complex technological networks, or the spreading of an epidemic in a human population. We have considered two types of protection resources: (1) *Preventive* resources able to "immunize" nodes against the spreading (e.g., vaccines), and (2) *corrective* resources able to neutralize the spreading after it has reached a node (e.g., antidotes). We assume that protection resources have an associated cost and have then studied the *budget-constrained allocation problem*, in which we find the optimal allocation of resources to contain the spreading, given a fixed budget. We have solved this optimal resource allocation problem in *weighted and directed* networks of *nonidentical* agents in polynomial time using GP. Furthermore, the framework herein proposed allows *simultaneous* optimization over both preventive and corrective resources, even in the case of cost functions being node-dependent.

We have illustrated our approach by designing an optimal protection strategy for a real air transportation network. We have limited our study to the network of the world's busiest airports by passenger traffic. For this transportation network, we have computed the optimal distribution of protecting resources to contain the spread of a hypothetical worldwide pandemic. Our simulations show that the optimal distribution of protecting resources follows nontrivial patterns that cannot, in general, be described using simple heuristics based on traditional network centrality measures.

We then presented the following recent extensions on this work: (1) a generalized framework to cover more realistic epidemic models, (2) a novel data-driven framework able to handle network uncertainties, and (3) an analysis tool that allows us to study non-Poissonian transmission and recovery rates. We concluded this chapter with a comparison between our results and common heuristics used in the literature.

References

[1] N. Bailey, *The Mathematical Theory of Infectious Diseases and Its Applications*. London: Charles Griffin & Company Ltd., 1975.

[2] M. Garetto, W. Gong, and D. Towsley, "Modeling malware spreading dynamics," in *IEEE INFOCOM 2003*, vol. 3, 2003, pp. 1869–1879.

[3] J. Leskovec, L. A. Adamic, and B. A. Huberman, "The dynamics of viral marketing," *ACM Trans. Web*, vol. 1, no. 1, p. 5, 2007.

[4] S. Roy, M. Xue, and S. Das, "Security and discoverability of spread dynamics in cyber-physical networks," *IEEE Trans. Parallel Distrib. Syst.*, vol. 23, no. 9, pp. 1694–1707, 2012.

[5] R. M. Anderson, R. M. May, and B. Anderson, *Infectious Diseases of Humans: Dynamics and Control*. Chichester: Wiley, 1992.

[6] M. J. Keeling and P. Rohani, *Modeling Infectious Diseases in Humans and Animals*. Princeton, NJ: Princeton University Press, 2007.

[7] P. van den Driessche and J.Watmourgh, "Reproduction numbers and sub-threshold endemic equilibria for compartmental models of disease transmission," *Math. Biosci.*, vol. 180, no. 1–2, pp. 29–48, 2002.

[8] S. Funk, E. Gilad, and V. A. A. Jansen, "Endemic disease, awareness, and local behavioural response," *J. Theor. Biol.*, vol. 264, no. 2, pp. 501–509, 2010.

[9] I. Z. Kiss, J. Cassell, M. Recker, and P. L. Simon, "The impact of information transmission on epidemic outbreaks," *Math. Biosci.*, vol. 225, no. 1, pp. 1–10, 2010.

[10] N. Perra, D. Balcan, B. Gonasalves, and A. Vespignani, "Towards a characterization of behavior–disease models," *PLoS ONE*, vol. 6, no. 8, p. e23084, Aug. 2011.

[11] P. Poletti, B. Caprile, M. Ajelli, A. Pugliese, and S. Merler, "Spontaneous behavioural changes in response to epidemics," *J. Theor. Biol.*, vol. 260, no. 1, pp. 31–40, 2009.

[12] D. Breda, O. Diekmann, W. F. de Graaf, A. Pugliese, and R. Vermiglio, "On the formulation of epidemic models (an appraisal of Kermack and McKendrick)," *J. Biol. Dyn.*, vol. 6, no. 2, pp. 103–117, 2012.

[13] M. M. Hikal, "Dynamic properties for a general SEIV epidemic model," *SIAM Rev.*, vol. 2, no. 1, pp. 26–36, 2014.

[14] H. W. Hethcote, "The mathematics of infectious diseases," *SIAM Rev.*, vol. 42, no. 4, pp. 599–653, 2000.

[15] J. Li, Y. Yang, and Y. Zhou, "Global stability of an epidemic model with latent stage and vaccination," *Nonlinear Anal. Real World Appl.*, vol. 12, no. 4, pp. 2163–2173, 2011.

[16] A. Korobeinikov and G. C. Wake, "Lyapunov functions and global stability for SIR, SIRS, and SIS epidemiological models," *Appl. Math. Lett.*, vol. 15, no. 8, pp. 955–960, 2002.

[17] A. Lahrouz, L. Omari, and D. Kiouach, "Global analysis of a deterministic and stochastic nonlinear SIRS epidemic model," *Nonlinear Anal. Model Control*, vol. 16, no. 1, pp. 59–76, 2011.

[18] H. W. Hethcote, H. W. Stech, and P. van den Driessche, "Stability analysis for models of diseases without immunity," *J. Math. Biol.*, vol. 13, no. 2, pp. 185–198, 1981.

[19] Y. Wang, D. Chakrabarti, C. Wang, and C. Faloutsos, "Epidemic spreading in real networks: An eigenvalue viewpoint," in *Proc. 22nd Int. Symp. Reliable Distributed Systems*, 2003, pp. 25–34.

[20] A. Ganesh, L. Massoulie, and D. Towsley, "The effect of network topology on the spread of epidemics," in *IEEE INFOCOM 2005*, vol. 2, 2005, pp. 1455–1466.

[21] P. Van Mieghem, J. Omic, and R. Kooij, "Virus spread in networks," *IEEE/ACM Trans. Networking*, vol. 17, no. 1, pp. 1–14, 2009.

[22] R. Cohen, S. Havlin, and D. Ben-Avraham, "Efficient immunization strategies for computer networks and populations," *Phys. Rev. Lett.*, vol. 91, no. 24, p. 247901, 2003.

[23] C. Borgs, J. Chayes, A. Ganesh, and A. Saberi, "How to distribute antidote to control epidemics," *Random Struct. Algor.*, vol. 37, no. 2, pp. 204–222, 2010.

[24] F. Chung, P. Horn, and A. Tsiatas, "Distributing antidote using PageRank vectors," *Internet Math.*, vol. 6, no. 2, pp. 237–254, 2009.

[25] Y. Wan, S. Roy, and A. Saberi, "Network design problems for controlling virus spread," *46th IEEE Conf. Decision and* Control, 2007, pp. 3925–3932.

[26] Y. Wan, S. Roy, and A. Saberi, "Designing spatially heterogeneous strategies for control of virus spread," *IET Syst. Biol.*, vol. 2, no. 4, pp. 184–201, 2008.

[27] V. M. Preciado and A. Jadbabaie, "Spectral analysis of virus spreading in random geometric networks," *Proc. 48th IEEE Conf. Decision and Control (CDC)*, 2009, pp. 4802–4807.

[28] E. Gourdin, J. Omic, and P. Van Mieghem, "Optimization of network protection against virus spread," in *8th Int. Workshop on the Design of Reliable Communication Networks*, 2011, pp. 86–93.

[29] V. Preciado, F. Darabi Sahneh, and C. Scoglio, "A convex framework for optimal investment on disease awareness in social networks," in *IEEE Global Conf. Signal and Information Processing*, 2013, pp. 851–854.

[30] V. Preciado, M. Zargham, C. Enyioha, A. Jadbabaie, and G. Pappas, "Optimal vaccine allocation to control epidemic outbreaks in arbitrary networks," in *IEEE Conf. Decision and Control*, 2013, pp. 7486–7491.

[31] V. Preciado, M. Zargham, C. Enyioha, A. Jadbabaie, and G. Pappas, "Optimal resource allocation for network protection against spreading processes," *IEEE Trans. Control Network Syst.*, vol. 1, no. 1, pp. 99–108, Mar. 2014.

[32] V. Preciado and M. Zargham, "Traffic optimization to control epidemic outbreaks in metapopulation models," in *IEEE Global Conf. Signal and Information Processing*, 2013, pp. 847–850.

[33] V. M. Preciado, M. Zargham, and D. Sun, "Traffic control for network protection against spreading processes," in Conference on Information Sciences and Systems, Princeton, NJ, 2014.

[34] M. Zargham and V. M. Preciado, "Worst-case scenarios for greedy, centrality-based network protection strategies," in *Conf. Information Sciences and Systems*, 2014, pp. 1–6.

[35] C. Nowzari, V. Preciado, and G. Pappas, "Optimal resource allocation in generalized epidemic models," *IEEE Trans. Control Network Syst.* vol., 4, no. 2, pp. 159–169, 2017.

[36] X. Chen and V. Preciado, "Optimal coinfection control of competitive epidemics in multilayer networks," *Proc. IEEE Conf. Decision and Control*, 2014, pp. 6209–6214.

[37] N. Watkins, C. Nowzari, V. Preciado, and G. Pappas, "Optimal resource allocation for competing epidemics over arbitrary networks," *Proc. IEEE American Control Conf.*, 2015, pp. 1381–1386.

[38] V. P. N. Watkins, C. Nowzari, and G. Pappas, "Optimal resource allocation for competitive spreading processes on bilayer networks," *IEEE Trans. Control Network Syst.*, in press.

[39] M. Ogura and V. Preciado, "Disease spread over randomly switched large-scale networks," *Proc. IEEE American Control Conference*, 2015, pp. 1782–1787.

[40] M. Ogura and V. Preciado, "Stability of spreading processes over time-varying large-scale networks," *IEEE Trans. Network Sci. Eng.*, vol. 3, no. 1, pp. 44–57, 2016.

[41] M. Ogura and V. Preciado, "Epidemic processes over adaptive state-dependent networks," *Phys. Rev. E*, vol. 93, 2016.

[42] M. Ogura and V. Preciado, "Spreading processes over networks with phase-type transmission and recovery times," *arXiv preprint arXiv:1502.07384v1*, 2015.

[43] M. Ogura and V. Preciado, "Spreading processes with general transmission and recovery times," *arXiv 1606.08518*, 2016.

[44] S. Han, V. Preciado, C. Nowzari, and G. Pappas, "Data-driven allocation of vaccines for controlling epidemic outbreaks," *IEEE Trans. Network Sci. Eng.*, vol. 2, no. 4, pp. 127–138, 2015.

[45] E. Ramirez-Llanos and S. Martinez, "A distributed algorithm for virus spread minimization," *2014 American Control Conf.*, 2014, pp. 184–189.

[46] A. Khanafer, T. Başar, and B. Gharesifard, "Stability of epidemic models over arbitrary graphs: A positive systems approach," *Automatica*, submitted.

[47] M. Rausand and A. Høyland, *System Reliability Theory: Models, Statistical Methods, and Applications*. Chichester: Wiley, 2004.

[48] B. R. Iyer, L. Donatiello, and P. Heidelberger, "Analysis of performability for stochastic models of fault-tolerant systems," *IEEE Trans. Comput.*, vol. 100, no. 10, pp. 902–907, 1986.

[49] A. L. Reibman and M. Veeraraghavan, "Reliability modeling: An overview for system designers," *Computer*, vol. 24, no. 4, pp. 49–57, 1991.

[50] A. D. Domínguez-García, "An integrated methodology for the performance and reliability evaluation of fault-tolerant systems," Ph.D. dissertation, Massachusetts Institute of Technology, 2007.

[51] V. Preciado and A. Jadbabaie, "Moment-based analysis of spreading processes from network structural information," *IEEE Trans. Autom. Control*, vol. 21, no. 2, pp. 373–382, 2013.

[52] V. Preciado, M. Draief, and A. Jadbabaie, "Structural analysis of viral spreading processes in social and communication networks using egonets," *arXiv preprint arXiv:1209.0341*, 2013.

[53] V. M. Preciado, M. Zargham, and D. Sun, "A convex framework to control spreading processes in directed networks," in *48th Ann. Conf. Information Sciences and Systems (CISS)*, March 2014, pp. 1–6.

[54] N. Biggs, *Algebraic Graph Theory*. Cambridge: Cambridge University Press, 1993.

[55] C. Meyer, *Matrix Analysis and Applied Linear Algebra*. Philadelphia, PA: SIAM, 2000.

[56] A. Lajmanovich and J. A. Yorke, "A deterministic model for gonorrhea in a nonhomogeneous population," *Math. Biosci.*, vol. 28, no. 3, pp. 221–236, 1976.

[57] D. Chakrabarti, Y. Wang, C. Wang, J. Leskovec, and C. Faloutsos, "Epidemic thresholds in real networks," *ACM Trans. Inf. Syst. Secur.*, vol. 10, no. 4, pp. 1–26, 2008.

[58] J. O. Kephart and S. R. White, "Directed-graph epidemiological models of computer viruses," in *IEEE Symp. Research in Security and Privacy*, 1991, pp. 343–359.

[59] J. O. Kephart and S. R. White, "Measuring and modeling computer virus prevalence," in *IEEE Symp. Research in Security and Privacy*, 993, pp. 2–15.

[60] C. Wang, J. C. Knight, and M. C. Elder, "On computer viral infection and the effect of immunization," in *Proc. 16th Annual Computer Security Applications Conf.*, 2000, pp. 246–256.

[61] M. Garetto, W. Gong, and D. Towsley, "Modeling malware spreading dynamics," in *INFOCOM Joint Conf. IEEE Computer and Communications Societies*, 2003, pp. 1869–1879.

[62] P. Van Mieghem, *Performance Analysis of Communications Networks and Systems*. Cambridge: Cambridge University Press, 2006.

[63] S. Boyd and L. Vandenberghe, *Convex Optimization*. Cambridge: Cambridge University Press, 2004.

[64] S. Boyd, S.-J. Kim, L. Vandenberghe, and A. Hassibi, "A tutorial on geometric programming," *Optim. Eng.*, vol. 8, no. 1, pp. 67–127, 2007.

[65] C. Schneider, T. Mihaljev, S. Havlin, and H. Herrmann, "Suppressing epidemics with a limited amount of immunization units," *Phys. Rev. E*, vol. 84, no. 6, p. 061911, 2011.

[66] M. Newman, *Networks: An Introduction*. Cambridge: Cambridge University Press, 2010.

[67] M. Zargham and V. Preciado, "Worst-case scenarios for greedy, centrality-based network protection strategies," in *Conf. Information Sciences and Systems*, 2014, pp. 1–6.

[68] V. P. C. Nowzari and G. Pappas, "Analysis and control of epidemics: A survey of spreading processes on complex networks," *IEEE Trans. Network Sci. Eng.*, vol. 36, no. 1, pp. 26–46, 2016.

[69] B. A. Prakash, D. Chakrabarti, M. Faloutsos, N. Valler, and C. Faloutsos, "Got the flu (or mumps)? Check the eigenvalue!" *arXiv:1004.0060*, 2010.

[70] V. Chandrasekaran and P. Shah, "Conic geometric programming," in *48th Ann. Conf. Information Sciences and Systems (CISS)*, 2014, pp. 1–4.

[71] E. Limpert, W. A. Stahel, and M. Abbt, "Log-normal distributions across the sciences: Keys and clues," *BioScience*, vol. 51, no. 5, 2001.

[72] A. L. Lloyd, "Realistic distributions of infectious periods in epidemic models: Changing patterns of persistence and dynamics," *Theor. Popul. Biol.*, vol. 60, no. 1, pp. 59–71, 2001.

[73] A. L. Lloyd, "Destabilization of epidemic models with the inclusion of realistic distributions of infectious periods," *Proc. Roy. Soc. Lond. B*, vol. 268, no. 1470, pp. 985–993, 2001.

[74] S. Blythe and R. Anderson, "Variable infectiousness in HFV transmission models," *Math. Med. Biol.*, vol. 5, no. 3, pp. 181–200, 1988.

[75] K. Lerman and R. Ghosh, "Information contagion: An empirical study of the spread of news on Digg and Twitter social networks," in *ICWSM*, vol. 10, 2010, pp. 90–97.

[76] P. Van Mieghem, N. Blenn, and C. Doerr, "Lognormal distribution in the Digg online social network," *Eur. Phys. J. B*, vol. 83, no. 2, pp. 251–261, 2011.

[77] P. Van Mieghem and R. Van de Bovenkamp, "Non-Markovian infection spread dramatically alters the susceptible-infected-susceptible epidemic threshold in networks," *Phys. Rev. Lett.*, vol. 110, no. 10, p. 108701, 2013.

[78] H.-H. Jo, J. I. Perotti, K. Kaski, and J. Kertész, "Analytically solvable model of spreading dynamics with non-Poissonian processes," *Phys. Rev. X*, vol. 4, no. 1, p. 011041, 2014.

[79] E. Cator, R. Van de Bovenkamp, and P. Van Mieghem, "Susceptible–infected–susceptible epidemics on networks with general infection and cure times," *Phys. Rev. E*, vol. 87, no. 6, p. 062816, 2013.

[80] S. Asmussen, O. Nerman, and M. Olsson, "Fitting phase-type distributions via the EM algorithm," *Scand. J. Stat.*, vol. 23, pp. 419–441, 1996.

[81] D. R. Cox, "A use of complex probabilities in the theory of stochastic processes," in *Math. Proc. Cambridge Philos. Soc.*, vol. 51, no. 2. 1955, pp. 313–319.

[82] R. Brockett, "Optimal control of observable continuous time Markov chains," in *47th IEEE Conf. Decision and Control*, 2008, pp. 4269–4274.

[83] Y. Hayashi, M. Minoura, and J. Matsukubo, "Recoverable prevalence in growing scale-free networks and the effective immunization," *arXiv preprint cond-mat/0305549*, 2003.

[84] H. Tong, B. A. Prakash, C. Tsourakakis, et al., "On the vulnerability of large graphs," in *2010 IEEE 10th Int. Conf. Data Mining (ICDM)*, 2010, pp. 1091–1096.

[85] G. Giakkoupis, A. Gionis, E. Terzi, and P. Tsaparas, "Models and algorithms for network immunization," Department of Computer Science, University of Helsinki, Tech. Rep., 2005.

[86] T. I. H. T. B. Aditya Prakash, L. Adamic, and C. Faloutsos, "Fractional immunization in networks," *Proc. SDM*, 2013.

[87] M. Grant, S. Boyd, and Y. Ye, "CVX: MATLAB software for disciplined convex programming," 2008. [Online]. Available: http://cvxr.com/cvx.

Part III Exercises

1. Distinguish the notions of confidentiality, security, and resilience as they relate to cyber-physical systems.
2. Here we step through an illustrative example using a two-generator, six-bus, seven-line power system model to highlight the steps for masking a DCOPF problem.

   The vectors and matrices needed to describe this model in the form of (8.1) are given as

$$\mathbf{A}_{\text{inc}} = \begin{bmatrix} 1 & -1 & 0 & 0 & 0 & 0 \\ 1 & 0 & -1 & 0 & 0 & 0 \\ 0 & 1 & -1 & 0 & 0 & 0 \\ 0 & 1 & 0 & 0 & -1 & 0 \\ 0 & 0 & 1 & -1 & 0 & 0 \\ 0 & 0 & 1 & 0 & -1 & 0 \\ 0 & 0 & 0 & 1 & 0 & -1 \end{bmatrix} \qquad \mathbf{b}_{\text{br}} = \begin{bmatrix} 1 \\ 1 \\ 1 \\ 1 \\ 1 \\ 1 \\ 1 \end{bmatrix}$$

$$\begin{aligned} \mathbf{B} &= {\mathbf{A}_{\text{inc}}}^T \operatorname{diag}(\mathbf{b}_{\text{br}}) \mathbf{A}_{\text{inc}} \\ \mathbf{P}_{\text{flow, max}} &= \begin{bmatrix} 2 & 6 & 6 & 6 & 6 & 6 & 6 \end{bmatrix}^T \\ \mathbf{P}_L &= \begin{bmatrix} 0 & 1 & 1 & 1 & 1 & 1 \end{bmatrix}^T \\ \mathbf{P}_{\text{g, min}} &= \begin{bmatrix} 0 & 0 & 0 & 0 & 0 & 0 \end{bmatrix}^T \\ \mathbf{P}_{\text{g, max}} &= \begin{bmatrix} 6 & 6 & 0 & 0 & 0 & 0 \end{bmatrix}^T \\ \mathbf{c}^T &= \begin{bmatrix} 3 & 4 & 0 & 0 & 0 & 0 \end{bmatrix} \end{aligned}$$

   For reference, the DCOPF (8.1) is expressed as

   Minimize $\mathbf{c}^T \mathbf{P}_g$
   Subject to:

$$\begin{aligned} -\mathbf{P}_g + \mathbf{B}\delta &= -\mathbf{P}_L \\ -\mathbf{P}_{flow,max} &\leq \operatorname{diag}(\mathbf{b}_{br}) \mathbf{A}_{\text{inc}} \delta \leq \mathbf{P}_{flow,max} \\ \mathbf{P}_{g,min} &\leq \mathbf{P}_g \leq \mathbf{P}_{g,max} \\ \delta_{ref} &= 0 \end{aligned}$$

a. For the defined system of matrices and vectors, convert the DCOPF model form to the standard form (8.2). To ensure all variables take positive values, define and use new variable vectors δ^+ and δ^-, such that $\delta = \delta^+ - \delta^-$. The representation can be made more compact by noting that there are only two generation variables, P_{g1} and P_{g2}, both of which are defined to be positive in the problem formulation. Define

$$x = \begin{bmatrix} P_{g1} \\ P_{g2} \\ \delta^+ \\ \delta^- \end{bmatrix}$$

and put into standard form.

b. Solve the optimization problem using MATLAB or any similar tool. The solution should yield $P_{g1} = 4.6$ and $P_{g2} = 0.4$.

c. Follow the transformation section of Chapter 8 to obtain a model of the form given in (8.7). Specifically, choose **Q** as any positive monomial matrix. Each row and column of **Q** will contain a single positive entry (a positive weighted permutation matrix). Choose **r** as any positive-valued vector. Add slack variables to convert the inequality constraints into equality constraints. Solve this transformed model and show that the solution is consistent with the solution in the previous step.

d. Obtain the fully transformed and masked problem in the form (8.2) using any nonsingular **P** matrix. Solve this transformed model and show that its solution is consistent with the optimal solution of the original using (8.11).

e. The problem formulation presents a single snapshot of a power system model. Suppose we wished to solve a new problem for the same network, generators, and cost functions, but with a different loading vector $\mathbf{P}_L$. Using the same transformations, how will (8.9) change? Will the transformation back to original variables (8.11) still map to an optimal solution with the new loading?

f. In the transformed system (8.9), the cost at the optimal solution may be a different value than that of the original model (8.1). If we want to compare the costs for different loading profiles, will the costs from the transformed system preserve an ordered ranking of costs from the original system?

g. (Advanced.) As described in this set of problems, the only restrictions on the transformations are that the nonzero elements of **Q** and r must be positive, and matrix **P** must be nonsingular. In practice we are interested in preserving "nice" properties of the system for the purposes of optimization. These include sparsity of the model, as discussed in Chapter 8, and ensuring the transformed model is not ill-conditioned. Discuss these issues with examples. (For example, try transformations such as **Q** and r with extremely varied values, and **P** sparse and full, also with widely varied eigenvalues.)

3. Chapter 8 introduces a process for masking power-system models so that sensitive information remains confidential, yet the models can still be used for system evaluation or research purposes. What are other infrastructures that may require model masking? What changes may need to be made to the masking procedure for these other potential applications?
4. One innovation of the method introduced in Chapter 8 is that the masked model retains the sparsity structure of the original model. Why is maintaining sparsity so important for power-systems applications?
5. For a small-scale example, implement in MATLAB the masking procedure introduced in Chapter 8.
6. Chapter 10 advances the π-calculus as a formal modeling tool for cyber-physical processes. Briefly explain what the π-calculus is, and discuss possible advantages and drawbacks of modeling CPS using this construct.
7. Answer the following problems related to Chapter 10. (Some of these are "back of the envelope calculations" and others are research problems.)

    a. In the example of Section 10.6, how can Fred ensure that his actions are always nondeducible? Is this economically feasible?
    b. How many possible settings are there for the pipeline system of Section 10.4 if the flow is discretized into three flow levels (1, 2, 3)? Is it possible to make the observable flow settings nondeducible? How many with seven flow levels?
    c. Modify the paint separator system to prevent the virus from disrupting information flow and show that the resulting system is not MSDND. (Hint: There are several ways to do this.)
    d. Draw multiple security partitions around the FREEDM system components. Where should MSDND be maintained and where should it be broken?
    e. Extend your model of multiple security partitions around the FREEDM system components to include a centralized type of monitoring and control system. Model this in MSDND. What new vulnerabilities are introduced by doing this? (Hint: Use the paint separator model to uncover potential problems.)

8. Explain how design-time and runtime efforts may be used in tandem to achieve security in a CPS.
9. What strategies can be used to protect a CPS from attack? What makes protection of CPSs difficult compared to traditional isolated cyber or physical systems?
10. Define the notion of trust for a CPS.
11. Chapter 11 poses the problem of allocating protection resources in a cyber-physical network as an optimization problem with a budget constraint. What are some real-world settings where this optimization framework is suitable?

12. Review the approaches that have been proposed to address resource allocation in dynamic networks. What are advantages and disadvantages of the bio-inspired approach proposed in Chapter 11?
13. Consider the following three networks with n nodes:

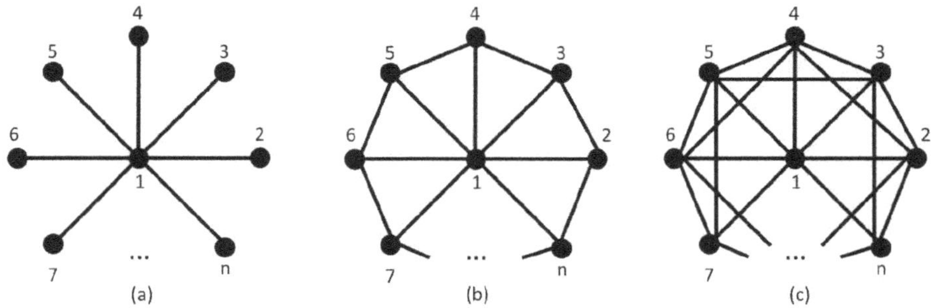

    a. Compute the largest eigenvalue of the adjacency matrices of the graphs in figures (a), (b), and (c) as a function of n.
    b. Consider the SIS model of spreading with $\beta = 0.1$ and $n = 100$. For what values of δ does an epidemic die out in each one of the three networks? (Reminder: The epidemic dies out when $\lambda_1 < \delta/\beta$.)
    c. Imagine you work for a health agency responsible for controlling an epidemic taking place in the above networks. Assume you can tune the spreading rates of the edges within a feasible interval $[\underline{\beta}, \overline{\beta}]$. Assume the cost associated with tuning β is given by $f_{ij}(\beta) = 1/\beta$. Write the associated optimization problem for each one of the above networks. (Hint: Your answer should look like (11.5)–(11.8).)
    d. Transform the optimization problem in (c) into a standard geometric program. (Hint: Your answer should look like (11.11)–(11.14).)
    e. Implement the geometric program from (d) using MATLAB's CVX Toolbox.
14. Consider a ground transportation CPS, such as the example pursued in Chapter 11. What are different types of attacks that may be enacted on the communication network for this CPS? What are possible impacts on the physical aspects of the CPS? What tools are available for evaluating the integrative physical-world impacts of cyber-attacks?

Part IV

Frameworks and Testbeds

The first three parts of the book were focused on new techniques for the operation, analysis, and design of cyber-physical systems (CPSs). In tandem with these methodological advances, new frameworks and testbeds are also needed for experimental as well as conceptual evaluation of CPSs. This part of the book is focused on developing frameworks and testbeds for the study of CPSs. The three chapters in this part are each narrowly focused on specific applications and/or aspects of CPS operations. What is important, however, is that they introduce representations for generic features and characteristics of CPSs, and in this sense broadly advance the state-of-the-art in modeling CPSs. In particular, the three chapters all address representation of complex time-varying and networked processes, which are prototypical of CPSs. Additionally, the chapters explicitly discuss the intertwining of dynamics in the physical world and information flow in the cyber world. The introduced frameworks and testbeds also acknowledge a wide range of performance characteristics related to CPSs, which are connected to structure, control, security, performance, uncertainty management, etc.

The frameworks and testbeds developed in this section are important because they provide concrete contexts for evaluating CPS principles and methods. In a broad sense, these frameworks and testbeds illustrate how core principles of CPS operations – such as the thrusts described in Parts I to III – combine to influence the holistic behavior and management of CPSs. We envision that these testbeds, and many other analogous testbeds/frameworks for other CPS components, will serve to contextualize and connect emerging CPS research, and also guide further research on CPSs.

The chapters in Part IV are organized as follows. Chapter 12, by Chejerla et al., provides a framework for resource management in cloud-accessed and -assisted CPSs. Chapter 13, by Dor and Lu, provides a framework for the middleware layer of CPSs. Finally, Chapter 14 introduces the concept that tandem physical- and cyber-world models (or CPS models) can be used to study complex biological processes, in this case neuronal networks that display sleep-wake-like oscillations.

12 Resource Management in Cloud-Assisted Cyber-Physical Systems

Brijesh Kashyap Chejerla, Sanjay Madria, and Sajal K. Das

12.1 Introduction

Cyber-physical systems (CPSs) are rapidly becoming the cornerstone of modern-day hybrid systems, with a physical subsystem and a cyber subsystem working in tandem to produce a complex distributed system. The hybrid nature of CPSs lies in integrating the physical component that yields observable outputs with a cyber component that processes the information obtained from it.

Real-world examples of CPSs, such as smart grids, transportation systems, healthcare, and robotics, are rapidly evolving from applications with limited quality of service (QoS) requirements to ones with complex interconnected QoS requirements, as shown in Figure 12.1. Such rapidly evolving QoS requirements warrant an improvement in the computational capabilities of the components serving the needs of the physical system of the CPS. However, the QoS requirements such as network latency, data processing, and data management for each of these applications vary with CPS application. Hence, in order to uphold such QoS requirements, effective management schemes for CPS resources, such as network, energy, and computation platform, have to be devised.

Cyber-physical systems are evolving into complex and hybrid systems at an exponential rate owing to the increase in the number of "actors". An actor in a CPS can be one of the various components of the application, such as a service provider and end-user of the cyber system, various actuating subcomponents of the physical system (e.g., embedded sensor systems), and the human who uses the entire CPS. We do not consider physical subsystems such as transformers and transmission lines which form a part of the topological setting of the CPS. With increasingly newer components being added to the existing mold or topology of a CPS, there is an increase in the number of actors that generate and use CPS data. Such massive amounts of data fall into the category of Big Data. Research into Big Data analytics has gathered unprecedented momentum. In smart grids, information about the real-time state data, test data, system information, data about defects and faults, inspection records, users' information storage, third-party information storage, end-users' usage data require massive distributed and heterogeneous processing capabilities. Resource demands for an actor in a smart grid application vary over time, with peak operations occurring during the day and a slowdown of information processing at night. Cloud scalability services such needs easily. It must be noted that in spite of the effectiveness of existing

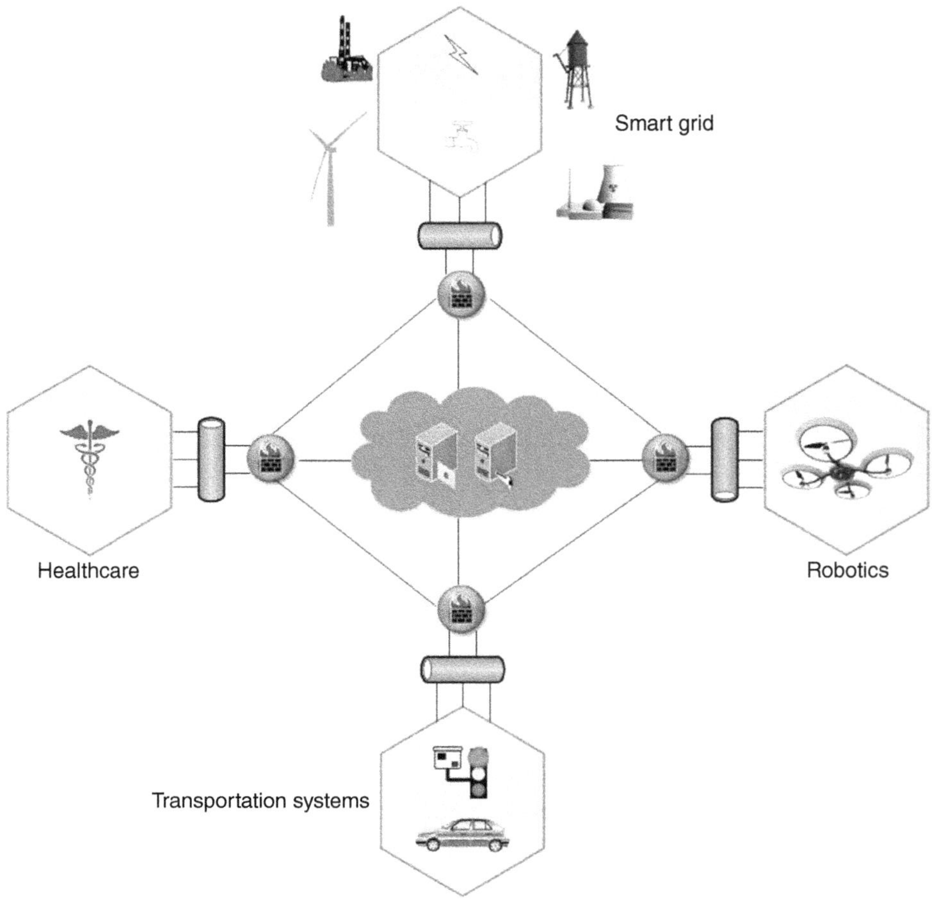

Figure 12.1 Different CPSs connected to the cloud.

algorithms for Big Data management, these cannot be directly used in the CPS domain because of the complexity of satisfying QoS requirements.

Big Data requires vast computational and data storage resources and usage of cloud platforms' elastic computing and storage capabilities is the only logical choice. Since cloud computing is in itself an independent entity, coupling it to a physical system of the CPS to complete the feedback loop has many challenges. The major challenges pertain to satisfying the QoS requirements of CPSs by effective management of computation and communication resources of both the physical and cyber components. This challenge also has to account for scaling of the CPS subsystems (actors). In order to solve such resource-sharing problems, many scheduling and optimization schemes have been proposed in the context of heterogeneous computing. This chapter aims to reflect state-of-the-art resource management methods and algorithms at various stages of data generation, communication, processing, analysis, and decision-making. The chapter attempts to provide a thorough understanding of the need for resource management required by each actor in the CPS.

This chapter is focused on the resource management of a cloud-integrated CPS. Management of resources as provisioned by a cloud environment requires identification and categorical classification. Therefore, based on our understanding, in this chapter, we classify resource management in cloud-assisted CPSs into five functional domains:

- global scheduling of virtualized versions of physical embedded counterparts;
- resource utilization estimation based on QoS;
- local scheduling of cloud resources and scheduling and placement of virtual machines;
- application scaling and provisioning of virtualized resources; and
- workload management.

Discussion about state-of-the-art techniques and methods is centered on these five functional domains. In the subsequent sections, we provide a comprehensive understanding of resource management of static and dynamic CPSs. In the same section, we provide the user with enough information about the scheduling methods used in embedded systems to construct a large-scale and distributed CPS. Aspects of resource management such as cooperative resource allocation and allocation under constraints are discussed to provide the reader with an understanding that would help to see the bigger picture this chapter inches toward. Thereafter, we will provide insights into how Big Data plays a major role in resource management of cloud-assisted CPSs. Data collected from the embedded components can grow manyfold over a short period of time and can quickly fall under the gamut of Big Data. Consequently, we will provide insights into how and where such resource-management schemes fit into the gamut of CPS applications requiring elastic storage and computing for Big Data. Finally, we will conclude the chapter by providing insights from our understanding of the existing resource-management schemes for their use in clouds-integrated CPSs by using a case study of an automobile CPS. An automobile CPS is perhaps the best example of how CPSs are evolving. It has subcomponents that comprise interconnected embedded and physical (mechanical) components. While the resource management of a single automobile is complex in itself, multiple automobiles can form a local group and interact/communicate with each other. This increases the complexity of the overall distributed system exponentially. This chapter lists the state-of-the-art technologies used in automobile CPSs and describes how integrating the cloud within such a dynamic CPS can be used at the various stages, setting a precedent for the need to develop resource management algorithms and methods.

12.2 Resource Management in Static and Dynamic CPSs

The theory of real-time systems traditionally concerns the temporal aspects of resource management, concerning the need for resource allocation. With the introduction of multiprocessor and multicore computing, the notion of locality is introduced to the problem. As systems spread out even further, connecting components through system buses or network protocols, spatial effects grow stronger. Distributed systems can arise by two primary mechanisms: decomposition of large systems or composition of individual parts. What separates a part in a distributed system from another does not have to be the Euclidian space but the two parts can be considered separate if they do not share the same resources. Cyber-physical systems are distributed systems that fall under two topological settings: static and dynamic.

12.2.1 Systems on Static Topologies

A static topology system is one in which the structure is explicitly defined at the start of the system design and remains unchanged. The physical connections among the components of the topology are well defined and do not change over time. An example of static topology is the networked control system (NCS) in which the scheduling and processing can be done using task graphs and queue systems. Task graph scheduling for a system is a popular and widely researched formulation of software components where, for each task, an assignment and a schedule are determined. The objective is to map a set of tasks with precedence constraints onto a set of identical processors, a problem proven to be NP-complete [1]. This formulation has resulted in numerous heuristic designs, most of them static in nature, which makes them infeasible for scenarios with high variability in resource availability. Recent work has moved toward more dynamic formulations, acknowledging the need to consider more uncertain situations [2].

Networked control systems such as industrial control systems are a common type of static topology system, where controllers, actuators, and sensors are physically separate nodes in the network but connected through some communication channel. Since control system performance is strongly dependent on timely measurements and actions, managing resources is key to a successful design. Applying control theory to the joint problem of managing the resources and the plant is a natural step, studied by [3]. This would later give rise to the event-based control structures discussed by [4, 5]. Given the similarities between scheduling of processor time and queue theory, there exist static topology resource management problems referred to as congestion control [6, 7]. With the ever-increasing demands for cost efficiency, it is expected that more and more dynamic methods will be applied in order to refrain from over-provisioning of available resources, as exemplified in [8].

12.2.2 Systems on Dynamic Topologies

A dynamic topology system is one in which the structures change over time, either autonomously or driven by decisions. Ad-hoc networks and overlay networks are typical examples of systems with dynamic topologies. In an autonomous vehicle CPS, such types of networks are present.

Peer-to-peer (P2P) and peer-assisted solutions have been used in wired and a few wireless networks to reduce infrastructure load. While traditional P2P has proven effective for cooperative resource sharing, the solutions provided are sometimes deemed system agnostic. This gives rise to further complications in areas of scheduling network traffic, which greatly increases congestion and consumption of other resources such as energy. In this subsection, we mention the effects of some scheduling schemes on energy management/consumption. However, we do not delve deeply into this particular aspect in order to focus more on resource management of the cloud-integrated aspects of CPSs.

Studies of local cooperative techniques that account for the energy cost of sharing data have recently been done by several independent groups. Common among them is that they leverage the lower energy consumption of short-range communication and point out the additional benefit of local schemes in that they offload the long-range infrastructure [9]. Authors in [10] introduce a framework for cooperative sharing, starting with an altruistic scheme and then introducing guard conditions to prevent participants from spending too much energy on serving content. The approach is based on a Markov decision process (MDP) formulation, which is studied in detailed in [11], and is validated through an implementation on Android-based phones showing substantial energy savings. One issue with the experimental setup is that the clients involved are assumed to be in contact with each other indefinitely. Additionally, the approach does not guarantee that participants will actually benefit from the scheme. Authors in [12] consider a game-theoretical approach for a streaming service. The theoretical framework includes both power and bandwidth considerations.

Authors in [13] also approached the problem using a game-theoretical basis, but explicitly addressed the energy issue and fairness. The proposed scheme permits selecting utility in a variety of ways, including greedy but unfair minimization of the total energy used by the system and minimizing the maximum energy spent by any client.

Based on this classification, in order to state the design problems associated with embedded systems in resource management terms, access to hardware and software resources must be exposed to the system designers as a divisible resource. Resource access in computer systems has traditionally been posed as a scheduling problem, but through virtualization and reservation-based scheduling (RBS) techniques, the access problem can be restated in resource allocation terms applied to CPS applications.

12.2.3 Resource Management in Embedded CPSs

This section introduces the prior research related to RBS that is necessary to understand the structures for the competitive, collaborative, and cooperative resource management strategies discussed in this chapter with respect to CPS applications. Central concepts are introduced, along with a brief historical survey of RBS, which is important for understanding when it is used in CPSs. The theory is derived from both its real-time scheduling roots and its queuing theory counterpart. Using the RBS formulation, a resource management policy can be posed as the solution to an optimization problem. These techniques are feed-forward techniques in which the allocations are calculated based on models rather than online measurements. The following are some of the most important considerations to be handled for effective resource management in embedded CPSs.

Temporal Isolation

A highly desirable property of the RBS approach is that a task that has reserved a specific amount of a resource should have access to this regardless of what other tasks are running on the system. This is called temporal isolation and makes very good sense for both continuous media and control-type applications, such as a robot monitoring an environment or a mission-critical networked control system. In both the CPS applications mentioned here, task priority and precedence is of the highest order in most cases due to the sensitivity of the data collected. Therefore, memory protection ensures that applications will behave functionally correctly even in the presence of other, potentially malfunctioning, software. Temporal isolation therefore guarantees temporal correctness.

Components and Compositions

In order to handle the complexity of large CPS systems, it is extremely crucial to build complex components from simpler components and couple them tightly. Each component generates tasks that run as threads with different priorities. Threads are the building blocks of traditional operating systems, which do not compose easily for tightly coupled systems [14]. As mentioned previously, the use of hierarchical RBS techniques enforces temporal isolation to create groups of threads with essentially the same outside properties as the atomic thread. This enables component-wise testing and verification, and also removes the need to explicitly know the structure of third-party software. Every CPS application depicted in Figure 12.1 generates tasks with different priorities. Tasks range from control-related computations to communication of information flow. Tasks are prioritized based on their execution significance in real-time, with the execution significance varying under different conditions such as faulty system behavior and security attacks.

Timing-Sensitive Applications

In real-time situations for CPS applications, the timely completion of the tasks mentioned above is important. Normally, if a task has a hard real-time deadline, it is assumed to function nominally if the deadline is met and fail if the deadline is missed. In soft real-time problems, the severity of this condition is relaxed and occasional deadline miss is allowed. For different CPS applications it is interesting to discuss how this affects the performance. Timing-sensitive CPS applications such as automobile CPS are examples where different soft and hard deadlines exist among the various components.

Graceful QoS Degradation

There may be an admission policy that denies tasks, which prevents the scheduler sustaining reservations. This might not always be desirable from the perspective of a user, but it is still viable with regard to the maintenance of system stability to have a graceful degradation of the performance as opposed to sudden system collapse. This becomes even more evident in embedded systems where resources are scarce. Consider systems such as NCSs; here, the embedded components that actuate the systems operate in a strict near real-time scenario and on most occasions adhere strictly to providing stability and operability with high reliability. However, under uncertainties such as system faults, sensor faults, and attacks, it is better to achieve conditional operability as opposed to complete system shutdown when deadlines of certain tasks are not met. This is called graceful degradation while the QoS of the CPS application is met.

Adaptive Reservations

A problem of the RBS is that execution time for periodic tasks may vary over time. Some deadlines are bound to be missed, as it is undesirable to base the calculations on the worst case. While the constant bandwidth server (CBS) [15] scheme can handle transient overruns, non-transient changes will lead to eventually infinite deadlines, which causes instability. One way to remedy this would be to dynamically set the budget for a server based on prior overrun statistics in a feedback control manner, commonly referred to as adaptivity. In [16], Abeni and Buttazzo introduce a metric called the scheduling error. For a periodic task τ_i with period T_i, the scheduling error ε_s is defined as

$$\varepsilon_s = d_s - (r_{i,j} + T_i) \tag{12.1}$$

i.e., the difference between the server deadline d_s and the soft deadline of the task. Scheduling error ε_s is driven toward 0 by the feedback control signal using the server budget Q_s.

A few of the design techniques for such a controller are discussed in [17]. A restriction is imposed such that even under the availability of extra unused bandwidth, a task t_i scheduled by a CBS will only receive the bandwidth Qs, and hence it is so-called hard reservation. Assume that t_i is a periodic task being served with a CBS, divided into jobs $J_{i,j}$ with the corresponding release times $r_{i,j}$. This gives $r_{i,j} + 1 = r_{i,j} + T_i$, where T_i is the task period. Each job is associated with a soft deadline $d_{i,j} = r_{i,j} + T_i$, that is $d_{i,j} = r_{i,j} + 1$. Let $f_{i,j}$ be the actual finish time for $J_{i,j}$ and $v_{i,j}$ be the finish time had t_i been running alone on a CPU with the fraction $b_i = Q_s = T_s$ of the actual CPU speed, i.e., the virtual finish time. We consider a modified definition of the scheduling error compared to (12.1):

$$E_{i,j} = (f_{i,j-1} - d_{i,j-1})/T_i \tag{12.2}$$

which is the scheduling error experienced for $J_{i,j-1}$. Since hard reservation is being used, having both $\varepsilon_j > 0$ and $\varepsilon_j < 0$ is undesirable since the task would either be missing deadlines or wasting bandwidth. The relation

$$v_{i,j} - \delta \leq f_{i,j} \leq v_{i,j} + \delta \tag{12.3}$$

where $\delta = (1 - b_i)T_s$ conveys that the CBS approximates the general processor sharing (GPS) algorithm by letting T_s go toward 0. It is reasonable to assume the use of (12.2) and (12.3) to approximate the scheduling error with

$$e_{i,j} = (v_{i,j-1} - d_{i,j-1})/T_i \tag{12.4}$$

A *difference* equation for the evolution of the scheduler error is then presented in [18]. This information is useful in determining the way reservations can be made for any CPS application in which deadline misses have to be handled effectively. For example, in the case of robotics, deadline misses are common when there is an abundance of data upon the occurrence of an event. Hence, the above-mentioned scheduling policies can be used to adaptively manage the resources for the various tasks.

12.2.4 Allocation under Resource Constraints

The purpose of resource management is to maximize system performance. Especially in the case of a nontrivial task, under resource constraints it is hard to run all components of the CPS at nominal execution rates. The error rates of all components cannot be kept at zero, and therefore a compromise has to be made in this regard to determine the most acceptable error rate, which can be determined using a global performance metric. For the set of independent components, a natural choice would be an aggregate of the individual utility functions, the selection of which is an important design parameter used to achieve a suitable compromise to the primary system use cases. For example, it is more meaningful to prioritize system services over other components or to minimize the worst-case rate error.

Convex Allocation Problems

In order to solve the allocation problem without human supervision, the utility function to minimize the worst-case error rate needs to be restricted. By limiting the component contributions J_i to differentiable convex functions, the resulting form will qualify as a convex optimization problem [19]. This formulation has several attractive properties:

- it is always a feasible problem with a unique solution;
- it holds a powerful theory for designing solvers; and
- it allows for general forms of system utility functions.

Steady-state rate error can be defined as

$$e_i(u_i) = r_i - y_i = r_i - k_i u_i \tag{12.5}$$

where e_i is the steady-state error, u_i is the input; r_i is the rate set point, y_i is the rate and k_i is the gain. Since (12.5) constitutes an affine mapping, any contribution from the component I, i.e., J_i, is taken as the composition of a convex function and $e_i(u_i)$ will also be convex [19]. For example, $e_i(u_i)$ and $|e_i(u_i)|$ are viable choices for J_i. This would mean that the problem is with rate error distribution and not resource distribution. However, eventually rate error distribution would affect resource distribution as rate error governs the amount of additional resources required to bring it down to 0.

When designing the objective function, some attention should be given to its sensitivity to parameter changes. Since parameters will be estimated online under noisy circumstances, adopting a linear programming-style objective can lead to sudden jumps in the solution. The authors of [20] provide an in-depth discussion on the merits of linear and quadratic objectives. This can be applied to the case of a CPS application employing embedded systems to actuate its control elements. The sensors that send information to the control unit or actuators send data with different data rates. The data sent are mission critical and thus the error rate must be kept at the minimum for which such convex allocation problems can be effectively used.

Measuring Time and Resource Consumption

The entire CPS framework requires the components to send data to the estimation algorithm in the form of cycle completion events. These events comprise time stamps and the accumulated CPU time for the component.

The resource measurement is performed at each time stamp owing to the sending and processing latencies. The cost of resource measurement is factored into all of the components, making it more distributed than centralized, reducing the computational burden on the estimation algorithm. Resource consumption measurements are done using system calls as it provides better precision. Alternatively, a CPU subsystem available with control groups can be chosen, but the system call is both simpler to use and more efficient than parsing text files.

Reservations with CFS Group Scheduling

The completely fair scheduler (CFS) group scheduling functionality offers good soft reservation capabilities. It is more intensive as the allocation theory assumes hard reservations. Resource allocation shares are calculated as a percentage of the total CPU capacity. The shares are translated into an integer number in such a way that the desired share equals the integer weight or sum of all weights. Quantization can be avoided by using a higher number of shares. The CFS scheduler does not choose the reservation period of each individual task, disallowing additional parameters to be set apart from the shares. Such a methodology can be used in the case of a smart grid CPS application where each actor who is trying to gain access to or push data into the centralized processing unit can be guaranteed a fair scheduling. The data about the actors can vary from private information to information representing the system state. In such cases, it is a nontrivial task to effectively determine the overall impact of choosing one set of users' data over the others. Hence, CFS group scheduling can be very effective in this case.

12.2.5 Cooperative Resource Management

Game theory has shown that two rational decision-makers will not cooperate if there is the possibility of one of them taking advantage of the other. This case is often referred to as the prisoners' dilemma [21] and illustrates the importance of trust between parties in a voluntary cooperation scenario. This scenario can be translated into the terms of

Table 12.1 Prisoners' dilemma for the sharing scenario

Cost for (A,B)	B cooperative	B uncooperative
A cooperative	$(w_l + 2w_s, w_l + 2w_s)$	$(2w_l + w_s, w_l + w_s)$
A uncooperative	$(w_l + w_s; 2w_l + w_s)$	$(2w_l ;2w_l)$

a cooperative file-retrieval scenario in a CPS. If two geographically colocated mobile robots, A and B, seek to access data which both the robots want from a database or a server or a cloud component, they could do so by cooperating to share the costs incurred for energy and communication. Let w_l denote the cost of accessing the data from a remote source and w_s the cost of communicating with a colocated source. Accessing data requires energy expenditure by sender and receiver, meaning that one local access incurs a cost of w_s for both robots. Assume also that $w_l \gg w_s$. A and B can now independently choose to either be cooperative or uncooperative based on the decisions made. The outcome of the possible decisions to retrieve both objects in terms of cost is listed in Table 12.1.

If both choose to cooperate, the costs will consist of one database access to fetch the data followed by two short-range accesses, where one is to deliver the object to the other robot and the other one is to get the other object from the database. In case only one chooses to cooperate, both will start out fetching one object from the database, but only one will be able to fetch from its peer. The other will then have to perform another database access to complete the set. The Nash equilibrium [22] of this game is that both A and B choose to be uncooperative. The two strategies that would resolve this problem are:

- w_s must be reduced to 0, to enable cooperation albeit with risks, or
- the off-diagonal choices must be eliminated, making it impossible for one party to exploit the other.

The first strategy can be seen in the behavior of traditional peer-to-peer networks, where cost of sharing doesn't exist [21]. Even a greedy participant's win doesn't incur any cost. In the applications of mobile CPSs, such assumptions become unrealistic owing to the constraints of battery and communication. Therefore, it is necessary that the system provides actors with the guarantee that the advantages outweigh the costs. A simpler scenario can be achieved by enforcing that all the transactions are bilaterally either cooperative or uncooperative. Research has shown that multicast transfers improve the efficiency of cooperation [23], but there is the risk that actors will opt out of the exchange system and just listen to multicasts. This will again reintroduce a prisoners' dilemma situation.

12.3 Big Data in CPSs

Up to now we have seen how different strategies exist in the context of pure CPS resource management. However, owing to recent trends in the evolution of CPSs and an exponential increase in the amount of data being generated for processing, Big Data analysis plays a major role in effective resource management of CPSs. Therefore, from our understanding, we categorize that there are four key differences that make Big Data modeling and analytics for CPSs different from similar issues in the general-purpose computing domain, where scalability of the cloud may directly be used for processing Big Data:

1. the tight coupling of computational elements with the physical elements through feedback control loops; and
2. a stricter and more rigorous engineering process for both hard and soft real-time CPSs as opposed to the process used in standard software engineering processes.

The above two lead to a formalized model-based design [24, 25] demonstrated by a variety of engineering tools, such as Simulink/Stateflow, Ptolemy [26], and more:

3. CPS systems comprise of many more operational modules as compared to IT systems. Such IT systems have a small number of operational models and they can be processed by combining domain knowledge and brute-force search to determine the state changes. However, such approaches do not fit systems with several discrete states that are hard to distinguish, such as in the case of complex CPSs [27].
4. The role and interpretation of data analytics is different in CPSs. In IT systems, the data can be processed with standard machine-learning techniques by increasing the state space. This cannot be applied directly to the case of a CPS owing to the data diversity, and thus it requires more complex data analytics.

For example, feature extraction can be used on the incoming stream of data on the first and second derivatives. Extracting the features of CPS data will enable us to distinguish between different operational modes. However, it is still a significant challenge to determine the most important variable for data analytics, given the constraints of cost and computation. Ad-hoc analytics cannot perform well because of the need for rigorous modeling of a CPS. Based on these differences, Big Data modeling and analytics for CPSs merits further research. We next describe the two major challenges in the research of Big Data analytics for CPSs.

Partial Knowledge of Input–Output Relations

The basic assumption is that the input and output variables for each subsystem of a CPS are known. This assumption is not always valid in a real-time scenario. Sensor data of a CPS usually require time series analysis and comprise thousands of metrics [28]. In such cases, only partial information about the input–output relations exists and even domain experts and operators may not be able to draw complete information. In a distributed computing systems environment such as a CPS, several factors make it impossible for humans to understand the entire system. Some of the limitations in understanding the system are the large scale of the application and complex interactions across components. In such environments, it is challenging to deploy a heterogeneous class of applications using the limited resources while still maintaining a high-quality service. CPS data comprise continuous and analogous data in applications such as a large manufacturing plant, transportation networks, and power grids. The data deluge caused by continuous sensing in a CPS is a hard problem to solve. There is a need to improve the ability to measure finer spatial and temporal granularity [29] and hence resource management on a scalable platform is of utmost importance.

Online Processing of High-Dimensional, Low-Quality Data
The authors in [29] demonstrate the need to tackle the analysis of massive amounts of high-dimensional data generated from the low-information content collected by sensors embedded into a real-time CPS. With the rise of industrial Big Data [30], and smart infrastructure in cities, transportation, homes, etc., a new challenge for modeling CPSs across multiple domains has arisen. Novel algorithms have to be developed to tackle this challenge, which requires an interdisciplinary approach that combines performance modeling and system building. For instance, a D-Streams programming model [31] deployed over Spark [32] is used to decompose the incoming stream of data into small groups for better analysis. The application constituted the use of an expectation maximization (EM) algorithm to estimate travel times on a network of roads from noisy GPS data in real time.

12.4 Scope of CPS Cloud Computing Resource Management

With this understanding for the need of a scalable computing resource to handle complex Big Data problems arising in CPSs, a cloud-based cyber component can be used effectively. Therefore, in this section we outline the scope of cyber resource-management in a cloud-assisted CPS. We specify the actor(s) and their role(s) in the resource management process along with their management objectives. Then, we discuss the various resources in a cloud and the technologies that are required to manage such resources. By the end of this section, we expect that the reader will understand the current cloud practices that can be used as is, and ones that need algorithmic modifications to cater to the QoS requirements of a CPS.

12.4.1 Actors and Cloud Application Provisioning Models

Cloud service providers provide users with three layers of resources: infrastructure as a service (IaaS), platform as a service (PaaS), and software as a service (SaaS). However, due to lack of a standardization technique in the cloud environment, the interconnections among cloud platforms and their roles and services are not concretely defined. In order to alleviate such inconsistency, recent standardization efforts, notably the OGF's Open Cloud Computing Interface (OCCI) [33], the DMTF's Cloud Infrastructure Management Interface (CIMI) [34], and Amazon Web Services [35], have provided a more technical understanding about the three layers of the cloud. Such strong technical understanding of each layer helps in providing a clear delineation of the layers in private, public, and hybrid clouds. As outlined by Armbrust et al. [36], we can assume that the following set of actors can present in a cloud environment: cloud provider, cloud user, and end-user. Depending upon the usage of the different layers of the cloud, these actors can be composed of a group of organizations using a public cloud or a single organization having its own private cloud. A brief description of the roles and responsibilities of each of the actors is given below.

Cloud provider: A cloud provider maintains the data center hardware and software. The cloud provider usually provides IaaS and/or PaaS resource abstractions to the cloud user or the end-user in a public or hybrid cloud. In a private cloud, the cloud provider provides SaaS along with the scope of maintaining IaaS abstractions to the end-user. A cloud provider acts according to the agreed upon service level agreements (SLAs) to deliver and meet the required QoS and resource management objectives.

Cloud user: A cloud user can be a single organization that uses the public cloud as provided by the cloud provider to offer its services of PaaS or SaaS to the end-user. The cloud user is responsible for meeting SLAs agreed with end-users. The objective of a cloud user is to minimize costs and maximize profits while ensuring that the scale of the resources as requested by the end-user is met by leasing the right amount of resources from a cloud provider. An example of the cloud user can be explained in the case of a smart grid, where intermediary power distributors or homeowners are the actors who provide services to the end-user.

End-user: An end-user simply uses the services as provided by the cloud provider or the cloud user. The end-user does not manage the resources of the cloud, but dictates the resource management by the cloud provider and the cloud user by generating workloads from the application for which the cloud resources are requested. End-users make use of the computational servers, storage devices, and networking equipment as the resources of the cloud. Applications are often run on specialized software platforms that provide software abstractions and application provisioning functionality. End-users in the context of a CPS can be the physical systems that need access to the cloud.

12.4.2 Management Objectives

Actors in a cloud-computing environment achieve objectives by configuring or influencing the entire resource management process. Of the different types of resource management scenarios, we provide one in which the cloud user leases resources from the cloud provider at an IaaS interface.

The primary objective of a cloud provider in the context of providing IaaS to the cloud user is to satisfy the SLA for the provisioning of a virtual infrastructure. An SLA is a formal agreement between the cloud provider and the cloud user, defining in quantitative terms the functional and non-functional aspects of the service being offered. The scope, specificity, and granularity of the SLAs vary widely, encompassing various aspects such as availability, performance, security and privacy, data access, fault tolerance, resource management, and dispute mediation. Quantifying these services doesn't have a standard and can be agreed upon in the SLAs. Among these management objectives, availability (e.g., minimum uptime of the services) and performance (average maximum processing delay) are directly related to resource management. An example of a typical SLA for an IaaS service is that of the Amazon EC2 service [37], which defines availability as the monthly uptime percentage of at least 99.95 percent for Amazon EC2 and Amazon EBS. A European Commission report [38] provides a detailed discussion about SLAs and the ensuing research on them. The cloud provider offers different levels of service to its customers based on their access priority for resources and may allocate the resources based on the terms of SLAs.

In addition to management objectives based purely on the SLAs, the cloud provider will ensure that its infrastructure management is better handled. Objectives include: (1) load balancing, where resources are provisioned in such a manner that the incoming data (in this particular case the CPS application) is fairly balanced across the cloud provider's infrastructure to optimize the overall utilization; (2) fault tolerance, where resource allocation is done in a manner that reduces the impact of a system failure on the system performance; and (3) energy minimization, where resource allocation is done in such a manner that the overall energy to execute a workload from the CPS application is reduced. The cloud provider may utilize a number of optimization techniques, as this entire process is nontrivial and fairly difficult to achieve given the various constraints imposed by the SLAs and QoS requirements. To solve this problem, the cloud user would use the elasticity property which Herbst et al. [39] discuss in detail and propose an alternative definition. The cloud user may thus formulate management objectives that reflect its approach to resource reservation either by conservatively over-provisioning resources (at extra cost) to accommodate demand surges, or aggressively minimizing cost by closely matching the level of leased resources with demand, but running the risk of violating SLAs with end-users.

12.4.3 Resource Types

This section details the main resources of a cloud that play a pivotal role in servicing the CPS application.

Computation Resources

Computation resources are the collection of physical machines (PMs), each composed of one or more processors, memory, network interface, and local I/O, which together provide the computational capacity of a cloud environment. Typically, PMs have deployed on them virtualization software that allows them to host a number of virtual machines (VMs) that are isolated from each other and that may run different operating systems, platforms, and applications. In the literature, most researchers model VMs and PMs as being constrained by their processing capacity and memory availability. However, recent work [40, 41] highlights the impact of contention between VMs caused by shared processor caches and other micro-architectural resources, suggesting that resource management process may benefit from more detailed models of computation resources.

Networking Resources

Computation resources (on PMs) within a data center are packaged into racks and are typically organized as clusters of thousands of hosts for resource allocation purposes. Critically, these PMs must be interconnected with a high-bandwidth network, typically built upon gigabit Ethernet or InfiniBand technologies. For all cloud-hosted applications, particularly those realized by a parallel computation, communication overhead imposed by the data center networking technologies and networking protocols constrain overall performance. There are two critical aspects. The first is network topology, the design of which significantly impacts performance and fault tolerance. Current data center network topologies are based on hierarchical, tree-like topologies similar to those used in early telephony networks, although a number of alternative topologies including proposals based on fat trees [42], hyper-cubes [43], and randomized small-world topologies [44] have emerged. In all cases, a key goal is engineering a scalable topology in which increasing the number of ports in the network should linearly increase the delivered bisection bandwidth [45]. The second key aspect is more directly tied to resource management: It is how to provide predictable latency and bandwidth in a data center network in the face of varying traffic patterns. Traditionally, the solution has been network over-provisioning, but this is prohibitively expensive in large-scale data centers and is inherently difficult due to lack of detailed traffic models. Given this, there has been a move toward implementing service differentiation via QoS policies that segregate traffic for performance isolation, so permitting high-level traffic engineering. A natural extension of this approach toward technologies enabling the virtualization of data center networks is currently gaining attention [46]. The creation of virtual data center networks offers the opportunity to deploy custom network addressing schemes and networking protocols, the latter being a particular concern given the problems associated with the TCP incast behavior in the data center context [47].

12.4.4 Resource Management Functions

In this section, we outline the main functionalities of the resource management systems for cloud-based CPS environments. These functional elements should coordinate to provide a complete resource management solution that monitors and controls different (physical and virtualized) resource types in line with management objectives. The cloud provider is responsible for monitoring the utilization of computation, networking, storage, and power resources and for controlling this utilization via global and local scheduling processes. The cloud user monitors and controls the deployment of its application modules on the infrastructure it leases from the cloud provider, and controls the workload it receives from end-users of its application(s). Both cloud provider and cloud user may have the ability to dynamically alter the prices they charge for leasing of infrastructure and usage of applications, respectively. The responsibilities of the cloud user are broadly similar to the tasks involved in managing dedicated infrastructure, but with the additional flexibility to request or release resources in response to changing application demands, or indeed, to changes in prices offered by the cloud provider. Finally, the end-user has limited responsibility for resource management; they generate workload requests (possibly influenced by dynamic prices quoted by the cloud user) and in some cases exert control of where and when workloads are placed. The framework is depicted from an IaaS perspective. However, it is also applicable to the PaaS and SaaS, where perspectives of the functional elements remain the same, but responsibility for provision of more of them rests with the cloud provider. In the PaaS case, the role of the cloud user is split into platform provider and application provider, where the degree of resource allocation responsibility falling on each varies depending on the scope of the provided platform. For example, with Microsoft Azure, application providers are responsible for controlling the admission and dispatch of workload requests, while with Google App Engine, the platform automatically handles demand scaling, launching application instances transparently as workload volumes increase. In the SaaS case, the platform and application provider are typically the same organization, and often that organization is also the cloud provider; such an organization would then have responsibility for all resource management functionality. We now introduce the scope of the functional entities in more detail; later, we survey the state-of-the-art in techniques for their realization.

Global Scheduling of Virtualized Resources

The cloud provider must support resources either directly leased by clouds users (for IaaS clouds) or utilized to support cloud-hosted applications (for PaaS/SaaS clouds). Global scheduling relates to the system-wide monitoring and control of virtualized and underlying physical resources in line with the cloud provider's management objectives. It encompasses admission control of requests from IaaS cloud users for deployment, the initial placement of physical infrastructure, and the adaptation of this placement over time.

Resource Utilization Estimation

Resource management decisions by the cloud provider and the cloud user require accurate estimations of the state of the physical and virtualized resources required to deliver the QoS needs of the applications. State estimation for computation, network, storage, and power resources are provided by a resource utilization estimation element which provides input into cloud monitoring and resource scheduling processes.

Local Scheduling of Virtualized Resources

Physical resources employ local resource management and scheduling functions to decide how to share access between the virtualized resources placed on them. For computation resources a hypervisor deployed on a machine will control the sharing of CPU, local I/O, and network interfaces between the VMs it hosts. Depending on hypervisor capabilities, local provisioning or scheduling functional elements may have the ability to dynamically adjust the share of resources allocated to VMs. Similarly, controllers of distributed storage systems will control access to storage capacity, network admission controllers and multi-path forwarding controllers share network resources, and power management systems can control power utilization through device control and server consolidation. In all cases the operation of local provisioning/scheduling functional elements should be coordinated so that they collectively satisfy management objectives.

Application Scaling and Provisioning

Cloud users who deploy applications leased from an IaaS cloud provider typically need to control how the software modules that comprise their application(s) are deployed over this, while in PaaS/SaaS clouds the cloud provider directly assumes this responsibility. For applications composed of multiple software modules, and for groups of applications that are closely coupled, their initial placement must be ascertained. In particular, for computation-intensive applications, such as MapReduce implementations, software modules may need to be configured for optimal resource usage given the characteristics of the VM they will execute on. Furthermore, as application demand changes, the placement and configuration of application modules may be dynamically changed – a process whose success is contingent of accurate estimation of future demand. This functionality is realized by the application scaling and provisioning functional element. In some cases, cloud users may offer end-users the capability of deploying bespoke software modules on their own; these users will need to make the decision as to whether they have sufficient resources available to deploy such modules. Finally, in response to changing application demand, the amount leased to support the application will need to be dynamically scaled.

Workload Management

Once application modules have been placed on the VM a cloud user must exert control over the workload requests generated by end-users. Initially, the cloud user must decide whether a given workload request can be processed in line with the cloud user's management objectives. Such management objectives may state, for example, that all accepted workloads should satisfy specified availability and performance SLAs. Once a workload is accepted, a related task is to dispatch it to one of the potentially many instantiations of the software modules that manage its execution, again in line with the cloud user's management objectives.

Cloud Management Systems

Management systems are used by the cloud providers and sometimes by the cloud users to provide the reporting and command-and-control interface used by human system administrators. They are used to define management metrics that provide feedback regarding the operation of the resource management system. These metrics will use the resource utilization estimates and demand profiles, together with other system information, all of which can be aggregated and fused to provide meaningful monitoring information. In particular, of critical importance is monitoring of the degree to which the system meets its SLAs. For the cloud provider these SLAs will be with its customers (the cloud users), whereas the cloud user may choose to monitor the performance of the cloud provider in meeting SLAs and/or may monitor its own adherence to the SLAs it has agreed with its own customers (the end-users). As discussed in Section 12.2, management objectives define the strategies according to which cloud resources are allocated to CPS applications.

12.5 Scheduling of the Cloud Resources to Satisfy the QoS

This section discusses scheduling and provisioning of virtual resources in servicing the CPS application.

12.5.1 Global Scheduling of Virtualized Resources

Global scheduling involves a system-wide perspective on the allocation of the physical and virtualized resources that comprise a cloud environment. Recent years have seen a proliferation of proposals for frameworks and techniques for global scheduling. Most proposals originate from academic researchers; notable exceptions include the discussion by Gulati et al. [48] of the VMware Distributed Resource Scheduler and Distributed Power Management systems, which provide centralized control of a cluster of virtualized servers, performing VM placement, as well as load balancing and VM consolidation for live migration. Frameworks from the academic community that incorporate controllers directly or indirectly cooperating to achieve global scheduling include: vManage [49], which loosely couples controllers for power and cooling management with controllers for virtualization management; the system proposed by Beloglazov and Buyya [50], which comprises controllers that cooperatively control VM placement and live migration; and TROPIC [51], which performs transactional cloud resource orchestrations that enforce safety, provide concurrency, and increase system robustness. The majority of proposed frameworks utilize resource controllers that are centralized in the sense that they have full control of the allocation of a given set of resources. Such solutions can reasonably be expected to scale only to systems incorporated on the order of thousands of servers. To meet the scalability challenge, some frameworks [52, 53] propose hierarchical organization of controllers. A small number of frameworks incorporate decentralized controllers as proposed by Wuhib et al. [53, 54] and Feller et al. [55].

VM Placement and Provisioning

Whenever a request for the provisioning of VMs is made by a cloud user, the cloud provider's resource management system schedules the VMs by placing them onto PMs. Such a decision should help achieve the cloud provider's current management objective(s). Initial work on the VM placement problem typically assumed that VMs are assigned static shares of a PM's managed resources (typically CPU and/or memory). Given this, the placement of VMs onto PMs is related to the vector bin packing problem, which can be used to model static resource allocation problems where the resources used by each item are additive. To achieve server consolidation, for instance, the optimal placement is one where the items (i.e., VMs) are packed into a minimum number of bins (i.e., PMs), such that the vector sum of the items received by any bin does not exceed the bin's (resource) limit. The vector bin packing problem and its variants are NP-hard problems; thus, heuristic algorithms are commonly proposed and evaluated.

Most proposed heuristics are based on greedy algorithms using simple rules, such as first fit decreasing (FFD) and best fit decreasing. Panigrahy et al. [56] study variants of the FFD algorithm and propose a new geometric heuristic algorithm, which scales to large data centers without a significant decrease in performance. Wilcox et al. [57] propose a new generic algorithm, the Reordering Grouping Genetic Algorithm (RGGA), which they apply to VM placement. Other works proposing bin packing heuristics for placement of VMs on PMs include [58–60]. Many variants of the VM placement problem have been studied. Cardosa et al. [61] propose a solution for VM placement that accounts for the ability of some hypervisors to dynamically allocate a pool of unreserved CPU and memory resources between a set of contending VMs. Swaminathan et al. [62] provide a solution for private clouds in which VMs are placed on those PMs that already host VMs with complementary workloads. Meng et al. [63] propose a VM placement solution in which multiple VMs are consolidated and provisioned together, in order to exploit the statistical multiplexing among the workload patterns of the individual VMs. More recent works recognize that, increasingly, provisioning requests from cloud users involve sets of VMs, rather than single VMs. In such cases, placement of the set of VMs is not only constrained by the collated resource requirements, but also by a range of possible placement constraints indicated by the cloud user.

Breitgand and Epstein [64] consider elastic services realized by a fixed constellation of VMs, but where the number of VM instances and their size varies in line with demand. They define an elastic services placement problem (ESPP) that extends the generalized assignment problem (GAP), but which is harder to solve directly as it does not admit a constant factor approximation. Instead, they transform ESPP into an analogous multi-unit combinatorial auction, which they solve using a column-generation method that provides good solutions for realistic-sized placement problems in reasonable time. A different perspective is taken by De and Roy [65], who outline a solution that focuses on the resiliency of VMs comprising a multi-tiered application, by seeking to spread the VMs across as many PMs as possible. Bin et al. [66] addressed a similar problem – that of making VMs resilient to k PM failures, so that if up to k such failures occur, there is a guarantee that this VM is relocated to another PM without relocating other VMs.

Dynamic VM Placement

Following initial placement, VMs can be rescaled (in a limited way) as demand from the CPS changes. However, from a global scheduling perspective a more powerful tool is live migration of VMs [67], whereby running VMs are paused, serialized, and transferred to a different PM, where they are once again scheduled for execution. Live migration offers many benefits, including improving fault management, simplifying system maintenance, and, from the resource management perspective, enabling global scheduling objectives, such as balancing load across PMs and consolidating

VMs on a minimal number of PMs. Many studies have assessed the overhead of VM live migration proposals for optimizing the process [68]. Bobro et al. [69] propose a first heuristic that dynamically remaps VMs to PMs in a manner that minimizes the number of PMs required to support a workload at a specified allowable rate of SLA violations; they present an extensive evaluation of the approach based on production traces. Sharma et al. [70, 71] propose a set of techniques for VM rescaling, replication, and live migration, in which the problems are formulated as integer linear programming problems for which greedy heuristic solutions are proposed. Zhang et al. [72] address control of VM migrations in scenarios in which the cloud provider intentionally overcommits their PM resources to VMs; their technique minimizes the number of VM migrations and migrates VMs with strong utilization correlation to different PMs in order to minimize the risk of future PM overload. From a more practical perspective, researchers have observed [73] that in public clouds the same VM instance types can exhibit significantly different performance levels depending on characteristics (processor, I/O speed, etc.) of the hosting PM; heterogeneity should thus be taken into account by VM migration controllers. Guo et al. [74] address the use of VM migration in the context of cloud bursting for enterprise applications, focusing in particular on the decision regarding which VMs to migrate to an external cloud environment and when. Shifrin et al. [75] also address cloud bursting; they analyze a more generalized task scheduling problem formulated as a Markov decision process and show that a threshold-based scheduling policy is optimal. Foster et al. [76] also address VM migration and VM consolidation, but from the perspective of how these are used to satisfy dynamically changing management objectives that are enforced via policies. Their study addresses how to change between the conflicting goals of minimizing SLA violations (by spreading VMs across more PMs) and minimizing power consumption (by consolidating VMs on fewer PMs). While the above solutions are centralized in nature, other works propose decentralized algorithms to control the live migration process. Yanggratoke et al. [77] propose a decentralized solution, utilizing a gossip protocol that realizes an efficient heuristic that again minimizes the number of migrations; their approach is shown to scale to problem sizes above 350,000 VMs. The results from this study are very useful in understanding how Big Data analytics resource management in CPSs can be handled. These works focus on calculating new VM placement plans that meet changing demand patterns. While some authors take into account the desire to minimize the number of VM migrations required to generate the new placement plan, few have focused on the challenge of planning the sequence of migrations in a manner that does not lead to violation of performance and security constraints, which can be applied to a CPS application. Ghorbani and Caeser [78] provide a solution for determining the order of VM migrations considering that there is limited bandwidth available for the migration process. Al-Haj and Al-Shaer [79] formulate VM migration planning as a constraints-satisfaction problem, which they solve using a satisfiability modulo theory (SMT) solver. This approach allows them to take into account a wide range of constraints relating to the migration process itself and the resulting VM placement plan. However, the computational overhead of using an SMT solver may mean that the approach would only be feasible for relatively small problem sizes and may not fit a dynamically changing large CPS application such as vehicular ad-hoc networks or smart grids. Zhang et al. [80] investigate dynamic VM placement in geo-distributed clouds. The goal is to minimize hosting cost while achieving SLA objectives under changing demand patterns and resource costs. Their solution is based on control- and game-theoretic models.

Network-Aware VM Placement

Virtual machines access a PM's network interface to communicate with other applications and system components. Hypervisors typically treat a PM's network interfaces as unmanaged resources. They do not provide a guaranteed allocation to individual VMs, relying on statistical multiplexing and the fact that VMs are unlikely to simultaneously maximize their use of their nominally assigned bandwidth [81]. However, this means that there is potential for VMs to affect each other's performance due to competition for network interface resources [82]. Recent works have addressed how to mitigate this risk by taking this competition into account when placing VMs. Wang et al. [83] apply the stochastic bin packing problem for bandwidth allocation to bursty network traffic flows to VM placement. They treat a VM's demand as a random variable and propose an online VM placement algorithm in which the number of PMs required on which to place a set of VMs is within $(1 + \varepsilon)(1 + \sqrt{2})$ of the optimum for any $\varepsilon > 0$. Breitgand and Epstein [84] improve the result of Wang et al. by an algorithm with a competitive ratio of $(2 + \varepsilon)$ for any $\varepsilon > 0$. Intuitively, it is beneficial to place VMs that interact with each other in close proximity in the data center network topology. Meng et al. [85] formulate heuristics that seek to minimize the aggregate traffic rates perceived by switches in the data center network by placing VMs with large mutual bandwidth in close proximity to each other. Evaluation results using production traces show a significant performance improvement. Similar approaches to minimizing data center traffic have been proposed by Jayasinghe et al. [86], Shrivastra et al. [87], and Wen et al. [88]. Biran et al. [89] address network-aware VM placement; their min cut ratio-aware VM placement (MCRVMP) formulation and heuristics to solve it incorporate constraints evolving from complex network topologies and dynamic routing schemes, but also take into account the time-varying nature of traffic demands in a manner that finds placements that have sufficient spare capacity across the network to absorb unpredicted traffic bursts.

12.5.2 Resource Utilization Estimation

Resource management processes require utilization estimates for two main purposes: first, to assess the level of free resources of given types within the system; and, second, to profile the resource utilization patterns of individual workloads so that impact of allocating resources to these workloads can be assessed. The latter is especially relevant for workloads exhibiting burstiness in resource utilization, where resource management processes often over-commit resources to workloads relying on statistical multiplexing to smooth out the average resource usage. The majority of cloud resource management research papers assume that instantaneous and/or historical resource usage measurements are accurate and readily available. Indeed, commonly used tools such as Nagios [90] and Ganglia [91] provide the capability to monitor a wide range of monitored metrics, including computation, network, and storage resource utilizations. One notable exception is Gulati et al. [92], who describe the particular CPU and memory related utilization metrics used by VMware's Distributed Power Management (DPM) to trigger management actions including VM migration and PM power-on. Recently, researchers have started to address the issue of estimating the utilization of micro-architectural resources such as shared processor caches, competition for which has been shown to negatively impact the performance of consolidated VMs [93]. Profiling individual application or workload resource utilization profiles is a more complex task, which has been generally addressed as part of resource demand profiling (see Section 12.2). While this is primarily done for resource allocation purposes, it is also important for assessing the amount of resources applications use as inputs into cost models used for pricing cloud services; to this end, Gmach et al. [94] assess a number of approaches for apportioning costs based on resource utilization estimation.

12.5.3 Local Scheduling of Virtualized Resources

In this subsection, we provide an overview of a range of techniques developed to control the localized scheduling of access by VMs to computation, network, and storage resources.

Local Scheduling of VMs

This section outlines how users can specify reservations (minimal allocations of a resource), limits (maximum allocations of a resource), and shares (weights governing the actual allocation when the resource is over-subscribed and other VMs are contending for the allocation). Solutions of this form seek to provide an appropriate balance between static, user-specified resource allocations and automated allocation of resources on VMs. In contrast, researchers tended to focus on providing fully automated solutions, in which continuously monitored VM resource demands are inputs to a resource controller, which dynamically allocates resources to contending VMs. For example, Padala et al. [95] propose a control-theoretic approach involving a two-layer multi-input, multi-output (MIMO) controller that facilitates the use of application priorities for service differentiation under overload. Xu et al. [96] address a similar problem, but apply fuzzy-logic approaches (both fuzzy modeling and fuzzy prediction) to provide input to a resource controller. Two more recent works addressing variants of this problem are by Rao et al. [97], who apply reinforcement learning techniques, and Han et al. [98], who propose what they term as a lightweight resource-scaling approach. Other researchers have addressed the allocation of PM resources to VMs from an alternative perspective, namely the role this process can play in minimizing data center energy usage; we describe two notable examples of such works. Cardosa et al. [99] remark that most work on power consolidation did not take into account the ability, as described above, for hypervisors to dynamically allocate resources.

Urgaonkar et al. [100] describe a solution involving the use of Lyapunov optimization to jointly maximize the utility of average application throughput and energy costs. The works discussed above all address the explicit local allocation of resources to VMs. However, recently researchers have started to address other aspects that influence VM performance. Sukwong et al. [101] address the sequence of VM execution, which they show can significantly impact the response time for requests that arrive concurrently for the different VMs hosted on a PM. They propose a local scheduler combining both compute resource allocation and control of VM execution sequencing.

Local Scheduling of Network Access

As outlined in Section 12.2.4, providing predictable latency and bandwidth is a significant challenge in the cloud. From a local scheduling perspective, TCP operates sub-optimally in virtualized environments. One well known issue is the TCP incast pattern in which high-bandwidth, low-latency conditions lead TCP implementations to behave in a pathological manner in which applications see goodput that is orders of magnitude lower than the link capacity [102]. Gamage et al. [103] report on findings that indicate that high levels of VM consolidation on a PM increase CPU scheduling latencies, which in turn greatly increases the round-trip times for TCP connections. Based on this and the related TCP shortcomings when used in the data center, a number of alternative transport protocols have been proposed and evaluated, e.g., multi-path TCP [104] and D3, a deadline-aware control protocol [105]. Curtis et al. [106] outline Mahout, a solution in which flows are detected at the PM and a central, OpenFlow-like controller provides traffic management functions to optimize the network to transfer these high-bandwidth flows. While the issues outlined above illustrate the potential effect of the network on VM performance, the central task of apportioning available network resources to co-hosted VMs remains. As outlined by Popa et al. [107], sharing network resources is challenging because the network allocation for a VM depends not only on the VMs running on the machine, but also on the other VMs in communication. Jeyakumar et al. [108] also address the network-sharing problem, proposing EyeQ, a solution in which a portion of access link bandwidth is not allocated, providing a simple local solution that offers statistical guarantees on allocation of bandwidth to contending VMs. Kumar et al. [109] focus on data parallel applications and argue that network resources should be shared in a manner cognizant of data transfer patterns that can be adopted by parallel applications. Shieh et al. [110] propose Seawall, a system in which bandwidth is shared among contending VMs to achieve max–min fairness via the tunneling of traffic through congestion-controlled, point-to-multipoint tunnels.

12.5.4 Application Scaling and Provisioning

In the scope of a CPS application, a cloud user that leases VMs from an IaaS cloud provider must monitor demand for its application(s), decide how many instances of the applications to deploy, and on which VMs, all the while observing constraints relating to which applications can/cannot be placed on given VMs. Given the bursty nature of demand from some CPS applications, this task has received considerable attention in recent years. Urgaonkar et al. [111] was one of the first works to propose a solution in which applications can share machines, although they consider only the dynamic modification of the number of instances deployed of a given application. Tang et al. [112] address the complete problem considering the allocation of multiple resource types, dynamic modification of the number of instances, and of placement of individual instances, as well as placement constraints. They formulate a class-constrained multiple-knapsack problem in which the objectives are to maximize the satisfied demand for the collection of applications, to minimize the number of application instance starts and stops, and to maximize the degree to which resource utilization is balanced across the machines. Gmach et al. [113] propose a fuzzy-logic-based controller that dynamically migrates or replicates application instances in response to changing demand.

Some researchers have sought to overcome the scalability limitations inherent in centralized application-scaling controllers. Adam and Stadler [114] describe a service middleware that implements decentralized algorithms for application placement and request routing that operates over a self-organizing management overlay. Their application placement algorithm seeks to maximize a utility function that captures the benefit of allocating computation resources to an application, given the relative importance of different applications. It operates by performing a local optimization on each server, based on state information from that server's neighborhood. Evaluation results show that the solution scales both in terms of number of servers and in the number of supported applications. Wuhib et al. [115] also propose a decentralized solution for application placement, using a round-based gossip protocol; in each round, a server probabilistically selects another server to exchange state information with information both servers then use to adjust their allocation of resources to applications. Moens et al. [116] argue that a hierarchical model in which centralized controllers (placing applications for a number of servers) are organized in a tree-like hierarchy can combine the benefits of close-to-optimal placements by centralized controllers with the scalability of a decentralized solution. They propose a scheme in which the hierarchy is structured in a manner inspired by B-Tree data structures and compare it to a modified version of the centralized algorithm proposed by Tang et al. [112]; however, they do not provide a comparison to a decentralized solution.

Recently, researchers have studied variants of the application provisioning and scaling problem described above. Carrera et al. [117] present an application placement algorithm which addresses fairness between applications. Their formulation seeks to achieve max–min fairness by maximizing the minimum utility value among all applications to which resources need to be allocated. In contrast to previous solutions, their formulation is a nonlinear optimization; they argue that the increased solution complexity is justified, given the improvement in fairness. Ishakian and Bestavros [118] also address fairness between applications, but argue that the primary goal should be to allocate resources to applications so that the degree to which SLAs are satisfied is maximized.

12.5.5 Workload Management

In CPSs, workloads are the application data in the usage sessions that are processed using cloud resources. Requests for these sessions must be admitted or blocked, and must be dispatched to one among a number of application instances that can process them. In this section we discuss works addressing a number of aspects: load balancing, workload control in data analytics frameworks, workload management of storage resources, and workload management in geo-distributed clouds.

Load Balancing

Given that elasticity is one of the main benefits associated with cloud hosting, the focus has been on efficient and scalable load balancing (also referred to as load sharing), sometimes incorporating admission control. Load balancing has been studied in many contexts and over many years; surveys of classical approaches are provided by Wang and Morris [119] and by Kremien and Kramer [120]. In the computing context, load-balancing algorithms seek to distribute workloads across a number of servers in such a manner that the average time taken to complete execution of those workloads is minimized, which typically results in server utilization being maximized and balanced. Load-balancing schemes can be classified as being static or dynamic. In static schemes the current state of the servers is not considered when dispatching workloads; examples of such schemes include random selection of servers and round robin (where workload is sent to servers in a statically defined, periodic sequence). On the other hand, dynamic schemes involve direct notification of or indirect inference of server state by the load balancer. A second classification distinguishes between preemptive and non-preemptive load balancers: in preemptive schemes, workloads, either queued for processing at a server or being processed, can be passed to another server for completion; in non-preemptive schemes, workloads are assigned to a server upon arrival at the load balancer and cannot be subsequently reassigned. Because of the dynamic nature of demands for the majority of cloud-hosted applications and the inefficiencies associated with transferring workloads between servers, most researchers have focused on the development and evaluation of dynamic, non-preemptive load balancing schemes. One of the well-known load-balancing schemes is termed the power-of-d algorithm, which is based on the supermarket model [121]. In this scheme, d servers are sampled, and the workload is dispatched to the server with the lowest utilization. This approach is shown to provide exponentially improved workload execution times (in the number of servers) in comparison to random server selection. One issue with this and similar schemes is the overhead associated with transferring server state to the load balancer. Breitgand et al. [122] address this point and propose an extended supermarket model, for which they show that there is an optimal number of servers that should be monitored to obtain minimal average service time at a given cost. Based on this result, self-adaptive load-balancing schemes can be developed for the cloud-assisted CPS.

Data Analytics Frameworks

An area of workload management that has generated considerable interest in recent years is the allocation of resources to data-intensive applications such as those based on the MapReduce model. Data analytics frameworks like Hadoop, which implement MapReduce, are now widely used in cloud-hosted applications, so how to optimize their performance is of significant interest. Zaharia et al. [123] showed that Hadoop's default task scheduler implementation performed badly when deployed on a public IaaS cloud due to its implicit assumption that computation resources available to it are homogeneous and that tasks make progress linearly. They propose a new task-scheduling algorithm called longest approximate time to end (LATE), which they show is robust to resource heterogeneity and thus leads to significantly improved Hadoop performance. Other researchers have also addressed the optimization of MapReduce implementations. Sandholm and Lai [124] address MapReduce platforms supporting execution of multiple applications; they propose the use of user-assigned priorities to dynamically adjust the allocation of resources to applications. Tan et al. [125] argue that, since map and reduce phases exhibit different characteristics and have a complex and tight interdependency, a scheduler should couple their progress rather than treat them separately. They use an analytic model and simulations to show that their coupling scheduler can outperform the standard fair scheduler by up to an order of magnitude. Recently, a more radical approach to solving these issues has been proposed by Valapalli et al. [126], who describe YARN, an analytics platform in which resource management functions are separated from the programming model: Many scheduling-related functions are delegated to per-job components. This separation provides flexibility in the choice of programming framework, so that the platform can support not only MapReduce, but also other models such as Dryad and Storm.

Besides improving the performances of the schedulers that are central to data analytics frameworks, researchers have also addressed how to provision these frameworks with resources during the execution of jobs. Lee et al. [127] address the issue that the heterogeneity of resources within a typical data center cluster environment means that certain jobs will execute more efficiently on certain machines. They propose a new metric they term "resource share," which captures the contribution of each resource to the progress of a job, showing how it can be used to determine the scheduling of resources in an architecture in which additional computation is required on a shorter-term basis. Hindman et al. [128] recognize that different data analytics frameworks have emerged that are optimized for different applications, and that such frameworks are often simultaneously run in the same cluster. They propose "Mesos," a thin resource-sharing layer that facilitates resource sharing across data analytic frameworks, through providing a standard interface to the frameworks to offer them resources. The frameworks themselves decide which resources to accept and how to use them for their computations. In [129], the authors address the problem of fair resource allocation where resources are heterogeneous, proposing a new resource allocation policy termed dominant resource fairness, which generalizes (weighted) max–min fairness to the multiple-resource scenario. More recently, this approach has been further enhanced to address hierarchical scheduling of multiple resources, in which nodes in a hierarchy receive at least their fair share of resources, regardless of the behavior of other nodes [130]. The study of these techniques has to be done with regard to the data generation in a CPS under the QoS constraints. It is fair enough to assume the above-mentioned works can fit the CPS model to a fair degree, but changes in the details are inevitable for each CPS application.

12.6 Case Study of an Automobile CPS

In order to give readers an understanding of the practical implications and limitations of a real-time CPS, in this section we describe a case study of the automobile CPS.

The case study considered in this chapter was inspired by the automotive industry. There is a growing trend toward complete autonomous systems in automobiles. Although the current technology limits the vehicle to intra-vehicular sensory communications, it still requires efficient resource management. The various onboard sensors share resources such as communication, computation, and storage. Each sensor generates workload requests, with different tasks having varying degrees of priority based on the information collected. The allocation of resources based on their criticality toward the safety of the car is of utmost importance. In order to give an overall picture of the resource allocation requirements of various components of a car, we describe a scaled down version of the car.

12.6.1 Functional View of the Automobile CPS

The model being described is a scaled down version of the actual car, in which the physical components are connected to resources for its functionality. These physical components interact with the physical environment, such as the road, weather, etc. Physical components comprise sensors, which collect information such as position, altitude, and velocity of the car, while actuators are the means that the system has for intra-vehicular environment manipulation, such as the engine and steering wheel. Sensors and actuators require computational activities whose output closes the feedback control loops for system stabilization, which enables the system to follow safety requirements. This setup requires a planner, which ensures that the pre-specified goals and objectives are met. As an example, the planner takes care of the amount of acceleration at multiple points of the overall route of the vehicle. The various decisions taken by the planner are based on several conditions such as saving time, saving fuel, overtaking slower vehicles, avoiding fixed obstacles, and maintaining traffic rules. The lowest level controllers' input information largely determines such behavior. From a high-level perspective, the system is composed of a set of nested control loops of the lower-level controllers where each control loop is activated periodically and has a different frequency. This allocation of communication resources is required to ensure that the sensors do not vie for the same frequency channel, which has the risk of some controllers' data going unnoticed. Obviously, in such situations where each controller has its own frequency, scheduling of network resources is unwanted. However, the information that is sent from the controllers that govern the stability and operational safety of the car needs quick data analysis and feedback for which scheduling on the computation resources is essential.

In this case study, we provide a bottom-up approach where, at the low level, the prototype of the car is equipped with two servos: one reflecting the engine used to move the car, and the other for controlling the steering angle. The two servos are controlled by a pulse width modulation (PWM) signal. Each wheel has a relative encoder attached for monitoring the speed. A potentiometer is mounted on the front of the car in order to get feedback on the steering position. In real-world systems, this accounts for the light detection and ranging (LIDAR) technology currently being used in prototype self-driving cars to steer the car in the designated lanes. LIDAR collects millions of data points every second, which need efficient processing at a very low latency. Hence, the information collected from such sensors is always given highest priority in scheduling mechanisms. Currently, the technology is being developed in such a way that the LIDAR has its own computation resources. However, in the future, the concept of shared resources will creep into the automobile CPS and effective scheduling and execution mechanisms must be developed that can process very high data amounts in low latency times. PID controllers are used to regulate the speed and the turning angle.

Other controllers such as gyros and accelerometers complete the set of low-level sensors. These are used to improve the estimate of the car position. The positioning of the car is always compared with its ideal trajectory, which is done by allocating sufficient compute resources to analyze the data. Information from the LIDAR sensor is used as feedback for the gyros and accelerometers. Deep learning algorithms [131] are employed to make meaningful and accurate representations of the information gathered through the LIDAR or a camera mounted on top of the car. The deep learning algorithms enable the car to follow the lanes on the road. Inputs to the deep learning techniques and the speed controller receive points from another controller that decides the sequence of maneuvers. This controller monitors the execution of the algorithm using encoders described previously, and the gyros and accelerometers to estimate the position of the car with respect to the road. This entire process requires extremely low-latency and high-accuracy computation. In order to allow this to happen, the on-board computational power must be very high. However, owing to the need to reduce the overall weight of the car and to consider design issues, such high computational power cannot be a viable and feasible solution at all times. In order to alleviate such a problem, the use of edge computing and cloud computing is a viable alternative. The use of such technology can only increase the latency, but recent improvements in LTE technology [132] can be made effective in such mission-critical and low-latency QoS requirements of the CPS.

In addition to the deep learning algorithms, vision algorithms can also be used for the mounted cameras by the use of a combination of randomized algorithms (RANSAC) and Kalman filtering [133]. Such sensing activities generate a widely changing computing workload, a situation difficult to manage with the standard tools of digital control, thus necessitating the need for scalable architectures.

It is possible to increase the performance of algorithms, such as described in this case study. The choice of using microprocessors to define the level of sensor and actuator interaction is reinforced; further, to process high-level intelligence systems, higher-level processors can be used. A common bus between the processing units facilitates the data flow for various tasks in the system.

12.6.2 Software Architecture and Mapping

Model-based approaches are used to model the design needs to fine-tune the designer of the control algorithms, schedule algorithms, and inferential algorithms along with the planner, which in turn maintains tight regulation of the pre-specified goals. In addition to highlighting essential mathematical and physical aspects related to the stability of the system and to the correctness of the design, the models composing the functional diagram also suggest a possible decomposition into subsystems, a definition of their relations and of the timing constraints for their execution.

Algorithms are used to generate the computational entities that will be used in the resource-allocation procedure after the system functionalities and time constraints are thoroughly defined. This refinement step produces a set of concurrent tasks, in which each task has a period related to the control loop to which the task belongs. These sets of tasks are represented using a directed acyclic graph (DAG). In the DAG, the nodes correspond to the tasks and the edges correspond to the communications. An edge is associated with a weight, specifying the amount of data to be sent based on the bitrate requirements. Every functional block (controllers) of the automobile CPS component generates a task, which communicates with the tasks from the other functional blocks involved in the same control loop. According to need, a task can be defined into subtasks for better and more efficient processing. For instance, the LIDAR or camera task could be split in two tasks: one for frame capture and one for image processing. The splitting automatically increases the complexity of the scheduling problem. However, such task splitting could be beneficial to the overall computational power utilization. Therefore, a balance has to be struck between power usage and latency. This problem can be alleviated by using the cloud algorithms described for VM allocation under resource constraints in Section 12.5.2.

Once the system has been refined into tasks, these tasks should be mapped into the hardware architecture of the vehicle as described in the functional view in order to take advantage of the physical parallelism enabled by the different computation units and by their interconnection buses. An optimized task allocation scheme that guarantees execution of all the tasks in their deadlines has to be designed. Rajkumar et al. [134] propose an architecture in this regard using parallel threading for an automobile CPS. Their scheme divides the tasks into subtasks and proposes a parallel scheduling scheme to satisfy the basic QoS requirements of the automobile CPS, such as speed acceleration and deceleration, braking, etc. Their scheme uses eight functional tasks

[134]: “BehaviorTask, MissionPlannerTask, OnRoadMotionPlannerTask, PrePlannerTask, RoadBlockageDetector, RobotClient, ServerTask and SimpleControllerTask.”

The BehaviorTask decides what to do when turning, intersection handling, and lane-changing occurs. The MissionPlannerTask interacts with the stored map to decide where to go. The OnRoadMotionPlannerTask and the PrePlannerTask send trajectories to the vehicle controller. The RoadBlockageDetector works with the BehaviorTask so that the vehicle detects a blocked road and finds an alternative route when needed. The SimpleControllerTask receives the actuator commands and directly interfaces with the vehicle hardware such as the accelerator, the brake, and the steering wheel. On the simulation cluster, this task operates in simulation mode, and the ServerTask and the RobotClient behave as the vehicle hardware. In this chapter, our focus is on the OnRoadPlannerTask running the motion planning algorithm with OpenMP enabled. The task generates curvature and velocity profiles for the vehicle hardware, so the lack of resources will affect the control algorithm, making the car drive in an unstable manner. If the planning algorithm does not meet the deadline, the steering wheel, for example, jerks and the car goes to an unexpected place, which can cause an accident.

In addition to satisfying this scheduling requirement, the traffic generated by task communication should be sustainable by the system. For this reason, the allocation of tasks to a specific computing unit should consider both the computational power of the CPU and also the bandwidth required. As a solution to such optimal allocation, tasks that need to communicate must be allocated in the same time window so that they will not increase the traffic in the system. This is because the data exchange will happen internally at the computational unit, and this communication power and cost is low compared to the traditional method of task communication and vying for resources.

By the use of this case study, we have discussed the various issues of resource allocation with regard to each component of the automobile CPS. The case study discusses intra-vehicle resource allocation; however, as the concept of connected cars becomes more and more viable, the resource allocation problems will only multiply. In order to effectively tackle such issues, robust cloud architectures have to be designed that can tightly integrate the various components of the automobile CPS.

12.7 Conclusions

In this chapter we explored the various types of resources-management schemes that have been proposed for embedded systems, which can be applied with some modification to the domain of CPSs. Also, we have identified the various problems that exist with the evolving trend of Big Data in CPSs. In order to satisfy the analytics problem of Big Data in CPSs, we listed various works in the literature that have proposed solutions to similar problems in cloud computing and mentioned how they can be extended to solving the problems of CPS data analytics. Finally, we concluded the chapter by presenting a case study of the automobile CPS to describe from the bottom up the various aspects that require effective resource-management schemes. We sincerely hope that the discussion in this chapter will help the many who are doing research in this area or building some CPS applications.

References

[1] Y. Robert, "Task graph scheduling," in *Encyclopedia of Parallel Computing*. New York: Springer, 2011, pp. 2013–2025.

[2] P. Choudhury, P. Chakrabarti, and R. Kumar, "Online scheduling of dynamic task graphs with communication and contention for multiprocessors," *IEEE Trans. Parallel Distrib. Syst*, vol. 23, no. 1, pp. 126–133, 2012.

[3] M. Branicky, S. Phillips, and W. Zhang, "Scheduling and feedback codesign for networked control systems," in *Proc. 41st IEEE Conf., Decision and Control*, 2002, vol. 2, pp. 1211–1217.

[4] A. Cervin and T. Henningsson, "Scheduling of event-triggered controllers on a shared network," in *Proc. 47th IEEE Conf. Decision and Control*, 2008, pp. 3601–3606.

[5] P. Tabuada, "Event-triggered real-time scheduling of stabilizing control tasks," *IEEE Trans. Autom. Control*, vol. 52, no. 9, pp. 1680–1685, 2007.

[6] F. I. Akyildiz, G. Morabito, and S. Palazzo, "TCP-Peach: a new congestion control scheme for satellite IP networks," *IEEE/ACM Trans. Netw.* vol. 9, no. 3, pp. 307–321, 2001.

[7] D. Katabi, M. Handley, and C. Rohrs, "Congestion control for high bandwidth-delay product networks," in *Proc. 2002 Conf. Applications, Technologies, Architectures, and Protocols for Computer Communications*, 2002, pp. 89–102.

[8] H. Wang, Z. Tian, and Q. Zhang, "Self-tuning price-based congestion control supporting TCP networks," in *Proc. 19th Int'l Conf. Computer Communications and Networks*, 2010, pp. 1–6.

[9] G. Kreitz and F. Niemela, "Spotify: large scale, low latency, p2p music-on demand streaming," in *Proc. 10th IEEE Int'l Conf. Peer-to-Peer Computing (P2P)*, 2010, pp. 1–10.

[10] L. McNamara, C. Mascolo, and L. Capra, "Media sharing based on colocation prediction in urban transport," in *Proc. 14th ACM Int'l Conf. Mobile Computing and Networking*, 2008, pp. 58–69.

[11] E. Jung, Y. Wang, I. Prilepov, et al., "User-profile driven collaborative bandwidth sharing on mobile phones," in *Proc. 1st ACM Workshop on Mobile Cloud Computing & Services: Social Networks and Beyond*, 2010, pp. 2:1–2:9.

[12] X. Jin and Y. K. Kwok, "Cloud assisted p2p media streaming for bandwidth constrained mobile subscribers," in *IEEE 16th Int'l Conf. Parallel and Distributed Systems (ICPADS)*, 2010, pp. 800–805.

[13] E. Yaacoub, L. Al-Kanj, Z. Dawy, et al., "A utility minimization approach for energy-aware cooperative content distribution with fairness constraints," *Trans. Emerg. Telecommun. Technol.* vol. 23, no. 4, pp. 378–392, 2012.

[14] E.A. Lee, "The problem with threads," *IEEE Comput.* vol. 39, no. 5, pp. 33–42, 2006.

[15] M. Lindberg, "From competitive to cooperative resource management for cyber-physical systems," 2014. [Online]. Available: http://lup.lub.lu.se/record/4437762.

[16] L. Abeni and G. C. Buttazzo, "Adaptive bandwidth reservation for multimedia computing," in *Proc. 6th Int'l Conf. Real-Time Computing Systems and Applications*, 1999, pp. 70–77.
[17] L. Abeni, L. Palopoli, S. Superiore, and J. Walpole, "Analysis of a reservation-based feedback scheduler," in *Proc. 23rd IEEE Real-Time Systems Symp. (RTSS)*, 2002, pp. 71–80.
[18] C. P. Marzario, T. Cucinotta, L. Palopoli, L. Marzario, and G. Lipari, "Adaptive reservations in a Linux environment," in *Proc. IEEE Real-Time and Embedded Technology and Applications Symp.*, 2004, pp. 238–245.
[19] S. P. Boyd and L. Vandenberghe, *Convex Optimization*. Cambridge: Cambridge University Press, 2004.
[20] J. M. Maciejowski, *Predictive Control: With Constraints*. Harlow: Pearson Education, 2002.
[21] R. Axelrod, *Evolution of Cooperation*. New York: Basic Books, 2006.
[22] F. Nash, "Equilibrium points in n-person games," *PNAS*, vol. 36, no. 1, pp. 48–49, 1950.
[23] S. Androutsellis-Theotokis and D. Spinellis, "A survey of peer-to-peer content distribution technologies," *ACM Comput. Surv.* vol. 36, no. 4, pp. 335–371, 2004.
[24] J. Sztipanovtis and G. Karsai, "Model-integrated computing," *IEEE Comput.*, vol. 34, no. 4, 1997.
[25] B. Selic, "The pragmatics of model-driven development." *IEEE Softw.*, vol. 20, no. 5, 2003.
[26] The Ptolemy Project. [Online]. Available: http://http://ptolemy.eecs.berkeley.edu.
[27] P. Derler, E. A. Lee, and A. S. Vincentelli, "Modeling cyber-physical systems," in *Proc. IEEE*, vol. 100, no. 1, 2012.
[28] S. Das, B. L. Matthews, A. N. Srivastava, and N. C. Oza, "Multiple kernel learning for heterogeneous anomaly detection: Algorithm and aviation safety case study," in *Proc. KDD,* 2010, pp. 47–56.
[29] T. Hunter, T. Das, M. Zaharia, P. Addeel, and A. M. Bayen, "Large scale estimation in cyber physical systems using streaming data: A case study with smartphone traces," *arXiv:1212.3393v1*, 2012.
[30] "The Rise of Industrial Big Data." [Online]. Available: www.geip.com/library/detail/13170.
[31] M. Zaharia, T. Das, H. Li, S. Shenker, and I. Stoica, "Discretized streams: An efficient and fault-tolerant model for stream processing on large clusters," in *Proc. HotCloud*, 2010, pp. 1–10.
[32] "Spark: Lightning-Fast Cluster Computing." [Online]. Available: http://spark-project.org.
[33] A. Edmonds, T. Metsch, A. Papaspyrou, and A. Richardson, "Toward an open cloud standard," *IEEE Internet Comput.*, vol. 16, no. 4, pp. 15–25, 2012.
[34] DMTF Group, "Cloud infrastructure management interface (CIMI) model and REST interface over HTTP specification," 2012. [Online]. Available: http://dmtf.org/standards/cmwg.
[35] Amazon Inc., "Amazon web services," 2012. [Online]. Available: http://aws.amazon.com.
[36] M. Armbrust, A. Fox, R. Griffith, et al., "A view of cloud computing," *Commun. ACM* vol. 53, no. 4, pp. 50–58, 2010.
[37] Amazon, "EC2 FAQs," 2013. [Online]. Available: http://aws.amazon.com/ec2/faqs.

[38] D. Kyriazis: "Cloud computing service level agreements: Exploitation of research results," European Commission, Tech. Rep., 2013.

[39] N.R. Herbst, S. Kounev, R. Ruessner, "Elasticity in cloud computing: What it is, and what it is not," in *Proc. 10th Int'l Conf. Autonomic Computing (ICAC)*, 2013, pp. 23–27.

[40] S. Govindan, J. Liu, A. Kansal, and A. Sivasubramaniam, "Cuanta: Quantifying effects of shared on-chip resource interference for consolidated virtual machines," in *Proc. 2nd ACM Symp. on Cloud Computing (SoCC 2011)*, 2011, pp. 22:1–22:14.

[41] J. Ahn, C. Kim, Y. R. Choi, J. Huh, "Dynamic virtual machine scheduling in clouds for architectural shared resources," in *Proc. 4th USENIX Workshop on Hot Topics in Cloud Computing*, 2012, pp. 1–19.

[42] M. Al-Fares, A. Loukissas, and A. Vahdat, "A scalable, commodity data center network architecture," in *Proc. ACM SIGCOMM Conference on Data Communication*, 2008, pp. 63–74.

[43] C. Guo, G. Lu, D. Li, et al., "BCube: A high performance, server-centric network architecture for modular data centers," *SIGCOMM Comput. Commun. Rev.*, vol. 39, no. 4, pp. 63–74, 2009.

[44] J. Y. Shin, B. Wong, and E. G. Sirer, "Small-world datacenters," in *Proc. 2nd ACM Symp. on Cloud Computing (SoCC)*, 2011, pp. 2:1–2:13.

[45] D. Abts and B. Felderman, "A guided tour of data-center networking," *Commun. ACM*, vol. 55, no. 6, pp. 44–51, 2012.

[46] M. F. Bari, R. Boutaba, R. Esteves, et al., "Data center network virtualization: A survey," *IEEE Commun. Surv.*, vol. 15, no. 2, pp. 909–928, 2013.

[47] Y. Chen, R. Griffith, J. Liu, R. H. Katz, and A. D. Joseph, "Understanding TCP incast throughput collapse in datacenter networks," in *Proc. 1st ACM workshop on Research on Enterprise Networking (WREN 2009)*, 2009, pp. 73–82.

[48] S. Kumar, V. Talwar, V. Kumar, P. Ranganathan, and K. Schwan, "Loosely coupled coordinated management in virtualized data centers," *Cluster Comput.*, vol. 14, no. 3, pp. 259–274, 2011.

[49] G. Jung, M. Hiltunen, K. Joshi, R. Schlichting, and C. Pu, "Mistral: Dynamically managing power, performance, and adaptation cost in cloud infrastructures," in *Proc. IEEE 30th Int'l Conf. Distributed Computing Systems (ICDCS)*, 2010 pp. 62–73.

[50] H. Moens, J. Famaey, S. Latrffie, B. Dhoedt, and F. De Turck, "Design and evaluation of a hierarchical application placement algorithm in large scale clouds," in *Proc. 12th IFIP/IEEE Int'l Symp. on Integrated Network Management (IM)*, 2011, pp. 137–144.

[51] F. Wuhib, R. Stadler, and H. Lindgren, "Dynamic resource allocation with management objectives implementation for an OpenStack cloud," in *Proc. 8th Int'l Conf. Network and Services Management (CNSM)*, pp. 309–315, 2012.

[52] F. Wuhib, R. Yanggratoke, and R. Stadler, "Allocating compute and network resources under management objectives in large-scale clouds," *J. Netw. Syst. Manage.*, vol. 23, pp. 111–136, 2013.

[53] E. Feller, L. Rilling, and C. Morin, "Snooze: A scalable and autonomic virtual machine management framework for private clouds," in *Proc. 12th IEEE/ACM Int'l Symp. Cluster, Cloud and Grid Computing (CCGrid)*, 2012, pp. 482–489.

[54] R. Panigrahy, K. Talwar, L. Uyeda, and U. Wieder, "Heuristics for vector bin packing," Microsoft Research, Tech. Rep., 2011.

[55] D. Wilcox, A. McNabb, and K. Seppi, "Solving virtual machine packing with a reordering grouping genetic algorithm," in *Proc. IEEE Congress on Evolutionary Computation (CEC)*, 2011, pp. 362–369.

[56] G. Jung, K. Joshi, M. Hiltunen, R. Schlichting, and C. Pu, "Generating adaptation policies for multi-tier applications in consolidated server environments," in *Proc. 2008 Int'l Conf. Autonomic Computing (ICAC)*, 2008, pp. 23–32.

[57] R. Gupta, S. Bose, S. Sundarrajan, M. Chebiyam, and A. Chakrabarti, "A two stage heuristic algorithm for solving the server consolidation problem with item–item and bin–item incompatibility constraints," in *Proc. 2008 IEEE Int'l Conf. Services Computing (SCC)*, 2008, vol. 2, pp. 39–46.

[58] B. Li, J. Li, J. Huai, et al., "EnaCloud: An energy-saving application live placement approach for cloud computing environments," in *Proc. 2009 IEEE Int'l Conf. Cloud Computing (CLOUD)*, 2009, pp. 17–24.

[59] M. Cardosa, M. Korupolu, and A. Singh, "Shares and utilities based power consolidation in virtualized server environments," in *Proc. 11th IFIP/IEEE Int'l Symp. Integrated Network Management* (IM), 2009, pp. 327–334.

[60] B. Viswanathan, A. Verma, and S. Dutta, "CloudMap: Workload-aware placement in private heterogeneous clouds," in *Proc. 13th IEEE/IFIP Network Operations and Management Symp. (NOMS)*, 2012, pp. 9–16.

[61] X. Meng, C. Isci, J. Kephart, et al., "Efficient resource provisioning in compute clouds via VM multiplexing," in *Proc. 7th Int'l Conf. Autonomic Computing (ICAC)*, 2010, pp. 11–20.

[62] L. Shi, B. Butler, D. Botvich, and B. Jennings, "Provisioning of requests for virtual machine sets with placement constraints in IaaS clouds," in *Proc. 12th IFIP/IEEE Int'l Symp. on Integrated Network Management (IM)*, 2013, pp. 499–505.

[63] D. Jayasinghe, C. Pu, T. Eilam, et al., "Improving performance and availability of services hosted on IaaS clouds with structural constraint-aware virtual machine placement," in *Proc. 2011 IEEE Int'l Conf. Services Computing (SCC)*, 2011, pp. 72–79.

[64] K. Konstanteli, T. Cucinotta, K. Psychas, and T. Varvarigou, "Admission control for elastic cloud services," in *Proc. 5th IEEE Int'l Conf. Cloud Computing (CLOUD)*, 2013, pp. 41–48.

[65] D. Breitgand and A. Epstein, "SLA-aware placement of multi-virtual machine elastic services in compute clouds," in *Proc. 12th IFIP/IEEE Symp. on Integrated Network Management (IM)*, 2011, pp. 161–168.

[66] P. De and S. Roy "VMSpreader: Multi-tier application resiliency through virtual machine striping," in *Proc. 12th IFIP/IEEE Int'l Symp. on Integrated Network Management (IM)*, 2011, pp. 185–192.

[67] E. Bin, O. Biran, O. Boni, E. Hadad, et al., "Guaranteeing high availability goals for virtual machine placement," in *Proc. 31st IEEE Int'l Conf. Distributed Computing Systems (ICDCS)*, 2011, pp. 700–709.

[68] C. Clark, K. Fraser, S. Hand, et al. "Warfield: Live migration of virtual machines," in *Proc. 2nd Conf. Symp. on Networked Systems Design & Implementation (NSDI)*, 2005, pp. 273–286.

[69] T. Wood, P. Shenoy, A. Venkataramani, and M. Yousif, "Black-box and gray-box strategies for virtual machine migration," in *Proc. 4th USENIX Symp. on Networked Systems Design & Implementation (NSDI)*, 2007, vol. 7, pp. 229–242.

[70] N. Bobroff, A. Kochut, and K. Beaty, "Dynamic placement of virtual machines for managing SLA violations," in *Proc. 10th IFIP/IEEE Int'l Symp. on Integrated Network Management (IM 2007)*, 2007, pp. 119–128.

[71] U. Sharma, P. Shenoy, S. Sahu, and A. Shaikh, "A cost-aware elasticity provisioning system for the cloud," in *Proc. 31st IEEE Int'l Conf. Distributed Computing Systems (ICDCS)*, 2011, pp. 559–570.

[72] U. Sharma, P. Shenoy, S. Sahu, and A. Shaikh, "Kingfisher: Cost-aware elasticity in the cloud," in *Proc. 2011 IEEE Int'l Conf. Computer Communications (Infocom)*, 2011, pp. 206–210.

[73] X. Zhang, Z. Y. Shae, S. Zheng, and H. Jamjoom, "Virtual machine migration in an overcommitted cloud," in *Proc. 13th IEEE/IFIP Network Operations and Management Symp. (NOMS)*, 2012, pp. 196–203.

[74] H. Xu and B. Li, "Egalitarian stable matching for VM migration in cloud computing," in *Proc. 2011 IEEE Conf. Computer Communications Workshops (Infocom Workshops)*, 2011, pp. 631–636.

[75] D. Gale and L. S. Shapley, "College admissions and the stability of marriage," *Am. Math. Month.* vol. 69, no. 1, pp. 9–15, 1962.

[76] J. Schad, J. Dittrich, and J. A. Quian-Ruiz, "Runtime measurements in the cloud: Observing, analyzing, and reducing variance," *Proc. VLDB Endow.*, vol. 3, no. 1–2, pp. 460–471, 2010.

[77] T. Guo, U. Sharma, T. Wood, S. Sahu, and P. Shenoy, "Seagull: Intelligent cloud bursting for enterprise applications," in *Proc. 2012 USENIX Annual Technical Conference (ATC)*, 2012, pp. 1–6.

[78] M. Shifrin, R. Atar, and I. Cidon, "Optimal scheduling in the hybrid-cloud," in *Proc. 2013 IFIP/IEEE Int'l Symp. on Integrated Network Management (IM)*, 2013, pp. 51–59.

[79] R. Yanggratoke, F. Wuhib, and R. Stadler, "Gossip-based resource allocation for green computing in large clouds," in *Proc. 7th Int'l Conf. Network and Service Management (CNSM)*, 2011, pp. 1–9.

[80] G. Foster, G. Keller, M. Tighe, and H. Lutffyya, "The right tool for the job: Switching data centre management strategies at runtime," in *Proc. 2013 IFIP/IEEE Int'l Symp. Integrated Network Management (IM)*, 2013, pp. 151–159.

[81] S. Ghorbani and M. Caesar, "Walk the line: Consistent network updates with bandwidth guarantees," in *Proc. 1st Workshop on Hot Topics in Software Defined Networks (HotSDN)*, 2012, pp. 67–72.

[82] S. Al-Haj and E. Al-Shaer, "A formal approach for virtual machine migration planning," in *Proc. 9th Int'l Conf. Network and Service Management (CNSM)*, 2013, pp. 51–58.

[83] Q. Zhang, Q. Zhu, M. Zhani, and R. Boutaba, "Dynamic service placement in geographically distributed clouds," in *Proc. 32nd IEEE Int'l Conf. Distributed Computing Systems (ICDCS)*, 2012, pp. 526–535.

[84] D. Breitgand and A. Epstein, "Improving consolidation of virtual machines with risk-aware bandwidth oversubscription in compute clouds," in *Proc. 2012 IEEE Int'l Conf. Computer Communications (Infocom)*, 2012, pp. 2861–2865.

[85] J. Schad, J. Dittrich, and J. A. Quian-Ruiz, "Runtime measurements in the cloud: Observing, analyzing, and reducing variance," *Proc. VLDB Endow.*, vol. 3, no. 1–2, pp. 460–471, 2010.

[86] M. Wang, X. Meng, and L. Zhang, "Consolidating virtual machines with dynamic bandwidth demand in data centers," in *Proc. 2011 IEEE Int'l Conf. Computer Communications (Infocom)*, 2011, pp. 71–75.

[87] J. Kleinberg, Y. Rabani, and E. Tardos, "Allocating bandwidth for bursty connections," in *Proc. 29th Annual ACM Symp. on Theory of Computing (STOC)*, 1997, pp. 664–673.

[88] D. Breitgand and A. Epstein, "Improving consolidation of virtual machines with risk-aware bandwidth oversubscription in compute clouds," in *Proc. IEEE Int'l Conf. Computer Communications (Infocom 2012)*, 2012, pp. 2861–2865.

[89] D. Jayasinghe, C. Pu, T. Eilam, et al., "Improving performance and availability of services hosted on IaaS clouds with structural constraint-aware virtual machine placement," in *Proc. 2011 IEEE Int'l Conf. Services Computing (SCC)*, 2011, pp. 72–79.

[90] V. Shrivastava, P. Zerfos, K.W. Lee, et al., "Application aware virtual machine migration in data centers," in *Proc. 2011 IEEE Int'l Conf. Computer Communications (Infocom)*, 2011, pp. 66–70.

[91] X. Wen, K. Chen, Y. Chen, et al., "VirtualKnotter: Online virtual machine shuffling for congestion resolving in virtualized datacenter," in *Proc. 32nd IEEE Int'l Conf. Distributed Computing Systems (ICDCS)*, 2012, pp. 12–21.

[92] O. Biran, A. Corradi, M. Fanelli, et al., "A stable network-aware VM placement for cloud systems," in *Proc. 12th IEEE/ACM Int'l Symp. on Cluster, Cloud and Grid Computing (CCGrid)*, 2012, pp. 498–506.

[93] J. W. Jiang, T. Lan, S. Ha, M. Chen, and M. Chiang, "Joint VM placement and routing for data center traffic engineering," in *Proc. IEEE Int'l Conf. Computer Communications (Infocom 2012)*, 2012, pp. 2876–2880.

[94] "Nagios the industry standard in IT infrastructure monitoring," 2015. [Online]. Available: www.nagios.org.

[95] "Ganglia monitoring system," 2015. [Online]. Available: http://ganglia.sourceforge.net.

[96] A. Gulati, A. Holler, M. Ji, et al., "VMware distributed resource management: Design, implementation, and lessons learned," *VMware Tech. J.*, vol. 1, no. 1, pp. 45–64, 2012.

[97] S. Govindan, J. Liu, A. Kansal, and A. Sivasubramaniam, "Cuanta: Quantifying effects of shared on-chip resource interference for consolidated virtual machines," in *Proc. 2nd ACM Symp. on Cloud Computing (SoCC)*, pp. 22:1–22:14, 2011.

[98] D. Gmach, J. Rolia, and L. Cherkasova, "Chargeback model for resource pools in the cloud," in *Proc. 12th IFIP/IEEE Int'l Symp. on Integrated Network Management (IM)*, 2011, pp. 622–625.

[99] P. Padala, K. Y. Hou, K. G. Shin, et al. "Merchant: Automated control of multiple virtualized resources," in *Proc. 4th ACM European Conf. Computer Systems (EuroSys)*, 2009, pp. 13–26.

[100] J. Xu, M. Zhao, J. Fortes, R. Carpenter, and M. Yousif, "Autonomic resource management in virtualized data centers using fuzzy logic-based approaches," *Cluster Comput.* vol. 11, no. 3, pp. 213–227, 2008.

[101] J. Rao, X. Bu, K. Wang, and C. Z. Xu, "Self-adaptive provisioning of virtualized resources in cloud computing," in *Proc. 2011 ACM SIGMETRICS Joint Int'l Conf. Measurement and Modeling of Computer Systems (SIGMETRICS)*, 2011, pp. 129–130.

[102] R. Han, L. Guo, M. Ghanem, and Y. Guo, "Lightweight resource scaling for cloud applications," in *Proc. 12th IEEE/ACM Int'l Symp. on Cluster, Cloud and Grid Computing (CCGrid)*, 2012, pp. 644–651.

[103] M. Cardosa, M. Korupolu, and A. Singh, "Shares and utilities based power consolidation in virtualized server environments," in *Proc. 11th IFIP/IEEE Int'l Symp. Integrated Network Management (IM)*, 2009, pp. 327–334.

[104] R. Urgaonkar, U. Kozat, K. Igarashi, and M. Neely, "Dynamic resource allocation and power management in virtualized data centers," in *Proc. 12th IEEE/IFIP Network Operations and Management Symp. (NOMS)*, 2010, pp. 479–486.

[105] Y. Chen, R. Griffith, J. Liu, R.H. Katz, and A.D. Joseph, "Understanding TCP incast throughput collapse in datacenter networks," in *Proc. 1st ACM Workshop on Research on Enterprise Networking (WREN)*, 2009, pp. 73–82.

[106] S. Gamage, A. Kangarlou, R. R. Kompella, and D. Xu, "Opportunistic flooding to improve TCP transmit performance in virtualized clouds," in *Proc. 2nd ACM Symp. on Cloud Computing (SoCC)*, 2011, pp. 24:1–24:14.

[107] L. Popa, A. Krishnamurthy, S. Ratnasamy, and I. Stoica, "FairCloud: Sharing the network in cloud computing," in *Proc. 10th ACM Workshop on Hot Topics in Networks (HotNets-X)*, 2012, pp. 22:1–22:6.

[108] V. Jeyakumar, M. Alizadeh, D. Mazieres, B. Prabhakar, and C. Kim, "EyeQ: Practical network performance isolation for the multi-tenant cloud," in *Proc. 4th USENIX Workshop on Hot Topics in Cloud Computing (HotCloud)*, 2012, pp. 1–6.

[109] G. Kumar, M. Chowdhury, S. Ratnasamy, and I. Stoica, "A case for performance-centric network allocation," in *Proc. 4th USENIX Workshop on Hot Topics in Cloud Computing (HotCloud)*, 2012, pp. 1–6.

[110] M. Lin, A. Wierman, L. Andrew, and E. Thereska, "Dynamic right-sizing for power-proportional data centers," in *Proc. 2011 IEEE Int'l Conf. Computer Communications (Infocom)*, 2011, pp. 1098–1106.

[111] B. Urgaonkar, P. Shenoy, and T. Roscoe, "Resource overbooking and application profiling in shared hosting platforms," in *SIGOPS Operating Systems Review 36(SI)*, 2002, pp. 239–254.

[112] C. Tang, M. Steinder, M. Spreitzer, and G. Paciffici, "A scalable application placement controller for enterprise data centers," in *Proc. 16th Int'l Conf. World Wide Web (WWW)*, 2007, pp. 331–340.

[113] G. Gmach, S. Krompass, A. Scholz, M. Wimmer, and A. Kemper, "Adaptive quality of service management for enterprise services," *ACM Trans. Web*, vol. 2, no. 1, pp. 8:1–8:46, 2008.

[114] C. Adam and R. Stadler, "Service middleware for self-managing large-scale systems," *IEEE Trans. Netw. Serv. Manage.*, vol. 4, no. 3, pp. 50–64, 2007.

[115] F. Wuhib, R. Stadler, and M. Spreitzer, "Gossip-based resource management for cloud environments," in *Proc. 6th IEEE Int'l Conf. Network and Service Management (CNSM)*, 2010, pp. 1–8.

[116] H. Moens, J. Famaey, S. Latre, B. Dhoedt, and F. De Turck, "Design and evaluation of a hierarchical application placement algorithm in large scale clouds," in *Proc. 12th IFIP/IEEE Int'l Symp. on Integrated Network Management (IM)*, 2011, pp. 137–144.

[117] V. Ishakian and A. Bestavros, "MORPHOSYS: Efficient colocation of QoS-constrained workloads in the cloud," in *Proc. 12th IEEE/ACM Int'l Symp. Cluster, Cloud and Grid Computing (CCGrid)*, 2012, pp. 90–97.

[118] Y. T. Wang and R. Morris, "Load sharing in distributed systems," *IEEE Trans. Comput. C*, vol. 34, no. 3, pp. 204–217, 1985.

[119] O. Kremien and J. Kramer, "Methodical analysis of adaptive load sharing algorithms," *IEEE Trans. Parallel Distrib. Syst.*, vol. 3, no. 6, pp. 747–760, 1992.

[120] M. Mitzenmacher, "The power of two choices in randomized load balancing," *IEEE Trans. Parallel Distrib. Syst.*, vol. 12, no. 10, pp. 1094–1104, 2001.

[121] D. Breitgand, R. Cohen, A. Nahir, and D. Raz, "On cost-aware monitoring for self-adaptive load sharing," *IEEE J. Sel. Areas Commun.*, vol. 28, no. 1, pp. 70–83, 2010.

[122] M. Zaharia, A. Konwinski, A. D. Joseph, R. Katz, and I. Stoica, "Improving MapReduce performance in heterogeneous environments" In *Proc. 8th USENIX Conf. Operating Systems Design and Implementation (OSDI)*, 2008, pp. 29–42.

[123] T. Sandholm and K. Lai, "MapReduce optimization using regulated dynamic prioritization," in *Proc. 11th Int'l Joint Conf. Measurement and Modeling of Computer Systems (SIGMETRICS)*, 2009, pp. 299–310.

[124] J. Tan, X. Meng, and L. Zhang, "Performance analysis of coupling scheduler for MapReduce/Hadoop," in *Proc. 2012 IEEE Int'l Conf. Computer Communications (Infocom)*, 2012, pp. 2586–2590.

[125] M. Schwarzkopf, D. G. Murray, and S. Hand, "The seven deadly sins of cloud computing research," in *Proc. 4th USENIX Workshop on Hot Topics in Cloud Computing (HotCloud)*, 2012, pp. 1–5.

[126] V. K. Vavilapalli, A. C. Murthy, C. Douglas, et al., "Apache Hadoop YARN: Yet another resource negotiator," in *Proc. 4th ACM Symp. on Cloud Computing (SoCC)*, 2013, pp. 1–16.

[127] G. Lee, B. G. Chun, and R. H. Katz, "Heterogeneity-aware resource allocation and scheduling in the cloud," in *Proc. 3rd USENIX Workshop on Hot Topics in Cloud Computing (HotCloud)*, 2011, pp. 1–5.

[128] M. R. Hines and K. Gopalan, "Post-copy based live virtual machine migration using adaptive pre-paging and dynamic self-ballooning," in *Proc. 2009 ACM SIGPLAN/SIGOPS Int'l Conf. Virtual Execution Environments (VEE)*, 2009, pp. 51–60.

[129] A. Ghodsi, M. Zaharia, B. Hindman, et al., "Dominant resource fairness: Fair allocation of multiple resource types," in *Proc. 8th USENIX Conf. Networked Systems Design and Implementation (NSDI)*, 2011, pp. 24–24.

[130] A. A. Bhattacharya, D. Culler, E. Friedman, et al., "Hierarchical scheduling for diverse datacenter workloads," in *Proc. 4th ACM Symp. on Cloud Computing (SoCC)*, 2013, pp. 1–15.

[131] K .Selyunin, D. Ratasich, E. Bartocci, and R. Grosu, "Deep neural programs for adaptive control in cyber-physical systems," *arXiv preprint arXiv:1502.04013*, 2015.

[132] T. Nakamura, S. Nagata, A. Benjebbour, et al., "Trends in small cell enhancements in LTE advanced," *Commun. Mag.*, vol. 51, no. 2, pp. 98–105, 2013.

[133] F. Moro, D. Fontanelli, and L. Palopoli, "Vision-based robust localization for vehicles," in *Instrumentation and Measurement Technology Conf. (I2MTC)*, 2012 pp. 553–558.

[134] J. Kim, H. Kim, K. Lakshmanan, and R. Rajkumar, "Parallel scheduling for cyber-physical systems: Analysis and case study on a self-driving car," in *Proc. ACM/IEEE 4th Int'l Conf. Cyber-Physical Systems (ICCPS '13)*, 2013, pp. 31–40.

13 Middleware for the Internet of Things

Rahav Dor and Chenyang Lu

The *Internet*, the network of networks, was not only a technological change. It affected human minds and changed how we operate as a collaborative society. In tandem with changing how we live, it affected how we think. We grew accustomed to sharing data and information that was once considered private, proprietary, or of no interest to others. This level of connectivity and wealth of information enabled a long list of useful applications and services that made the Internet and our lives what they are today. Another technological wave is coming and it may be larger in terms of connectivity and novel applications we will be able to build. The human endeavor is about to embark on the next disruptive technology – the *Internet of Things* (IoT).

We say that the IoT will be larger because, while the Internet connected people to information (predominately produced by people), the IoT will connect people and machines to data produced by everyday objects, and there are many more everyday objects than people. From inception, the IoT will enable machines to talk to other machines and produce new datastreams. Possibly the cardinal difference is that machines and people alike will not only consume data but will be able to control and activate everyday objects. When the IoT achieves popularity, it will surpass the Internet along the following key dimensions:

- **Scale:** The IoT will connect far more sources of data and everyday physical objects. This would produce more datastreams, either raw or refined.
- **Participants:** Autonomous machines and software processes, along with people, will natively participate in interactions.
- **Closed-loop:** Data will be comprehensible by people and machines and this will enable either or both to make decisions and command objects.
- **Diversity:** The objects will be excessively heterogeneous.

The IoT will give machines and people alike access to data and myriad everyday objects at a greater scale than the Internet does. In the IoT lingo these objects are named *Things*. Together with the networks and applications that connect Things, the IoT will be formed. Our toothbrush, the dog's water bowl, temperature sensors, motion sensors, our homes, roads, buildings, cities, and the world – these will all be producing data that someone, and something, can consume. Moreover, the IoT is bidirectional and will allow machines and people to issue commands to (actuate) connected Things. Applications (apps) would be composed and be informed by data emanating from Things, and these apps would actuate our world through Things. The impetus for creating this emerging technology is to better our lives through the IoT.

13.1 Motivation for an IoT Middleware

This chapter is focused on an enabling technology that is essential to realize the IoT. The world is full of Things, diverse in their computing and networking facilities. Many Things are not designed with such facilities, and these will be bolted onto them, or they will be remotely sensed to become a part of the IoT ecosystem. This presents substantial variety in what can be sensed or actuated, and in how communication can be accomplished. How could we possibly interconnect such diverse Things as computer programs, in-network processes, and data of varied structure and amounts? How can we enable application developers to compose smart apps using the world as a data source, without requiring expertise in the intricacies and oddities of each and every sensor, actuator, or datastream? The answers critically reside with the technology discussed in this chapter.

We view the IoT as a cyber-physical system (CPS) on a global scale, where everyday *physical* Things are enhanced with *cyber* resources (sensing, computing, networking, and actuating). In this CPS, in-network processes can perform computation on datastreams as the data traverses the network, and create additional streams and services. This CPS extends to the cloud, where historical data and additional software processes and services run.

In this chapter we provide an overview of one technology that will be important in enabling this vision – a software layer that will help app developers interface with such myriad Things. Such a software layer is generically referred to as *middleware*. It is used in many contexts and for many purposes. A middleware abstracts low-level details of what lies below it (here, Things, computers, streams, services, and the network) and provides useful services to what lies above it (here, apps). The useful services of our middleware can be, for example, finding and communicating with Things in a uniform way. This is of added value because most IoT apps will want to connect to many Things, so considerable design and coding efforts will be duplicated unless something provides that functionality for all. We will introduce three different middleware, each with its unique approach, advantages, and disadvantages.

It would require more than a single chapter to host a complete representation of competing middleware that could power the IoT. In the struggle between providing meaningful information about a few and a high-level survey, we choose the former. Our selected three options allow us to discuss middleware systems with very different architectures, or even philosophies, and in doing so we provide the reader with what we hope is a broad perspective. This choice should not suggest that other technologies, such as the work of [1–12], are less likely to find their place in the IoT.

The middleware we choose are the following: (1) The *AllJoyn* framework [13], which follows the peer-to-peer (P2P) network model and uses remote procedure call (RPC) from one peer (app) to another to invoke services or change state. In a P2P network, participating peers share tasks and perform work for each other. RPC is an inter process communication technique that allows one process to invoke services provided by another process, where the developer is not required to deal with the details of the remote invocation. AllJoyn services are stateful and interactions occur within sessions that apps join. This architecture is of value for apps that need some sort of continuity as time progresses or a state of the world due to past interactions. It also allows AllJoyn to manage network dynamics such as device disconnection caused by wireless interference or mobility. We will see, though, that this architecture choice makes the framework challenging to run on resource-constrained Things.

(2) CoAP is a standard [14] developed by the Internet Engineering Task Force (IETF). It is developed around the idea that since internet technologies and standards are so successful, they would be equally fruitful for enabling the IoT. Importantly, where CoAP is adapted, seamless Internet–IoT interoperation will be enabled. CoAP changes internet technologies to fit the requirements of resource-constrained Things by defining compact packet structures to lower computation demands and network bandwidth. It depends on the unreliable UDP transport for the same reason, and it touts the HTTP/REST architecture. Thus, CoAP is stateless and fits well onto resource-constrained Things because hosts do not need to maintain state in memory. This design, though, is not the best fit for apps whose future behavior depends on the past, and they would like to delegate hysteresis management to the server.

(3) The Building Operating System Services (BOSS) [15] provide operating system abstractions for Things in smart buildings. As an operating system, we will see that BOSS provides services surpassing AllJoyn and CoAP. Depending on the app that one intends to build, these services may be a blessing or introduce unneeded overhead.

13.2 The Services We May Want from a Middleware

We can begin to appreciate why we would want a middleware by asking what added-value services we may want it to provide. We would like to identify services that have little to do with apps' core functionality and are general enough that most IoT apps would need them. A well-designed middleware will allow app developers to author only their app's core logic and delegate all other operations to the middleware.

We start by noting that a good middleware will "wrap the IoT" by abstracting the diversity of Things within it and provide easy access to the resources inside. Since we view the IoT as a CPS, the middleware should provide access to both physical Things such as sensors and actuators, and cyber Things such as services and data that the network and other apps make available.

Consider as an example an app we wish to author, named *OurApp*, which needs to know the temperature at a location. Suppose that the temperature can be provided by a local public sensor *tSensor*, or by an app *EnvApp* that already runs in the network. *EnvApp* is designed to use *tSensor* when it is functioning, accessible, and not busy with other activities. It can also ask a nearby private temperature sensor *tDorSensor* for a deference reading. Otherwise *EnvApp* will revert to querying the Internet for the temperature in that location. We should ask ourselves a few questions:

- How would we, or worse *OurApp* without human-infused knowledge, know that there is a sensor, or an app providing a temperature reading service?
- How would *OurApp* know about the services provided by *tSensor* or *EnvApp*?
- What are the interfaces to access *tSensor* and *EnvApp*? What parameters should be passed? What data and in which structure will be returned?
- Can we submit a service request to either using a unified interface?
- How would *OurApp* know which is functioning correctly at the time of service request?
- What is the reliability and trustworthiness of those providing services? What is their accuracy and tolerance?

An added-value middleware would enable us to discover that such services exist and provide a unified way of accessing their services. A better middleware will address concerns such as the last two questions as well.

Take as another example an app that identifies location and a person as they approach their home. The app may glean the required information by proximity perimeter arrival, followed by a home entrance event, and then locate the person in a particular room. Suppose that the core functionality of this app is to play the person's favorite music as they enter a set of rooms and transfer the audio as they roam around the house. Another app, built by a different developer, needs similar location and person identification in order to turn the lights to a favorite settings. Yet a third app, built by a coffee aficionado app developer, brews coffee to a personalized temperature and has it ready at the appropriate time. If the coffee app senses that the coffee cup was not lifted off the coffee maker, it would gently flicker the lights and talk through the audio system in the room the person is in. All of these apps, and likely many others, would need access to a (similar) set of services to enable each app to accomplish its (different) core functionality. Here are some of the services we already identified:

- person identification service
- proximity perimeter arrival service
- entrance service
- room ID service
- audio actuation service
- light actuation service
- coffee machine actuation
- coffee machine temperature reading.

It would be beneficial if access to those services were unified by a middleware that abstracts the individual sensors', and actuators', particulars. It would be an engineering marvel if each app could include only the logic pertaining to its core functionality.

This concludes the motivation part of the chapter. Next, we review in more detail the three middleware.

13.3 AllJoyn Case Study

The first middleware we review is the AllJoyn framework [13]. Being a framework, it is a collection of libraries and classes providing a certain architecture, using which software can be built. The architecture in this case is a P2P network that enables direct app-to-app communication over a local network. Qualcomm released AllJoyn in 2011 to connect apps with nearby devices. In December 2013 the framework was contributed to the AllSeen Alliance, an open-source project of the Linux Foundation. The core services that AllJoyn provide are:

1. A process named *Router* that manages interactions between apps and handles networking. It works with the underlying network layers, managing devices that leave and rejoin the network, serializing method calls, and delivering them to the remote app that will run the method.
2. Manage an abstraction named *BUS*. Each host on the network that runs an AllJoyn Router can be connected to another host by an AllJoyn BUS-segment. The BUS is a concatenation of all BUS segments and its main function is to maintain routes connecting multiple distributed hosts.
3. Discovery of nearby devices.
4. Manage RPCs.
5. Represent resource-limited devices by proxy entities, making them appear as equal members on the distributed BUS. These entities will appear on Routers, which always run on resource-rich hosts.

Architecturally, the Router and the BUS provide the AllJoyn middleware with the mechanics of connecting hosts and routing traffic. Different middleware may choose other ways of achieving this machinery. RPC is another architectural choice; it is how AllJoyn choose to invoke remote services. The proxy is AllJoyn's approach to connecting Things to the middleware. Things connect via TCP to an AllJoyn Router, which acts as their proxy. The impact of this architectural choice is that while TCP guarantees packet delivery, it requires more memory and computation resources than UDP.

When an app developer uses AllJoyn, she or he is free from worrying about networking issues. AllJoyn will handle link failures and deal with the consequences that occur when apps running on mobile devices temporarily disconnect, for example because the device moved out of range. AllJoyn manages such dynamics, enabling connections to persist as devices disappear and appear.

13.3.1 System Requirements

AllJoyn libraries are available for C, C++, C#, Java, and Objective-C programming languages; it runs on Android, iOS, OS-X, Linux, Windows, and OpenWrt [16] operating systems.

The networks supported by AllJoyn are Wi-Fi, Powerline, and Ethernet. Peers need to be on the same local network because the framework does not support communication between subnets. AllJoyn is still evolving and not all programming languages and network combinations are supported on all operating systems.

AllJoyn core services are provided by a software component called the AllJoyn *Router*, which must run on each physical host on which apps run. The Java-based Router daemon consumes over 3 MB of memory and the C daemon about one-quarter of this. With this memory requirement the daemon must be deployed on at least μServer (e.g., [17, 18]) class of machines or a modern smartphone.

The framework's solution to connecting resource-constrained Things is the AllJoyn *Thin Client* (AJTC). AJTC core service is a TCP connection from a Thing to an AllJoyn Router running on a resource-rich host. The Router acts as a proxy for the limited device, which appears as an equal AllJoyn host on the network. In AllJoyn, the thin client is a BUS attachment connecting a Thing to the rest of the BUS. To have a small footprint and for portability, AJTC is written purely in C and provides a limited set of AllJoyn APIs (application programming interfaces). These APIs are quite different (only the major abstractions are preserved) from the standard AllJoyn APIs, which is a limitation on the usability of the middleware because app developers need to learn two different APIs.

AllJoyn's original target was not the IoT, but one that is focused on proximity-based apps running on smartphones. Still today, the AllJoyn standard Thing is the Arduino Due [19] with an Ethernet shield, making the system a wired-network IoT. We did not find rigorous empirical studies of AJTC; however, publications of the AllSeen Alliance report that the smallest AJTC implementation consumes about 25 KB of memory, a reasonable requirement that makes it fit on many modern Things.

13.3.2 Using AllJoyn: An Overview

An application wishing to connect to AllJoyn follows these steps:

1. The app instantiates a BusAttachment object from the framework and uses it to connect to an AllJoyn Router.
2. AllJoyn will create a uniqueName for the app, which the app will get by calling the *getUniqueName()* method.
3. The app associates the uniqueName with a uniqueWellKnownName of its choice.
4. The app calls the *requestName(String name, ...)* method to tell AllJoyn about its uniqueWellKnownName.

Apps or Things that connect to the AllJoyn middleware are known only by their uniqueWellKnownName, and therefore it should be chosen to advertise their purpose. We will learn later that some middleware, for example BOSS, allows us to specify additional searchable tags that can be used to find services.

Following the setup phase, the app has two paths it can follow. It can find other AllJoyn apps and join their session(s), or create a session into which other apps may join.

Applications communicate by invoking each other's methods, which appear to be local. Behind the scenes, AllJoyn uses RPC to invoke methods. Methods may get values (e.g., read from sensors), activate a remote Thing (actuate), and use other app methods as supported by that app developer.

13.3.3 AllJoyn Router

The Router is the main module providing AllJoyn core services. Routers may run as a dedicated service or run on the same device alongside an app. Each Router forms its own BUS attachment segment on the host it is running on and communicates with other Routers to form the AllJoyn distributed BUS:

- Apps connect to the local BUS attachment on the host they are running.
- Things connect to a Router by using AJTC.

A Router serving AJTC can be running on a dedicated host or on a Wi-Fi access point (AP). This serves as a proxy for the Things connected to it, presenting them as equal peers in the middleware.

AllJoyn only handles local communication. It does not support bridging between subnets and therefore only apps within the same Wi-Fi network range or subnet can communicate.

Hosts must run an AllJoyn Router to connect to the middleware and the Routers are connected to each other as a basic networking service provided by AllJoyn – the formation of a distributed BUS. The BUS and the apps attached to it on each host form a distributed system, with the AllJoyn framework providing the appearance (from the app's point of view) that they have access to each other's methods as if they were all local. This simplifies app development, leaving the low-level details to the middleware. The Routers are responsible for maintaining the distributed BUS, the discovery process discussed below, and low-level services such as serializing and transporting method invocations and the returned data across the BUS.

13.3.4 Connecting to AllJoyn

To participate in AllJoyn communication, an app instantiates a persistent singleton object of type BusAttachment from a class provided by the framework. This object is the app connection to the BUS.

As part of the middleware, the app determines its wellKnownUniqueName. As a matter of practice, AllJoyn recommends the use of the reverse domain name notation for this name, followed by some unique ID. The application developer is responsible for making sure that this name is unique among all apps that could be present on the distributed BUS, an imposition on developers that a middleware should avoid.

Once this process is complete, the app is connected to the AllJoyn middleware and can continue to the next step, which is advertising itself or discovering other applications.

13.3.5 App Advertising and Discovery

Apps with useful services should tell other apps that they are available – for example, to process data from sensors or to participate in a multi-host game. To advertise an app calls the *advertiseName(String name, ...)* method, which will register it with the AllJoyn Router to which the app is attached via its own BusAttachment.

AllJoyn will prompt any other app that will go into discovery mode about the availability of this app, as well as other apps that previously advertised themselves with any AllJoyn Router on any host. It is the responsibility of the Routers to communicate, such that apps do not need to worry about on which host other apps run. It is of great advantage to app developers that all services appear as a local method call.

To go into discovery mode, an app creates a BusListener object from the framework. The framework libraries and classes we discussed so far provide complete implementations of the architecture AllJoyn designers had in mind. The BusListener class, though, requires some specialization to customize its methods such that they are more relevant to the app's needs or core functionality. Once this object is instantiated, the app registers it with its BusAttachment. With this process complete, the app can query for any wellKnownUniqueName string or a prefix of it by calling the *findAdvertisedName(String namePrefix)* method. This is a non-blocking method, so the app is free to perform other operations while it waits for the discovery to complete. If AllJoyn Routers find an app matching the specified query string, a call-back will be invoked on the BusListener object, triggering the *foundAdvertisedName(String name, ...)* method.

Apps can remove their name from the advertised list AllJoyn Routers maintain by calling *cancelAdvertiseName(String name, ...)*. The Routers will immediately notify other apps by triggering the *lostAdvertisedName(String name, ...)* method of the other apps, BusListener objects.

13.3.6 Starting to Communicate

After the connection to the BUS (Section 13.3.4) and discovery (Section 13.3.5) processes are complete, apps need to take one more step to make their services available or use other apps' services. In AllJoyn, apps can talk to each other only after they join the same session. Any app can create a session and other apps can join it. Apps can create and join more than one session, with each session grouping apps with a shared purpose or benefiting from logically related services. An app starting a session can set the following restrictions on it:

- which physical layer is supported (e.g., Wi-Fi);
- proximity of apps that can join – apps running on any host on the network or only apps running on the localhost host; and
- number of apps in the session – a pair of peers or any number of them.

The app defines these restrictions by instantiating a SessionOpts object, which it passes to the BusAttachment along with a designated port parameter and a SessionPortListener object. Both classes' implementations are provided by the framework.

The abstraction AllJoyn calls *port* is a value that other apps need to know in order to join the session. AllJoyn assumes that other apps wishing to connect with a particular app will know a priori the correct port number. There is no "port discovery" process provided by AllJoyn. This is another drawback of the AllJoyn middleware, in particular for apps that are made by different vendors.

The *session* abstraction in the AllJoyn architecture now contains all of the necessary parts for communication and control:

- port, the address of a specific session;
- SessionOpts object, which holds the session configuration; and
- various methods that will be triggered when important events occur.

This list embodies one of the major advantages of AllJoyn, which is rooted in the philosophy behind this middleware. AllJoyn does not see its role simply as providing services when asked for; it actively helps apps by managing the dynamic that could occur due to other apps' actions or network conditions. For example, before an app will be allowed to join a session, the app owning the session will be triggered, allowing it to decide whether to accept or reject the request to join. This is above and beyond passing the restrictions discussed earlier. AllJoyn will also aid the app trying to join by sending it a status object with the value OK and invoking its *sessionJoined()* method if joining will be allowed. Hence both the passive session owner and session joiner will be triggered when needed. This saves a significant amount of code that app developers do not need to author, or ever see in their code, allowing them to focus on their app's core functionality.

There are additional services that become available as soon as a session is formed. For example, the *sessionLost()* method will be triggered if the session is broken for some reason, and any app in the session will have its *sessionMemberAdded()* and *sessionMemberRemoved()* methods trigger as apps join or leave the session.

13.3.7 Dynamic Network Conditions

As mobile devices move about, their wireless network strength changes or the device can roam beyond the network's coverage area. When this occurs the middleware will trigger the *lostAdvertisedName(String name, ...)* method on all other applications' BusListener objects that are part of the session that was lost. AllJoyn guarantees that this trigger will occur within 120 seconds at the latest. The delay is purposeful; within this period AllJoyn does not expunge the state of the lost app. If it returns within this period it will not need to go through all of the connection steps to rejoin the sessions it was a part of.

13.3.8 Apps' Specific Functionality

Networked apps need to communicate with each other. In AllJoyn the first communication step is for apps that have services to offer to start session(s), which other apps can join. This was discussed in Section 13.3.6. After a session is formed, communication occurs through remote method invocations (RPC). But how do apps know about the specific services of other apps?

To allow apps to specify and publicize the services they provide, AllJoyn uses an abstraction it names BusObject. For the middleware this is only a known object name it can reference, with no methods or services defined in it. It is the app developer who will implement methods (services) in their own instance of BusObject, to fit the app's needs and the services it wants to make available to callers. To make the names and signature of these methods known, the developer prepares an XML file named BusInterface describing them.

A BusInterface file must accompany each BusObject. Apps desiring to communicate with app A instantiate an object called ProxyBusObject, which knows how to read the BusInterface description file and expose the services described in it as "local" methods available to the app. De facto the ProxyBusObject is a local representation of a remote BusObject.

13.3.9 AllJoyn Summary

AllJoyn provides services that allow app developers to focus on their app's core functionality and abstract away the complexity of enabling communication between apps. Apps advertise their presence and the services they want to provide using the middleware. AllJoyn allows apps to discover each other as long as they are on the same local network. App A can call apps' {B, C, ... } methods and any app in {B, C, ... } can call A's methods. All apps {A, B, C, ... } see each other's methods as if they were local to them, whether they run on the same or remote hosts. This locality abstraction is of great value to developers. With AllJoyn, developers are saved from a lot of network communication details and code that they do not need to author. They instantiate a few objects, call methods in the correct sequence, and author two files: implementation (code) of their app services (authored in the BusObject class) and the specification in XML of these services (authored in the BusInterface file). However, the bootstrapping process a developer needs to go through to connect to the middleware, the number of objects, and sequencing of calls, is intricate relative to the other middleware we will discuss.

Each connected app on the AllJoyn middleware holds a proxy representation of all other apps it is connected to. Each proxy is an object holding the representation of the other app. This is a rather resource-intensive architectural choice, but, as many design alternatives, has its advantages – for example, apps can be dynamically alerted to important events occurring in the middleware.

AllJoyn manages the network dynamics. Apps can drop out of network range and rejoin the session with no work imposed on the app developer. This is a real situation, in particular when a host moves and the network is wireless. It is further pronounced when one considers low-power wireless networks that are sensitive to noise, interference from nearby networks, and even the presence of people.

AllJoyn's solution for Things, the thin client (AJTC), yields a relatively large runnable that does not fit on many Things. The AJTC lack of support for wireless technologies such as Bluetooth or Zigbee is an additional limitation in its current implementation.

AllJoyn is limited to supported local apps because it does not support inter-network communication. With this limitation it cannot support more than a local IoT, limited to a single network such as Wi-Fi. As we write this chapter, the AllSeen Alliance is discussing a gateway agent that will remedy this.

13.4 CoAP Case Study

In this section we examine a middleware that was forged for the IoT and designed by an internet standard organization. The core idea behind the standard is adapting internet technologies as the basis for an IoT middleware. The Constrained Application Protocol (CoAP) [14] is by the IETF. The IETF is responsible for many of the internet standards and its guidelines play an important role in our net-based society. The CoAP first draft was created at the end of 2010 as a data transfer protocol for constrained networks and nodes (e.g., 8-bit micro-controllers). The current standard is from 2014 and follows the original principles of using internet technologies to build an IoT as open and scalable as the Internet.

As in other sections of this chapter, our objective is to examine design alternatives of CoAP abstractions and not to explain every aspect of the standard.

13.4.1 Internet Technologies

There are many technologies that define and enable the Internet. At the time of writing, IP is used at the network layer, TCP or UDP at the transport layer, and HTTP at the application layer. HTTP is used as a request–response protocol (e.g., requesting services by using the GET or POST verbs and receiving a response). Many services operate in the application layer and can use each other's services or services from the lower layers. Many internet apps communicate using the REST [20] architecture, use URI [21] to identify resources such as servers or their services, and exchange XML or JSON objects as containers for describing requests or data. The structure of XML and JSON objects may be specified by schemas [22, 23]. WADL [24], or a similar language, can be used to specify/describe the services' APIs. To be concise in the rest of the chapter we will refer to this collection of technologies as the "Internet Technologies Stack," or ITS for short.

13.4.2 Compatibility with Internet Technologies

CoAP architects knew that a simple deployment of ITS on Things would not work. Things do not have sufficient computing power, memory, or network bandwidth to run them. There are also other reasons that ITS would not be applicable. For example, constrained nodes cannot handle the relatively large packets of the Internet. The CoAP philosophy is to look for ways to borrow ITS principles, support a good set of services that make sense for the IoT, and make everything require less. If the implementations are kept interoperable, then CoAP may enable integration between the Internet and IoT that is as seamless as one could hope for.

CoAP relies on existing protocols tailored for resource-constrained devices, and its primary contribution is at the application layer: The physical and data link layers are the responsibility of the network chips and firmware implemented on Things. At the network layer it relies on low power wireless personal area network (6LoWPAN [25]) instead of IPv6 or IPv4, which are used on the Internet. 6LoWPAN is a specialized flavor of the IP version 6 protocol that minimizes packet size and was designed to work on wireless sensor networks. At the transport layer, CoAP uses UDP exclusively, but allows apps to request a guaranteed delivery of some messages. The realization of reliable transport becomes the app's responsibility and not that of the network, as is the case with ITS. The session and presentation layers are rarely used in modern implementations of the OSI model.

13.4.3 The Application Layer

In ITS, HTTP runs in the application layer. It became the de facto protocol used by internet apps for requesting and receiving services. Many apps also adhere to the Representational State Transfer (REST) architecture. Such success and proliferation should be considered when it comes to designing the IoT software stack. Indeed, CoAP is designed to realize REST on constrained nodes, dubbed the Constrained RESTful Environments (CoRE). Its design takes into consideration limitations on packet size, higher probability of packet loss, Things' duty cycle, and their computing, memory, network throughput, and power constraints.

The most important objective for CoAP is to provide request–response operations. In CoAP, requests are denoted inside CoAP messages by including a *method code*. Responses carry the returned content and a *response code*.

Upon arrival at the server, the *method* within the request is applied to the *resource* on the server. All that an app needs to know to request a service is the resource URI and to send it a CoAP message. In AllJoyn we had to go through an intricate object-creation sequence and any two communicating apps were almost entangled (remember that each held objects representing the other app) to be able to request services from each other. This is one of the great benefits of adapting the REST architecture. CoAP apps are decoupled and very little setup is needed to consume services.

CoAP specifies a stateless (REST) architecture. A client requests a service, and the server responds. The server does not hold the state for the request; each request–response process stands on its own.

This is different from AllJoyn and brings about a problem that we did not face before. CoAP asynchronous responses can arrive in any order, so how can apps match response to request when multiple requests are made? For example, when the following requests are sent: "send current heart rate," "send average temperature," "send current blood oxygenation." CoAP architects addressed this challenge with an abstraction they named a *token*. The client generates the token at the time it sends the request message, and the server will return the token with the response, which the client can match.

As in ITS, resources are identified by their URI, and apps ask for a service by sending a CoAP message to a URI. The message contains a verb naming the method that the server should apply on the resource. CoAP defines the following methods with semantics very similar to HTTP:

- **GET:** Returns a representation (e.g., the current sensor reading) of the resource.
- **PUT:** Asks the server to create or update the resource with the given representation (e.g., turn ON).
- **POST:** Similarly to PUT, asks the resource to handle the request. As a result the representation of the resource may change.
- **DELETE:** Asks the server to delete a resource.

The response message will contain the representation of the resource after the method was applied to it. It will also include a return code: Success, Client Error (malformed request syntax or some other reason due to the fault of the consumer), or Server Error.

We see that the CoAP specification makes it very simple to consume services. But with so little coupling between apps and services, how do apps know what services are available? We will answer this important question in the next two sections.

13.4.4 URIs

Borrowing from HTTP, CoAP identifies or locates resources by URI. The naming scheme is *coap* or *coaps*, corresponding to *http* and *https* respectively (the latter being the secure version). The resource identified by the URI is operated on by a method provided with the CoAP message.

13.4.5 Discovery

To allow us to use proper internet terms we need to introduce a few new abstractions before we embark on discussing discovery and answering the question we posed above.

Hosts are servers, clients, or Things. Basically, any machine or device on the network that can run code is a host.

The abstraction named */.well-known/* is used by ITS to retrieve the *resources* of a host. Its typical use is to obtain meta information such as the services available on a host. Correspondingly, in CoAP, */.well-known/core* is the default entry point for requesting the list of services rendered by a Thing.

In internet jargon, "discovery" is called "web discovery" and the description of relations between resources is named "web linking" (link is another name for URI). Using web discovery, we find out about a host URI and then use web linking to find out about its services.

Web Application Description Language (WADL), or a similar language, can be used to describe the API of an application. These languages describe the services in formats that can be easily parsed by machines. For example, the set of resources, relationships between resources (how they web link from each other), the methods that can be applied to each resource, and so on.

With these terms in mind, we are ready to discuss how discovery works in CoAP.

With energy consideration as a prime directive, CoAP architects decided to make web discovery optional. A host chooses to support resource discovery. Since a host may be handling multiple resources, such as heart rate reading, blood oxygenation reading, medicine actuation – each identified by a unique URI (a URI equals a service) – then the host can selectively decide which (if at all) are discoverable.

Discovering services is similar to climbing a tree. If we can reach the lowest tree branch, we can start climbing; we then look for the next reachable branch. If we can find out which hosts are connected on the network, we can ask each host what resources it hosts. Since we don't even know which hosts are available, we send the following CoAP message:

```
/.well-known/core
```

to a *multicast address* of the network itself. This multicast address can cover anything from the local to global segments of the network. The responses will be a list of hosts, which will allow us to then query the hosts for their services.

But CoAP architects still had a problem to solve. The CoAP standard defines that the port on which CoAP hosts listen is configurable. AllJoyn had a similar challenge and there it was decided that an app port had to be known a priori. If we do not know anything about the hosts, how do we know which port they are listening to? CoAP deals with this design challenge by requiring that hosts that choose to support discovery join one or more of an *all-CoAP-nodes* multicast address and that they listen for discovery requests on the default CoAP port number, as well as on IPv4 and IPv6 multicast addresses.

This solution continues to provide decoupling between apps and services and renders apps with discovery for hosts on the network, followed by discovery of their services.

The next design question we would like to address is which entity should do the discovery filtering, hosts or apps? The number of hosts on the network and the number of services on each host can be large, producing a lot of traffic if hosts do not filter their discovery responses. On the other hand, in the IoT many hosts are Things, which should be very careful with their energy usage. With both benefits and drawbacks, CoAP architects left it up to the hosts to decide whether to support filtering or not. The design choice made by CoAP architects is apt. If a host can afford it, it can choose to support filtering; when it cannot, this design choice did not rob IoT systems from filtering as it can be performed by more powerful hosts or the apps.

An app can simplify the discovery when a host name is known a priori by using DNS [26, 27] to obtain the host IP address. For example, there may be a generally accepted agreement that all home medical system hosts are named *medihome*. Once an app has the IP address of the local medihome host, it can send a direct (unicast) discovery message to that address to query its available resources.

Upon a discovery query, a host will return the resource type, interface description, and media type (when applicable) of its available services. As an example, we may get the following response from the host

```
</sensors/medical/heart-rate>;if="sensor",
</sensors/medical/blood-oxygenation>;if="sensor"
```

This is a terse response, which is actually more terse than what we have shown here because it will be returned as a long string and not broken into separate lines as we have depicted here for readability. In the response, the comma separates the resources, and the semicolon does the same for the parameters of each resource. This response indicates one interface named *sensor*; two resources, *heart-rate* and *blood-oxygenation*, are available from this interface; and the URI of each is given to us. Having this information now allows the app to interact directly with the services.

Let's look at an example in which the host is asked to filter its response. Here is a CoAP message asking the host to return information only about the *heart-rate* sensor:

```
GET /.well-known/core?rt=heart-rate
```

The response will be

```
</sensors/medical/heart-rate>;rt="heart-rate";if="sensor"
```

This cryptic way of specifying queries is common on the Internet, where often URIs include the request parameters. As long as the overhead is negligible, we hypothesize that passing JSON or XML objects may facilitate effective communication that is easier to work with.

13.4.6 CoAP Summary

There are a few libraries that implement the CoAP standard in Java or C. A library named **libcoap** [28] is available for both Contiki [29] and TinyOS [30]. It is an implementation of version 3 of the standard (we are now in version 18). Libcoap consumes about 12 KB on Contiki and a limited version (among other omissions only GET and PUT methods were implemented) consumes approximately 7 KB on TinyOS. A typical request (for data or actuation such as "turn LED ON") sends about 115 bytes. A typical response (an answer to "get temperature" request) returns about 220 bytes. However, when using a service that returns three modalities (temperature, humidity, and voltage) about 230 bytes were transmitted [31], so only 10 additional bytes for two additional modalities. In this example the content portion takes about 15 bytes and the overhead such as headers 215 bytes, suggesting that CoAP and the other protocols it relies on in the lower levels are still chatty. This is worthy of further research.

CoAP offers good separation between services and apps as opposed to the coupled approach we saw in AllJoyn. As always, this comes with a price tag. Any reasonable implementation of AllJoyn that chooses meaningful method names provides context for developers and maintainers of an app. It will be harder to wrap context around the terse GETs and PUTs of CoAP. How to enjoy both worlds is an area for further research.

We treated the stateless nature of CoAP as an advantage thus far, but stateful architecture has advantages as well. Any object of AllJoyn is a container holding some state of the world it represents. Programming paradigms that use objects enjoy the benefit of state without any extra work on behalf of the apps. There are many settings in which state is advantageous. Take, for example, a multiplayer game. It is very natural for an app to instantiate an object for each of the players, endowing the app with state about it own player and of the others. Trying to do that with a RESTful architecture would be hard and would result in more complicated app code.

We did not like the cryptic nature of URI-based queries, but this is nothing new; we don't like it on the Internet either. More than technical, this is a cultural problem. Designers think in certain ways when their total real-estate is a URI string. The result may be limited filtering, for example. With the objective of connecting the world, expressive discovery and filtering is important and the research community should look for ways beyond "communicating on the URI line."

We have learned about a middleware that was conceived by borrowing from Internet technologies and adapted for the IoT. For an internet browser to be able to connect all the way to a Thing, while using a uniform software stack (that is, without adding too many intermediaries, bridges, gateways, or other translators) would be a great advantage. This exemplifies why we think CoAP is a good piece of innovation – not because it invented a new wheel, but exactly because it did not.

13.5 BOSS Case Study

BOSS [15] is a research endeavor from the University of California, Berkeley. It was not designed as a general-purpose IoT middleware, but as a distributed CPS for controlling buildings. We review BOSS in this chapter because we believe that some of its core modules and services can be repurposed as a general IoT middleware, some of them novel and absolutely useful.

Buildings today are controlled by numerous isolated and closed systems. For example, the HVAC system cannot readily access data from the alarm motion sensors to infer occupancy. BOSS aims to integrate these isolated systems and manage the building under one OS. Because BOSS exposes all sensing and actuation points of the building, apps not envisioned before can be written. BOSS includes mechanisms to allow for such app composition, possibly by a third party, while still keeping the operation of the building safe. The operation of the building is considered safe at two levels: under the overall supervision of BOSS, and at the default safety control loops of the original equipment manufacturers.

BOSS is a distributed OS with modules running on machines that make the most sense from the perspective of safe building control. The main modules are:

- component abstraction layer
- naming and semantic modeling
- semantic data querying
- real-time data processing and management
- transaction management
- authorization
- running apps.

BOSS is implemented mostly in Python, with performance-sensitive parts in C. The system follows service-oriented design. Most services communicate using REST architecture and exchange JSON objects. This is an improvement over the terse messaging of CoAP.

13.5.1 Component Abstraction Layer

It is difficult to acquire data from sensors or send commands to actuators because of device (Thing) diversity. The objective is uniform communication, but Things are built by so many manufacturers who use diverse protocols and offer or need data in various encodings and formats. To make matters worse, Things have little computation power so it is hard to run modern software on them and so it is hard to abstract the lack of uniformity. To overcome these challenges, BOSS uses the Simple Measurement and Actuation Profile (sMAP) [32] to present and access all Things in a uniform way. sMAP is a specification for accessing data feeds emanating from diverse sources. On the output side of sMAP, all Things live in a shared namespace, from which all Things respond to REST calls.

13.5.2 sMAP

The core abstraction in sMAP is the *timeseries* tuple *(time, value)* which is named a *point*. Points are made unique by assigning a universally unique identifier (UUID) to each, and are made expressive when apps add metadata to them. Metadata are key–value tuples that describe each point, tagging it with any information that apps deem important. Standard metadata includes what is being measured, units, calibration constants, sampling rate, and how much buffering is desired. sMap provides the services to read and set points, subscribe to changes, receive periodic notifications, and define or retrieve key value metadata describing the point. sMAP also supports the aggregation of a few sensors' data into a single new point.

sMAP deals with the heterogeneity of Things by using *drivers* to connect to Things. Drivers are Things-specific code fitted to the intricacies of sensors, actuators, and other devices to make them accessible to sMAP. By definition, drivers provide base services such as reliable delivery, failover, buffering of sensory data, metadata management, and secure actuation (using SSL). Buffering allows sMAP to save data local to a Thing when its connectivity to the rest of the network is temporarily cut.

Things often measure more than one modality of the world. A pulse-oximeter, for example, has two modalities. It measures both the heart rate and blood oxygenation of a person. In sMAP, each device (Thing) is named a *point* and each measured modality is named a *channel*. Points publish into one or more sMAP channels. All channels are exposed at URIs and are accessible over HTTP with either TCP or UDP. One of the advantages of using HTTP is that devices on low-bandwidth links may use proxies to cache their sMAP data. TCP can only be used to read points that are already in sMAP namespace on a μServer-grade machine, not directly from Things. This is the same TCP–UDP design trade-off that CoAP had to deal with.

We discussed earlier in the chapter whether a middleware can provide the services we expect while being efficient enough to run on resource-constrained devices. The designers of sMAP decided to implement the IP stack, as well as use JSON objects as data containers. This design decision has implications because both the IP stack and JSON introduce an overhead that makes them hard or impossible to implement on Things, at least on today's micro-controllers. To address this challenge, sMAP designers used specialized versions of some layers of the IP stack: Code on Things uses UDP instead of TCP, embedded binary HTTP (EBHTTP) [33] instead of HTTP, and encoded JSON instead of JSON. The result, an implementation of sMAP for 802.15.4 networks, takes just over 19 KB of code and consumes 10 percent of the bandwidth [32]. These figures are for sMAP only, not BOSS, but they are important in and of themselves because in our opinion sMAP can play a key part in enabling the IoT with or without BOSS. sMAP was successfully deployed over regular computers, wireless gateways, and 802.15.4 motes. Drivers exist for HTTP, OPC-DA, 6LoWPAN, RS-485, and BACnet/IP [32]. The recommended approach is to run sMAP on the devices it monitors or controls, and when this is not possible, on the next closest gateway or proxy.

13.5.3 Semantic Modeling and Query

After sMAP does its work, all Things are present in a common namespace. BOSS then takes the next innovative step and simplifies discovery well beyond what we have seen so far. BOSS takes a unique approach to finding components. Because buildings are complicated and require domain knowledge, BOSS designers integrated a proximate query language as a core service [34]. Apps specify the components they are interested in in terms of their relationship to other components, and search for devices based on their functional or spatial relationships. BOSS uses an approximate query language and the $<$ and $>$ operators, with $A > B$ meaning that A supplies or feeds into B. The end result of this technology is that apps might not lose their usefulness and context when they are deployed in different buildings. For example, queries such as the following (taken from [32]) can be issued:

```
lights in room 410
lights on the fourth floor
cannot dim the lights below 50%
access is only provided at night
```

The query language uses spatial, electrical, HVAC, and lighting views of the building, as well as a GIS database as metadata to inform these queries.

Services can also be found by using their canonical DNS names when the flexibility of the query language is not needed. Either way, such a query language clearly cannot be handled by Things. The BOSS solution is to run modules on any server that has a strong enough processor. To achieve that, BOSS was designed as a distributed OS: sMAP is run on Things or as close as possible to them. Things are then exposed as points and channels in a shared namespace. The rest of the BOSS modules, databases, query language, and so on can run on any machine, which can be either inside the organization or even in the cloud.

13.5.4 Time Series Data Processing

For apps to control buildings in any meaningful way, they need access to the state of the building. It is certainly beneficial to have access to the current state, but many algorithms would benefit from historical data as well. The complication emerges when we realize that real-time streams and historical data are in different places. The former emanates from Things, and the latter is in a database. BOSS offers a time series module that provides uniform access to both.

The Time Series Service (TSS) contains three components. A database called *readingdb* that stores historical data, a query language, and data-cleansing operators. Points are time-stamped and stored in the database, where each stream is treated as a time series. TSS is nicely integrated with the points' metadata, which can be used for querying.

13.5.5 Transaction Management

A transaction is a sequence of operations that takes a system (for BOSS this is the building) from a starting state to an end state, where both states are defined as correct. "Correct" can carry different semantics, such as safe, coherent, or valid. During the transaction the system is allowed to be in an incorrect state. The role of a transaction management system is to guarantee that a transaction is either carried through to completion, from a starting correct state c_1 to an ending correct state c_n, or the system will be rolled back to its starting state c_1.

This is a nice concept, but not so easy to guarantee when the system is distributed. If a transaction actuates Things that sit on different network domains, network failures can prevent the transaction management module from yielding its promised correctness. To solve this design challenge, BOSS defines a fault domain, outside of which transactions are no longer guaranteed. The lowest fault domain is a sensor or actuator. sMAP is responsible for that level, and it, or the device internal fallback control logic, is responsible for the device's basic (safe) operation if the device is cut from the rest of the network. The best way to achieve this is by running sMAP as a colocated service on the device itself. But as noted earlier, this is often not possible for either computing requirements or procedural issues. The question remains, then, how can the fault domain be expanded? BOSS addresses this challenge with the following abstractions:

- **Lease:** Associated with an actuator. Defines the time that an operation is valid. If the lease expires and the transaction had not yet been completed, the action will be rolled back.
- **Revert sequence:** Specifies how to undo each action that is part of the transaction.
- **Error policy:** Defines what to do in case of a partial failure.

Values for these abstractions are specified by apps when they request transactions. BOSS will run the revert sequence if the transaction had not run to its completeness and the lease expired. If the error policy specifies so, actions may be reverted on partial failures as well.

13.5.6 BOSS Summary

BOSS packs many features beyond a basic IoT middleware. It provides rich metadata for each point and makes these tags, as well as semantic data, available for querying. Apps can discover Things based on the relationships between Things and their physical environment ("lights on the fourth floor"). This makes apps transferable between locations. BOSS discovery is much more flexible and robust than we have seen in AllJoyn or CoAP.

sMAP decouples Things from apps' logic by conforming to the request–response paradigm. This is the same approach that CoAP takes. An app running on BOSS sends asynchronous requests, which BOSS will respond to. A BOSS app only needs to know the name of a service it wants to use, and to learn these names they can be queried without any prior knowledge. This is a major difference from AllJoyn, in which we had to follow intricate steps for one app to know about another, or know about sensors and actuators. Custom classes had to be authored and described in XML. Proxy objects representing remote apps had to be created to allow remote method invocation. While sMAP drivers and sMAP specification are still a custom coding exercise for each device, they are being developed with separation of concerns. sMAP objects are not a part of any app. This eases software development and is a welcome design choice.

Discovery in BOSS brings about an interesting advantage by means of its semantic query, but it introduces a great burden of configuration. There will be no semantics in BOSS without defining the entire spatial, electrical, HVAC, and lighting views. It would be an interesting research to try to infer the semantics with less configuration.

Transaction management in the context of actions performed on a sequence of actuators was a new concept we learned about in this section. It adds safety and peace of mind for CPSs.

For data transfer, we much prefer the use of JSON (in BOSS) or XML versus remote method invocation (in AllJoyn). Although JSON may be too descriptive for embedded systems, we believe that being expressive triumphs the bandwidth challenge. Moreover, BOSS use of encoded JSON demonstrates that the bandwidth challenge may be addressed.

The request–response paradigm has an additional advantage. Data producers and consumers are decoupled. They can run on different machines and no app needs to maintain hooks into another app to receive data or operate an actuator. At the same time, for a large body of apps this is a disadvantage. A multiplayer game server, for example, would benefit from maintaining the state of all participants. AllJoyn will excel at such stateful situations, and games servers are not the only type that would benefit from maintaining states.

BOSS bases its architecture on existing standards used on the general Internet, while specializing them. The BOSS designers have shown that despite using standards such as HTTP and JSON (in their compact forms), the end result is a solution that requires a reasonable amount of memory (just over 19 KB in one case). Deployments of BOSS exists on a number of embedded systems. For full disclosure, though, many of them are not running on Things, but on the closest computer connected to the Things. Therefore, it is still an open question whether truly resource-constrained devices can afford sMAP.

13.6 A CPS Middleware for the IoT: Summary and Observations

We started by thinking on what services a middleware for the IoT will need to render and continued with an analysis of three existing middleware. The middleware we visited present different approaches.

AllJoyn is an object-oriented middleware. We believe that this is the preferred approach for apps that need to maintain state or apps that can benefit from hysteresis (imagine a service that can become better over time if it remembers what happened before). Many types of services are of this type. Right now, AllJoyn operates within the confinement of a single Wi-Fi network, but if it were to go beyond this, it would need to deal with the issue of scale. Services that hold state require more memory as the number of clients they serve increases. At an IoT scale this eventually becomes impractical. AllJoyn specializes in managing the network dynamics, a feature that we have not seen provided by the other two middleware.

CoAP adapts the technologies of the Internet to the IoT. This offers great advantages, such as the promise of seamless Internet–IoT interoperability, scalability, and services that can run inside the network (for example, proxies that can perform computations on passing data). Proxies also improve the lifetime of Things by responding on behalf of a Thing, providing data that is still fresh in the proxy memory.

BOSS (like CoAP) follows the request–response paradigm and uses internet technologies. It adds a database, a module to describe relationships between Things, semantic querying beyond basic discovery, data series services that treats real-time and historical data uniformly, and transaction management.

Discovery is limited in AllJoyn. No matter how creative and descriptive a service name is made, it is not going to be descriptive enough. CoAP improves by adding parameters to the query string (such as the resource type), but neither are a match for

BOSS metadata, use of JSON, and the ability of the apps themselves to add key–value pairs to the metadata.

AllJoyn is the only middleware that manages network dynamics, allowing devices to temporarily roam out of the network range without losing their session configuration. Both CoAP and BOSS may be lacking in this area. This is the result of being stateless – they do not have the session abstraction. But they present a different architectural philosophy. For them, there is really no need to manage leaving and joining. A BOSS channel is either there or not, temporarily. For intermittent network conditions, sMAP offers buffering, storing sensory data until the point is connected again. But still, for many types of apps, AllJoyn will be a much better choice here, for example, apps that would enjoy being automatically notified about other apps being present (or not) on the network.

We have been discussing different protocols or middleware that offer different advantages. Their design inherently embodies different trade-offs. We believe that a successful IoT architecture will eventually run a set of protocols; there isn't and there shouldn't be one "winner" protocol or middleware. As this chapter tried to demonstrate, a middleware that excels in some of its features will lack in others. A set of viable middleware should therefore be used and interoperate to optimize the overall yield of the IoT infrastructure.

For example, given their contrasting nature, CoAP and AllJoyn are suitable for different use cases: CoAP for large-scale, loosely coupled applications, and AllJoyn for the local area in which a number of apps need to be aware of each other. CoAP would be better for accessing resource-constrained Things in terms of energy and computation demands. On the other hand, in the setting of a single IoT location such as a home or a city park, where different apps may need to interact or at least know about each other's presence, AllJoyn provides better services. BOSS provides an appealing platform for buildings, but work remains to generalize it to other types of IoT applications.

The middleware reviewed in this chapter – as well as others that weren't – remain heavyweight for too many Things. Work remains in developing lightweight protocols and services to seamlessly incorporate small Things into IoT networks.

IoT networks present researchers with an opportunity to look again at security and privacy, an area that is not handled well on the Internet. The discussion about "who owns the data" should involve many disciplines, including philosophy, law, and business. Until that occurs, computer scientists, through middleware and tools, can articulate solutions that turn the current practices on their heads, such as by having the protocols and middleware support configuration by owners of Things that allow them to limit who (and which device) can use the data.

Finally, we believe that the network itself should be capable of doing more complex computation as the data traverses nodes on the path. How much computation is done and on which node will be determined at runtime according to energy availability, the capabilities of the nodes, security and privacy, and concurrent operations. We do not believe that doing everything in the cloud (as the trend strongly favors today) is realistic or architecturally beneficial.

References

[1] J. M. Schlesselman, G. Pardo-Castellote, and B. Farabaugh, "OMG Data-Distribution Service (DDS): Architectural update," in *Military Communications Conf.*, 2004, pp. 961–967.

[2] D. Locke, "MQ Telemetry Transport (MQTT) v3.1 protocol specification," IBM developerWorks Technical Library, Tech. Rep., 2010.

[3] S. Vinoski, "Advanced message queuing protocol," *IEEE Internet Comput.*, vol. 10, no. 6, pp. 87–89, 2006.

[4] L. L. V. Mäntysaari, "Extensible Messaging and Presence Protocol (XMPP)," *Sem. Instant Messaging and Presence Architectures in the Internet*, 2005, pp. 1–9.

[5] Microsoft, "Service Bus," 2015. [Online]. Available: http://msdn.microsoft.com/en-us/library/ee732537.aspx.

[6] OpenIoT Consortium, "Open source solution for the Internet of Things into the cloud," 2012. [Online]. Available: https://github.com/OpenIotOrg/openiot/wiki/Documentation.

[7] D. Guinard and V. Trifa, "Towards the Web of Things: Web mashups for embedded devices," in *Workshop on Mashups, Enterprise Mashups and Lightweight Composition on the Web*, vol. 15, 2009, pp. 1–8.

[8] C.-L. Fok, G.-C. Roman, and C. Lu, "Servilla: A flexible service provisioning middleware for heterogeneous sensor networks," *Sci. Comput. Program.*, vol. 77, no. 6, pp. 663–684, 2012.

[9] C.-L. Fok, G.-C. Roman, and C. Lu, "Adaptive service provisioning for enhanced energy efficiency and flexibility in wireless sensor networks," *Sci. Comput. Program.*, vol. 78, no. 2, pp. 195–217, Feb. 2013.

[10] G. Hackmann, C.-L. Fok, G.-C. Roman, and C. Lu, "Agimone: Middleware support for seamless integration of sensor and IP networks," in *Distributed Computing in Sensor Systems*, 2006, pp. 101–118.

[11] C.-L. Fok, G.-C. Roman, and C. Lu, "Agilla: A mobile agent middleware for self-adaptive wireless sensor networks," *ACM Trans. Autonom. Adapt. Syst.*, vol. 4, no. 3, p. 16, 2009.

[12] C. Dixon, R. Mahajan, S. Agarwal, et al. "An operating system for the home," in *Proc. 9th Conf., Networked Systems Design and Implementation*, 2012, pp. 25–25.

[13] AllSeen Alliance, "Open source IoT for the Internet of everything," 2015. [Online]. Available: https://allseenalliance.org.

[14] Z. Shelby, K. Hartke, and C. Bormann, "The Constrained Application Protocol (CoAP)," Internet Engineering Task Force, Tech. Rep., 2014.

[15] S. Dawson-Haggerty, A. Krioukov, J. Taneja, et al. "BOSS: Building Operating System Services," in *Proc. 10th Conf. Networked Systems Design and Implementation*, 2013, pp. 443–457.

[16] OpenWRT, "OpenWRT wireless freedom," 2015. [Online]. Available: https://openwrt.org.

[17] Raspberry Pi Foundation, "Raspberry Pi," 2015. [Online]. Available: https://www.raspberrypi.org.

[18] BeagleBoard Foundation, "BeagleBoard," 2015. [Online]. Available: http://beagleboard.org.

[19] Arduino, "Arduino Due," 2015 [Online]. Available: https://store.arduino.cc/usa/due.

[20] R. T. Fielding, "Architectural styles and the design of network-based software architectures," Ph.D. dissertation, University of California, 2000.

[21] World Wide Web Consortium, "Naming and addressing: URIs, URLs, ..." 2006. [Online]. Available: www.w3.org/Addressing.
[22] D. Lee and W. W. Chu, "Comparative analysis of six XML schema languages," *Sigmod Record*, vol. 29, no. 3, pp. 76–87, 2000.
[23] K. Zyp and F. Galiegue, "JSON schema: core definitions and terminology," Internet Engineering Task Force, Tech. Rep., Jan. 2013.
[24] M. J. Hadley, "Web Application Description Language (WADL)," Sun Microsystems, Inc., Tech. Rep., 2006.
[25] J. Hui and D. Culler, "Extending IP to Low-Power, Wireless Personal Area Networks," *IEEE Internet Comput.*, vol. 12, no. 4, pp. 37–45, 2008.
[26] P. V. Mockapetris, "Domain names: Concepts and facilities," ISI Network Working Group, RFC 1034, Tech. Rep., 1987.
[27] P. V. Mockapetris, "Domain names: Implementation and specification," ISI Network Working Group, RFC 883, Tech. Rep., 1983.
[28] O. Bergmann, "Libcoap: C-implementation of CoAP," 2013. [Online]. Available: https://libcoap.net.
[29] Contiki, "Contiki: The open source OS for the Internet of Things," 2015. [Online]. Available: http://www.contiki-os.org.
[30] P. Levis, S. Madden, J. Polastre, et al., "TinyOS: An operating system for sensor networks," in *10th Int'l Symp. Ambient Intelligence*. 2005, pp. 115–148.
[31] K. Kuladinithi, O. Bergmann, T. Pötsch, M. Becker, and C. Görg, "Implementation of CoAP and its application in transport logistics," in *Proc. IP+SN*, 2011, pp. 1–7.
[32] S. Dawson-Haggerty, X. Jiang, G. Tolle, J. Ortiz, and D. Culler, "sMAP: A simple measurement and actuation profile for physical information," in *Proc. 8th Conf. Embedded Networked Sensor Systems*, 2010, pp. 197–210.
[33] G. Tolle, "Embedded Binary HTTP (EBHTTP)," Internet Engineering Task Force, Tech. Rep., 2010.
[34] A. Krioukov, G. Fierro, N. Kitaev, and D. Culler, "Building Application Stack (BAS)," in *Proc. 4th Workshop on Embedded Sensing Systems for Energy-Efficiency in Buildings*, 2012, pp. 72–79.

14 *In Silico* and *In Vitro* Modeling Platforms for Stimulation of Emergent Processes in Neuronal Networks

Sandip Roy, Patrick Corrigan, and James Krueger

14.1 Introduction

The emergence of synchronized dynamics, including rhythmic behaviors, among loosely coupled or semi-autonomous units is common in nature (e.g., [1–3]). Meanwhile, complex engineered networks increasingly require algorithms for coordination or synchronization of network components [4]. Motivated by the need for modeling synchronization in nature and designing synchronization in engineered systems, scientists have developed an extensive analytical theory explaining the emergence of synchronization (and other global dynamics) in interconnected systems as a consequence of simple local interactions among units. These efforts to explain synchronization and other emergent phenomena in networks using mathematical models originate from such diverse fields as circuits engineering, particle physics, and sleep neuroscience, and together form the core of the "science of networks" that has developed over the last 20 years or so. The studies of synchronization provide a compelling verification that simple and extremely limited units can coordinate to achieve highly complex functions, and hence they are of fundamental importance both to understanding complexity in nature and to engineering complex network dynamics.

Yet, as synchronization processes in nature are more thoroughly studied, it is increasingly becoming clear that emergent behaviors should be viewed not only as a natural consequence of coordination among intrinsically independent units, but also as regulatory tools that in fact can often outperform regulation by a central authority. Specifically, experimental evidence suggests that distributed regulatory processes are intrinsically suited for processing multiple afferent inputs and actuating multiple consequences, in a way that is flexible, robust, and cheap. As the experimental evidence showing the possible benefits of distributed regulation grows, there is increasing interest in characterizing flexibility and robustness of synchronization processes from descriptive models. Indeed, a few recent studies have considered performance measures for synchronization (e.g., robustness), with a particular focus on understanding the role of the network topology in robustness to subunit and communication failures (e.g., [5–7]). While these works are beginning to expose the performance benefits and limitations of distributed regulation, a comprehensive study requires characterizing synchronization dynamics *that are subject to realistic external stimulation*. Specifically, it is crucial to understand how distributed regulation processes can integrate stimulation or input signals, and how the processes are modulated by these inputs.

While studies on emergence bear on diverse natural and engineered networks, understanding emergence in brain networks is a keystone challenge – both because of the profound implications on human health of such understanding, and because of the amazing complexity of brain networks compared to other network dynamical processes. Evidence is mounting that core brain functions, including sleep and memory, emerge through bottom-up coordination or regulation of brain units. Brain circuits, and hence these core brain functions, are modulated by myriad sensory inputs. Characterizing the impact of such stimulation or activity or *use* on brain circuits is crucial to understanding distributed regulation of core functions like sleep.

The original evidence that brain processes are distributed stems from *in vivo* experimentation, primarily using animal models. However, characterizing the mechanisms and consequences of distributed regulation *in vivo* is extremely difficult. Such characterization begs for modeling platforms that allow easy manipulation and stimulation of brain circuits, multi-location measurement of response characteristics, and formal analysis of emergent properties. Our group has been pursuing development of mathematical and cell-culture modeling platforms for emergent behaviors in the brain, with a particular focus on models related to sleep regulation. These models are designed to permit analysis of emergence in brain circuits under multimodal stimulation, and hence to inform understanding of the mechanisms and consequences of distributed regulation. While our analysis using these models is very preliminary, the models appear to be promising tools for understanding brain processes at multiple scales and resolutions. This chapter reports on three models for emergence under sensory stimulation in brain circuits, and presents a few preliminary analyses using the models. Our focus here is to trace the development of increasingly complex models for synchronization/emergence in the brain. We begin with an abstract mathematical model that derives from classical models for emergence, then introduce an *in silico* model that captures local function state evolutions underlying sleep regulation (albeit abstractly), and finally describe an *in vitro* platform for characterizing distributed mechanisms for sleep. The *in vitro* platform permits sensing, actuation, and real-time computational analysis of electrical signals generated by networks of neurons grown in culture, using multi-electrode-array technology. We discuss how the *in silico* and *in vitro* platforms can be used in tandem to gain insight into brain processes.

An important consequence of our modeling effort is that mathematical, *in silico* and *in vitro* models are being used in tandem to understand coordination of brain units. Interfacing these cyber and physical world models is important, since the models together can permit characterization of brain functions (such as sleep) across temporal and spatial scales in the brain. Specifically, the high-resolution mechanistic analyses allowed by the cell-culture model can be used to improve the abstract models for interactions among network components in the mathematical and simulation models. Conversely, the predictions of global emergence and stimulation dependence provided by the simulation models can guide experimentation in culture. Additionally, measurements made in the culture platform must be recorded, mined, and analyzed using cyber-tools. In these senses, our modeling efforts are leading to a *cyber-physical platform* for understanding synchronization in brain networks. We believe that cyber-physical modeling tools of this sort may prove valuable for understanding a range of complex biological processes, and hence their development and analysis may constitute an important focus in cyber-physical systems (CPS) research.

We believe that the research presented here introduces an interesting direction of work on CPS testbeds. In particular, it introduces the idea that CPSs can be used as *models* for perhaps the most complex system in existence, the human brain (which itself may philosophically be viewed as a CPS with physical-world and information-processing functions). Unlike other CPS testbeds, our effort here does not attempt to replicate exactly an operational CPS in small scale, but rather focuses on building a new CPS – consisting of a cyber-physical technology for *in vitro* analysis of neuronal networks, as well as coupled computational models of brain networks – that models aspects of brain function. The research presented here, in this sense, introduces that CPSs can be built to develop new scientific understanding of the brain. This study also addresses essential questions related to actuation and sensing of emergent network processes, which are fundamental to diverse large-scale CPSs like electric power systems and the Internet-of-Things (IoT).

A key finding of our study is that interaction topology and interaction strengths among the subunits in a network synchronization process can play a critical role in the network's ability to (1) maintain synchronization in the presence of stochastic external stimulation; and (2) robustly yet quickly adapt to local stimulation-input trends. Based on this finding, we make the argument that efficient distributed regulation often requires rather special interaction topologies and intermediate interaction strengths among network subunits. These analyses support a couple of important hypotheses of the distributed sleep-regulation theory, including that: (1) an appropriately structured distributed regulation process can permit robust maintenance of a global state while permitting modulation of this state by local activity inputs; and (2) stabilization of the neuronal network topology (which is plastic) is of critical importance to maintaining distributed synchronization.

The remainder of this chapter is organized as follows. First, a classical linear model for network synchronization to represent stochastic stimulation inputs (Section 14.2). For this model, analytical methods are used to tie the network topology and interaction strengths with spatial and temporal characteristics of the synchronization dynamics. This analysis illustrates that special topologies and intermediate interaction strengths are needed for robust synchronization and sensitivity to sensory inputs (Section 14.3). Second, we study synchronization under sensory stimulation for a nonlinear model for distributed, use-dependent cortical sleep regulation (Section 14.4). Through simulations, we illustrate that the network topology and interaction strengths again significantly modulate the impact of stochastic stimulations (or use of activity) on the synchronization dynamics. Third, a cell-culture model for distributed sleep regulation is overviewed, in which neuronal networks grown on a multi-electrode array are subject to stimulation and measurement (Section 14.5).

14.2 A Linear Model for Network Synchronization under Stimulation

An abstract model for network synchronization under stimulation is formulated (Section 14.2.1) by enhancing a classical liner network model for synchronization or consensus. Formal analysis and simulation of the model are undertaken to characterize the model's dynamics under stochastic stimulation (Section 14.2.2).

14.2.1 Model Formulation

Synchronization phenomena are observed both in nature and in engineered network systems [1, 2, 8], and hence there has been an extensive effort to model and analyze synchronization. Many of these modeling studies focus on: (1) developing synchronization models that are descriptive of the internal dynamics of a particular system; (2) determining whether or not the system becomes synchronized using the model; and/or (3) characterizing the speed at which the system becomes synchronized. Beyond these basic studies, a wide range of further characterizations of synchronization or synchronization-like network dynamics have been pursued, including robustness analysis of synchronization to component failures and design of synchronized tracking, among many others (e.g., [9]). A common theme in this research is to tie characteristics of the synchronization dynamics to the network's underlying graph structure. While a wide family of synchronization models has been characterized in this way, there has been a particular focus on several canonical models, including especially linear network models (with network components that may have either scalar or vector states) defined on a graph. These linear network models, in which network components are at their essence averaging states of graphical neighbors, permit extensive analysis of dynamics and often serve as apt approximations for more complex models.

Our focus here is on developing graphical characterizations for synchronization dynamics that are subject to stochastic stimulation. Many synchronization processes – including the sleep-regulation dynamics of interest to us and power system swing dynamics, among others – are naturally subject to uncertain stimulations, which (whether desired or not) serve to modulate the synchronization dynamics. We first use a canonical linear network model to study synchronization processes that are subject to stochastic stimulations. This simple model is not meant to capture the specific mechanisms underlying sleep regulation or other distributed brain functions. Rather, it provides a simplified context for understanding the consequences of stochastic stimulation on emergent phenomena in networks. The linear model permits us to analytically determine how the network's graph topology affects the synchronization dynamics. Precisely, we can show that networks with intermediate and heterogeneous interaction strengths (1) maintain strong synchronization in the face of stochastic stimulations, while also (2) permitting certain critical stimulation inputs to modulate the synchronization state. The results from this analysis are valuable to us for two reasons: first, because they provide an explicit characterization of synchronization under stimulation for a model that has wide application; and, second, because they provide insight into, and guide simulation and analysis of, more sophisticated models of emergence, such as the distributed sleep model described in Section 14.3.

Formally, let us consider a network with n subunits or components (labeled $i = 1, \ldots, n$), each of which has a scalar state $x_i[k]$ that evolves along a discrete time axis $k = 0, 1, 2, \ldots$. We model each network component as determining its next state by averaging its current state with those of certain neighboring components (as specified by a weighted and directed graph). Formally, we model the network state $x[k]$ (which comprises the states of the n subunits together, i.e., $x[k] = \begin{bmatrix} x_1[k] \\ \vdots \\ x_n[k] \end{bmatrix}$) as being governed by the following linear time-invariant differential equation:

$$x[k+1] = Ax[k] + v[k]$$

where the following apply.

The **topology matrix** $A \in R^{nxn}$ has nonnegative entries and satisfies $\sum_j^n A_{ij} = 1$, $\forall\, i$, i.e., each row of A sums to 1. The topology matrix specifies how each network component updates its state through (weighted) averaging of other components' states. Specifically, the entry A_{ij} (at row i and column j) indicates the impact of the current state of component j on the next state of component i.

The **stimulation vector** $v[k] \in R^{nx1}$ captures external stimulation inputs to each of the network components. We will consider several models for the stimulation vector. First, we will consider impulsive stimulations at single components, i.e., the stimulation vector $v[0]$ at the initial time is a 0–1 indicator vector while $v[k] = \vec{0}$ for $k \neq 0$. Also, with the motivation that stimulation inputs to synchronization dynamics are often stochastic in nature (as in sleep regulation), stochastic-process models for the stimulation vector will also be considered. In particular, Poisson arrival process models will be considered as representations of stochastic driving stimulations that can modulate the synchronization state. Precisely, for the Poisson process model, the stimulation vector at each time is assumed to capture the number of arrivals in a Poisson process with a specified arrival rate; thus, the stimulation input $v_i[k]$ at component i at time k is assumed to be a Poisson random variable with mean $\lambda_i[k]$, that is independent of the stimulation input at any other component at time k and of all inputs at other times. We will also consider a zero-mean Gaussian white noise model for the stimulation vector as a representation of noise impacting the synchronization dynamics, focusing especially on the case that the noise inputs at each component are independent and identically distributed.

It is also natural for us to define a network graph from the topology matrix, which captures how each component's current state affects the next state of other components. Specifically, we define a weighted and directed **network graph** G with n vertices, which represent the n components in the network. An edge is drawn from vertex j to vertex i, if and only if $A_{ij} \neq 0$ (i.e., the current state of component j impacts the next state of component i). Further, for these edges, we define the weight of the edge as A_{ij}.

The synchronization model described above has been extensively studied in the case that external stimulations are excluded (i.e., $v[k] = 0$). For this case, the connection between the network's graph topology G and characteristics of the synchronization dynamics, including the presence or absence of synchronization and the rate of synchronization, is well understood [8]. Of particular note, it is well known that synchronization is achieved (i.e., the component's states asymptotically approach a common value) under broad connectivity conditions on the network topology. Our aim here is to characterize the impact of the graph topology, as well as the specifics of the stimulation input, on the synchronization dynamics in the case where stimulation is present. With a little effort, it is easy to see that transient stimulation inputs (such as the impulsive inputs described above) will not prevent synchronization in the asymptote, but may both modify the synchronous state and cause transient desynchronization: Our aim is to relate these modifications of synchronization to the location of the transient stimulation and the network topology. Meanwhile, when persistent stochastic inputs drive the synchronization dynamics, perfect synchronization typically will be lost but (under some conditions) the states of the network agents will remain close or will be *imperfectly synchronized.* In this case, our aim is to characterize the *strength* or *extent* of synchronization in terms of the network topology and specifics of the stochastic inputs. Through these analyses, we will aim to understand how networks maintain synchronization in the presence of external stimulation, while also modulating the synchronization dynamics in reaction to important stimulation inputs.

In order to characterize the extent or strength of synchronization for stochastic stimulation inputs, we require analytical measures for the extent of synchronization. A natural measure for the extent of synchronization is the expected squared error between pairs of network components. We will focus on this error measure, i.e., we will aim to characterize the measure

$$error\,(i,j) = E\left[\left(x_i\,[k] - x_j\,[k]\right)^2\right] = E\left[e^T x\,[k]\,x^T\,[k]\,e\right]$$

(where $e_i = 1$, $e_j = -1$ and all other values in $e \in R^{1xn}$ are zero) as an indication of the desynchronization between components i and j.

14.2.2 Model Analysis

The focus of our analysis is to understand how the network's topological structure (its connectivity) implicates its dynamics. For the linear network model, significant insight into this relationship can be obtained via formal analysis along with limited simulation. Two particular studies have been undertaken toward this aim, and are described here. First, the strength of synchronization is characterized in terms of the network topology for stochastic stimulation inputs. Second, the impact of anomalous stimulation at individual network components on the synchronization dynamics is studied, with the aim of understanding the adaptiveness of the dynamics to local stimulation. In developing these results, we will at some points limit the analysis to symmetric network topologies (or at least distinguish between the symmetric and asymmetric case) for the sake of tractability. Noting that many network interactions leading to synchronization (including some interactions in neuronal networks) are symmetric, symmetrizable through diagonal (network-preserving) transformations, or at least bidirectional, the results for the symmetric case are useful in their own right. Further, they provide a stepping stone toward the analysis in the general case.

Strength-of-Synchronization Analysis

Our aim is to characterize the error measure $error\ (i, j)$ defined for the linear synchronization model in terms of (1) properties of the network matrix A and corresponding network graph Γ; and (2) the statistics of the stochastic stimulation input. In particular, we assume that the stimulation input $v\,[k]$ model is subject to a stationary zero mean white stochastic input signal $v\,[k]$ with covariance matrix $\sigma^2 I$, and aim to characterize the asymptotics of $error\ (i, j)$ (for pairs of components, or summed over pairs) in terms of Γ and σ^2. We note that the characterizations only depend on the statistics (specifically, second moment) of the stochastic input signal, not the particulars of the distribution. We also assume throughout that the network graph is connected, and assume the initial states of the network components are nil (without loss of generality, since only the asymptotic dynamics are of interest).

As a preliminary step in the strength-of-synchronization analysis, let us develop a recursion for the covariance matrix of state, which equals $E\left[x\,[k]\,x^T\,[k]\right]$ since $x\,[k]$ has zero mean at each time step. For convenience in this analysis, let us define $M\,[k] = x\,[k]\,x^T\,[k]$. We note that $M\,[k+1]$ can be expressed in terms of the time-k state, as follows:

$$M\,[k+1] = x\,[k+1]\,x^T\,[k+1] = Ax\,[k]\,x^T\,[k]\,A^T + Ax\,[k]\,v^T\,[k] \\ + v\,[k]\,x^T\,[k]\,A^T + v\,[k]\,v^T\,[k]$$

where we have used the state equation of the synchronization model: $x\,[k+1] = Ax\,[k] + v\,[k]$. Taking expectations of this expression, we find:

$$\begin{aligned} E\,[M\,[k+1]] &= E\left[Ax\,[k]\,x^T[k]\,A^T + Ax\,[k]\,v^T[k] + v\,[k]\,x^T[k]\,A^T + v\,[k]\,v^T[k]\right] \\ &= E\left[Ax\,[k]\,x^T\,[k]\,A^T + v\,[k]\,v^T\,[k]\right] \\ &= AE\left[x\,[k]\,x^T\,[k]\right]A^T + E\left[[k]\,v^T\,[k]\right] \\ &= AE\left[x\,[k]\,x^T\,[k]\right]A^T + I = AE\,[M\,[k]]\,A^T + I \end{aligned}$$

The recursion for the covariance matrix, along with the assumption that the state is initially nil, yields an explicit expression for the covariance. Specifically, after a little algebra, we obtain:

$$(E\,[M\,[N]]) = I + AA^T + A^2\left(A^T\right)^2 + A^3\left(A^T\right)^3 + \cdots + A^{N-1}\left(A^T\right)^{N-1}$$

In order to develop graph-theoretic characterizations for the state's covariance (and in turn the error metric), it is helpful for us to express the covariance matrix in terms of the spectrum (eigenvalues and eigenvectors) of the topology matrix A. Such a spectral characterization is especially tractable in the case that A is symmetric, so let us first consider this special case. Let us also further assume that the network graph G is connected and aperiodic. In this case, we note that A has a single unity eigenvalue, while the remaining eigenvalues are real and in the open interval $(-1,1)$. Let us denote the eigenvalues of A, in decreasing order, as $1 = \lambda_1 > \lambda_2 > \ldots > \lambda_n > -1$.

For symmetric A, we note that $A = A^T \rightarrow AA^T = A^2$. Using the spectral decomposition of A as $A = V\Lambda V^{-1}$, where the eigenvalues in Λ are in decreasing order WLOG:

$$\begin{aligned} E\left[M\left[N\right]\right] &= I + A^2 + (A)^4 + A^6 + \cdots + (A)^{2(N-1)} \\ &= V\left(I + \Lambda^2 + (\Lambda)^4 + (\Lambda)^6 + \cdots + (\Lambda)^{2(N-1)}\right) V^{-1} \end{aligned}$$

To further simplify the covariance matrix, let us consider the diagonal matrix $\left(I + \Lambda^2 + (\Lambda)^4 + (\Lambda)^6 + \cdots + (\Lambda)^{2(N-1)}\right)$. The ith diagonal entry in this matrix is given by $\sum_{n=0}^{N-1} (\lambda_i^2)^n$, which simplifies to: $\sum_{n=0}^{N-1} (\lambda_i^2)^n = \frac{1-(\lambda_i^2)^N}{1-\lambda_i^2}$ for any $\lambda_i \neq 1$ and $\sum_{n=0}^{N-1} (\lambda_i^2)^n = N$ for $\lambda_i = 1$.

The covariance matrix can thus be written as $E\left[M\left[N\right]\right] = V\ \Gamma\ V^{-1}$, where $\Gamma = \begin{bmatrix} N & 0 & & 0 \\ 0 & \frac{1-(\lambda_2^2)^N}{1-\lambda_2^2} & & 0 \\ & & \ddots & \\ 0 & \cdots & & \frac{1-(\lambda_n^2)^N}{1-\lambda_n^2} \end{bmatrix}$, $V = \begin{bmatrix} \uparrow & \uparrow & & \uparrow \\ 1 & v_2^T & \cdots & v_n^T \\ \downarrow & \downarrow & & \downarrow \end{bmatrix}$, $V^{-1} = \begin{bmatrix} \leftarrow & 1^T & \rightarrow \\ \leftarrow & v_2 & \rightarrow \\ & \vdots & \\ \leftarrow & v_n & \rightarrow \end{bmatrix}$ and $v_2, \ldots, v_n$ are the right eigenvectors of A associated with non-unity eigenvalues. Here, we have used the fact that a nonnegative matrix with unit row sums whose associated graph is connected has a strictly dominant eigenvalue at 1 with corresponding eigenvector **1**. We have also used that the left and right eigenvectors of a symmetric matrix are identical.

Next, noting that the error measure at time N is given by $error\,(i,j) = e^T E\left[M\left[N\right]\right] e$, we recover: $error\,(i,j) = \left(v_{2,i} - v_{2,j}\right)^2 \left(\frac{1-(\lambda_2^2)^N}{1-\lambda_2^2}\right) + \left(v_{3,i} - v_{3,j}\right)^2 \left(\frac{1-(\lambda_3^2)^N}{1-\lambda_3^2}\right) + \cdots \left(v_{n,i} - v_{n,j}\right)^2 \left(\frac{1-(\lambda_N^2)^N}{1-\lambda_n^2}\right)$, where we have used the fact that $e^T \vec{1}$. The asymptotic value of the error measure is of particular interest to us, and can be computed easily from the general expression:

$$\begin{aligned} \lim_{N\to\infty} error\,(i,j) &= \left(v_{2,i} - v_{2,j}\right)^2 \left(\frac{1}{1-\lambda_2^2}\right) + \left(v_{3,i} - v_{3,j}\right)^2 \left(\frac{1}{1-\lambda_3^2}\right) \\ &\quad + \cdots \left(v_{n,i} - v_{n,j}\right)^2 \left(\frac{1}{1-\lambda_n^2}\right). \end{aligned}$$

We note here that, although $E[M[N]]$ does not reach a limit, the error is bounded. The above expressions for the error permit us to develop several spectral and graph-theoretic characterizations of synchronization errors in the linear synchronization model, and hence to provide insight into the strength and characteristics of synchronization in the model. These various characterizations broadly follow from the following observation regarding the error expressions above: eigenvalues of the topology matrix A near 1 whose ith and jth eigenvector components are far apart contribute significantly to $error\,(i,j)$. With some effort, this basic insight can be translated into simpler spectral bounds on the error, and in turn into graph-theoretic characterizations of the error. Here, we focus on presenting the results and using them to inform understanding of synchronization strength/properties under stochastic stimulation; we omit proofs of the results. Here is the list of structural and graph-theoretic results:

(a) $2\left(\frac{1-(\lambda_{min}^2)^N}{1-\lambda_{min}^2}\right) < error\,(i,j) < 2\left(\frac{1-(\lambda_{max}^2)^N}{1-\lambda_{max}^2}\right)$, where λ_{max} is the non-unity eigenvalue of A with largest magnitude, and λ_{min}^2 is the non-unity eigenvalue with smallest magnitude. In the asymptotic case, the expression reduces to $2\left(\frac{1}{1-\lambda_{min}^2}\right) < error\,(i,j) < 2\left(\frac{1}{1-\lambda_{max}^2}\right)$. Noting that the upper bound is a monotonically increasing function of the largest-magnitude non-unity eigenvalue, we see that the error can be upper-bounded based on the distance of this magnitude from 1. In fact, noting that the topology matrix is a symmetric *stochastic matrix*, numerous well-known results from *algebraic graph theory* can be brought to bear to bound the largest magnitude eigenvalue away from unit circle, and hence bound the error. We omit detailed results, but note that strong bounds are obtained for strongly and uniformly connected network graphs. In particular, networks where short, high-weight paths exist between each pair of components can be guaranteed to have small errors. Thus, we see that synchronization strength is increased through uniform connectivity among the network components. Such analysis also indicates the particular benefit of long-range connections in networks (such as are present in "small-world graphs") in achieving strong synchronization.

(b) Consider the total error, $\sum_{i,j} error\,(i,j)$. The total error is a monotonically decreasing function of network connectivity, in the following sense: Increasing edge weights between two network components i and j while correspondingly decreasing the self-loop weights at these two components will always reduce the error. Again, the result indicates that strong and uniformly connected networks achieve strong synchronization (albeit at the cost of requiring high interconnectivity).

(c) Within a particular network, strongly connected clusters of components will tend to be strongly synchronized, while components in different clusters (which are weakly interconnected with each other) are more apt to be desynchronized. This result suggests that networks subject to stochastic stimulation may exhibit strong regional synchronization patterns, even when some global desynchronization is present. The result also highlights that weak cuts in the network graph lead to desynchronization

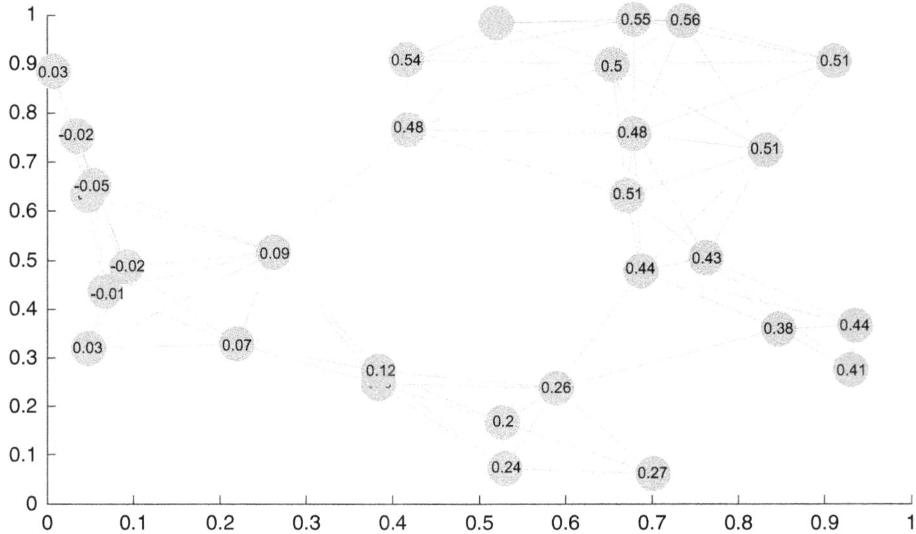

Figure 14.1 The linear synchronization model with added stochastic noise is analyzed, for a network with 30 nodes. Sample correlation coefficients between the state at the unmarked node (near the top middle of the figure) and at every other node are shown. Strong correlations with nearby nodes are observed, while correlations with distant nodes are much weaker.

among regions of the network. This topology-dependence of the synchronization is illustrated in Figure 14.1.

Let us also briefly discuss the strength-of-synchronization analysis for the case that the state matrix A is asymmetric. In contrast to the symmetric case, exact expressions for error (i, j) in the asymmetric case are rather complex; however, the error can again be systematically bounded in terms of the subdominant eigenvalue of A and hence connected to the network's graph topology. The bound is similar in structure to that for the symmetric case (depending on the distance of the subdominant eigenvalue from unit circle), and requires some amount of technical machinery to develop, so we omit the details. However, a brief outline of how the bound can be developed is worthwhile. To begin, we recall that $error\,(i, j) = e^T\,(E\,[M\,[N]])\,e = e^T\left(I + AA^T + A^2\left(A^T\right)^2 + A^3\left(A^T\right)^3 + \ldots + A^{N-1}\left(A^T\right)^{N-1}\right)e = e^2 + A^T e^2 + \ldots + \left(A^T\right)^{N-1} e^2$. To continue, we note that A is a stochastic matrix, and hence that the columns of $\left(A^T\right)^r$ can be interpreted as containing time-N state-occupancy probabilities of a Markov chain, for each possible initial state of the Markov chain. Thus, $\left(A^T\right)^r e$ captures the difference in state occupancy probabilities resulting for two different initial Markov chain states. Thus, assuming A is ergodic, it follows immediately that the entries in $\left(A^T\right)^r e$ can be upper bounded by decaying exponential functions of r, since the state occupancy probabilities from each initial state approach a common asymptote exponentially. In fact, numerous exponential bounds have been developed for deviations from the asymptote in terms of the subdominant eigenvalue of A, that automatically yield bounds on the entries of $\left(A^T\right)^r e$. Upper bounds on *error*(i, j) in terms of the subdominant eigenvalue then follow readily. As with the symmetric case, these upper bounds on *error*(i, j) indicate that densely and uniformly connected networks will be strongly synchronized.

Adaptability to Stimulation

Naively, the above characterizations of synchronization strength seem to suggest that densely and uniformly connected networks achieve strong and robust synchronization for the linear network model. Strongly connected networks are able to withstand stochasticity in stimulation inputs, to maintain close synchronization among network components. However, several factors other than synchronization strength may drive the formation/design of network synchronization processes, which engender trade-offs between synchronization strength and other performance metrics. First, we note that strongly connected networks are often costly to build (or even impossible to build, if space/dimension constraints are present); thus, a design cost trade-off may limit network connectivity. Perhaps of even greater significance, synchronization processes are often required to be reactive to particular important stimulation inputs while tempering the impact of other, less critical, stimulations. Although the synchronization dynamics should be able to withstand stochasticities in stimulation in general, the process *should* differentially react to stimulation trends or impulses in certain critical network components. This need to incorporate stimulation inputs is pronounced for brain networks, where regulatory mechanisms must account for and react to stimulation or use. For instance, sleep regulation is well-known to be use-dependent: Activity or use of brain circuits play a key role in sleep duration and characteristics.

Here, let us study the impact of stimulation inputs at various components in the linear synchronization model, on both asymptotic and transient characteristics of the synchronization dynamics. Noting that the model is both linear and time-invariant, we see that the impacts of stimulations at various times superpose, and further that the impacts are invariant to stimulation time. Thus, it is sufficient to study the impact of impulsive stimulations at time 0 at particular network components ($v_i\,[0] = 1$, where i is the node being stimulated, is the only nonzero value for the stimulation input), to characterize the impact of stimulation characteristics on the synchronization dynamics. Here, we characterize the impact of such impulsive stimulations, assuming that the network is synchronized at the time of stimulation. Broadly, an impulsive stimulation to a network component modifies the synchronization dynamics in two ways: first, it modifies the synchronous state of the network, i.e., it drives the states of all the network components to a new equilibrium. Second, the stimulation produces a transient response that yields the transition from the old to the new synchronous state. Here, we briefly discuss both the asymptotic and transient modulations of the synchronous state upon impulsive stimulation of single network components. Specifically, we study (1) the extent of modification of the asymptotic synchronous state, and (2) the transient response time for the dynamics. A key finding is that, for critical network components to achieve significant and differential impact on the synchronization dynamics, nonuniformity in connectivity throughout the network is necessary. Thus, for a strong but receptive synchronization, an intermediate and non-uniform connectivity is desirable.

Let us briefly discuss the asymptotic and transient analyses/simulations.

Asymptotic analysis: The new synchronous equilibrium state of the linear network model after impulsive stimulation differs from the original equilibrium. The change in the equilibrium can be found as follows: (1) The left eigenvector of A associated with its unity eigenvalue should be found and normalized so that the entries in the vector sum to 1 and (2) the ith entry of this vector is the change in equilibrium state effected by the impulsive stimulation. This spectral characterization of the change in the equilibrium immediately provides graph-theoretic insight into the network components that can strongly impact the asymptotic state. First, for symmetric A, we recall that the left eigenvector (normalized as described above) has identical entries of $1/n$. Thus, the change in equilibrium state due to an impulsive stimulation at any network component is identical, and relatively small. In contrast, for asymmetric network topologies, the change in equilibrium can be dependent on the stimulation location. In particular, as can be verified using majorization arguments [10], centrally located network components that strongly influence other components but are not strongly influenced themselves (i.e., whose corresponding vertices in the graph have large out-degree and small in-degree) will cause large equilibrium changes. Thus, networks require non-uniform and directional topologies to achieve significant and differential impact through stimulation at particular locations.

Transient Analysis: Through simulations, we have determined that transient dynamics upon local impulsive stimulation are also strongly dependent on the local network topology. In particular, we have pursued characterization of the settling time to the new equilibrium upon impulsive stimulation at each network component for symmetric network topologies. The settling time displays a dependence on (1) the degree of the graph vertex associated with the stimulated component; and, especially, (2) the centrality of the vertex in the network. In particular, stimulations at components that are centrally located and have high connectivity rapidly achieve transition to the new equilibrium, while transition due to stimulations at other locations is slow. These results show that networks with strongly and weakly connected components naturally permit stimulations at certain components to achieve rapid modulation of the synchronous state, while stimulations at other components have much delayed impact. Such differential impact in networks with weak and strong connections can be verified analytically, using the slow-coherency theory [11]; we omit the details here. This dependence suggests that non-uniform connectivity provides a mechanism for differentially weighting the impact of multiple stimulation inputs. Such non-uniform connectivity may result from deliberate design of the full network, or from evolutionary modification of the network through stimulation.

14.3 *In Silico* Synchronization Network Model for Distributed Sleep Regulation

Sleep plays a critical and holistic role in animal function. Despite extensive research on sleep (including experimental, theoretical, and modeling efforts), the purpose of sleep in many ways remains mysterious. Recently, several research groups have introduced a new paradigm that recognizes sleep as a distributed and local-use-dependent phenomenon [12–16]. In particular, this new theory posits that neuronal assemblies (small groups of neurons in the brain) maintain local sleep states. These local sleep states evolve in time, in a way that is modulated by (1) local external stimulation or activity or *use* of afferent neuronal circuits; (2) loose couplings among the neuronal assemblies; and (3) some amount of global or hierarchical regulation. This new paradigm holds significant promise for informing our understanding of sleep function, in part because it automatically captures the intrinsic self-adaptation and distributed regulation that is so central to understanding brain function.

During the last 10 years or so, significant experimental evidence has been obtained verifying aspects of the distributed sleep theory, and identifying electrical and biochemical mechanisms underlying local sleep. Despite these advances, understanding of distributed sleep and its consequences remains incomplete, in part because of the complexity inherent to *in vivo* distributed stimulation and measurement of neuronal systems. To overcome this challenge, our group is pursuing the development of both mathematical and cell-culture *models* for distributed and use-dependent neuronal oscillations. Of importance to our development here, we introduced in [15] a mathematical model for distributed sleep that views neuronal groups as *activity integrators* that are governed by local stimulation, loose couplings with other neuronal groups, and global regulation via biological clock signals. We have shown (through both simulations and analytical methods) that this activity-integrator network model qualitatively captures several key properties of distributed sleep, including the ability of local functional states to synchronize without global regulation, and the receptiveness of the oscillatory synchronous state to local overstimulation. However, the preliminary analysis does not take the critical step of relating these sleep properties with the topology of the network of couplings among neuronal groups. Understanding this dependence could potentially lend significant insight into the function of sleep, by identifying neuronal network topologies that both robustly achieve regulation and are differentially adaptive to stimulation: It is postulated that sleep plays a critical role in *stabilizing* neuronal network topologies, and such robust yet adaptive topologies may be the ones achieved by this stabilization role.

The model for sleep regulation that we have introduced in fact can be viewed as a network synchronization model that is impacted by stochastic stimulation inputs. In particular, both activity and functional states of neuronal groups – which are the components in the network model – synchronize, and this synchronization dynamics is impacted by use or external stimulation. This synchronization-under-stimulation viewpoint on the sleep model suggests that the topology may play a critical role in the regulatory role and adaptiveness of the network, in analogy to the linear synchronization model. Here, we argue through simulation and limited mathematical analysis that the distributed sleep model achieves robust yet adaptive synchronization for only a range of intermediate weights (coupling strengths), and some special graph structures. This analysis informs understanding of stabilization of neuronal circuits. In addition, this analysis (in conjunction with the analogous results for linear network models) provides some interesting insight into network synchronization processes that are subject to external stimulation. In these ways, the *in silico* model provides a platform for understanding stimulation-integration in brain function in general, and specifically in sleep regulation. While the model is far abstracted from the detailed mechanisms underlying brain unit coordination, it captures abstractly the interfaces among neuronal groups and the integration of stimulation/activity that lead to global emergence of brain functions. Here, we review the *in silico* model for distributed sleep regulation, and study integration of stimulation inputs in sleep regulation using the model.

14.3.1 Model Formulation

Here, we only briefly review the dynamics of the model for distributed sleep introduced in [15], and kindly ask the reader to see [15] for a careful justification of the model. Briefly, the model captures the dynamics of two state variables for each network component (neuronal assembly) $i \in 1,\ldots,n$: a binary *sleep state variable* that identifies the functional state of the neuronal assemblies (specifically as "Awake" or "Asleep"); and an *integrated activity state* that captures integrated stimulation (due to local activity, network coupling, or global regulation) since the previous sleep state transition. The evolution of the components' states is defined separately depending on whether the neuronal assembly is awake or asleep (i.e., depending on its sleep state variable). The dynamics in each case are presented next, with the associated notation defined just afterwards:

$$
\begin{aligned}
&\mathit{Wake}\ (S_i\ (t) = 0):\\
&\qquad \dot{x}_i(t) = +u_i(t) + \sum_{j=1}^{n} S_j(t) step(x_j(t) - E_j) w_{ji}\\
&\qquad S_i\ (t) \to 1\ if\ x_i\ (t) - E_i > T_i\\
&\mathit{Sleep}\ (S_i\ (t) = 1):\\
&\qquad \dot{x}_i(t) = -r(t) + \sum_{j=1}^{n} (1 - S_j(t)) step(E_j - x_j(t)) w_{ji}\\
&\qquad S_i\ (t) \to 0\ if\ E_i - x_i\ (t) > T_i
\end{aligned}
$$

where:

- $x\ (t) \in R^{nx1}$, is the activity state vector, containing the integrated activity states of the n components (neuronal assembly).
- $u\ (t) \in R^{nx1}$ captures the local activity input for each neuronal assembly. This input reflects external stimulation or use of the neurons in the assembly (due to sensory stimulation, motor movements, etc.) during waking. We typically model the local activity input as a structured stochastic process – for instance, a time-varying Poisson or Gaussian white noise process – to capture varying local activity. Poisson and Gaussian models for neuronal stimulation have been extensively justified (e.g., [17]). While more intricate models for neuronal stimulation have been developed, these simple models are sufficient for us to capture the time-varying use of neuronal assemblies that drives sleep dynamics. As in [15], we will also consider deterministic overstimulation of neuronal assemblies.
- $r\ (t) \in R^{nx1}$ captures the local recovery inputs for each neuronal assembly during sleep. In contrast with the waking period, an organism is not receptive to sensory

stimulation during sleep; however, neuronal activity continues throughout the sleep period. The recovery input captures such activity. Again, a stochastic-process model is assumed for the recovery input.

- $S \in R^{nx1}$ is a sleep state vector that captures the sleep state variables of the neuronal assemblies. Specifically, if the jth neuronal assembly is awake then $S_j = 0$, and if the jth neuronal assembly is asleep then $S_j = 1$.
- $W \in R^{nxn}$ is the topology matrix, which captures the couplings among the neuronal assemblies that permit synchronization. In particular, if $w_{ji} = 0$ then neuronal assembly j does not modulate the dynamics of neuronal assembly i. If, however, $w_{ji} \neq 0$, then neuronal assembly j influences the dynamics of neuronal assembly i, with the strength of the impact dependent on the weight.
- $E \in R^{nx1}$ is an abstract representation of the total energy or activity-integration capability available during the awake portion of the cycle, and analogously $E \in R^{nx1}$ is the energy replenishment required during the sleep stage.
- $T \in R^{nx1}$ is the threshold that actuates change of the sleep state. Specifically, if the integrated activity exceeds the available energy by T_i during waking, then sleep results; correspondingly, if the rejuvenation input exceeds the requirement by the threshold amount during sleep, the neuronal assembly awakens.

The model is illustrated in Figure 14.2.

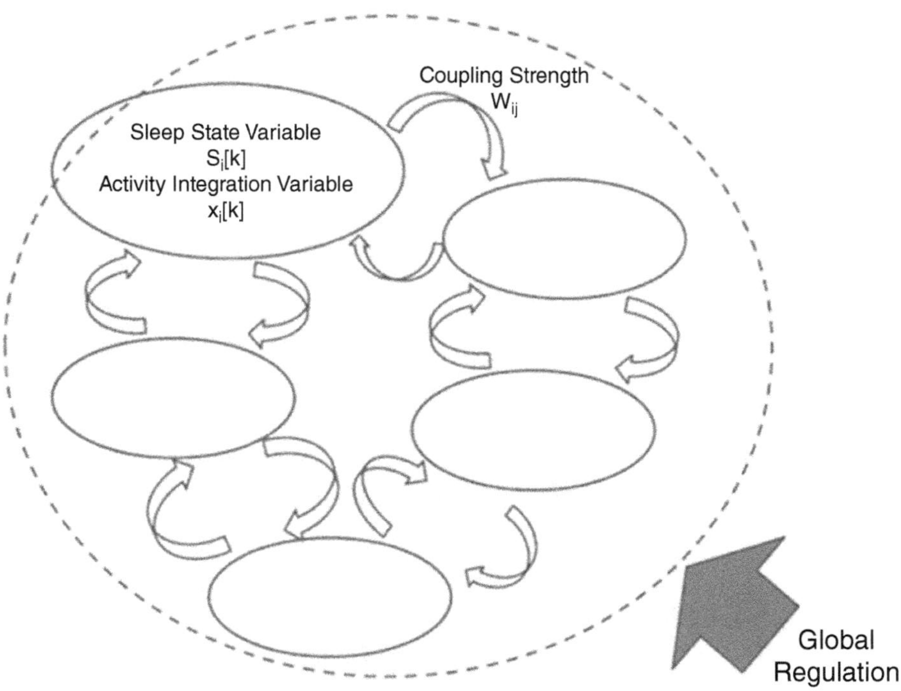

Figure 14.2 Diagram of the *in silico* sleep model. Activity-integration and sleep state variables associated with neuronal assemblies are modulated by local activity inputs, couplings with other neuronal assemblies, and some global regulation.

In our previous work, we have shown that the above model for distributed sleep self-synchronizes if commensurate constant activity inputs are applied at the neuronal assemblies, under broad connectivity conditions on the network topology [15]. In particular, both the activity-integration and sleep state variables for each of the neuronal assemblies oscillate in synchrony. Further, we have demonstrated that deterministic overstimulation of a single neuronal assembly modulates the global sleep state. Here, our primary focus is on characterizing the role of the network topology in the dynamics, and characterizing the synchronization of the model using realistic stochastic models for the stimulation inputs. Through such analysis, we aim to identify network topologies that permit robust yet adaptive synchronization in a stochastic-stimulation regime.

As with the linear model, we will require appropriate error measures to characterize the strength of synchronization. We will use the correlation and squared-error measures defined for the linear network model. In addition, we will also measure strength of synchronization by summing the absolute value of the difference between the two activity integration variables over a time interval: $error_1(i, j) = \sum_{k=1}^{N} |x_i[k] - x_j[k]|$, where i and j are the neuronal assemblies being compared. We consider this alternate absolute-difference or 1-norm error measure to permit comparison of our analyses here with those given in [15].

14.3.2 Simulation and Analysis

Like the linear network model, the distributed sleep model introduced in [15] represents a synchronization dynamics that is subject to stochastic stimulation. Both *in vivo* and *in vitro* experiments have shown that (1) a global sleep-correlated synchronized functional-state oscillation emerges in neuronal networks under broad conditions; and (2) the sleep oscillation is use (or activity or stimulation) dependent. Given these characteristics, we would expect a predictive mathematical model for sleep to display a strong synchronization that is nevertheless adaptive to trends in important stimulation inputs. In [15], we determined that, for a broad class of network topologies, the sleep model achieves asymptotic synchronization for deterministic (constant) activity inputs. We also noted the use dependence of the synchronous state. However, we did not study the network dynamics under realistic stochastic stimulation. Here, we study how the strength of synchronization and adaptiveness of the sleep-model dynamics depends on the model's network (coupling) topology, by subjecting the dynamics to stochastic stimulation inputs.

Strength-of-Synchronization Analysis

Through simulation, we have studied the dependence of synchronization strength/extent (as defined from the activity-integration states) on sleep-model parameters, including especially the characteristics of the stochastic activity inputs and the network topology of the model. Since several parameters affect the network dynamics, we find it most illustrative to present these studies as a series of comparative experiments, highlighting key observations and outcomes for each experiment.

Experiment 1: Dependence on Coupling Strengths. The total extent of synchronization achieved by the sleep-model dynamics has been compared for different coupling strengths (weight values). Precisely, we compare the model dynamics as the nonzero weights in the network matrix are scaled (while other parameters are held constant); that is, we consider the change in the dynamics assuming that the network structure and local model characteristics are held fixed, but the interaction strengths in the model are scaled up or down. In these experiments, each neuronal assembly is assumed to be driven by a stochastic activity input (specifically, a Poisson-process input) during waking. Given the stochastic inputs, the network components do not maintain exact synchronization, and the interplay between the stochastic drive and the network couplings governs strength/extent of synchronization. Our aim is to characterize the strength of synchronization. These experiments show that the neuronal assemblies (network components) are most strongly synchronized for intermediate coupling strengths. This characteristic reflects *in vivo* sleep, wherein connectivities outside an intermediate range are known to correlate with pathologies.

Experiment 2: Dependence on Activity-Input Characteristics. The total strength of synchronization achieved by the distributed sleep model was compared for three different activity-input signal regimes: (1) deterministic (in fact, constant) stimulation; (2) Poisson-process stimulation; and (3) Gaussian stimulation. For a fair comparison, the activity-input signal for each neuronal assembly is chosen to have the same (and constant) mean for the three cases. As expected, the deterministic stimulation allows perfect synchronization among the network components, while the stochastic stimulations achieve only imperfect synchronization. The simulations suggest that the strength of synchronization for these stochastic activity inputs is tied to the variability of the activity input: The more variable Poisson input signal yields higher average desynchronization compared to the Gaussian input. The dependence of synchronization strength on activity-input variability also correlates with *in vivo* measurements of brain function.

Experiment 3: Synchronization Characteristics across a Network. We have compared the strength of synchronization among different pairs of components within a network (again assuming a Poisson activity input at each network component). In particular, the correlation between one component's activity state and those of other components has been plotted on the network graph. These results highlight that strongly connected subnetworks display strong synchronization, while distant and/or weakly connected components tend to be more desynchronized.

The results of the three experiments are relevant to the ongoing study of distributed sleep in several ways:

1. They give further justification that the distributed sleep model is predictive, at a qualitative level, of aspects of *in vivo* sleep.
2. They provide some insight on neuronal network topologies that facilitate synchronization. Specifically, topologies in which all components are connected

by short/strong paths, and which have intermediate coupling strengths, can achieve strong synchronization in the face of stochastic stimulation.

3. They relate the dynamics of the distributed sleep model with the dynamics of the simple linear synchronization model. In particular, both models display a strong topology dependence of synchronization dynamics under stochastic stimulation, with particular network types and connectivities showing stronger synchronization compared to others. The tight cohesion between the models may prove valuable in developing further characterizations of the sleep model, and they also hint that the topology-dependence of activity-driven regulation may be a more fundamental property of networks.

Sleep Model: Receptiveness to Activity Inputs

The introductory paper on the distributed sleep model [15] has already verified that the synchronization dynamics are modulated by trends in activity inputs (e.g., over-stimulation of a neuronal assembly). We would like to better understand how this impact on the synchronization dynamics depends on the network topology and on the location of the modified stimulation in the network. To this end, we have studied how this impact on the synchronization dynamics depends on the network topology and on the location of the modified stimulation in the network, through simulation. Specifically, we consider overstimulation of a single network component for a half-day duration (assuming synchronization initially), and characterize the extent of disruption to the synchronized activity-state dynamics. In analogy with the linear model, we find that stimulations at central locations with high local connectivity tend to have large impacts compared to stimulations at weakly connected components. In particular, stimulations at central, highly connected locations disrupt synchronization for longer periods, and enact a more significant change in the synchronized state in the long term. In fact, for sufficiently strong overstimulations at centrally located components, long-term desynchronization can result; however, over a wide range of overstimulations, synchronization is achieved again, albeit at a slower rate when a centrally located component is stimulated.

14.4 A Cell-Culture Model for Distributed Sleep Regulation

In tandem with the mathematical modeling efforts, our group has also been pursuing cell-culture modeling of neuronal networks using multi-electrode-array technology. Specifically, this technology permits growth of dissociated neuronal networks in culture, and distributed stimulation and measurement of the cultured networks. We believe that such a cell-culture model is a compelling platform for understanding sleep regulation, for two reasons. First, the model intrinsically tests a core hypothesis of the sleep theory, that any group or network of neurons exhibit functional states that are correlative with sleep and waking. Second, the model may provide a flexible platform for understanding the multi-time-scale biochemical and electrical mechanisms underlying sleep regulation, by permitting easy manipulation of neuronal circuits and targeted stimulation and measurement of the circuits in isolation from global regulatory drives (e.g., the circadian clock). Our development and use of the cell-culture is still in its preliminary stages, and much remains to be done to understand to what extent the model can capture characteristics of sleep and other whole-brain functions. We point the reader toward the concurrent works [18, 19] for characterization of sleep- and wake-relevant states in the model. Our focus here is to describe the cell-culture model, which is itself CPS slow-wave, and to present some initial studies of the model to illustrate its use.

The cell-culture model has three technical aspects: (1) a basic capability to measure electrical activity at multiple nodes (locations) from mixed neuronal/glial networks grown on MEA dishes; (2) data storage capabilities as well as an algorithm suite that mines recordings for multimodal sleep correlates and spatiotemporal activity patterns; and (3) a capability to modulate network responses via electrical and optical stimulation, as well as via modification of sleep-related pathways through biochemical and genetic means. The model implicates a close interface of physical systems (the neuronal network itself, the multi-electrode array) and cyber capabilities (data recording and storage, algorithms for extracting signal characteristics). Following is a brief description of the three technical aspects.

14.4.1 Multi-location Electrical Measurement from *In Vitro* Networks Grown on MEA Dishes

Mixed cultures of neurons and glia from the whole brain of one day old mice are grown on multi-electrode array (MEA) dishes that contain either 60 electrodes in a single well embedded in the bottom of the dish, or six wells containing nine electrodes each [20–22]. Similar testbeds are widely used for characterization of firing dynamics of neurons in culture [23–26]. Neurons grow a well-connected network in culture. These cultures survive for over 60 days *in vitro* and stay electrically active. Our analysis of developmental data indicates that several relevant correlates (burstiness index, cross-channel correlation coefficients, delta-wave power; see discussion of sleep correlates below) stabilize after 10–14 days, which matches with the literature. Thus, >14-day-old cultures are used for experiments. We record spontaneous electrical signals (field potentials or *voltages*) at a 20 kHz frequency from neurons at multiple electrodes (multiple network nodes) using a commercially available recording system (Multi-Channel System, Germany), which applies a wide-band filter (0.1 Hz–3 kHz). Electrical field potentials recorded at multiple MEA electrodes are shown. These field potentials show complex temporal and spatial patterns, and show phase lag correlation indicative of signal propagation. The setup also allows growth of cultures with varying neuron-to-astrocyte ratios.

14.4.2 Data Analysis for Sleep/Waking Correlates

The recorded field potentials capture action potentials of single neurons as well as slower regional depolarizations, both of which are representative of *in vivo* neuronal measurements. To study correlates of sleep or other emergent phenomena, features of interest at time scales ranging from microseconds to tens of hours may need to be considered. Thus, we record at a high frequency (10 or 20 kHz) and over a long duration (6 hours or more, and additionally on the next day). This necessitates extensive data storage and makes computation taxing. A suite of supporting software is under development that processes and filters electrical recordings from the MEA. The data analysis approach is shown in Figure 14.3. Briefly, the low- and high-frequency components of the recorded signal are isolated through digital filtration. The faster component (>200 Hz) is used for spike analysis and the slower component (<100 Hz) for analysis (see Figure 14.3 legend for details). Initial data analyses were focused on deriving simple features from these signals that are indicative of state, including action potential (AP) counts and burstiness indices for 1- and 5-minute intervals during a recording. These analyses, for the nominal unstimulated case, match the literature showing that cultured neurons are nominally in a sleep-like state with low AP density and high burstiness. The initial software development also permits analysis of additional sleep/waking correlates, including cross-channel correlation for the slow-wave signal, low-frequency (delta wave) content in individual channels, and cross-spectral content. While similar correlation and spectral analyses have been developed in the literature [27, 28], the focus here is to analyze the time-course progression of these correlates for stimulation patterns and biochemical modifications that are relevant to sleep and cognitive tasks. An initial analysis of spatial patterns in cross-channel correlation has also been undertaken, as described below. Finally, in describing the required cybercapabilities, we note that data storage and retrieval is a nontrivial task: multi-hour recording at 20 kHz frequency is required for each experiment, which yields hundreds of megabytes to gigabytes of data per experiment. Appropriate storage, retrieval, and organizational tools are needed to analyze these data.

14.4.3 Stimulation and Biochemical Modulation

Mixed cultures of neurons and glia have been stimulated electrically (for single-node or spatially patterned stimulation) and optically (for stimulation of the whole network). Noninvasive optical stimulation of the network is achieved using optogenetics [29, 30]. Optical (blue light, 470 nm LED) stimulation of neurons in the *in vitro* network was achieved through targeted expression of channelrhodopsin-2 in neurons [30]. Immediately post-stimulation, multiple correlates showed significant change relative to pre-stimulation, returning to baseline the day after. The multimodal state change has characteristics of waking *in vivo*, including decreased burstiness, slow-wave correlation, and low-frequency signal power. Further, the response characteristics were strongly dependent on the stimulation pattern (bursty vs. periodic), duration, and frequency. Additionally, as *in vivo*, stimulation induced expression of c-fos [30, 31] and the sleep cytokine tumor necrosis factor TNFα at 30–90 minutes post-stimulation, and increased low-frequency signal power (correlative with *in vivo* sleep).

Mixed cultures of neurons and glia have also been treated with TNFα which is known to alter local network states *in vivo* [32, 33]. Preliminary data [21, 22, 31, 34]

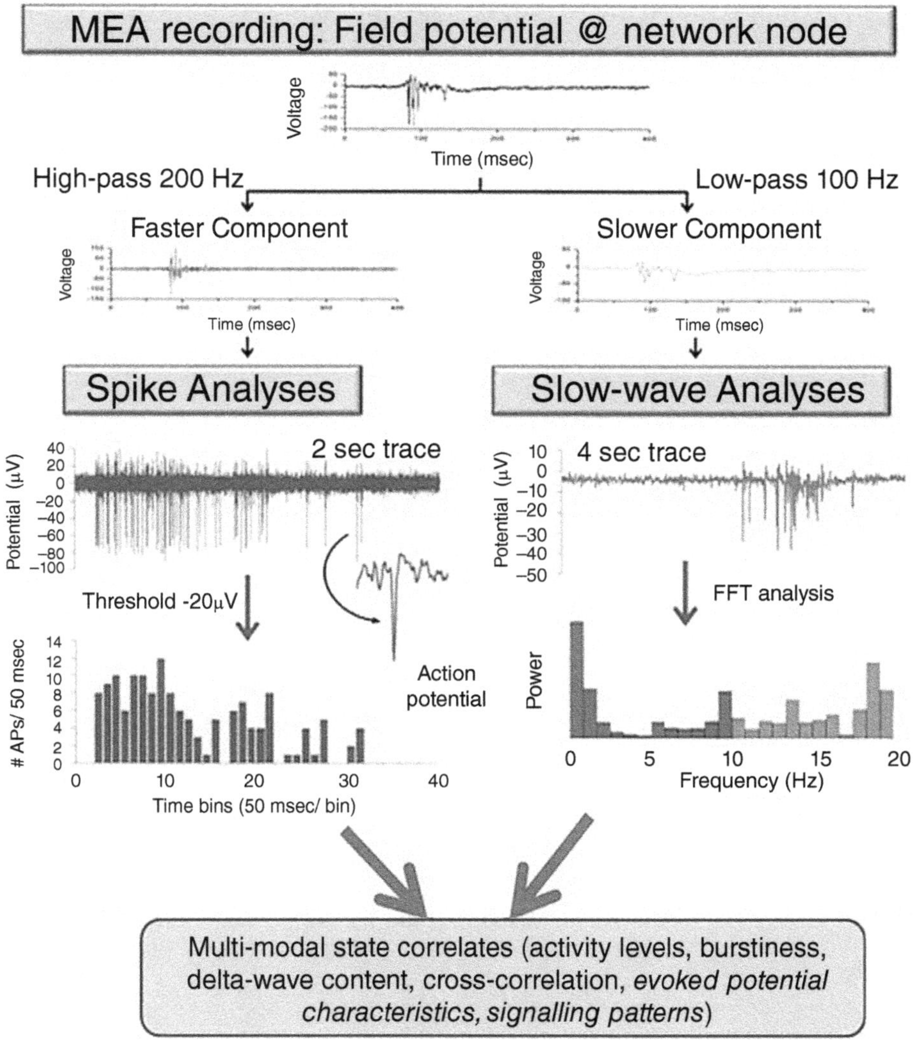

Figure 14.3 MEA data analysis. Field potentials recorded at each network node are digitally filtered to obtain high-frequency (filter: high-pass 200 Hz, Butterworth second order) and low-frequency (filter: low-pass 100 Hz, Butterworth second order) signal components. The faster component is used for spike analyses and the slower component is used for slow-wave analyses. For spike analyses, APs are counted for a time bin (50 ms for this example) with a threshold potential (−20 μV for this example, calculated as 5× of the standard deviation of a 1 s recording trace without any AP in it). For slow-wave analyses, frequency content is also obtained using fast Fourier transform (FFT) analysis. Time courses of multi-modal state correlates (including sleep/waking indicative correlates) are obtained from these two signals.

show that TNFα alters the firing dynamics of the neurons in culture: The total number of AP/0.5 h intervals is reduced, the low-frequency peak is enhanced, and spatial correlation is enhanced. This change is reversible: 10 hours after addition of 6 nM TNFα, neuronal firing patterns recover to baseline. The state changes are correlative of those observed *in vivo* during sleep. Further, it is consistent with *in vivo* results showing that neuronal activity enhances TNF expression and application of TNF to a cortical column increases the probability of the column going into the sleep-like state [33].

14.4.4 Using the Models Together: Correlation Analysis

The cell-culture model development is still in its early stages, but preliminary results are promising in that they appear to show emergence of states, whose properties correlate with those observed in the sleep–wake cycle *in vivo* and which can be manipulated via stimulation and biochemical modification. We believe that the culture model can be used in tandem with the mathematical models, to understand distributed brain processes (including specifically sleep regulation) at multiple time scales. For instance, data from the cell-culture model can be subjected to a cross-channel correlation analysis. Such an analysis gives an indication of the fast time scale electrical signaling among neuronal groups that reflects and underlies the global sleep–wake state. In Figure 14.4 the spatial dependence of correlation (synchronization strength) between network components is shown for the recorded electrical data.

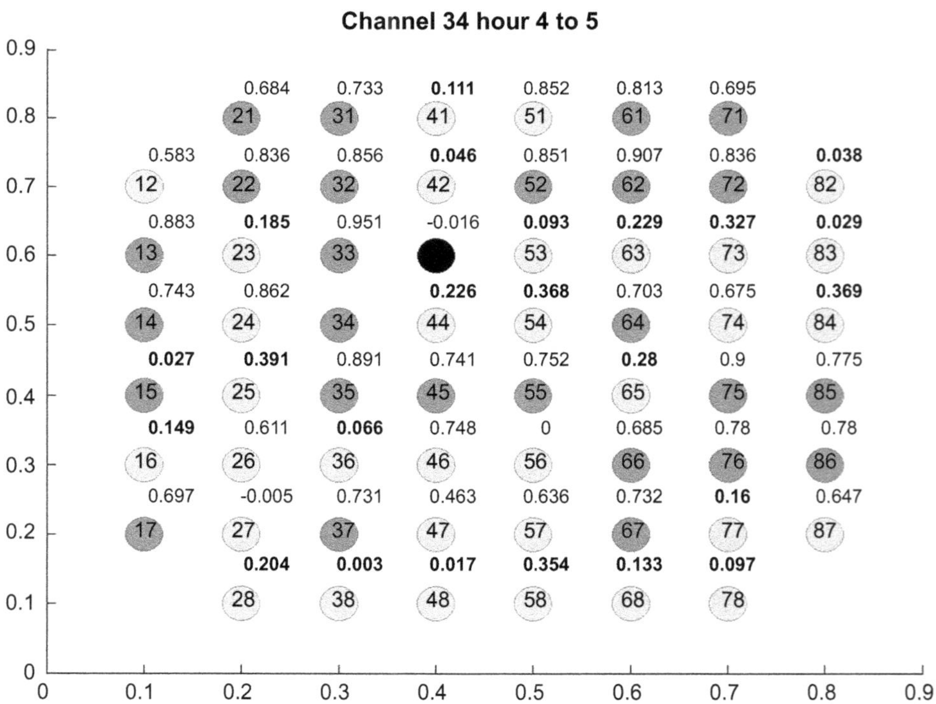

Figure 14.4 Correlation coefficient mapping for testbed data. The number directly above each node (electrode in the culture testbed) is the correlation coefficient with respect to node 34. Notice that node 43 corresponds to a grounding electrode.

The similarities between the correlation pattern in the recorded data and those predicted by the mathematical models are remarkable, with pairs of components within tightly connected subnetworks displaying strong correlation relative to other pairs in each model. This comparison provides some support for the fact that the mathematical models are evocative of real neuronal network dynamics, and conversely exposes faster time scale dynamics that are not captured in the mathematical models. Also, the comparison potentially suggests a means for tuning the mathematical models: The mathematical model parameters can be selected so that their spatial dependencies match those obtained for the cell-culture model. We expect to address this parametrization task in future work.

References

[1] S. Strogatz, *Sync: The Emerging Science of Spontaneous Order*. New York: Hyperion, 2003.

[2] S. H. Strogatz, "Spontaneous synchronization in nature", presented at the *Proc. 1997 IEEE Int'l Frequency Control Symposium*, Orlando, FL, May 1997.

[3] R. E. Mirollo and S. H. Strogatz, "Synchronization of pulse-coupled biological oscillators," *SIAM J. Appl. Math.*, vol. 50, no. 6, pp. 1645–1662, Dec., 1990.

[4] A. Pogromsky, G. Santoboni, and H. Nijmeijer, "Synchronization and partial synchronization in a network of identical systems," presented at the *Proc. 10th Workshop on the Nonlinear Dynamics of Electronic Systems*, Izmir, Turkey, Jun. 2002.

[5] F. Zhang, W. Chen, and Y. Xi, "Motion synchronization in mobile robot networks: robustness," in *Int'l Conf. Intelligent Robots and Systems*, 2006, pp. 5570–5575.

[6] G. Q. Zhong, K. T. Ko, and K. F. Man, "Robustness of synchronization in coupled Chua's circuits," in *Industrial Electronics*, 1998, pp. 436–440.

[7] X. F. Wang and G. Chen, "Synchronization in scale-free dynamical networks: Robustness and fragility," *arXiv cond-mat/0105014*, May 2001.

[8] C. W. Wu and L. O. Chua, "Application of Kronecker products to the analysis of systems with uniform linear couplings," *IEEE Trans. Circuits Syst. Video. Technol.*, vol. 42, no. 10, pp. 775–778, Oct. 1995.

[9] L. Chen, S. Roy, and A. Saberi, "On the information flow required for tracking control in networks of mobile sensing agents," *IEEE Trans. Mob. Comput.*, vol. 10, no. 4, pp. 519–531, 2011.

[10] S. Roy, A. Saberi, and Y. Wan, "Majorizations for the dominant eigenvector of a nonnegative matrix," in *Proc. 2008 American Control Conf.*, 2008, pp. 1965–1966.

[11] J. H. Chow, *Time-Scale Modeling of Dynamic Networks with Applications to Power Systems*. New York: Springer, 1982.

[12] J. M. Krueger, D. M. Rector, S. Roy, et al., "Sleep as a fundamental property of neuronal assemblies," *Nat. Rev. Neurosci.*, vol. 9, no. 12, pp. 910–919, 2008.

[13] J. M. Krueger and F. Obal Jr, "A neuronal group theory of sleep function," *J. Sleep Res.*, vol. 2, pp. 63–69, 1993.

[14] P. Sengupta, S. Roy, and J. M. Krueger, "The ATP–cytokine–adenosine hypothesis: How the brain translates past activity into sleep," *Sleep Biol. Rhythms*, Feb. 2011. DOI: 10.1111/j.1479-8425.2010.00463.x

[15] S. Roy, J. M. Krueger, D. M. Rector, and Y. Wan. "Network models for activity-dependent sleep regulation," *J. Theor. Biol.*, vol. 253, pp. 462–468, 2008.

[16] D. M. Rector, J. L. Schei, H. P. A. Van Dongen, G. Belenky, and J. M. Krueger, "Physiological markers of localized sleep," *Eur. J. Neurosci*, vol. 29, pp. 1771–1778, 2009.

[17] M. Bazhenov, I. Timofeev, M. Steriade, and T. Sejnowski, "Model of thalamocortical slow-wave sleep oscillations and transitions to activated states," *J Neurosci.*, vol. 22, no. 19, pp. 8691–8701, Oct. 2002.

[18] P. Sengupta, K. A. Jewett, C. J. Davis, et al., "Somnogenic cytokine TNFα induces a new functional state in dissociated cortical networks", presented at *SLEEP 2011*, Minneapolis, 2011.

[19] K. A. Jewett, S. Roy, C. J. Davis, and J. M. Krueger, "Electrophysiological signatures of sleep in vitro," *J Neurosci.*, submitted.

[20] P. Sengupta, S. Roy, and J. M. Krueger, "The ATP–cytokine–adenosine hypothesis: How the brain translates past activity into sleep," *Sleep Biol. Rhythms*, vol. 9, pp. 29–33, 2011.

[21] P. Sengupta, K. Jewett, J. Clinton, and J. M. Krueger, "Optogenetic stimulation enhances interleukin-1-beta levels in cultured neurons," *Biophys. J.*, vol. 100, no. 3, p. 95, 2011.

[22] H. P. A. Van Dongen, G. Belenky, and J. M. Krueger, "A local, bottom-up perspective on sleep deprivation and neuro-behavioral performance," *Curr. Top. Med. Chem.*, vol. 11, pp. 2414–2422, 2011.

[23] D. J. Bakkum, Z. C. Chao, and S. M. Potter, "Spatio-temporal electrical stimuli shape behavior of an embodied cortical network in a goal-directed learning task," *J. Neural Eng.*, vol. 5, no. 3, pp. 310–323, 2008.

[24] J. van Pelt, I. Vajda, P. S. Wolters, M. A. Corner, and G. J. Ramakers, "Dynamics and plasticity in developing neuronal networks in vitro", *Prog Brain Res.*, vol. 147, pp. 173–188, 2005.

[25] M. A. Corner, J. van Pelt, P. S. Wolters, R. E. Baker, and R. H. Nuytinck, "Physiological effects of sustained blockade of excitatory synaptic transmission on spontaneously active developing neuronal networks: An inquiry into the reciprocal linkage between intrinsic biorhythms and neuroplasticity in early ontogeny," *Neurosci. Biobehav. Rev.*, vol. 26, no. 2, pp. 127–185, 2002.

[26] M. A. Corner, "Spontaneous neuronal burst discharges as dependent and independent variables in the maturation of cerebral cortex tissue cultured in vitro: A review of activity-dependent studies in live 'model' systems for the development of intrinsically generated bioelectric slow-wave sleep patterns," *Brain Res. Rev.*, vol. 59, no. 1, pp. 221–244, 2008.

[27] D. A. Wagenaar, R. Madhavan, J. Pine, and S. M. Potter, "Controlling bursting in cortical cultures with closed-loop multi-electrode stimulation," *J. Neurosci.*, vol. 25, no. 3, pp. 680–688, Jan. 2005.

[28] A. E. Hady, G. Afshar, K. Broking, et al., "Optogenetic stimulation effectively enhances intrinsically generated network synchrony," *Neural Circuits*, vol. 7, no. 167, 2013. DOI: 10.3389/fncir.2013.00167

[29] K. Deisseroth, "Optogenetics", *Nat. Methods*, vol. 8, no. 1, pp. 26–29, 2011.

[30] F. Zhang, L. P. Wang, E. S. Boyden, and K. Deisseroth, "Channelrhodopsin-2 and optical control of excitable cells", *Nat. Methods*, vol. 3, no. 10, pp. 785–792, 2006.

[31] P. Sengupta, K. A. Jewett, J. Clinton, and J. M. Krueger, "Optogenetic stimulation enhances interleukin-1 beta levels in cultured neurons" *Biophysical J.*, vol. 100, no. 3, p. 95a, 2011.

[32] D. M. Rector, I. Topchiy, and M. Rojas, "Local cortical column activity states and localized delta wave differences," *Sleep*, vol. 28, no. A26, 2005.

[33] L. Churchill, D. M. Rector, K. Yasuda, et al., "Tumor necrosis factor: Activity dependent expression and promotion of cortical column sleep in rats," *Neuroscience*, vol. 156, pp. 71–80, 2008.

[34] P. Sengupta, K. Jewett, J. Clinton, and J. M. Krueger, "Optogenetic stimulation enhances interleukin-1-beta levels in cultured neurons," *Biophys. J.*, vol. 100, no. 3, p. 95, 2011.

Part IV Exercises

1. What are the major subdomains associated with resource management in cloud-assisted cyber-physical networks? Choose one of these domains and explain why new techniques are needed compared to: (a) traditional computing systems and (b) non-cloud-assisted CPSs.
2. Implementation of cloud-assisted CPS technologies requires cloud-based programming of CPSs. As a supplement to the research presented in Chapter 12, explore advances in cloud programming of CPSs (e.g., for Internet-of-Things applications).
3. Answer the following questions that are derived from Chapter 12:
    a. What kind of network traffic is generated by a CPS application? How is it different from any other network traffic? Why?
    b. Which type of data processing is more suited for CPSs, batch or stream? Why? How would that affect your choice of a cloud service provider?
    c. How does cloud computing fit into an evolving CPS? What are its advantages?
    d. List the different kinds of CPS applications that will need cloud integration and briefly describe the integration to each application.
    e. How is a CPS different and similar to an embedded system?
    f. Is the following statement true or false? State your reasons: "Integrating cloud computing with CPSs is counter-intuitive as the CPS data are kept as close to the CPS as possible."
    g. Describe the security issues in a cloud-integrated CPS.
    h. Based on your understanding of the fundamental principles of CPSs, what are the considerations to bear in mind while designing a scheduling algorithm for a CPS?
    i. Will the existing security measures for the cloud be sufficient to complete entire CPS security?
    j. What type of real-time system does a CPS fall under? Based on your classification, can you provide application examples?
    k. Since a primary understanding of the various subcomponents of a CPS and its resources has been given in the chapter, the reader is advised to do the following:
        i. Determine and list all the resources that need to be managed for a CPS to be stable.

    ii. Define the various tasks that a CPS consists of both at a high level (e.g., extracting features of the CPS traffic trace) and at a low level (e.g., integer and floating point computation).
    iii. Devise and implement a genetic algorithm that takes into account constrained resources and schedules the relevant tasks to manage resources efficiently. Note that genetic algorithms are heuristic approaches in nature and one solution/algorithm will be different from another.
    iv. Theoretically determine the runtime complexity of your algorithm (approximation of the solution). Computation complexity will be high for a genetic algorithm, so can you think of any other better scheduling algorithm? If yes, describe in detail how the other algorithm is better.

4. Discuss the connection between Big Data analytics and resource management in CPSs.
5. What is middleware? Review the development of middleware in computing, and note settings where middleware is commonly used.
6. What is the motivation for a middleware for the Internet of Things (IoT)? What are some special properties envisioned for this IoT middleware?
7. Answer the following detailed questions about the middleware testbeds for IoT introduced in Chapter 13:
    a. What must be true of a Wi-Fi access point for it to run AllJoyn?
    b. Why did AllJoyn designers recommend the use of reverse domain name notation (reverse-DNS) as a naming convention?
    c. Why does an app connection to a Router need to be persistent and a singleton?
    d. What are the limitations of AllJoyn in terms of scalability to IoT applications? What in the design of AllJoyn is the source of this limitation?
    e. What are the optimizations made in AllJoyn and CoAP in order to accommodate resource constraints of IoT devices?
    f. Before apps discover what services are available in an IoT location, they know nothing about what is available on the network. Critically missing is knowledge about which network port servers are listening to in order to answer a discovery query. Without the knowledge of the port number, how it is possible for CoAP apps to discover what is available on the network?
    g. Why does BOSS-style query language make apps transferable between IoT locations?
    h. What are the differences in the discovery schemes employed by CoAP, AllJoyn, and BOSS?
    i. You are asked to build a smartphone app that receives data from Things and other apps around a house. Which middleware do you think will be the best to use for this and why?

8. Is the brain a CPS? Play devil's advocate and argue both for and against.
9. The testbed introduced in Chapter 14 is rather different from the many CPS testbeds, in the sense that the CPS is a model for a different natural system (the brain) rather than a small-scale replica of an engineered system. Discuss the possible benefits of having a cyber-physical model for the brain.
10. Chapter 14 discusses both cyber (*in silico*) and physical (*in vitro*) models for brain network processes. Discuss how the two types of models could be used in a tightly integrated way to replicate or gain insight into brain processes and, specifically, sleep.

Part V

Conclusion

The 14 preceding chapters, subdivided into four parts, have introduced diverse concepts and techniques that are foundational to the analysis and design of cyber-physical systems (CPSs). These chapters together review and advance core directions of research that are essential for both students and practitioners in the field of CPSs. Given the enormous research effort that has taken place on CPSs, these chapters are far from a complete treatment of the field. However, we believe strongly that these studies from leading researchers in the field give voice to some of the key ideas, directions of thought, and methods that will underlie the operation of real-world CPSs in the near future.

The book contains one final chapter, authored by the editors, which is meant to synthesize the concepts and ideas covered in the book, and at the same time excite the readers to pursue research on CPSs in a direction with great need. The chapter focuses specifically on the management of large-scale cyber-physical infrastructures. Classically, the field of CPSs has been focused on real-time control using embedded computing in physical-world devices, not on strategic management of terrestrial-scale cyber-enabled infrastructures. Yet, these management problems are essential CPS problems, in that they require a tight integration of cyber- and physical-world functions to be solved. The final chapter posits ten factors that are central to strategic or wide-area management of infrastructures, and explains how methods for CPSs can be brought to bear on the grand challenge problem of wide-area management. In short, it is a call to action to CPS researchers to address wide-area management of large-scale infrastructures.

15 Wide-Area Management of Cyber-Physical Infrastructures: A Call to Action

Sandip Roy, Sajal K. Das, and Mengran Xue

15.1 Introduction

Terrestrial-scale infrastructures like road- and air-transportation systems, electric power grids, human social networks, and defense systems are foundational to our modern society and economy. The central importance of these large-scale infrastructures dictates that they must be managed in a way that is failsafe, efficient, and robust. Over time, however, the operations of large-scale infrastructures are becoming increasingly complex: pervasively integrating cyber technologies, spanning ever-wider scales, involving heterogeneous stakeholders and multifaceted functions, being subject to a wider array of cyber and physical world threats, becoming increasingly co-dependent, and allowing lower capacity and cost tolerances, among other challenges. Crucially, this evolution toward complexity means that local events are increasingly likely to implicate wide-area impacts across the heterogeneous cyber and physical subsystems in the infrastructure. The growing presence of such network-wide dynamics necessitates flexible wide-area management of these infrastructures. Wide-area management of terrestrial infrastructures has been, and still largely remains, a manual endeavor. For example, national-level air-traffic management initiatives in the United States are strategized via a teleconference involving stakeholders (e.g., the central command, regional offices, airlines), wide-area security and fault resolution in electric power grids involves manual input, and mitigation protocols for emerging infectious diseases are developed by government actors in an ad-hoc fashion. Limited decision-support capabilities are available for some of these management needs (e.g., airtraffic system operators have some weather-forecast products available), and automation has been developed to manage some faster-scale wide-area dynamics (e.g., remedial action schemes for the electric power grid). However, these limited automation and decision-support functions tend to be highly specialized, focusing on particular event trajectories (e.g., fog impacting San Francisco airport, or failure of a particular tie-line in the electric power grid), and often instigating hard-coded responses rather than flexibly shaping the network's dynamics. There is thus a growing need for new decision-support software and automation, for flexible and resilient wide-area management of infrastructure–network dynamics. In fact, in several domains, the development of integrative management strategies constitute *grand challenge problems* which may have profound impact on economic development, human health, and scientific progress. Consider, for example the overarching strategic management needs for the Next Generation Air Transportation System (NextGen), the drive to control emerging epidemics like ebola and chikungunya with limited control resources, and the challenge of integrating renewable generation in the electric power grid. This chapter is a *call-to-action to the cyber-physical system* (CPS) community, to approach wide-area management of infrastructures from a CPS perspective.

The field of CPSs has its roots in the embedded-systems community. It is natural, therefore, that most CPS research initially focused on individual physical-world devices with embedded computing, communication, and control (cyber) systems. From this start, the CPS community has begun to address an increasingly wider scope of problems, using a range of techniques that integrate computer science and engineering concepts, and lately social science. One particular focus in recent years has been on the Internet of Things (IoT), comprising platoons of physical-world devices with networked communications. This track of research is a natural generalization of the original embedded systems focus, from single devices with embedded cyber capabilities toward a network of such devices. In a separate direction, there has also been a growing recognition that modern infrastructure networks are CPSs, involving tightly intertwined physical-world dynamics and cyber technologies. A number of different problems related to the management of infrastructures are being studied under the CPS umbrella, with a primary focus on transportation and energy-related applications. The studied problems are rather varied in their scope, but many have to do with either: (1) exploiting new embedded cyber capabilities (including pervasive sensing) to achieve new functionalities in modernized infrastructures; and (2) addressing new challenges arising due to the growing integration of cyber systems, such as the possibility for cyber-security violations that propagate to the physical world. A limited subset of this effort has been concerned particularly with wide-area management problems. Developing generic CPS technologies for these infrastructure–network problems entails several challenges, however, and there are still gaps between the research being undertaken in the CPS community and implementation of specialized technologies in particular domains.

The focus of this chapter is on describing challenges and characteristics of CPS problems for infrastructure networks, particularly problems related to wide-area management of large-scale infrastructures. The chapter is organized as a list of 10 challenges or characteristics or *factors* that are central to managing modern cyber-physical infrastructures. For each, a conceptual discussion is given, specific examples are provided for illustration, and relevant ongoing work inside and outside the CPS community is discussed. Our hope is that, through the description of challenges and characteristics of wide-area infrastructure management problems, this chapter will guide the CPS community's research toward addressing grand challenges in infrastructure–network management. In this way, the chapter is meant to serve as a call-to-action for the CPS community, to address critical challenges in the wide-area management of large-scale infrastructure networks.

	Power System	Air Traffic Network	Veterinary Hospital
ms-min	Transient stability, contingency protection	Runway collision avoidance, surface operations	Intra-animal epidemiology
min-hr	Line-overheating resolution, hourly markets, demand response	Traffic control, tactical traffic management, security	Disease control through quarantine, treatment
hr-day	Unit commitment, transmission-system planning	Full-day strategic management, route planning	Policy changes for community health

Automatic to Manual

Growing Spatial Scale

Figure 15.1 Control and management problems at multiple time and spatial scales are diagrammed, for three cyber-physical infrastructures.

15.2 Ten Factors in the Management of Cyber-Physical Infrastructures

15.2.1 Factor 1: Decision-Making at Multiple Time Scales

Terrestrial-scale infrastructures require decision-making at multiple time and spatial scales, ranging from fast localized control actions (milliseconds to minutes) to wide-area strategic management for efficient operation (hours to a few days) and long-term planning and improvement (weeks to years). Figure 15.1 catalogs representative decision-making tasks at the three scales for three infrastructures, namely the electric power grid, the air transportation system, and the human infrastructure underlying disease spread and its management.

Fast time-scale controls in infrastructures are typically local and/or specialized for particular functions [1–5]. They are very often tasked with maintaining safety and operational fidelity using constrained actuation, in the face of specific, harsh disruptions. The need for a fast response dictates that they are automatic rather than manual. A diverse range of fast controls are already in place in infrastructures, but the need for new functionalities and the integration of new technologies are continuing to lead to the development of new and refined controls. For example, key fast controls of the electric power grid, such as automatic generation control (AGC) and remedial action schemes (RAS) are well-established. However, new communication capabilities and pervasive sensor integration (e.g., deployment of synchrophasors) are permitting more flexible responses to fault events. Also, many hardware control systems are being replaced by controls implemented in embedded cyber systems. At the same time, new requirements and constructs in power grid operations, such as demand–response problems and higher penetration of distributed generation, are necessitating new controls. These new control tasks crucially involve cyber systems, and many are attracting significant attention from the CPS community. Similar changes are occurring in diverse infrastructure networks.

At intermediate time horizons, decision-making in infrastructure networks remains a primarily manual endeavor, although some decision-support tools are available to assist the human operators (to differing extents depending on the infrastructure) [6–10]. These strategic decision-making or management problems typically entail scheduling resources across a wide area, and must consider the holistic network-wide impacts of the designed strategies. These strategies are designed with the infrastructure's mission in mind, which includes not only the performance of the engineered components, but also the financial ramifications on stakeholders. Compared to the fast control problems, these strategic management problems are hard not because of constraints and critical safety/time implications, but because of the following: (1) the number of decision variables, (2) the complexity and inherent stochasticity of network-wide responses in uncertain environments, and (3) the amorphousness and complexity of measuring and managing performance at a mission scale. The need for such wide-area strategic decision-making was limited in the past, and traditionally has been addressed by human operators in an ad hoc way. For instance, strategic air-traffic management (nationwide management over a 2–24 hour look-ahead) is completed manually by stakeholders (FAA national and regional offices, airlines) via an early-morning teleconference for strategizing management initiatives together with periodic manual corrections from this initial plan. However, infrastructure networks are growing in sophistication, to a point where manual strategizing is insufficient. Specifically, the networks are evolving toward increasingly diverse functionalities, wider-area interconnectivity, lower tolerance, more complex regulations, growing uncertainty, more pervasive communications and computations, etc., which all complicate management. Developing decision-support capabilities and automation for wide-area management of infrastructures is thus a central grand-challenge problem for a number of infrastructures. For instance, strategic traffic management is a key need for NextGen [11]; the Department of Energy's roadmap for the electric power grid stresses the need for nimble decision-making in the future renewables-rich grid, and management of emerging infectious diseases has come to the forefront during the last couple of years. We believe that these core management problems are at their essence CPS problems, which can and should be addressed by the CPS community; to date, the focus of the CPS community on these strategic decision-making problems has been relatively sparse compared to the fast-time controls. A primary goal of this chapter is to highlight that these strategic decision-making tasks are CPS problems, and characterize them in a way that fosters and directs research by the CPS community.

Decision-making at the longest (planning) time horizons is also largely manual [12, 13]. These decisions are often driven by changing operational needs or demands as well as regulatory and technological issues, which naturally are monitored manually and often require human ingenuity to resolve. These planning tasks will remain manual for some time to come, but will be aided by growing efficiency and automation at the faster time scales.

It is important to stress that the decision-making tasks at multiple time scales for an infrastructure network are inextricably linked. Solutions for fast control problems will influence strategy choices and long-term plans, while new strategies and plans may dictate how fast controls are tuned and used. Increasingly, different critical infrastructures are also becoming interdependent on each other [14]. Hence, control and management problems are becoming linked not only across temporal and spatial scales, but across multiple infrastructures. We believe that research by the CPS community will benefit from an integrative understanding of infrastructure control and management problems at multiple scales, across multiple infrastructures.

15.2.2 Factor 2: Working from Domain-Specific Models for Physical-World Processes

Physical-world infrastructures have a much longer history than cyber systems. For example, the United States' ground transportation system has developed over centuries, healthcare systems have been in place since the United States was settled, the electric power grid was built starting from the late 1800s, and even the air transportation infrastructure has a history of about a century; many infrastructures (e.g., ground transportation and water-distribution systems) have been in place since antiquity in some form or other. Furthermore, these infrastructures have played a primary role in economic and societal development, and hence their construction and efficient operation have long been of central importance. Given this long history, it is not surprising extensive domain-specific expertise has been developed for most infrastructures, in the form of operational know-how, models, tools, and understanding of challenges and needs. In fact, for most infrastructures, a catalog of models with varying resolutions and scopes are available, which are variously used for evaluation, control, management, and planning tasks. These models capture essential physics-based or empirically derived dynamics of the network at the proper resolution for a particular evaluation or decision task. The models are widely used by practitioners for design, evaluation, and forecasting.

Control and wide-area management of infrastructures requires leveraging appropriate models for the infrastructures' dynamics. For instance, strategic air-traffic management crucially depends on accurate models and forecasts for weather, traffic congestion, impacts of management initiatives, etc. (see [15] for an overview). Likewise, electric power system control and management are predicated on models for power flow solutions and swing dynamics, forecasts of renewable generation and loads, models for market mechanisms, etc. (e.g., [16]). We believe strongly that, as CPS researchers address infrastructure-management tasks, they must leverage domain-specific models for infrastructure network processes. Specifically, they must know the state-of-the-art in the application domain of interest, and be able to draw on models of the proper resolution and scope, for the particular management problem of interest. Some decision-making problems will undoubtedly require new models (e.g., ones that interface the cyber and physical processes involved), but these efforts must also be grounded in the existing modeling paradigms.

Recent advances in strategic air-traffic management provide a useful case study. During the last five years, several efforts involving academic, industrial, and operational partners have been focused on developing strategic management tools for the US airspace system [6, 17]. These tools are meant to leverage cyber capabilities to provide operators in the traffic network's central command (the Air Traffic Control

Strategic Command Center) with: (1) probabilistic weather-impact forecasts and (2) optimal schedules and rates for traffic management initiatives (flow restriction actions) in the face of uncertain severe weather. Research in this direction has been fruitful, and outcomes are nearing operational implementation in the airspace system. The management tools leverage two classes of established models for airspace traffic. First, they use flow-level queuing models that permit fast quantification of traffic flows under severe weather and flow restriction at a nationwide scale for strategy design. Second, for evaluation and validation, they use detailed models that track individual aircraft [18]. The tools also draw on established models for traffic demand, model-based weather-forecast products, and models for translating weather conditions to airspace and airport capacity impacts (e.g., [19–21]). Tool-development also required some new modeling capabilities (e.g., for representing traffic management initiatives [TMIs] in the queueing framework, and interpolating weather impacts); however, these models were also closely tied to existing models and operational know-how. It is crucial that further work in this space draws on the domain-specific models that have been developed.

In summary, modeling needs for infrastructure management problems contrast sharply from needs on other CPS applications because of the extensive history of modeling in these fields. Unlike emerging CPS applications, models do not need to be built from the ground up. This means, however, that modeling and design efforts must respect existing domain-specific models and constructs, including representations of physical-world processes, specialized notions of performance, etc., while also introducing new modeling constructs in targeted ways.

15.2.3 Factor 3: Cyber in the Problem, Cyber in the Solution

One main reason why infrastructure management is becoming increasingly challenging, and requiring automation rather than manual decision-making, is the pervasive integration of cyber systems into infrastructures. The transformation of large-scale infrastructures from largely physical systems to CPSs complicates management in several ways:

1. It enables increasingly complicated operational paradigms which rely on more high-dimensional data and long-distance data transfer. These new paradigms can permit significant performance improvement, but can also invoke increasingly wide-area dependencies in both the cyber and physical worlds. For instance, air-traffic management historically used limited telephone communications between regional control centers, but now increasingly relies on web-based data communications (e.g., to access weather data, inter-center communications by email, cloud computing). These communications implicate greater interdependency among control regions and greater reliance on external data in the cyber world. The higher traffic densities allowed by such coordination also lead to increased spatial and temporal interdependency in the physical-world network.

2. It exposes the infrastructures to new vulnerabilities. These include cyber-security issues (witness the various cyber-attacks on infrastructures such as Stuxnet); growing observability of network dynamics to external agents (e.g., power system operators post congestion information to the web, and market players are privy to much more information); and accidental intrusions like software failures, human–machine interface issues, overreliance on decision support, software bugs, etc.
3. It complicates equipment maintenance and life-cycle management.

While the integration of cyber systems is introducing new problems in infrastructure management, our viewpoint is that the cyber world is also instrumental to solving these problems. Cyber systems provide unique capabilities for seamless secure data transfer among stakeholders, fast computation, effective human–machine interfaces, pervasive sensing, and sensor-data fusion. These technologies can be used to great benefit in solving both fast control and strategic management problems in large-scale networks, provided that designers acknowledge and properly model the network's cyber and physical components, and address specific control/management problems of interest. Numerous exciting research projects are being undertaken toward such solutions in the computer science and controls engineering communities, as well as in domain-specific forums (e.g., toward addressing the cyber-security and vulnerability issues noted above). However, there is still a gap between operational practice and generic computing/controls methodologies, and much remains to be done in leveraging computing resources effectively. As one instance, the integration of phasor measurement units in the electric power grid can enable estimation of and response to fast transients in the electric power grid resulting from diverse disruptions. Controls engineers are developing cutting-edge techniques for estimation and control in dynamical networks that may support use of phasor measurement unit data [22, 23]; however, these techniques focus on models that are abstracted from current operational practice. Meanwhile, work in the power-engineering community in this direction tends to be more heuristic, but is yielding specific advances that are being integrated into operations [24, 25].

15.2.4 Factor 4: Sensing and Especially Actuation Are Sparse, Expensive, and Legacy

Wide-area strategic management of infrastructures depends on measurements of the ground truth (sensing) and control/management actions for modulating the network's responses (actuation). In applying CPS technologies to infrastructures, it is important to acknowledge that sensing and especially actuation capabilities are often very sparse, expensive, and legacy. Specifically, for wide-area management, typically only a limited set of control procedures are used which probe the physical-world dynamics in a parsimonious way. They are rather inflexible in their scope and are expensive to use. For instance, for the air-traffic system, strategic management is achieved using several specific types of TMIs [6, 7]: e.g., ground delay programs that control arrival rates at specific airports by holding flights at certain origin airports, airspace flow programs that restrict traffic flows across specified boundaries, and playbook reroutes. These TMIs act on traffic flows (rather than individual aircraft), and do so in very specific ways. Also, they can only be implemented at a limited number of locations in the airspace system (a few key airports, flows/routes, or region boundaries, for example), and depend on their propagative impacts to shape flows across the airspace system. Use of TMIs also incurs significant cost in terms of delays and fuel usage for airlines and management overhead, and policy implementation requires a significant lag. Likewise, wide-area control capabilities in the power grid are also limited and legacy, though new power electronics are permitting some further flexibility [26]. In comparison with actuation, sensing capabilities are typically more pervasive and flexible, but sensors also may be significantly constrained and subject to noise and bias. For instance, policy-makers in controlling disease spreads rely on infection counts based on doctors' reports, but such measurements are very noisy and also may differentially sample subpopulations (e.g., rural areas or lower-income neighborhoods may not be effectively sampled). In the power grid, sensors were also historically slow, few, and localized; phasor measurement units have permitted more pervasive sensing but suffer from noise, clock jitter, failure, etc., and remain expensive enough that they are still sparsely deployed. Further complicating management is the fact that an infrastructure's stakeholders may each only have access to a subset of its measurements and actuations, hence the system is intrinsically a distributed or decentralized one.

Cyber decision-support systems – which at their essence are feedback loops that use sensor data to set actuations – must work with the infrastructure's limited and legacy sensing and actuation capabilities. For instance, strategic management software for the air transportation industry must provide solutions in terms of existing TMIs which operate on flows in structured ways – solutions that schedule individual aircraft, for instance, will not be feasible for implementation. Exploiting existing actuation and sensing is challenging, however, precisely because these capabilities are parsimonious and legacy: this means that the limited sensing and actuation must be leveraged to indirectly manipulate the hardwired parts of the network to shape overall responses. We note that this challenge contrasts with autonomous-agent CPS applications (e.g. autonomous-vehicle control and sensor-networking applications), where there is often significant flexibility to guide the dynamics of each network component or agent; in these applications, computation and communications are often the key challenges, not the difficulty of manipulating a hardwired physical system. In specific arenas, such as air-traffic management, methods for designing such parsimonious management capabilities to shape overall network performance are being explored [6, 7, 27]. More broadly, there is a wide literature on decentralized and network control that bears on these feedback problems [28]. However, the classical literature on decentralized control typically seeks to make network components impervious to external impact rather than exploiting the network's structure, and hence often does not apply to the decision-making problems of interest. Very recently, a focus on network control with sparse measurement and actuation has emerged, which is concerned with exploiting the network's hardwired topology (e.g. [26, 29, 30]; see also Figure 15.2). We believe that these recent efforts on network control are crucial for developing

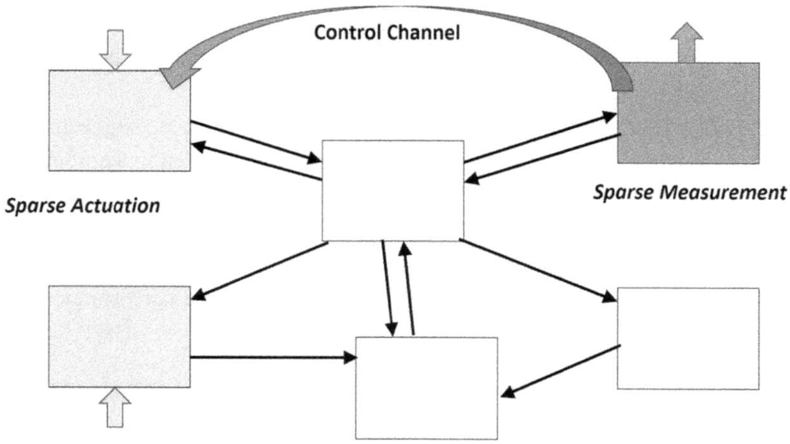

Figure 15.2 Infrastructures comprise many interconnected components, but typically sensing and actuation are sparse and legacy.

CPS solutions for infrastructure management. Figure 15.2 highlights the research on topology-exploiting network control with limited sensing and actuation. We stress that research in this direction is quite preliminary, and there is a rich space of open problems that may be of interest to the CPS community.

15.2.5 Factor 5: Respecting the Human in the Loop

As noted already, wide-area strategic management of large-scale infrastructures often involves humans in the decision loop. The human operators are typically government personnel or contractors who are based in regional or national control centers. For example, transmission system operators for the power grid manage the wholesale power markets and wide-area grid operations from their control room [31], while strategic air-traffic management is undertaken by managers in regional and national command centers [6], and major disease outbreaks may be managed in an ad hoc way by specialists at the Centers for Disease Control in collaboration with government agencies and medical-care facilities [32]. It is worth noting that some infrastructures (e.g., online social networks or the internet backbone) may be privately owned and managed, while others do not have dedicated control centers with human operators (e.g., much of the road traffic network). Regardless of these variations, human operators play a major role in strategic decision-making for nearly all infrastructures.

Most human operators of infrastructures greatly desire, and would benefit from, advanced decision-support tools. However, the intricacy inherent to infrastructure management, as well as economic and policy considerations, usually dictate that cyber decision-making tools should assist rather than replace human operators. To gain traction for human-in-the-loop operations, CPS approaches to decision-support should focus on:

1. developing user interfaces that allow managers to easily parse and visualize the provided data, and interpret suggested management strategies, at the proper time scales for decision-making;
2. providing operators with data that validate or evaluate suggested management strategies, as well as multiple options for effective management strategies [33];
3. allowing input from managers regarding on-the-ground realities, performance objectives, and special circumstances requiring deviations in strategies [33];
4. better understanding and acknowledging concerns that are specific to human operators, such as fatigue and time-on-task-related performance degradation [34].

Developing tools with these features will require close collaboration among infrastructure-network engineers, human-factors-and-cognition experts, and the managers themselves. CPS researchers are uniquely positioned to develop these multidisciplinary collaborators, and bring complementary know-how in several areas related to human-in-the-loop solutions (e.g., human–machine interfaces, game theory). We believe that the CPS community needs to pursue closer interactions with infrastructure managers and engineers toward implementing these technologies.

15.2.6 Factor 6: Modeling the Environment

Environmental factors crucially influence strategic management of terrestrial-scale infrastructures. For instance, wind and insolation characteristics determine renewable-generation levels, driving power grid operations; disease spreads show seasonal fluctuations due to weather features; and both air and road traffic systems are disrupted by severe weather. Environmental or natural factors such as electrical storms, environmental noise, and temperature profiles may also influence the cyber systems embedded in infrastructures. At the relevant time horizons for strategic decision-making, the environmental factors influencing infrastructure operations may have significant uncertainty. Appropriate models for environmental drivers and their uncertainties, as well as the impacts of these drivers on the infrastructure's dynamics, are needed.

Much work has been done on modeling and forecasting the environmental factors that influence common infrastructures' physical networks. For instance, many tools are available for modeling and forecasting severe weather and its impact on air-traffic system capacities (e.g., [21]). Similarly, wind and solar generation forecasting for the power grid have been extensively studied (e.g., [35]), and seasonal variations in disease spread have been modeled (e.g., [36]). Likewise, noise and environmental factors influencing cyber systems have been extensively modeled (e.g., [37]). While environmental factors that influence infrastructure performance have been widely modeled, using these models to address strategic decision-making remains challenging, for several reasons. First, the developed models sometimes do not have the appropriate resolution and scale for strategic decision-making. Second, the models vary widely in their mathematical details and sometimes lack tractability, which makes it difficult to integrate multiple environmental factors impacting an infrastructure, and to develop statistical analyses of possible impacts. Third, decision-making under uncertainty remains a very challenging problem because of the state-space explosion that results when analyzing nonlinear systems under uncertainty.

We believe that CPS researchers are developing techniques and tools that hold promise to address these needs in decision-making under environmental uncertainty. First, because CPS research integrates controls engineering and computer science techniques, it is leading to new stochastic modeling classes that share the tractability of computer science models (e.g., Markov random field models) with those of classical control-theory models (e.g., linear systems driven by white noise) [21, 38–40]. These models – such as the influence model – can enable common representation and integration of diverse environmental uncertainties and their impacts [27, 41], and also can support interpolation of forecasts to achieve the correct resolution and scale. Second, CPS researchers are developing new techniques for decision-making under uncertainty. In particular, scenario-based algorithms for decision-making, which leverage forecasts generated by new models for environmental uncertainties, are promising for providing high-performance and implementable solutions to decision-making problems.

15.2.7 Factor 7: The Growing Challenge of Cyber-Security

Cyber-security is a constant, and growing, concern in our modern world. Cyber-attacks on infrastructures are particularly concerning because their impacts can propagate to the infrastructure's physical components, potentially exacting significant cost and performance degradation. Recently, several cyber-attacks to infrastructures have occurred, which have effected a significant propagative impact in the physical world. The Stuxnet worm is a prominent example of a cyber-attack that had physical-world impacts, particularly on Iran's nuclear and energy systems. Beyond Stuxnet, several further attacks have been promulgated on CPS infrastructures (including manufacturing, water-distribution, and transportation infrastructures) which have had significant propagative impact in the physical world. As a whole, CPS infrastructures have a wide range of interconnected cyber components (e.g., control systems, communication links, networked computers) which are prone to diverse cyber-attacks (e.g., computer viruses and worms, denial-of-service and other network attacks, spoofing attacks, and various crude attacks such as destruction of equipment). The attack surface for infrastructures is quite large, and evidence suggests that attacks can have a propagative impact across a wide area. Beyond deliberate attacks, accidental disruptions may also incur significant propagative impact on infrastructure function. Thus, cyber-attacks on infrastructures are likely to become increasingly common, impactful, and challenging to address. New control and management tools for large-scale infrastructures must address this sophisticated emerging challenge.

There is a growing research and development effort on the security of CPSs, broadly defined [42–54]. Within this broad research effort, infrastructure–network security has been one primary focus area. The effort on cyber-physical infrastructure security has several thrusts. At the implementation level, there has been a concentrated effort to learn and apply cyber-security tools (e.g., firewalls) and best practices to infrastructures' cyber systems, with the intent of proactively preventing attack. This approach is appealing in that cyber-security know-how can be applied directly; however, such an approach cannot prevent all attacks, and may be prohibitively costly given the large attack surface and diversity of possible threats to CPSs (and particularly cyber-physical infrastructures). Very recently, CPS researchers and also control theorists have begun to approach cyber-security from an alternative perspective that is focused on *implications* rather than *exploits*. The main concept of this approach is to categorize and evaluate cyber-attacks in terms of their propagative impacts on the CPS, rather than based on the specific mechanisms of attacks. This approach thus abstracts away details of the attack vector, and instead depends on integrative models of the CPS (which may capture information flow, physical-world dynamics, etc.) to evaluate impacts. A main outcome of this implications-based perspective is that key vulnerabilities of the system can be identified, allowing parsimonious placement of defense resources to mitigate impacts. The implications-focused approach also permits a common framework for cyber-attacks and other disruptions (e.g., damage to physical-world components, environmental influences, or performance degradation due to human operators' mistakes). If successful, the approach can lead to simple insights into vulnerabilities in terms of the essential structure of the CPS.

While the implications-focused approach to CPS security is promising at a theoretical level, there are still some barriers to its implementation in real-world decision-support systems for infrastructures. First, the initial studies have focused on canonical models for CPS dynamics, which do not replicate the complexity of infrastructures' physical and cyber processes. Second, the initial studies largely do not assimilate expertise on cyber, physical, and human threat responses: Either the cyber-attacks are represented too abstractly, or the physical dynamics are not modeled at the appropriate resolution for evaluating attack impacts; and human elements are rarely represented at all in analyzing threats. Third, metrics for measuring the impact of a cyber-attack on an infrastructure have not been carefully defined. We advocate that the CPS community should address these barriers, toward a comprehensive methodology for mitigating wide-area impacts of cyber-attacks to large-scale infrastructures.

15.2.8 Factor 8: Developing Simple Graph-Based Rubrics

Engineers and computer scientists are obsessed with formal optimization: We understandably seek solutions that are provably successful, and also provably the "best." The focus on formal solutions and optimization is understandable, and would be ideal if this were a perfect world – i.e., if our models for physical and cyber systems were perfect, if model parameters were exactly known, if uncertainties were precisely modeled, if performance metrics were known and agreed upon, if the decision variables in the optimization could be enumerated and relevant ones selected, if the design problem didn't change from day to day, if we had sufficient data, if we had infinite computational power to solve the optimization, if the humans operators could fully interpret the optimal design and its performance under uncertainty, if the operators could actually deploy the optimal design, etc. However, these assumptions are never quite valid, and for infrastructures they are often especially invalid. It is no surprise, therefore, that control and management of infrastructures greatly depends on simple rubrics, human intuition, and suboptimal solutions, while "optimal" strategies fall by the wayside. As cyber systems are pervasively integrated in infrastructures, their operation is becoming yet more complicated, which makes simplicity in decision-making even more crucial. It is thus our strong belief that CPS research should seek simple insights into infrastructure management and control, which can be easily leveraged and deployed in wide-area decision-making. It is important to stress that this perspective is not meant to exclude formal analysis and optimization approaches. Indeed, model-driven analyses as well as formal controller-design and optimization approaches are crucial, but these solutions need to be translated to simple insights and rubrics that can be used readily by operators.

In this section, we advocate particularly for graph-based rubrics for managing cyber-physical infrastructures [23, 26, 55–57]. For many large-scale infrastructures, the network's topology or graph remains largely invariant, and essentially guides and limits the dynamics of the infrastructure. That is, the topology of the network essentially dictates its propagative wide-area dynamics, and hence crucially influences control and management, vulnerability/security analysis, etc. Further, the topology of the network is often well-known or measured accurately (even when other parameters are unknown), and infrastructure engineers traditionally rely on topological understandings to achieve control. For example, electric power engineers place sensing resources (e.g., synchrophasors) at the edges of the graphs, which are known to give cleaner signals. Likewise, air-traffic managers rely on a topological understanding of traffic flows and congestion to design traffic management initiatives. Typically, these graph-theoretic rubrics derive from experience rather than formal techniques, but the growing complexity of the infrastructures is making it harder to obtain such insights by experience. Thus, there is a growing need for deriving graph-based metrics for infrastructure management which are based on formal approaches.

Recently, an exciting research area has emerged in the controls engineering community, which seeks to tie propagative behaviors in networks and their control to the network's graph [23, 26, 55, 59]. This literature builds on the spectral or algebraic graph theory, which connects characteristics of a network's graph to the spectrum of various matrices (e.g., Laplacian) defined from the graph which govern relevant network dynamics. From the classical linear systems theory, it is known that the spectrum essentially dictates the network's dynamical response, and hence the dynamics can be tied to the graph structure in this way. The recent literature in the controls engineering community has sought to extend these graphical characterizations of dynamics, toward understanding input–output processes in networks such as observability and estimation from sparse sensors, controllability, robustness, zero structure, etc. These results provide a pathway toward gaining graph-theoretic rubrics on the management of infrastructure network dynamics. For instance, this theory can be used to ascertain that, indeed, sensors at the edges of a complex network permit the most effective estimation of the power-system swing dynamics, as recognized by electric power engineers, and further give additional insight into sensor placement (e.g., dependence of estimation performance on connectivity).

15.2.9 Factor 9: Exploiting Incidental Measurements

The sensing of infrastructure dynamics is undergoing a paradigm shift, from sparse dedicated sensing using custom-made devices toward pervasive ad hoc sensing. For example, smart phones are now routinely used for crowdsourcing to measure traffic congestion in road traffic networks, smart meters are being deployed across the electric power grid, and social media is being used to track infectious-disease spread. Furthermore, these diverse new sensing modalities are intrinsically networked, and hence the collected data can be rapidly transferred to control centers (whether centralized or distributed). These ad hoc sensors provide an incredible volume and diversity of new measurements of infrastructure processes, yet to date have only been used in a limited way to forecast and especially manage infrastructures. For example, cell-phone data are now commonly used to map congestion [60], but are not used to reschedule traffic signals or forecast future congestion patterns. Likewise, blog data are being leveraged to great effect to map disease spread [61], but have not been used systematically to develop management strategies (e.g., restrictions on transportation) during disease outbreaks. The hardware and software technologies needed to actuate infrastructure networks based on new sensing modalities are largely in place. The barrier to using the new sensors in management instead primarily lies in processing and interpreting the data from these new sensor types. Crucially, many of the new sensing modalities are "incidental" rather than dedicated, in the sense that the sensor's characteristics are primarily governed by factors other than the management task of interest. For instance, the mapping tools discussed above use such incidental measurements: Cell phones are able to measure traffic congestion simply because drivers keep them powered on in their cars for other reasons; individuals blog about infectious diseases in communicating with friends rather than to inform disease control authorities; and sensors embedded on miners may identify hazardous gas or pollutant levels, but their locations are governed by the miner's responsibilities which are unrelated to the sensing task. A key challenge is to exploit such incidental measurements to gain situational awareness of wide-area infrastructure dynamics, so as to support management. Also, the fact that the measurements are incidental tends to reduce stakeholders' trust in them, necessitating performance guarantees or bounds along with the estimates obtained from the data. Because the new sensing modalities are incidental and pervasive, the recorded data sets are often extremely highly dimensional but also incomplete and uninformative.

To date, incidental measurements have only been used in infrastructure networks for simple real-time monitoring tasks. There is a strong need for using these incidental measurements, together with traditional sensing and models, to forecast future dynamics and support management. The CPS community is well positioned to address this need, since CPS research intrinsically connects infrastructure modeling and analysis, sensing and communication system analysis, and formal techniques for analyzing sensing and feedback control (see [62]). Of interest, several recent works from the CPS community have made inroads toward exploiting incidental measurements for infrastructure management, including in the road transportation, environmental monitoring, and disease control spaces.

15.2.10 Factor 10: Economics Matter

Day-to-day operations of modern cyber-physical infrastructures are affected by the economic drives of stakeholders, and hence the performance of wide-area management and control schemes are directly influenced by economics. For instance, the unit commitment plans developed by power transmission system operators (under generation and load uncertainty) directly influence market operators, and influence whether profiteers can take advantage of the power markets to make money. Likewise, economic drives may compel airlines to fly around an airspace region constrained by traffic management initiatives rather than incur a delay; this economic incentive may be modified if the shape of the constrained airspace is changed. Additionally, the planning of air traffic management initiatives in the morning teleconference is strongly influenced by economic factors, since airlines are directly involved in this discussion. In a broad sense, since selfish profit-oriented stakeholders are involved in infrastructure management, it is inevitable that their financial motivations influence the management of the network. At longer planning time horizons, economic factors also clearly play a role in decision-making.

While human operators consider economic factors in wide-area management, this aspect of decision-making tends to be partially unstructured, and often is not based on formal models. Also, in some cases, there may be an information gap among stakeholders, which influences the efficacy of the decision-making process. It is our belief that CPS research on infrastructure management must acknowledge these diverse economic parameters within formal modeling, analysis, and design approaches. Of particular relevance, the integration of new cyber technologies is permitting finer control and management of infrastructure processes; as the CPS community seeks coordinated design of these management capabilities, researchers must understand and account for the economic factors involved in management. Particularly, CPS researchers should be able to leverage formal modeling capabilities as well as collaborative research with economists and social scientists, to evaluate the economic drivers and their impact on management.

CPS researchers have already made notable contributions connected to the economics of infrastructure management. For instance, the CPS community has been at the forefront of research on demand response programs in the power grid, leading to work that acknowledges the decision-making processes, physical-world dynamics (e.g., operation of HVAC systems), and economic factors that influence these programs [63–66]. Likewise, CPS researchers have been effective in bringing forth game-theoretic approaches to evaluate the performance of cyber-physical transportation systems [67]. These initial efforts have naturally been focused on specific new functionalities in infrastructures, rather than integrative wide-area management. However, we believe that the CPS community is well-positioned to comprehensively incorporate economic incentives in wide-area decision-making for infrastructures.

References

[1] P. Kundur, *Power System Stability and Control*. New York: McGraw-Hill, 1994.

[2] N. Jaleeli, L. S. Vansylyck, D. N. Ewart, et al., "Understanding automatic generation control," *IEEE Trans. Power Syst.*, vol. 7, no. 3, pp. 1106–1122, 1992.

[3] A.N. Venkat, I. A. Hiskens, J. B. Rawlings, and S. J. Wright, "Distributed MPC strategies with application to power system automatic generation control," *IEEE Trans. Control Syst. Technol.*, vol. 16, no. 6, pp. 1192–1206, 2008.

[4] J. S. Thorp and A. G. Phadke, "Protecting power systems in the post-restructuring era," *IEEE Comput. Appl. Power*, vol. 12, no. 1, pp. 33–37, 1999.

[5] H. Erzberger, T. J. Davis, and S. Green, "Design of center-TRACON automation system," in *Proc. AGARD Guidance and Control Symp. Machine Intelligence in Air Traffic Management*, Berlin, Germany, 1993.

[6] C. Taylor, C. Wanke, Y. Wan, and S. Roy, "A framework for flow contingency management," in *Proc. of* 11th *AIAA Aviation Technology, Integration, and Operations (ATIO) Conference*, Virginia Beach, VA, 2011.

[7] Y. Wan, C. Taylor, S. Roy, et al., "Dynamic queuing network model for flow contingency management," *IEEE Trans. Intell. Transp. Syst.*, vol. 14, no. 3, pp. 1380–1392, 2013.

[8] B. Sridhar, S. R. Grabbe, and A. Mukherjee, "Modeling and optimization in traffic flow management," *Proc. IEEE,* vol. 96, no. 12, pp. 2060–2080, 2008.

[9] N. P. Padhy, "Unit commitment: A bibliographical survey," *IEEE Trans. Power Syst.*, vol. 19, no. 2, pp. 1196–1205, 2004.

[10] A. J. Conejo, M. Carrión, and J. M. Morales, *Decision Making under Uncertainty in Electricity Markets*. New York: Springer, 2010.

[11] C. G. Souders, S. McGettigan, E. R. Dash, and J. May, "10.2 Weather integration concept of operations for transforming the national airspace system," *12th Conf. Aviation Range and Aerospace Meteorology*, Washington DC, 2006.

[12] B. Sridhar, K. S. Sheth, and S. Grabbe, "Airspace complexity and its application in air traffic management," in *2nd USA/Europe Air Traffic Management R&D Sem.*, 1998, pp. 1–9.

[13] R. Piwko, et al., "Wind energy delivery issues [transmission planning and competitive electricity market operation]," *IEEE Power Energy Mag.*, vol. 3, no. 6, pp. 47–56, 2005.

[14] V. Rosato, et al., "Modelling interdependent infrastructures using interacting dynamical models," *Int. J. Crit. Infrastruct.*, vol. 4, no. 1–2, pp. 63–79, 2008.

[15] C. D. Wickens, A. S. Mavor, R. Parasuraman, and J. P. McGee, eds. *The Future of Air Traffic Control: Human Operators and Automation*. Washington, DC: National Academies Press, 1998.

[16] P.M. Anderson and A. A. Fouad, *Power System Control and Stability*. Chichester: Wiley, 2008.

[17] M. O. Ball, R. Hoffman, and D. Knorr, "Assessing the benefits of collaborative decision making in air traffic management," *Prog. Astronautics Aeronautics*, vol. 193, pp. 239–252, 2001.

[18] K. D. Bilmoria, B. Sridhar, G. N. Chatterji, et al., "FACET: Future ATM concepts evaluation tool," *Air Traffic Contr. Q.*, vol. 9, pp. 1–20, 2001.

[19] C. Wanke, et al., "Measuring uncertainty in airspace demand predictions for traffic flow management applications," presented at the *AIAA Guidance, Navigation and Control Conf.*, Austin, TX, USA, 2003.

[20] L. Song, C. Wanke, and D. Greenbaum, "Predicting sector capacity for TFM decision support," in *6th AIAA Aviation, Integration, and Operations Conference*, Wichita, KS, 2006.

[21] M. Xue, S. Zobell, S. Roy, et al., "Using stochastic, dynamic weather-impact models in strategic traffic flow management," in *Proc. 91st American Meteorological Society Annual Meeting*, Seattle, WA, 2011.

[22] S. Nabavi and A. Chakrabortty, "Topology identification for dynamic equivalent models of large power system networks," in *American Control Conference (ACC)*, pp. 1138–1143, 2013.

[23] J. Valdez, X. Zun, J.A. Torres, and S. Roy, "Fast fault location in power transmission networks using transient signatures from sparsely-placed synchrophasors," in *North American Power Symp. (NAPS)*, 2014, pp. 1–6.

[24] D. Hu and V. M. Venkatasubramanian, "New wide-area algorithms for detection and mitigation of angle instability using synchrophasors," *Power Engineering Society General Meeting*, 2007, pp. 1–8.

[25] A. R. Borden and B. C. Lesieutre, "Variable projection method for power system modal identification," *IEEE Trans. Power Syst.*, vol. 29, no. 6, pp. 2613–2620, 2014.

[26] K. Koorehdavoudi, M. Halami, S. Roy, et al., "Input–output characteristics of the power transmission network's swing dynamics," *IEEE Conf. Decision and Control*, 2016, pp. 1846–1852.

[27] C. Taylor, T. Masek, C. R. Wanke, and S. Roy, "Designing traffic flow management strategies under uncertainty," in *FAA/Eurocontrol Air Traffic Management Research and Development Seminar and Exhibit*, 2015.

[28] D. D. Šiljak, *Large-Scale Dynamic Systems: Stability and Structure*, Vol. 2. Amsterdam: North Holland, 1978.

[29] J. Abad Torres and S. Roy, "Connecting network graph structure to linear-system zero structure," in *American Control Conference (ACC)*, 2013, pp. 6114–6119.

[30] J. Abad Torres and S. Roy, "Graph-theoretic characterisations of zeros for the input–output dynamics of complex network processes," *Int. J. Control*, vol. 87, no. 5, pp. 940–950, 2014.

[31] H. Singh, S. Hao, and A. Papalexopoulos, "Transmission congestion management in competitive electricity markets," *IEEE Trans. Power Syst.*, vol. 13, no. 2, pp. 672–680, 1998.

[32] National Center for Infectious Diseases (US), *Addressing Emerging Infectious Disease Threats: A Prevention Strategy for the United States*. Washington, DC: Centers for Disease Control and Prevention, 1994.

[33] C. Wanke and C. Taylor, "Exploring design trade-offs for strategic flow planning," in *AIAA Aviation*, 2013, pp. 12–14.

[34] B. C. Satterfield and H. P. A. Van Dongen, "Occupational fatigue, underlying sleep and circadian mechanisms, and approaches to fatigue risk management," *Fatigue Biomed. Health Behav.*, vol. 1, no. 3, pp. 118–136, 2013.

[35] A. M. Foley, P. G. Leahy, A. Marvuglia, and E. J. McKeogh, "Current methods and advances in forecasting of wind power generation," *Renewable Energy*, vol. 37, no. 1, pp. 1–8, 2012.

[36] E. Lofgren, N. H. Fefferman, Y. N. Naumov, et al., "Influenza seasonality: underlying causes and modeling theories," *J. Virol.* vol. 81, no. 11, pp. 5429–5436, 2007.

[37] P. Xie, J. H. Li, X. Ou, et al., "Using Bayesian networks for cyber security analysis," in *Int'l Conf., Dependable Systems and Networks (DSN)*, 2010, pp. 211–220.

[38] C. Asavathiratham, et al., "The influence model," *Control Syst.*, vol. 21, no. 6, pp. 52–64, 2001.

[39] M. Xue, S. Roy, S. Zobell, et al., "A stochastic spatiotemporal weather-impact simulator: Representative scenario selection," in *AIAA Aviation Technology, Integration, and Operations Conf.*, Virginia Beach, VA, 2011.

[40] R. Dhal and S. Roy., "Layered moment-linear network models as tools for strategic air traffic flow management," Proc. Infotech@Aerospace, 2012, pp. 1–23.

[41] Y. Zhou, et al., "Multivariate probabilistic collocation method for effective uncertainty evaluation with application to air traffic flow management," *IEEE Trans. Syst. Man. Cybern.,* vol. 44, no. 10, pp. 1347–1363, 2014.

[42] S. K. Das, K. Kant, and N. Zhang, *Handbook on Securing Cyber-Physical Critical Infrastructure: Foundations and Challenges*. New York: Morgan-Kauffman, 2012.

[43] M. Xue, S. Roy, Y. Wan, and S. K. Das, "Security and vulnerability of cyber-physical infrastructure networks: A control-theoretic approach," in *Handbook on Securing Cyber-Physical Critical Infrastructure: Foundations and Challenges*, S. K. Das, K. Kant, and N. Zhang, Eds. New York: Morgan-Kaufmann, 2012, pp. 1–30.

[44] X. Jin, M. Dan, N. Zhang, et al., "Game theory for infrastructure security: the power of intent-based adversary models," in *Handbook on Securing Cyber-Physical Critical Infrastructure: Foundations and Challenges,* S. K. Das, K. Kant and N. Zhang, Eds. New York: Morgan-Kaufmann, 2012, pp. 31–53.

[45] J.-W. Ho and S. K. Das, "Node compromise detection in wireless sensor networks," in *Handbook on Securing Cyber-Physical Critical Infrastructure: Foundations and Challenges,* S. K. Das, K. Kant and N. Zhang, Eds. New York: Morgan-Kaufmann, 2012, pp. 281–300.

[46] A. Sturaro, S. Silvestri, M. Conti, and S. K. Das, "Towards a realistic model for failure propagation in interdependent networks," *Proc. International Conference on Computing, Networking and Communications (ICNC 2016)*, Kauai, HI, Feb. 15–18, 2016.

[47] A. Cardenas, S. Amin, B. Sinopoli, et al., "Challenges for securing cyber physical systems," in *Workshop on Future Directions in Cyber-Physical Systems Security, DHS, 23,* 2009.

[48] J. A. Torres, D. Sahabandu, R. Dahl, and S. Roy, "Local open-and closed-loop manipulation of multi-agent networks," *Proc. ACM/IEEE 6th Int. Conf. Cyber-Physical Systems*, 2015, pp. 21–30.

[49] J.-W. Ho, M. Wright, and S. K. Das, "Fast detection of mobile replica node attacks in wireless sensor networks using sequential hypothesis testing," *IEEE Trans. Mob. Comput.*, vol. 10, no. 6, pp. 767–782, Jun. 2011.

[50] N. Zhang, W. Yu, X. Fu, and S. K. Das, "Maintaining the defender's reputation in anomaly detection against insider attacks," *IEEE Trans. Syst. Man Cybernet. B*, vol. 40, no. 3, pp. 597–611, Jun. 2010.

[51] P. De, Y. Liu and S. K. Das, "Deployment aware modeling of node compromise spread in wireless sensor networks," *ACM Trans. Sens. Netw.*, vol. 5, no. 3, pp. 413–425, May 2009.

[52] P. De, Y. Liu, and S. K. Das, "An epidemic theoretic framework for vulnerability analysis of broadcast protocols in wireless sensor networks," *IEEE Trans. Mob. Comput.*, vol. 8, no. 3. pp. 413–425, Mar. 2009.

[53] S. Siddharth, A. Hahn, and M. Govindarasu, "Cyber–physical system security for the electric power grid," *Proc. IEEE*, vol. 100, no. 1, pp. 210–224, 2012.

[54] M. Xue, W. Wang, and S. Roy, "Security concepts for the dynamics of autonomous vehicle networks," *Automatica*, vol. 50, no. 3, pp. 852–857, 2014.

[55] Y. Wan and S. Roy, "On inference of network time constants from impulse response data: Graph-theoretic Cramer-Rao bounds," in *Proc. 48th IEEE Conf. Decision and Control*, 2009, pp. 4111–4116.

[56] M. Xue, et al., "Initial-condition estimation in network synchronization processes: Algebraic and graphical characterizations of the estimator," in *AIAA Guidance, Navigation and Control Conf.*, Portland, OR, 2011.

[57] J. A. Torres and S. Roy, "A two-layer transformation for characterizing the zeros of a network input–output dynamics," in *IEEE 54th Ann. Conf. Decision and Control (CDC)*, 2015, pp. 902–907.

[58] F. Pasqualetti, S. Zampieri, and F. Bullo, "Controllability metrics, limitations and algorithms for complex networks," *IEEE Trans. Control Network Syst.*, vol. 1, no. 1, pp. 40–52, 2014.

[59] R. Dhal and S. Roy, "Vulnerability of continuous-time network synchronization processes: A minimum energy perspective," in *IEEE 52nd Ann. Conf. Decision and Control (CDC)*, 2013, pp. 823–828.

[60] S. Amin, S. Andews, S. Apte, et al., "Mobile century using GPS mobile phones as traffic sensors: A field experiment," *Proc. 15th World Congress on Intelligent Transportation Systems*, New York, 2008.

[61] A. Sadilek, H. A. Kautz, and V. Silenzio, "Predicting disease transmission from geo-tagged micro-blog data," *Proc. 26th AAAI Conference on Artificial Intelligence*, 2012, pp. 136–142..

[62] S. Roy and R. Dhal, "Situational awareness for dynamical network processes using incidental measurements," *IEEE J. Sel. Top. Signal Process.*, vol. 9, no. 2, pp. 304–316, 2015.

[63] M. Balandat, F. Oldewurtel, M. Chen, and C. Tomlin, "Contract design for frequency regulation by aggregations of commercial buildings," *52nd Ann. Allerton Conf. Communication, Control, and Computing*, 2014, pp. 38–45.

[64] I. Yang, D. S. Callaway, and C. J. Tomlin, "Indirect load control for electricity market risk management via risk-limiting dynamic contracts," in *American Control Conference (ACC)*, 2015, pp. 3025–3031.

[65] J. Mathieu, et al., "Quantifying changes in building electricity use, with application to demand response," *IEEE Trans. Smart Grid*, vol. 2, no. 3, pp. 507–518, 2011.

[66] J. Mathieu, D. S. Callaway, and S. Kiliccote, "Examining uncertainty in demand response baseline models and variability in automated responses to dynamic pricing," in *50th IEEE Conference on Decision and Control and European Control Conference (CDC-ECC)*, 2011, pp. 4332–4339.

[67] C. Tomlin, G. J. Pappas, and S. Sastry, "Conflict resolution for air traffic management: A study in multiagent hybrid systems," *IEEE Trans. Autom. Control*, vol. 43, no. 4, pp. 509–521, 1998.

Index

Lightning Source UK Ltd.
Milton Keynes UK
UKHW051953191120
373739UK00006B/83

9 781107 066618